Rough Sets

Advances in Soft Computing

Editor-in-chief
Prof. Janusz Kacprzyk
Systems Research Institute
Polish Academy of Sciences
ul. Newelska 6
01-447 Warsaw, Poland
E-mail: kacprzyk@ibspan.waw.pl
http://www.springer.de/cgi-bin/search-bock.pl?series=4240

Lech Polkowski

Rough Sets

Mathematical Foundations

With 20 Figures

Springer-Verlag Berlin Heidelberg GmbH

Prof. Dr. Lech Polkowski

Polish-Japanese Institute of Information Technology
ul. Koszykowa 86
02-008 Warsaw
Poland

and

Department of Mathematics and Computer Science
University of Wormia and Mazury
ul. Żofnierska 14a
10-561 Olsztyn
Poland
Lech.Polkowski@pjwstk.edu.pl

ISSN 1615-3871
ISBN 978-3-7908-1510-8   ISBN 978-3-7908-1776-8 (eBook)
DOI 10.1007/978-3-7908-1776-8

Library of Congress Cataloging-in-Publication Data applied for
Die Deutsche Bibliothek – CIP-Einheitsaufnahme
Polkowski, Lech: Rough sets: mathematical foundations / Lech Polkowski. – Heidelberg; New York: Physi-
ca-Verl., 2002
  (Advances in soft computing)
  ISBN 978-3-7908-1510-8

© Springer-Verlag Berlin Heidelberg 2002

Originally published by Physica-Verlag Heidelberg New York in 2002

Softcover Design: Erich Kirchner, Heidelberg

SPIN 10882886   88/2202-5 4 3 2 1 0 – Printed on acid-free paper

Dedicated to Professor Zdzisław Pawlak on his 75th birthday

Preface

Er wunderte sich, dass den Katzen gerade an der Stelle zwei Löcher in den Pelz geschnitten werden, wo sie die Augen hätten

Georg Christoph Lichtenberg, *Aphorismen*

Rough sets, similarly to fuzzy sets, have been invented to cope with uncertainty. Both paradigms address the phenomenon of non – crisp concepts, notions which, according to Frege, are characterized by a presence of a non – empty *boundary* which does encompass objects neither belonging with certainty to the given concept nor belonging with certainty to its complement.

A crisp concept $A \subseteq X$ induces on the universe X the partition $\{A, X \setminus A\}$ into objects belonging to A and objects belonging to $X \setminus A$ as the problem whether $x \in A$ is decidable on the basis of knowledge available to us.

The partition $\{A, X \setminus A\}$ can be rendered in functional language into the *characteristic function* χ_A which assumes the value 1 on A and the value 0 on $X \setminus A$.

Assume our knowledge increases due to our experience, new experiments etc. etc. In the language of partitions, this results in passing from the partition $\{A, X \setminus A\}$ to a partition $\mathcal{P}$ whose members are subsets of the universe containing objects indiscernible one to another on the basis of new knowledge.

In consequence, the formerly crisp notion A may become non – crisp: for some objects x the problem whether $x \in A$ may become undecidable: some objects indiscernible to x may be in A while some objects indiscernible to x may be not in A.

Either paradigm solves the description problem of A in its own way; rough sets approach this problem by considering *approximations* to A induced by the partition $\mathcal{P}$ viz. into the *lower approximation* A_P those objects fall whose whole class $[x]_P$ is contained in A. We can by all means conclude that such objects belong in A with *certainty* (or, *necessarily*). To the *upper approximation* A^P those objects fall whose class $[x]_P$ intersects A but is not contained in A; we can conclude that such objects belong *possibly* in A. The *boundary*

$B_P = A^P \setminus A_P$ consists of ambiguous objects possibly but not certainly in A.

Fuzzy sets approach the problem of description of A in the other way: the characteristic function $\chi_A : X \to \{0,1\}$ undergoes a modification to a *fuzzy characteristic* function $\mu_A : X \to [0,1]$. We can regard objects in the inverse $\mu_A^{-1}(1)$ as belonging in A with certainty while objects in $\mu_A^{-1}((0,1])$ belong in A possibly.

Both approaches are related; assuming e.g. the case of a finite universe X, and a partition $\mathcal{P}$, we can assign to each $x \in X$, a quotient $\mu_P(x) = \frac{|A \cap [x]_P|}{|[x]_P|}$ according to an idea of Pawlak and Skowron. The function μ_P is then a fuzzy characteristic function inducing same approximations as the partition P. Conversely, given a fuzzy characteristic function χ_A, we can induce a partition P_χ by letting

$$[x]_{P_\chi} = [y]_{P_\chi} \Leftrightarrow \chi_A(x) = \chi_A(y).$$

Then certainty and possibility regions induced by χ_A correspond to the lower and upper approximations induced by P_χ.

From this point both paradigms go their own ways. It is the aim of this book to present mathematical foundations and methods of the rough set theory. As one may expect, rough set theory with its notions of the lower approximation, the upper approximation and the boundary region is related to 3 – valued logics, while in fuzzy set theory various infinite valued logics are reflected. The topic of mutual relationships and influences of each to other is taken up in the end (cf. Chapter 14 where we outline the ideas of *rough fuzzy* and *fuzzy rough* sets leading to hybrid constructions).

Mathematical foundations of rough set theory are constructed from fragments of set theory, logic, topology and algebra. We aim at presenting a full picture of results obtained in rough set theory by set–theoretical, logical, topological and algebraic methods.

We intend at making a complete and self – contained exposition of the necessary material, so in addition to advanced topics presented in Chapters 10 – 12 on rough set – theoretic results, we include all necessary preliminaries from the concerned areas of mathematics so the reader may find in the book along with results on rough sets also all information from the relevant mathematical field. In accordance with this desire, we include as an interlude between Chapter 1 bringing the basic information about rough sets and intended as a self–contained tutorial, and Chapter 10, a series of Chapters devoted to respective areas of pure mathematics. Chapter 1 contains moreover a bibliography of 420 recent papers in rough set theory.

The reader will find in Chapter 2 a short but precise and rigorous course in the sentential logic covering the classical deductive theory as well as the Gentzen – style axiomatization in the version due to Kanger. Chapter 3 brings forth an exposition of Syllogistic: in the light of the Słupecki result on Syllogistic as the complete logical theory of containment and intersection i.e. the basic tools in defining rough set approximations, Syllogistic appears as one of the main logical reflexions of rough set theory.

In Chapter 4, many – valued finite – valued logics are encountered for the first time. We discuss the 3 – valued logic of Łukasiewicz and we introduce the Wajsberg axiomatization of this logic along with the proof of its completeness. Many–valued logics are discussed in the Rosser – Turquette spirit and finally 4 – valued logic of Łukasiewicz is introduced as a vehicle to present the basic modal logic.

Modal logics are discussed in Chapter 5 in which we also introduce the Kripke semantics and prove the completeness of basic logics: K, T, S4, and S5.

Set theory is exposed in Chapter 6. In addition to standard notions of naive set theory recalled, we discuss basic constructions of formal set theory among them Cartesian products, relations, orderings, and filters and essentially important notions of a lattice and of a Boolean algebra. A special attention is devoted to equivalence and tolerance relations as to constructs rough set theory deals with frequently and substantially.

Chapter 7 is devoted to topological structures and we discuss basic notions and most important results of set theoretic topology among them compactness and completeness. Important topics like regular sets, filters, Stone representation, are brought in there. As a preliminary to a discussion of rough set topologies in Chapter 11, we include a section on topologies on collections of closed sets containing a discussion of the Hausdorff – Pompéju metric.

A discussion of basic notions of lattice theory is carried out in Chapter 8. We highlight the notions of a distributive lattice, a Stone algebra, pseudo – Boolean (Heyting) algebra, including representation theorems for distributive lattices and Heyting algebras. We regard this Chapter as preparatory to Chapters 10 and 12 in which rough sets are discussed also in an abstract setting of lattices.

The last in this part Chapter 9 concerns predicate calculus. We give a naive introduction as well as a formalization of the predicate calculus both in the Rasiowa – Sikorski style and in the Gentzen style. In addition, intuitionistic logic is introduced as a preliminary to Chapter 10, and the Łukasiewicz calculus of fractional values in logic is recalled to be used in Chapter 10. As a preliminary to Chapter 12, Lindenbaum –Tarski algebras of logical calculi are introduced and studied in some detail.

After preliminaries are set out, we proceed with an advanced discussion of mathematical results pertaining to rough sets. In Chapter 10, we discuss independence issues, covering the Pawlak – Novotný theory of reducts and independence via semi – lattices. After that we discuss an abstract theory of approximations and the theory of partial dependencies referring to the Łukasiewicz calculus of fractional values (Chapter 9).

Chapter 11 is devoted to topological theory of rough sets. In the setting of information systems with countably many attributes, we construct topologies on rough sets and almost rough sets metrizable by metrics D, D^*, D' making rough sets into complete metric spaces. As an application, we introduce the approximate fixed point theorem on rough set spaces and we discuss for the first time in literature the notion of a fractal in information systems showing continuity of fractal dimension with respect to induced metrics.

In Chapter 12, a study of algebraic properties of rough sets is presented. We begin with the Lindenbaum – Tarski algebra of the 3–valued Łukasiewicz logic i.e. with the Wajsberg algebra and we introduce Łukasiewicz algebras. We include a proof of equivalence of Wajsberg algebras to 3 – Łukasiewicz algebras. Then we demonstrate that rough sets in disjoint representation may be given a structure of a Łukasiewicz algebra i.e. that they are related in a natural way to the 3 – valued logic of Łukasiewicz. Parallel representations of rough sets as a Heyting algebra, a double regular Stone algebra or a Post algebra due to Pagliani, Pomykala et al. are also introduced. After that we pursue the logical thread and we present the Rauszer results about the equivalence between calculus of independence and the fragment of intuitionistic logic. We conclude with an account of modalities in information systems presenting the information logic *IL* due to Vakarelov.

At this point our presentation of rough set mathematics *sensu stricto* ends. We come back to the starting idea in this preface viz. to relations between rough and fuzzy set theories.

In order to give the reader an insight into basic differences in technical approaches to either theory, we present in Chapter 13 an account of infinite valued logic, based on an axiomatics due to Łukasiewicz. We present a syntactic proof of its completeness after Rose–Rosser with some simplifications concerning matrix dichotomies which we replace with some simple algebraic lemmas.

After that we present an account of a fuzzy sentential logic due to Pavelka in which we introduce adjoint pairs being a logical counterpart of pairs (t – norm, residuated implication) so essential in fuzzy set calculi. Working with the Łukasiewicz adjoint pair, we include the proof of completeness of the fuzzy sentential logic in the Pavelka sense.

Chapter 13 introduces the reader to mathematical world associated with the fuzzy set theory and contrasted with the 3–valued world of rough set theory.

This contrast is underlined in Chapter 14 where the ideas of rough fuzzy set and fuzzy rough set are introduced. We begin with a discussion of t – norms and t – conorms along with induced residuated implications and we prove the representation theorem about t – norms and t – conorms from which we infer the remarkable fact that the Łukasiewicz adjoint pair is the unique up to equivalence adjoint pair making the fuzzy sentential logic complete.

Then we introduce rough fuzzy sets in the Dubois – Prade style followed by fuzzy rough sets. Here the notions of a fuzzy equivalence relation as well as a fuzzy partition are discussed after Zadeh et al., and a proposal for fuzzy rough sets is laid out. Finally, we return to the lattice setting and we discuss the Brouwer – Zadeh lattices of fuzzy sets with rough set style approximations equivalent to the certainty and possibility regions of a fuzzy sets mentioned in the introduction to this Preface.

It may be noticed from the outline of the content presented above that our book covers all essential results in mathematical theory of rough sets which require a deeper mathematical knowledge.

The book may be used as a text for a one – semester graduate course on rough set theory, exploiting Chapters 10 – 12 with a possible addition of excerpts from Chapters 13, 14.

Due to Chapters 2-9 the book may be also used as a text for one – semester undergraduate course covering less demanding topics of Chapters 10 – 12 along with fragments of Chapters 2–9 as a prerequisite.

Finally the book may be used as a text for a course on mathematical foundations of Computer Science as it covers the basic facts from logic, set theory, topology and lattice theory.

Over 320 exercises in Chapters 2-13 provided with careful hints allow the reader to gain additional information. The basic text along with exercises makes the book completely self – contained and the diligent reader will master the topic on its own without the need to refer to other sources if they are pleased to do so.

One more usage of the book will be as a reference to the area of mathematical foundations of rough set theory as each Chapter is provided with the essential bibliography. The author hopes that the book will provide many newcomers to rough set theory with the necessary background on the current state of art in this area and by this it will stimulate the further development of the theory.

The author has experienced the friendship and help of many during his work on rough sets.

The thanks should go first to the late Professor Helena Rasiowa, an eminent logician, a disciple of Jan Łukasiewicz whose legacy so permeates this book, who brought the author to the rough set theory, and to Professor Zdzisław Pawlak, the founder of the rough set theory for His kind help and support. This book is offered to him.

The author would like to thank Professor Andrzej Skowron for his help and support, and for many years of close and still going into future collaboration on rough sets which has made those years so pleasant.

Professor Janusz Kacprzyk has always supported the author in his editorial tasks which is acknowledged with gratitude.

Colleagues from the Polish–Japanese Institute of Information Technology has provided a friendly atmosphere and facilities which allowed for preparing of this book.

I am also thinking of my wife Maria and my son Martin in grateful acknowledgment of their support and love.

Finally the author thanks Dr Martina Bihn and her colleagues from Physica–Verlag/Springer–Verlag for their skilled help in making this book.

This work was supported by the Grant from the Polish State Committee for Scientific Research (KBN) No. 8T11C 024 17. A help from the Grant Nr 8T11C 02519 from the Polish State Committee for Scientific Research(KBN) is also acknowledged.

For the reader convenience a guide map of the content with suggestions for connections among Chapters is offered. A graduate course could cover 1-10-11-12-13-14, a less demanding undergraduate course could encompass the same without more technical proofs. A course on Mathematics for Computer Science could cover 2-3-4-5-6-7-8-9 or 2-3-4-5-9. A course on Fuzzy Set Theory devoted to foundational issues may be organized around Chapter 13 and Chapter 14.

The author does hope that this book will offer the rough set community a useful reference book and researchers from other areas of investigation, in particular from the field of fuzzy set theory, will find therein a basic information on rough set theory, its purposes, methods, and mathematical tools it does require as well as on relations between the two theories.

April, 2002   Lech Polkowski

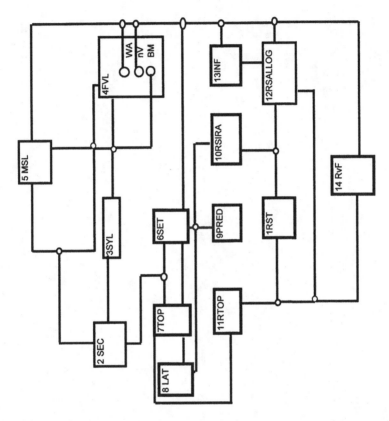

Figure 1: A guide map: WA is Sections 4.1-4.4, nV is Sections 4.5-6, BM is Section 4.7; 14RvF is Chapter 14

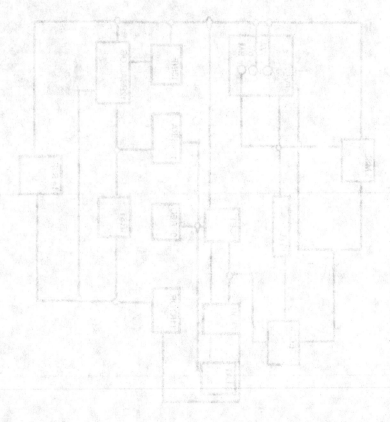

Contents

PART 2: PREREQUISITES

PART 3: MATHEMATICS OF ROUGH SETS

PART 4: ROUGH VS. FUZZY   411

13 Infinite–valued Logical Calculi   413

14 From Rough to Fuzzy   465

PART 1: ROUGH SETS

Chapter 1

Rough Set Theory: An Introduction

The human understanding is of its own nature prone to suppose the more order and regularity in the world than it finds. And though there be many things which are singular and unmatched, yet it devises for them parallels and conjugates and relatives which do not exist.

Francis Bacon, *Novum Organum*, I, 45

Part I: Basic notions and structures related to rough sets

1.1 Knowledge representation

In rough set theory, knowledge is interpreted as an ability to classify some objects (cf. [Pawlak82^a, 81^b]). These objects form a *set* called often a *universe of discourse* and their nature may vary from case to case: they may be e.g. medical patients, processes, participants in a conflict etc., etc.

Classification of objects consists in this theory (cf. op.cit., op.cit.) in finding a *set* of *subsets* of the universe of objects such that each object answers to at least one subset in the sense that it is an *element* of this subset. Subsets of the universe of objects are called often *categories* and we may say that classification of objects consists in finding a *covering* of the universe of objects by a set of categories.

The most clear, unambiguous, classification case happens when the covering of the universe of objects consists of *pair–wise disjoint* categories as in this case each object answers to a unique category thus classification acquires a *functional* character. We say that in this case categories included in classification form a *partition* of the universe of objects.

Such a disjoint classification may be looked upon also as a constraint on objects: given a partition C one may define a constraint R_C by letting $(x, y) \in R_C$ if and only if there is a category $c \in C$ such that $x \in c$ and $y \in c$. Here we apply the symbol $\in$ which is read *is an element of* and it is the membership symbol of set theory.

The constraint R_C is in mathematical language called a *relation* on the universe of objects; it is easy to see, by disjointness of categories in C, that the relation R_C satisfies for each triple x, y, z

$$
\begin{align}
&(i) \ (x, x) \in R_C \\
&(ii) \ if \ (x, y) \in R_C \ then \ (y, x) \in R_C \\
&(iii) \ if(x, y) \in R_C \ and \ (y, z) \in R_C \ then \ (x, z) \in R_C
\end{align}
\tag{1.1}
$$

(i) means *reflexivity*, (ii) means *symmetry*, (iii) means *transitivity*.

A relation which satisfies equation 1.1 is said to be an *equivalence relation*.

One may thus state that *elementary knowledge* is encoded in rough set theory in the form of a pair (U, R) where U is a universe of objects and R an equivalence relation on U. For such an elementary knowledge, we may form *elementary granules* of knowledge by looking at categories forming the corresponding partition: for each object $x \in U$, we may consider the set $[x]_R = \{y \in U : (x, y) \in R\}$ i.e. we collect in $[x]_R$ all objects which answer to the same category in the initial partition as x i.e. we restore the category whose element is x. Categories of the form $[x]_R$ are *elementary granules of knowledge*. From mathematical point of view they are *equivalence classes* of the equivalence R (cf. a discussion in [Pawlak91, Ch. 2]; for a discussion from the granularity of knowledge point of view cf. [Polkowski–Skowron01]).

We have witnessed the fact that discussion of knowledge in rough set theory does involve mathematical objects like sets, relations, equivalence classes etc.etc. For this reason, in Chapter 6 we include a thorough discussion of basic notions and results of Set Theory, including the theory of Relations.

The set of all elementary granules (categories) of knowledge is thus the set $U/R = \{[x]_R : x \in U\}$ of all equivalence classes of the relation R called the *quotient set* of the universe U by the equivalence R. As any two objects x, y with the property that $x \in [y]_R$ i.e. $[x]_R = [y]_R$ are not distinguishable from each other by means of knowledge expressed by means of R, one says often that x, y are *indiscernible* and R is said to be the *indiscernibility relation* (cf. [Pawlak91, 82^a]).

The reader will find therefore in Chapter 6 a throughout discussion of quotient structures as well.

Elementary granules (categories) of knowledge may be fused together by means of set–theoretical operations of the union, the intersection, and the complement, also discussed in Chapter 6, viz. given elementary granules $G_1, ..., G_k$ one may form their *union* $G = G_1 \cup G_2 \cup ... \cup G_k$, and having two such unions G, H it is possible to form their *intersection* $G \cap H$ as well as their difference $G \setminus H = H \cap H^c$ where H^c is the *complement* to H i.e. $H^c = U \setminus H$. In this way we obtain the set of *granules* (categories) of knowledge.

The fact that classification R is unambiguous allows for a functional rendering of knowledge viz. for each $x \in U$ we may define a value $a_R(x) = [x]_R$ uniquely, which defines a function a_R on the universe U with values in the quotient set U/R, in symbols, $a_R : U \longrightarrow U/R$. The function a_R may be interpreted as a *feature, attribute* of objects in U defined with respect to knowledge R. The indiscernibility of objects may be now expressed by means of the attribute a_R: for each pair x, y of objects, x, y are indiscernible if and only if $a_R(x) = a_R(y)$ i.e. if and only if $x \in a_R^{-1}(y)$ where $a_R^{-1}(y) = \{z \in U : a_R(z) = a_R(y)\}$ is the *fibre* of a_R defined by y. Elementary granules of knowledge are thus fibres of the attribute a_R and granules of knowledge (categories) are unions of those fibres i.e. *inverse images* $a_R^{-1}(Y)$ of (finite) subsets $Y \subseteq U$ of the universe U.

This rendering of basic notions related to knowledge in functional language demonstrates the importance of theory of functions on sets for the development of rough set theory. The reader will witness evidence for this claim in what follows and again in Chapter 6 we bring forth the basic notions and facts related to functional structures.

Finally, we call our attention to the fact that knowledge may be encoded in a more complex form, by considering a family of classifications (i.e. equivalences) $\mathcal{R} = \{R_i : i \in I\}$ on the universe U. In this more intricate case, a plethora of classifications arise viz. for each subset $\mathcal{R}_0 \subseteq \mathcal{R}$ of $\mathcal{R}$, the intersection $\bigcap \mathcal{R}_0$ is a classification in its own right.

In particular, the intersection $\bigcap \mathcal{R}$ of the family $\mathcal{R}$ of classifications is a classification, the finest classification possible over $\mathcal{R}$. The pair $(U, \mathcal{R})$ is also called a *knowledge base* (cf. [Pawlak91]).

We now proceed to a discussion of means for representing knowledge in rough set theory.

1.2 Information systems

The semantic aspect of knowledge as a classification on the universe of objects is in need of its syntactic counterpart. In rough set theory, this syntactic representation of knowledge is provided in the form of an *information system* (cf. [Pawlak 81^{a,b}]). The idea of an information system stems from the functional representation of a classification as an attribute of objects.

	a_1	a_2
x_1	1	3
x_2	1	0
x_3	3	1
x_4	3	1
x_5	4	2
x_6	1	2
x_7	4	2

Figure 1.1: An information system A1

Thus, an information system is a pair (U, A) where U is a set (a universe) of objects and A is a set of attributes; as we already know, each attribute $a \in A$ is a function $a : U \to V_a$ where the set V_a is the *value set* of the attribute a. The classification R_a associated with the attribute a is defined by the condition $(x, y) \in R_a$ if and only if $a(x) = a(y)$; one does interpret the relation R_a as the relation of *indiscernibility* with respect to the attribute a and for this reason, the relation R_a is called the a–*indiscernibility relation* and it is denoted with the symbol IND_a (or, $IND_{\{a\}}$).

It follows that the information system $\mathcal{A} = (U, A)$ does induce the classification $\mathcal{R}_{\mathcal{A}} = \{IND_a : a \in A\}$ as well as the knowledge base $(U, \mathcal{R}_{\mathcal{A}})$. Each set $B \subseteq A$ of attributes induces the classification $IND_B = \bigcap \{IND_a : a \in B\}$ called the B–*indiscernibility relation*.

We may look at the knowledge base induced by the information system A1 cf. Fig. 1.

Example 1.1. *Let us look at Fig. 1. The non-empty subsets of the attribute set A are $\{a_1\}$, $\{a_2\}$ and $\{a_1, a_2\}$.*

The indiscernibility relation $IND_{\{.\}}$ defines three non–trivial partitions of the universe

1. $IND_{\{a_1\}} = \{\{x_1, x_2, x_6\}, \{x_3, x_4\}, \{x_5, x_7\}\}$.

2. $IND_{\{a_2\}} = \{\{x_1\}, \{x_2\}, \{x_3, x_4\}, \{x_5, x_6, x_7\}\}$.

3. $IND_{\{a_1, a_2\}} = \{\{x_1\}, \{x_2\}, \{x_3, x_4\}, \{x_5, x_7\}, \{x_6\}\}$.

1.3   Exact sets, rough sets, approximations

Categories which are equivalence classes of the equivalence relation R on the set of objects U are called by us *elementary granules of knowledge* encoded by R. These categories are given to us once R is specified. For this reason, these categories are called R–*exact elementary categories* (cf. [Pawlak91]).

As elementary R–exact categories are pair–wise disjoint, the operation of union of families of sets $\bigcup$ (cf. Chapter 6) yields us all other categories which may be described in terms of R; these categories are called R–exact. We may observe that intersections as well as differences of R–exact categories are R–exact categories.

Any category $X \subseteq U$ which is not R–exact is said to be R–rough; in common usage, we speak rather of R–exact sets and R–rough sets than of R–exact or R–rough categories, and in what follows we adopt the former terminology.

R–exact sets may be used to approximate with them sets which are R–rough. For an R–rough set $Y \subseteq U$, we may define following [Pawlak82^a], two exact sets, called, respectively, the R–lower– and the R–upper–approximation. The idea of these approximations goes back to the general idea of approximation from below and from above, known for example in geometry or in the theory of integral.

One thus defines the R–lower and the R–upper approximations of Y, denoted $\underline{R}Y$ and $\overline{R}Y$, respectively, as follows

$$
\begin{aligned}
\underline{R}Y &= \{x \in U : [x]_R \subseteq Y\} \\
\overline{R}Y &= \{x : [x]_R \cap Y \neq \emptyset\}
\end{aligned}
\tag{1.2}
$$

Let us observe that lower and upper approximations are defined in equation 1.2 in terms of containment and non–empty intersection of classes. Therefore theory of approximations falls in the province of Syllogistic of Aristotle– a logical theory of containment and non–empty intersection of non-empty classes. For this reason, we discuss Syllogistic in Chapter 3 pointing to its complete logical rendering of the approximation calculus.

The reader will consult Fig. 2 below in which lower and upper approximations of a concept marked in grey are traced with thick contours. This is a picture one may have in mind when discussing these approximations.

R–lower and R–upper approximations obey certain basic laws (cf. [Pawlak 91]) which are discussed in Chapter 10 in which approximations are studied from set–theoretic point of view.

Let us observe here only that

$$
\underline{R}Y \subseteq Y \subseteq \overline{R}Y
\tag{1.3}
$$

for each category (set) $Y \subseteq U$ and that

$$
\underline{R}Y = \overline{R}Y
\tag{1.4}
$$

if and only if Y is R–exact.

In the context of an information system $\mathcal{A} = (U, A)$, approximations may

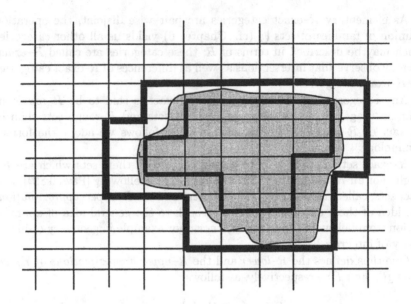

Figure 1.2: The lower and the upper approximations to the concept in grey are marked in thick contours on a grid of equivalence classes

be defined for each set $B \subseteq A$ of attributes viz.

$$\underline{B}Y = \{x \in U : [x]_{IND_B} \subseteq Y\}$$
$$\overline{B}Y = \{x : [x]_{IND_B} \cap Y \neq \emptyset\} \tag{1.5}$$

We may look at instances of approximations induced by the information system A1 discussed in Example 1.1.

Example 1.2. *We have for instance for* $X = \{x_1, x_2, x_3\}$:

1. $\underline{a_1}X = \emptyset$.

2. $\overline{a_1}X = \{x_1, x_2, x_3, x_4, x_6\}$.

3. $\underline{a_2}X = \{x_1, x_2\}$.

It follows on further examination, that the concept $X = \{x_1, x_2, x_3\}$ *is exact for neither of the three attribute sets while the concept* $Y = \{x_1, x_2\}$ *is both* $\{a_2\}$ *– and* A *–exact.*

The important consequence of introducing approximations in the knowledge base (U, R) is that each set $Y \subseteq U$ may be represented as the pair $(\underline{R}Y, \overline{R}Y)$ of R–exact sets; clearly, this representation is non–ambiguous only in case Y is R–exact, for an R–rough set Y, we may have other rough sets Z with the property that $\underline{R}Y = \underline{R}Z$ and $\overline{R}Y = \overline{R}Z$.

To disambiguate rough sets, we may resort again to equivalence relations and we may define a new relation $\equiv$ of *rough equality* (cf. [Novotný–Pawlak85[a,b,c]]) by letting $Y \equiv Z$ if and only if $\underline{R}Y = \underline{R}Z$ and $\overline{R}Y = \overline{R}Z$. It follows easily that $\equiv$ is an equivalence; we may call its equivalence classes *reduced R–rough sets*. Clearly, there is one–one correspondence between reduced R–rough sets and pairs of R–exact sets of the form $(\underline{R}Y, \overline{R}Y)$.

1.4 Set–algebraic structures

For a knowledge base $\mathcal{U} = (U, R)$, we may form the set $\mathcal{E}_\mathcal{U}$ of R–exact sets. It follows then from our discussion in the previous section that $\emptyset, U \in \mathcal{E}_\mathcal{U}$ and $\mathcal{E}_\mathcal{U}$ is closed with respect to set–theoretic operations of the union, the intersection, the complement and the difference: the set $\mathcal{E}_\mathcal{U}$ with constants $\emptyset, U$ and set–theoretic operations of the union, intersection and the complement is an instance of a structure called *Field of sets*, a particular case of a more general structure called *Boolean algebra*. For this reason, Boolean algebras fall in the province of our interest and we discuss in Chapter 6 structural properties of Boolean algebras and in Chapter 8 we consider Boolean algebras as set–algebraic structures.

On the other hand, R–rough sets do not form a field of sets as e.g. the union of two rough sets may be an exact set etc. etc. In search for an appropriate set–algebraic structure for rough sets, we represent rough sets (or, reduced rough sets) as pairs (I, C) of exact sets where $I \subseteq C$ and I, C satisfy some additional conditions discussed in Chapters 10, 12 or as pairs (I, D) of exact sets where $D = U \setminus C$ hence $I \cap D = \emptyset$ (this representation is thus called *disjoint*). It turns out that a number of well-known set–algebraic structures may be applied to represent rough sets, among them *Nelson algebras* (cf. [Pagliani98[a]]), *Heyting algebras* (cf. [Pagliani98[a]], [Obtulowicz85]), *Stone algebras* (cf. [Pomykala88], [Iwinski87]), *Post algebras* (cf. [Pagliani98[a]]), *Łukasiewicz algebras* (cf. [Pagliani98[a]]). These structures are discussed in Chapters 8 and 12 except for Łukasiewicz algebras whose discussion begins in Chapter 4 with 3–valued Łukasiewicz logical calculus and Wajsberg algebras and continues through Chapter 12 where Wajsberg algebras are shown to be equivalent to 3–Łukasiewicz algebras. Also in Chapter 12, the above mentioned representations of rough sets are constructed.

1.5 Topological structures

Continuing with the knowledge base $\mathcal{U} = (U, R)$, we may observe that the field of sets $\mathcal{E}_\mathcal{U}$ admits as well a topological interpretation (cf. [Skowron88]) viz. we declare as an *open set* any R–exact set and only such a set. As $\mathcal{E}_\mathcal{U}$

is closed on unions and intersection of sets, it becomes a topological space with open sets = R–exact sets. The particular features of this space, which need not be observed in an arbitrary topological space are: (i) any open set is a closed set (as the complement of an R–exact set is an R–exact set) (ii) the intersection of any family of open sets is an open set (as the family $\mathcal{E}_\mathcal{U}$ is closed on arbitrary intersections). We will see in Chapter 7 dedicated to topological structures that these properties characterize topological spaces of exact sets. We denote this topology with the symbol Π_R.

On the other hand, rough sets do not admit so simple topological description. We may recall the characterization of rough sets by means of exact approximations. Let us state here some additional properties of approximations (cf. [Pawlak91]).

The lower approximation operator $\underline{R}$ satisfies the following

$$\begin{aligned}
(i)\ &\underline{R}(X \cap Y) = \underline{R}X \cap \underline{R}Y \\
(ii)\ &\underline{R}X \subseteq X \\
(iii)\ &\underline{R}(\underline{R}X) = \underline{R}X \\
(iv)\ &\underline{R}\emptyset = \emptyset
\end{aligned} \tag{1.6}$$

which correspond closely to properties of the topological *interior operator* with respect to the topology Π_R.

Similarly, the upper approximation operator $\overline{R}$ satisfies dual properties

$$\begin{aligned}
(i)\ &\overline{R}(X \cup Y) = \overline{R}X \cup \overline{R}Y \\
(ii)\ &X \subseteq \overline{R}X \\
(iii)\ &\overline{R}(\overline{R}X) = \overline{R}X \\
(iv)\ &\overline{R}U = U
\end{aligned} \tag{1.7}$$

which correspond closely to properties of the topological *closure operator* with respect to the topology Π_R.

We obtain thus exact sets as interiors, respectively, closures of rough sets in the topology Π_R and we characterize approximation operators as interior, respectively, closure operators in the corresponding topology.

Chapter 7 brings the basic notions and results of set–theoretic topology many of which are applied in a study of topological structures on rough sets in Chapter 11. Rough sets are studied there in the context of a knowledge base induced by an information system $\mathcal{A} = (U, A)$ with a countable set A of attributes as a non–trivial case with respect to the topology Π_A induced by the family of all equivalence classes $[x]_{IND_a}$ where $x \in u$ and $a \in A$. It turns out that topologies on rough sets may be constructed (cf. [Polkowski93^{a,b}]) by applying suitably modified metric topologies of Hausdorff and Pompéju on sets of closed subsets of a topological space. This idea leads to metrics D, D^*, D' on families of rough sets which metrics make families of rough sets into complete metric spaces. A thorough discussion of all topological notions is carried out in Chapter 7 and in Chapter 11 topological spaces of rough sets are studied and some applications are discussed in particular the approximate Banach fixed–point theorem and fractal dimension in information systems.

1.6 Logical aspects of rough sets

Reasoning about knowledge for a given knowledge base $\mathcal{U} = (U, R)$ requires deductive or inductive mechanisms (logical calculi) for making inferences about objects in terms of attributes and their values. Logics to this end were proposed among others in [Orlowska–Pawlak84[a,b]], [Orlowska84,85,89, 90], [Pawlak87], [Rasiowa–Skowron86[a,b,c]], [Vakarelov89].

We will present here the knowledge base in the form of an information system $\mathcal{A} = (U, A)$.

For the sake of being specific, we briefly and informally recall here the logic DL (*decision logic*) following [Pawlak91]. Its elementary formulae are of the form $(\bar{a}, \bar{v})$ where $a \in A$ and $v \in V_a$. Formulae are constructed from elementary formulae by means of propositional functors $\vee$ (of alternation, or disjunction), $\wedge$ (of conjunction), $\neg$ (of negation), $\Rightarrow$ (of implication). Thus, if α, β are formulae then $\alpha \vee \beta$, $\alpha \wedge \beta$, $\neg \alpha$, $\alpha \Rightarrow \beta$ are formulae.

Formulae are interpreted in the universe U as follows. The acceptance symbol $\models$ will be used here. For an object $x \in U$ and a formula α, the symbol $x \models \alpha$ will denote that α is satisfied by x. The satisfaction relation is defined by induction on complexity of formulae measured by the number of occurrences of propositional functors. Thus

$$
\begin{aligned}
& x \models (\bar{a}, \bar{v}) \ iff \ a(x) = v \\
& x \models \alpha \vee \beta \ iff \ either \ x \models \alpha \ or \ x \models \beta \\
& x \models \alpha \wedge \beta \ iff \ x \models \alpha \ and \ x \models \beta \\
& x \models \neg \alpha \ iff \ not \ x \models \alpha \\
& x \models \alpha \Rightarrow \beta \ iff \ either \ x \models \neg \alpha \ or \ x \models \beta
\end{aligned}
\tag{1.8}
$$

A formula α is *true* if and only if $x \models \alpha$ for every $x \in U$. Let us observe that in case $B = \{a_1, ..., a_k\} \subseteq A$ and a formula $\alpha : \alpha_B(v_1, .., v_k)$ is of the form $(\overline{a_1}, \overline{v_1}) \wedge ... \wedge (\overline{a_k}, \overline{v_k})$, if $x \models \alpha$ and $(x, y) \in IND_B$ then $y \models \alpha$ and if it is not the case that $(x, y) \in IND_B$ then it is not the case that $y \models \alpha$.

We may introduce the notion of a *meaning* of a formula α as $[\alpha] = \{x : x \models \alpha\}$ and thus we have: the meaning $[\alpha_B(v_1, ..., v_k)]$ is identical to the equivalence class $[x]_{IND_B}$ where $a_i(x) = v_i$ for each $a_i \in B$. In this way we obtain a logical description of indiscernibility.

1.6.1 Dependencies

True formulae of the form

$$
\alpha_B(v_1, ..., v_k) \Rightarrow \alpha_C(w_1, ..., w_m)
\tag{1.9}
$$

are called *local dependencies*. Dependencies were introduced in [Pawlak85] and studied among others in [Novotný–Pawlak85^d, 88^a, 89] and [Pawlak–Rauszer85].

Let us observe that the truth of the formula 1.9 means that for every $x \in U$, we have

$$if \ x \models \alpha_B(v_1, ..., v_k) \ then \ x \models \alpha_C(w_1, ..., w_m) \qquad (1.10)$$

or, equivalently, $[x]_{IND_B} \subseteq [x]_{IND_C}$. Read informally, the last inclusion means that if values of B–attributes are fixed on the object x then C–attributes take uniquely determined values on that object. Looking from the functional point of view, we may say that C *depends functionally* on B at x.

Motivated by this observation, one may introduce the notion of a (B,C)–*dependence* (cf. op.cit., op.cit.) as a set $\mathcal{F} = \{\phi_i \Rightarrow \psi_i : i \in I\}$ of formulae of the form 1.9 such that the following requirements are fulfilled

$$\begin{aligned} &(i) \ [\phi_i \wedge \psi_i] \neq \emptyset \ for \ each \ i \in I \\ &(ii) \ for \ each \ x \in U, \ there \ is \ i \ \in I \ with \ x \models \phi_i \wedge \psi_i \end{aligned} \qquad (1.11)$$

In case each formula $\phi_i \Rightarrow \psi_i \in \mathcal{F}$ is true, the (B,C)–*dependence* is said to be a *functional dependence*. We know from our previous discussion of local dependencies that in the case of a functional dependence $\mathcal{F}$, we have $[x]_{IND_B} \subseteq [x]_{IND_C}$ for each $x \in U$, thus there exists a function which assigns to each set $\{a(x) : a \in B\}$ a unique set $\{a(x) : a \in C\}$.

In case there is no functional (B,C)–dependence, we say that C *depends partially* on B (cf. [Novotný–Pawlak88^a]). There is in this case a question of a measure of degree of partial dependency. We sketch a few strategies for defining such a measure.

First, we may invoke the idea of Łukasiewicz of assigning fractional truth values to formulae of unary predicate calculus viz. given a formula $\alpha(x)$ over a finite set U of objects, we may assign to α the truth value (weight) $w_\alpha = \frac{\|\alpha\|}{|U|}$ where $|X|$ denotes the number of elements of a finite set X (cardinality of X) and $[\alpha] = \{x \in U : x \models \alpha\}$ is the meaning of α. Then we have for an implication $\alpha \Rightarrow \beta$ that

$$w_{\alpha \Rightarrow \beta} = 1 - w_\alpha + w_{\alpha \wedge \beta} \qquad (1.12)$$

The idea now is to add weights of all formulae in $\mathcal{F}$ relative to the number of all formulae (i.e. cardinality of $\mathcal{F}$). This will give a fraction of objects on which the dependency is locally functional. To give a formula for this measure, let us consider formulae $r : \phi_i \Rightarrow \psi_i \in \mathcal{F}$ which are not true and let us denote by the symbol $\mathcal{F}_\phi$ the set of formulae $r : \chi_i \Rightarrow \psi_i \in \mathcal{F}$ for which

χ_i is ϕ and by κ_ϕ the cardinality of this set i.e. the number of rules of the form $\phi \Rightarrow \psi_i$ in $\mathcal{F}$. Then we have (assuming $\sum_{r \in \mathcal{F}_\phi} w_{\phi \wedge \psi_i} = w_\phi$)

$$\sum_{r \in \mathcal{F}_\phi} w_r = \sum_{r \in \mathcal{F}_\phi} (1 - w_\phi + w_{\phi \wedge \psi_i}) = \kappa_\phi - (\kappa_\phi - 1) w_{\phi_i} \qquad (1.13)$$

Let us observe that each true formula r in $\mathcal{F}$ yields $w_r = 1$ in the overall sum and thus the measure $\alpha_{B,C}$ is given by

$$\alpha_{B,C} = \frac{|\mathcal{F}_t| + \sum_\phi (\kappa_\phi - (\kappa_\phi - 1) w_\phi)}{|\mathcal{F}|} \qquad (1.14)$$

where $\mathcal{F}_t$ is the set of true formulae in $\mathcal{F}$, i.e.

$$\alpha_{B,C} = \frac{|\mathcal{F}| - \sum_\phi (\kappa_\phi - 1) w_\phi}{|\mathcal{F}|} \qquad (1.15)$$

Clearly, $0 \le \alpha_{B,C} \le 1$; in case all formulae are true i.e. the dependency $\mathcal{F}$ is functional, $\kappa_\phi = 1$ for each ϕ and thus $\alpha_{B,C} = 1$. Conversely, $\alpha_{B,C} = 1$ implies the existence of a functional dependency of C on B.

Other strategy may involve an interpretation of a formula $r : \alpha \Rightarrow \beta$ as a conditional i.e. according to Lukasiewicz (cf. the discussion of the Lukasiewicz fractional truth values in Chapter 9), we let $v_r = \frac{w_{\alpha \wedge \beta}}{w_\alpha}$ as the truth value of the implication r. Repeating the strategy of counting the sum of truth values of formulae in $\mathcal{F}$ relative to their number and using v_r we obtain the measure

$$\beta_{B,C} = \frac{|\mathcal{F}_t| + |\Gamma|}{|\mathcal{F}|} \qquad (1.16)$$

where $\Gamma = \{\mathcal{F}_\phi : \phi\}$ is the set of classes $\mathcal{F}_\phi$ of formulae in $\mathcal{F}$ with the common predecessor ϕ. It follows immediately that $\beta_{B,C} = 1$ if and only if C depends functionally on B.

There exist in rough set theory measures of partial dependence related to the notion of a *positive set* (cf. [Pawlak85,91], [Novotný–Pawlak88^a]). In case of a B, C–dependency $\mathcal{F}$, one defines the *positive set* $POS_\mathcal{F}$ as the set of true formulae and then the measure $\gamma_{B,C}$ is defined via

$$\gamma_{B,C} = \frac{|POS_\mathcal{F}|}{|\mathcal{F}|} \qquad (1.17)$$

Again, $\gamma_{B,C} = 1$ if and only if C depends functionally on B.

Finally, the notion of a positive set may be defined by means of the set U of objects; to this end, one may apply approximations viz. given $B, C \subseteq A$, one lets

$$POS_B C = \bigcup_x \underline{B}[x]_{IND_C} \tag{1.18}$$

i.e. $POS_B C$ (called the *positive region*) is the set of those elements x for which $[x]_{IND_B}$ is a subset of $[x]_{IND_C}$. Then we let

$$\delta_{B,C} = \frac{|POS_B C|}{|U|} \tag{1.19}$$

Either of measures defined above takes value 1 only in case of functional dependency; although their numerical values differ on particular information systems and sets of attributes yet they distinguish between functional and partial dependency.

Thus (functional) dependence is a notion which has a logical characterization in *DL*–logic.

Example 1.3. *For* A1, *Fig. 1.1, consider attribute sets* $\{a_1\}$ *and* $\{a_2\}$. *As the relation* IND_{a_1} *induces classes* $\{x_1, x_2, x_6\}, \{x_3, x_4\}, \{x_5, x_7\}$, *and the relation* IND_{a_2} *induces classes* $\{x_1\}, \{x_2\}, \{x_3, x_4\}, \{x_5, x_6, x_7\}$, *neither of those sets depends functionally on the other.*

We calculate measures of partial dependency of $\{a_2\}$ *on* $\{a_1\}$. *The* (a_1, a_2)– *dependence* $\mathcal{F}$ *consists of the following formulae*

1. $(a_1, 1) \Rightarrow (a_2, 3)$

2. $(a_1, 1) \Rightarrow (a_2, 0)$

3. $(a_1, 1) \Rightarrow (a_2, 2)$

4. $(a_1, 3) \Rightarrow (a_2, 1)$

5. $(a_1, 4) \Rightarrow (a_2, 2)$

of which formulae 4,5 are true i.e. $|POS_{\mathcal{F}}| = 2$. *For non-true formulae there is only one predecessor* ϕ *viz.* $(a_1, 1)$ *and thus* $\kappa_\phi = 3$ *and* $||[\phi]|| = 3$. *We thus have*

I. $\alpha_{a_1, a_2} = \frac{2 + 2 \cdot \frac{3}{7}}{5} = \frac{4}{7}$

II. $\beta_{a_1, a_2} = \frac{2 + 1}{5} = \frac{3}{5}$

III. $\gamma_{a_1, a_2} = \frac{2}{5}$

IV. $\delta_{a_1, a_2} = \frac{4}{7}$

as the positive region $POS_{a_1, a_2} = \{x_3, x_4, x_5, x_7\}$.

Dependence is a fundamental notion in rough set theory (cf. a discussion in [Pawlak91, Ch. 7]). We devote to this notion a substantial part of Chapter 10 where we discuss the abstract characterization of dependency due to Novotný and Pawlak and we recall an algebraic characterization of independence due to Rauszer. We also offer therein a closer look at calculus of partial dependencies.

As logics for reasoning about knowledge involve the sentential calculus as well as the predicate calculus, we include in Chapter 2 a discussion of the sentential calculus and in Chapter 9 we recapitulate basic facts about the predicate calculus along with the Łukasiewicz theory of fractional truth values.

1.6.2 Modal aspects of rough sets

The relational system $(U, \{IND_B : B \subseteq A\})$ induced by an information system $\mathcal{A} = (U, A)$ allows for *modalities*. Modal constructions are immanent in rough structures: given a set $X \subseteq U$, and a set $B \subseteq A$ of attributes, one may interpret the formula $x \in \underline{B}X$ as the statement that x belongs *necessarily* in X while the formula $x \in \overline{B}X$ may be interpreted as the statement that x belongs *possibly* in X. These modalities may be rendered formally in logical calculi, cf. [Orlowska84, 85, 89, 90], [Vakarelov89].

Given a formula α and a relation R on the universe U we may consider a formula $[R]\alpha$ with the interpretation $x \models [R]\alpha$ if and only if $y \models \alpha$ for every $y \in U$ with the property that $(x, y) \in R$. Then the operator $[R]$ does satisfy the basic postulate of modal logic concerning the *necessity operator* viz.

$$(K) \quad [R](\alpha \Rightarrow \beta) \Rightarrow ([R]\alpha \Rightarrow [R]\beta) \tag{1.20}$$

i.e. the formula (K) is true.

Richer relational systems may be defined in the context of generalized information systems when one admits the existence of many–valued attributes i.e. a *many–valued information system* $\mathcal{M}$ is a pair (U, A) such that for $a \in A$ we have $a : U \to 2^V$ where 2^V is the set of all subsets of the value set V. Thus, $a(x) \subseteq V$ for each pair a, x. In this context, some basic relations may be induced (cf. [Orlowska–Pawlak84[a,b]]) e.g. the *indiscernibility* relation I defined via $(x, y) \in I$ if and only if $a(x) = a(y)$ for each $a \in A$, the *containment* relation C defined via $(x, y) \in C$ if and only if $a(x) \subseteq a(y)$ for each $a \in A$, and the *tolerance* (similarity) relation T defined via $(x, y) \in T$ if and only if $a(x) \cap a(y) \neq \emptyset$ for each $a \in A$. Each of these relations $R \in \{I, C, T\}$ does induce the modal operator of necessity $[R]$. Logics dealing with these operators were constructed (e.g. as logics $DIL, NDIL, INDL$) in [Orlowska–Pawlak84[a,b]].

In Chapter 12, we present the *information logic IL* due to [Vakarelov89], a complete modal logic interpreted in many–valued information systems encompassing operators $[I], [C], [T]$.

Modal logics are discussed in Chapter 4, where a basic modal logic due to Łukasiewicz is presented and in Chapter 5 in which modal logics $K, T, S4, S5$ are discussed from the point of view of the Kripke (many–world) semantics.

1.6.3 Many–valued logics for rough sets

Recalling the representation of a rough set X as a pair (I, D) of exact sets, we may observe that the three induced sets viz. $I, U \setminus D, B = (U \setminus D) \setminus I$ do represent logical values of truth (I), falsity $(U \setminus D)$, and uncertainty (B). This poses a possibility of interpreting rough sets as a 3–valued algebra of Łukasiewicz, the Lindenbaum–Tarski algebra of the 3–valued Łukasiewicz logic in which truth values are $1, 0, \frac{1}{2}$. That it is indeed so is shown throughout following chapters: in Chapter 4, many–valued logical calculi are discussed and Wajsberg's axiomatization of the Łukasiewicz 3–logic is proved to be complete. In Chapter 12 the equivalence of Wajsberg algebras i.e. Lindenbaum–Tarski algebras of the Wajsberg sentential calculus and Łukasiewicz algebras is demonstrated and a representation of rough sets as a Łukasiewicz algebra is constructed (cf. [Pagliani98^a]). Lindenbaum–Tarski algebras are introduced in Chapter 9.

1.7 Decision systems

The category formation process described above may be regarded from Machine Learning point of view (cf. [Mitchell98]) as a case of *unsupervised learning*: given a data table, concepts are formed by the system itself on the basis of descriptors extracted from the data. In the most important applications, however, one has to deal with the *supervised case* when objects in the universe are pre–classified by an expert (by means of a distinguished attribute d not in A, called the *decision*) and the task of a rough set based algorithm is to induce categories which would approximate as closely as possible the given pre–classification by an expert's decision.

To discuss this problem in formal terms, we recall the notion of a *decision system* understood as a triple $\mathsf{A_d}=(U, A, d)$ where (U, A) is an information system as introduced above, and d is a distinguished attribute called the *decision*; as any other attribute, d is a function $d : U \to V_d$ on the universe U into the value set V_d. Decision systems were introduced in [Pawlak85^b] and studied in [Pawlak86,87^a].

The decision attribute d induces the partition of the universe U into equivalence classes of the relation IND_d of d–*indiscernibility*. Assuming, without any loss of generality, that $V_d = \{1, 2, ..., k(d)\}$, we obtain the par-

	a_1	a_2	d
x_1	1	3	0
x_2	1	0	0
x_3	3	1	1
x_4	3	1	1
x_5	4	2	1
x_6	1	2	0
x_7	4	2	1

Figure 1.3: A decision system $A2_d$.

tition $\{X_1, X_2, ..., X_{k(d)}\}$ of U into *decision classes*. These classes express the classification of objects done by an expert e.g. a medical practitioner, on the basis of their experience, often intuitive and/or based on long experience. This classification should now be modeled by a rough set algorithm in terms of the attribute–value formulae in DL–logic induced from *conditional* attributes in the attribute set A. There are few possibilities imposing themselves.

1. Each of decision classes $X_1, X_2, ..., X_{k(d)}$ may be approximated by its lower, resp. upper, approximations over a set $B \subseteq A$ of attributes. The resulting approximations $\underline{B}X_i$, $\overline{B}X_i$ provide a (local) description of decision classes.

2. Global approximation of decision classes may be produced e.g. by looking at B–indiscernibility classes providing a refinement of the partition by IND_d i.e. we are interested in those B–classes $[x]_B$ which satisfy the condition $[x]_B \subseteq X_i$ for some X_i. This idea leads to the notion of the *positive region of the decision* introduced above.

Given a decision system A_d, and a set $B \subseteq A$ of attributes, we let

$$POS_B(d) = \{x \in U : \exists i \in \{1, 2, .., k(d)\}.[x]_B \subseteq X_i\}.$$

The set $POS_B(d)$ is called the *B–positive region of d*.

Example 1.4. *Let us look at Fig. 1.3. Non–empty subsets of the set A are $\{a_1\}$, $\{a_2\}$ and $\{a_1, a_2\}$. There are two decision classes: $X_1 = \{x : d(x) = 0\} = \{x_1, x_2, x_6\}$ and $X_2 = \{x : d(x) = 1\} = \{x_3, x_4, x_5, x_7\}$. We can see that*

1. $POS_{\{a_1\}}(d) = U$

2. $POS_{\{a_2\}}(d) = \{x_1, x_2, x_3, x_4\}$

3. $POS_A(d) = U$

We may distinguish between the two cases: $POS_B(d) = U$ in which we say that the decision system A_d is B–*deterministic* (there is an exact description of d in terms of IND_B as d depends functionally on B) and $POS_B(d) \neq U$ in which we say that A_d is B–*non–deterministic* (we may only obtain an approximate description of d in terms of IND_B as d depends partially only on B).

In our example above, we have found that our decision system has been both $\{a_1\}$–deterministic and A–deterministic. This fact poses a problem of *information reduction*: some attributes may be redundant in the classification process and their removal reduces the amount of information encoded in data while preserving knowledge. This remark leads to the fundamental notion of a *reduct*.

1.8 Knowledge reduction

Optimization in a knowledge base $(U, \{IND_B : B \subseteq A\})$ consists in finding minimal sets of attributes which preserve knowledge represented in the original information system (U, A); in accordance with the MDL–principle (cf. [Rissanen83], [Mitchell98, Ch. 6]) such optimization leads to logical descriptors in DL–logic having formulae of the shortest possible length.

Optimization of knowledge may be started with any of functions $\eta \in \{\alpha, \beta, \gamma, \delta\}$ of pp.13, 14; given $\eta_{B,C}$ for $B, C \subseteq A$, we may search for a subset $D \subseteq B$ which is a minimal subset of B preserving the partial dependency value $\eta_{B,C}$ i.e.

$$
\begin{aligned}
(i) \; & \eta_{D,C} = \eta_{B,C} \\
(ii) \; & if \; E \subseteq D \; and \; E \neq D \; then \; \eta_{E,C} \neq \eta_{D,C}
\end{aligned}
\tag{1.21}
$$

A set D satisfying (i), (ii) in equation 1.21 is said to be a B, C, η–*reduct*. Let us observe that in case $\eta_{B,C} = 1$ i.e. when C depends functionally on B, the notion of a reduct does not depend on η and a B, C, η–reduct D is a minimal subset of B with respect to the condition that C depends functionally on D. In this case D is called a B, C–reduct.

In the yet more particular case when $B = C$, a B, C–reduct is a minimal subset D of B such that $IND_D = IND_B$; indeed, for any $D \subseteq B$ we have $IND_B \subseteq IND_D$ hence B may depend functionally on D only if $IND_B = IND_D$. In this case a B, D–reduct is called a *reduct of B*.

Reducts of A are said to be reducts of the information system (U, A).

In other words, a reduct is a minimal set of attributes from A that preserve the partitioning of the universe by IND_A, and hence the original classification. Finding a minimal reduct is computationally an NP-hard problem as shown in [Skowron89] and [Skowron–Rauszer92].

Fortunately, there exist good heuristics that e.g. compute sufficiently many reducts in an admissible time, unless the number of attributes is very high.

1.8.1 Reducts via boolean reasoning: discernibility approach

It was shown by Skowron [Skowron89] and Skowron and Rauszer (op. cit.) that the problem of finding reducts of a given information system may be solved as a case in *boolean reasoning*; the idea of boolean reasoning, going back to George Boole, is to represent a problem with a boolean function and to interpret its *prime implicants* (an *implicant* of a boolean function f is any conjunction of literals (variables or their negations) such that for each valuation v of variables, if the values of these literals are true under v then the value of the function f under v is also true; a *prime* implicant is a minimal implicant) as solutions to the problem. This approach to the reduct problem is implemented as follows.

Let A be an information system with n objects and k attributes. The *discernibility matrix* of A is a symmetric $n \times n$ matrix M_A with entries c_{ij} consisting of sets of attributes discerning between objects x_i and x_j where $i, j = 1, ..., n$:

$$c_{ij} = \{a \in A : a(x_i) \neq a(x_j)\} \tag{1.22}$$

Once the discernibility matrix M_A is found, the *discernibility function* f_A for the information system A, a Boolean function of k boolean variables $a_1^*, ..., a_k^*$ (corresponding to attributes $a_1, ..., a_k \in A$), is defined as below, where $c_{ij}^* = \{a^* : a \in c_{ij}\}$

$$f_A(a_1^*, ..., a_k^*) : \bigwedge \left\{ \bigvee c_{ij}^* : 1 \leq j, c_{ij} \neq \emptyset \right\} \tag{1.23}$$

Then, (op.cit., op. cit.), there is a one-to-one correspondence between the set $PRI - IMP(f_A)$ of prime implicants of the discernibility function f_A and the set $RED(A)$ of reducts of the information system A; specifically,

Proposition 1.1. *A conjunction $\wedge_{j=1}^{m} a_{i_j}^*$ is in $PRI - IMP(f_A)$ if and only if the set $\{a_{i_1}, a_{i_2}, ..., a_{i_m}\}$ is in $RED(A)$*

Indeed, consider a set $B \subseteq A$ of attributes, and define a valuation v_B via $v_B(a^*) = 1$ if $a \in B$ and $v_B(a^*) = 0$ if $a \notin B$. Assume that f_A takes on the value 1 under v_B. Then f_A in the dual, normal disjunctive form : $\bigvee_u \bigwedge d_u^*$ where $\bigwedge d_u^*$ is $a_{1,u}^* \wedge ... \wedge a_{k_u,u}^*$, takes on the value 1 as well, hence for some u we have the value 1 for $\bigwedge d_u^*$ meaning that $\{a_{1,u}, ..., a_{k_u,u}\} \subseteq B$. It follows that a minimal B with $IND_A = IND_B$ is of the form $\{a_{1,u}, ..., a_{k_u,u}\}$ for some u.

Example 1.5. *Look at the decision system shown in Fig. 1.4, below. We compute reducts in its conditional part* A

We may check that the discernibility function f_A is (d, e, f, r are boolean variables corresponding to a_1, a_2, a_3, a_4, respectively, and the $\wedge$ sign is omitted):

	a_1	a_2	a_3	a_4	Decision
x_1	1	1	1	2	yes
x_2	1	0	1	0	no
x_3	2	0	1	1	no
x_4	3	2	1	0	yes
x_5	3	1	1	0	no
x_6	3	2	1	2	yes
x_7	1	2	0	1	yes
x_8	2	0	0	2	no

Figure 1.4: A decision table $A3_d$

$$
\begin{aligned}
f_A(d,e,f,r) \;:\; &(e \vee r)(d \vee e \vee r)(d \vee e \vee r)(d \vee r)(d \vee e)(e \vee f \vee r)(d \vee e \vee f) \\
&(d \vee r)(d \vee e)(d \vee e)(d \vee e \vee r)(e \vee f \vee r)(d \vee f \vee r) \\
&(d \vee e \vee r)(d \vee e \vee r)(d \vee e \vee r)(d \vee e \vee f)(f \vee r) \\
&(e)(r)(d \vee f \vee r)(d \vee e \vee f \vee r) \\
&(e \vee r)(d \vee e \vee f \vee r)(d \vee e \vee f \vee r) \\
&(d \vee f \vee r)(d \vee e \vee f) \\
&(d \vee e \vee r)
\end{aligned}
$$

The function f_A can be brought to an equivalent simplified form after boolean absorption laws are applied:

$$ f_A(d,e,f,r) : er $$

It follows that the set $\{a_2, a_4\}$ of attributes is the reduct of the information system A.

Reducts and dependencies are discussed in Chapter 10, when we follow an abstract theory of reducts and dependencies due to Pawlak and Novotný and framed in the theory of semi–lattices, and where we also include an algebraic characterization of independence due to Rauszer.

1.8.2 Reducts in decision systems

The problem of reduction of the conditional attribute set in decision systems may be approached on similar lines. We resort here to the notion of the positive region of a decision, defined above.

Given a decision system A_d, a subset $B \subseteq A$ of the attribute set A is called a *relative reduct of* A_d if

$$
\begin{aligned}
&(i)\ POS_B(d) = POS_A(d) \\
&(ii)\ B\ \textit{is minimal with respect to } (i)
\end{aligned}
\tag{1.24}
$$

Relative reducts may be found along similar lines to those for reducts i.e. by an appropriate application of Boolean reasoning (cf. Skowron and Rauszer (op. cit.)). It suffices to modify the discernibility matrix:

$$M_{\mathsf{A_d}}^d = (c_{ij}^d) \; where \quad \begin{matrix} c_{ij}^d = \emptyset \; in \; case \; d(x_i) = d(x_j) \\ c_{ij}^d = c_{ij} - \{d\} \quad otherwise \end{matrix} \tag{1.25}$$

The matrix $M_{\mathsf{A_d}}^d$ is the *relative discernibility matrix of* $\mathsf{A_d}$.

From the relative discernibility matrix, the *relative discernibility function* $f_{\mathsf{A_d}}^d$ is constructed in the same way as the discernibility function was constructed from the discernibility matrix. Again, as shown by Skowron and Rauszer (op. cit.), the set $PRI - IMP(f_{\mathsf{A_d}}^d)$ of *prime implicants* of $f_{\mathsf{A_d}}^d$ and the set $REL - RED(\mathsf{A_d})$ of relative reducts of A_d are in one–to–one correspondence described above.

Example 1.6. *Relative reducts for the decision table* $A3_d$ *are:*
$REL - RED(\mathsf{A_d}) = \{\{a_1, a_2\}, \{a_2, a_4\}\}.$

1.8.3 Rough membership functions

While two measures introduced above may be regarded as global estimates of approximation quality, we may have (and feel the need for) a more precise, local measure indicating the roughness degree of a set X at particular objects. This purpose is served by the *rough membership function* due to [Pawlak–Skowron94]. In agreement with the rough set ideology, the rough membership function is information–dependent and a fortiori it is constant on indiscernibility classes.

Given a set $B \subseteq A$ of attributes, the rough membership function $\mu_X^B : U \to [0, 1]$ is defined as follows

$$\mu_X^B(x) = \frac{|[x]_B \cap X|}{|[x]_B|} \tag{1.26}$$

The rough membership function can be interpreted as an unbiased estimate of $\Pr(x \in X \mid x, B)$, the conditional probability that the object x belongs to the set X, given x is defined uniquely in terms of B (cf. [Pawlak01]).

Clearly, a set X is B–exact if and only if $\mu_X^B = \chi_X$ i.e. the rough membership function of X is identical with its *characteristic function* χ_X taking values $0, 1$ only ($\chi_X(x) = 1$ when $x \in X$ and 0, otherwise). Thus, rough set theory agrees with the classical set theory when only exact sets are concerned. We will meet rough membership functions in Appendix, Chapter 14, where they will be archetypical examples of exact fuzzy rough sets.

At this point we conclude our survey of basic notions of rough set theory accompanied by pointers to relevant mathematical structures induced in rough set environment and to their discussion in the following chapters. We would like to include now into our survey of rough sets a brief discussion of their present state which we hope would give the reader an insight into some of the topics currently studied in rough sets.

Part 2: Selected current trends in rough set research

The main application of decision rules is in classification of new objects, not included in the training decision system. Thus, decision rules adequate to this task should be robust i.e. stable with respect to "sufficiently small" perturbations of information sets of objects and noise–resistant i.e. they should preserve classification in the presence of noise.

It is therefore one of the main research tasks of rough set community to explore various ideas leading to induction of rules well predisposed to the task of classification of new yet unseen objects. These tasks encompass among others

1. Searching for adequate sets of attributes (reducts or approximations to them) which would determine predecessors of decision rules.

2. Searching for adequate techniques for discovery of proper patterns (meanings of rule predecessors) which would ensure the high classification rate by decision rules.

3. Searching for decomposition methods in a two–fold sense; first, in order to decompose (preprocess) complex attributes e.g. to discretize continuous attributes, and next, to decompose complex (large) decision systems into smaller ones, from which it would be computationally feasible to extract local decision rules and then to aggregate them (fuse) into global ones pertaining to the original table. Clearly, this point has close relations to 1 and 2.

We now present an outline of techniques and ideas developed in order to fulfill the program described in 1–3 above. The main purpose of this currently going research can be described as a search for decision–related features in the conditional part of the decision system which would induce classifiers of a satisfactory quality.

1.9   Similarity based techniques

Rough set community has found quite recently that one possible and promising way of obtaining good classifiers is to relax the indiscernibility relations to tolerance or similarity relations (cf. [Bazan 98], [Greco99], [Greco98], [Nguyen–Nguyen98^a], [Slowinski0x], [Skowron–Stepaniuk96], [Stepaniuk00]).

We will call in the sequel a *tolerance* relation any relation τ which satisfies for each pair $x, y \in U$ (cf. equation 1.1)

$$
\begin{aligned}
&(i) \ \tau(x, x) \\
&(ii) \ \tau(x, y) \implies \tau(y, x)
\end{aligned}
\tag{1.27}
$$

A *similarity* relation will be any relation τ which is merely reflexive.

We will outline a few ways by which tolerance or similarity approach enters rough set methods.

1.9.1 General approximation spaces

An abstract rendering of classical rough set approach has been proposed in the form of an *approximation space* i.e. a universe U endowed with operators l, u on the power set $P(U)$ of U whose properties reflect the properties of lower and upper approximations. A modification of this approach is defined as a *generalized approximation space*. A detailed analysis of generalized approximation spaces with applications to rule induction may be found in [Stepaniuk00].

A *generalized approximation space* is a system $GAS = (U, I, \nu)$, where

1. U is a universe (i.e. a non–empty set of objects).

2. $I : U \to P(U)$ is an *uncertainty function*.

3. $\nu : P(U) \times P(U) \to [0, 1]$ is a *rough inclusion function*.

The uncertainty function assigns to an object x the set $I(x)$ of objects similar in an assumed sense to x.

Practically, such functions may be constructed e.g. from metrics imposed on value sets of attributes:

$$
y \in I(x) \Leftrightarrow max_a\{dist_a(a(x), a(y))\} \leq \varepsilon
\tag{1.28}
$$

where $dist_a$ is a fixed metric on V_a and ε is a fixed threshold.

Values $I(x)$ of the uncertainty function define usually a covering of U (as it is usually true that $x \in I(x)$ for any x: uncertainty functions defined in rough set theory on the lines sketched above induce usually tolerance relations).

The rough inclusion function ν defines the degree of inclusion between two subsets of U.

An example of a rough inclusion is a *generalized rough membership function*

$$
\nu(X, Y) =
\begin{array}{l}
\frac{|X \cap Y|}{|X|} \ if \ X \neq \emptyset \\
1 \quad otherwise
\end{array}
\tag{1.29}
$$

The lower and the upper approximations of subsets of U in the generalized approximation space may be defined as follows.

For an approximation space $GAS = (U, I, \nu)$ and any subset $X \subseteq U$, lower and upper approximations are defined by

$$\underline{GAS}X = \{x \in U : \nu\,(I\,(x)\,, X) = 1\}$$
$$\overline{GAS}X = \{x \in U : \nu\,(I\,(x)\,, X) > 0\} \tag{1.30}$$

From both theoretical as well as practical points of view, it is desirable to allow I, ν to depend on some parameters, tuned in the process of optimalization of the decision classifier.

The analogy between classical rough set approximations and those in generalized approximation spaces may be carried out further e.g. the notion of a positive region may be rendered here as follows.

Given a classification $\{X_1, .., X_r\}$ of objects (i.e. $X_1, .., X_r \subseteq U$, $\bigcup_{i=1}^{r} X_i = U$ and $X_i \cap X_j = \emptyset$ for $i \neq j$, where $i, j = 1, \ldots, r$), the *positive region* of the classification $\{X_1, \ldots, X_r\}$ with respect to the approximation space GAS is defined by

$$POS\,(GAS, \{X_1, \ldots, X_r\}) = \bigcup_{i=1}^{r} \underline{GAS}X_i \tag{1.31}$$

The quality of approximation of the classification $\{X_1, \ldots, X_r\}$ and other related numerical factors may be then defined on the lines of classical rough set theory, with appropriate modifications.

The notions of a reduct and a relative reduct are preserved in generalized approximation spaces, their definitions adequately modified. For instance, the notion of a reduct may be defined in a generalized approximation space as follows.

A subset $B \subseteq A$ is called a *reduct* of A if and only if

$$\begin{array}{l} (i) \ for \ every \ x \in U,\ I_B\,(x) = I_A\,(x) \\ (ii)\ for\ every\ proper\ subset\ C \subset B\ (i)\ holds\ not \end{array} \tag{1.32}$$

where I_B is a partial uncertainty function obtained by restricting ourselves to attributes in B.

Similarly, one extends the definition of a relative reduct and a fortiori decision rules are induced by classical techniques adapted to this case. Algorithms based on Boolean reasoning presented above may be extended to the new case.

One may observe that due to the fact that in this new context we may operate with parameterized tolerance relations, we obtain a greater flexibility in modeling the classification by the decision via conditional attributes; in consequence, we have prospects for better classification algorithms.

1.9.2   Rough mereology

In generalized approximation spaces, the search is for tolerance relations, induced from value sets of attributes as ingredients of uncertainty functions and such that decision rules defined from these relations have sufficient quality.

Such approach has obvious advantages of being flexible and practical.

One may look however at this situation with a theoretician's eye and notice that from an abstract point of view this approach actually is about establishing for any two objects (or, concepts) a degree in which one of them is the "part" of the other; the notion of a "part in degree" being non–symmetrical would lead in general to similarity (and non–tolerance) relations. Clearly, one may adapt to this new situation all algorithms for reduct and rule generation presented above.

In addition, a judiciously defined notion of " a part in degree" would lead to a new paradigm having close relations to theories for approximate reasoning based on the primitive notion of a "part" i.e. mereological and meronymical theories; also this paradigm would have affinities with fuzzy set theory based on the notion of being an element in a degree. Rough mereology fulfills these expectations.

Rough mereology has been proposed as a paradigm for approximate reasoning in complex information systems (cf. [Polkowski–Skowron01] and entries in the Selected Bibliography). Its primitive notion is that of a "rough inclusion" functor which gives for any two entities of discourse the degree in which one of them is a part of the other. Rough mereology may be regarded as an extension of rough set theory as it proposes to argue in terms of similarity relations induced from a rough inclusion instead of reasoning in terms of indiscernibility relations; it also proposes an extension of mereology as it replaces the mereological primitive functor of being a part with a more general functor of being a part in a degree. Rough mereology has deep relations to fuzzy set theory as it proposes to study the properties of partial containment which is also the fundamental subject of study for fuzzy set theory. Rough mereology may be regarded as an independent first order theory but it may be also formalized in the traditional mereological scheme proposed first by Stanisław Leśniewski where mereology is regarded and formalized within ontology i.e. theory of names (concepts). We regard this approach as particularly suited for rough set theory which is also primarily concerned with concept approximation in information systems.

We formalize here the functor of "part in degree" as a predicate μ_r with $X\mu_r Y$ meaning that X is a part of Y in degree r. We recall that a mereological theory of sets may be based on the primitive notion of an *element el* satisfying the following

$$
\begin{aligned}
&(i) \ X el X \\
&(ii) \ X el Y \wedge Y el Z \Longrightarrow X el Z \\
&(iii) \ X el Y \wedge Y el X \Longrightarrow X = Y
\end{aligned}
\tag{1.33}
$$

The functor *pt* of *part* is defined as follows

$$
X pt Y \Longleftrightarrow X el Y \wedge non(X = Y)
\tag{1.34}
$$

Rough mereology is constructed in such way as to extend mereology. The following is the list of basic postulates of rough mereology.

$$(RM1)\ X\mu_1 Y \iff XelY$$
$$(RM2)\ X\mu_1 Y \implies \forall Z.(Z\mu_r X \implies Z\mu_r Y)$$
$$(RM3)\ X = Y \wedge X\mu_r Z \implies Y\mu_r Z \tag{1.35}$$
$$(RM4)\ X\mu_r Y \wedge s \leq r \implies X\mu_s Y$$

(RM1) means that being a part in degree 1 is equivalent to being an element; in this way a connection is established between rough mereology and mereology: the latter is a sub–theory of the former.

(RM2) is meaning the monotonicity property: any object Z is a part of Y in degree not smaller than that of being a part in X whenever X is an element of Y.

(RM3) means that the identity of individuals is a congruence with respect to μ.

(RM4) does establish the meaning " a part in degree at least r".

We may observe that $X\mu_r Y$ may be regarded as the statement that X is similar to Y in degree r i.e. as a parameterized similarity relation. We call the functor μ a *rough inclusion*.

One may ask about procedures for defining rough inclusions in information systems. The following simple procedure defines a rough inclusion in an information system (U, A) in case we would like to start with rows of the information table.

1. Consider a partition $P = \{A_1, A_2, ..., A_k\}$ of A.

2. Select a convex family of coefficients: $W = \{w_1, w_2, ..., w_k\}$ (i.e. $w_i \geq 0$, $\sum_{i=1}^{k} w_i = 1$).

3. Define $IND(A_i)(x, x') = \{a \in A_i : a(x) = a(x')\}$.

4. Let $r = \sum_{i=1}^{k} w_i \frac{card(IND(A_i)(x,x'))}{card(A_i)}$.

5. Let $x\mu_r x'$.

The rough inclusion defined according to this simple recipe is symmetric: $x\mu_r x' \iff x'\mu_r x$.

However, taking an additional care, one may find a bit more intricate recipes leading to similarities not being tolerances.

1.9.3   Rough mereology in complex information systems: in search of features in distributed systems

Rough mereology is well–suited to the task of concept approximation in complex information systems e.g. distributed systems. We outline here its workings in the case of a distributed/multi–agent system.

Distributed systems of agents

We assume that a pair (Inv, Ag) is given where Inv is an *inventory of elementary objects* and Ag is a set of intelligent computing units called shortly *agents*. We consider an agent $ag \in Ag$. The agent ag is endowed with tools for reasoning and communicating about objects in its scope; these tools are defined by components of the agent label.

The *label of the agent ag* is the tuple

$$lab(ag) = (\mathsf{A}(ag), \mu(ag), L(ag), Link(ag), O(ag), St(ag)),$$

$$Unc - rel(ag), Unc - rule(ag), Dec - rule(ag))$$

where

1. $\mathsf{A}(ag)$ is an information system of the agent ag.

2. $\mu(ag)$ is a functor of part in a degree at ag.

3. $L(ag)$ is a set of unary predicates (properties of objects) in a predicate calculus interpreted in the set $U(ag)$ (e.g. in the descriptor logic $L_{A(ag)}$).

4. $St(ag) = \{st(ag)_1, ..., st(ag)_n\} \subset U(ag)$ is the set of *standard objects* at ag.

5. $Link(ag)$ is a collection of strings of the form $ag_1 ag_2 ... ag_k ag$ which are elementary teams of agents; we denote by the symbol $Link$ the union of the family $\{Link(ag) : ag \in Ag\}$.

6. $O(ag)$ is the set of *operations* at ag; any $o \in O(ag)$ is a mapping of the Cartesian product $U(ag_1) \times U(ag_2) \times ... \times U(ag_k)$ into the universe $U(ag)$ where $ag_1 ag_2 ... ag_k ag \in Link(ag)$.

7. $Unc - rel(ag)$ is the set of parameterized *uncertainty relations* $\rho_i = \rho_i(o_i(ag), st(ag_1)_i, st(ag_2)_i, ..., st(ag_k)_i, st(ag))$ where

$$ag_1 ag_2 ... ag_k ag \in Link(ag), o_i(ag) \in O(ag)$$

are such that

$$\rho_i((x_1, \varepsilon_1), (x_2, \varepsilon_2), ., (x_k, \varepsilon_k), (x, \varepsilon))$$

holds for $x_1 \in U(ag_1), x_2 \in U(ag_2), .., x_k \in U(ag_k)$ and $\varepsilon, \varepsilon_1, \varepsilon_2, .., \varepsilon_k \in [0, 1]$ if and only if

$$\begin{aligned} &(i) \ x_j \mu(ag_j)_{\varepsilon_j} st(ag_j) \ (j = 1, 2, ..., k) \\ &(ii) \ x \mu(ag)_\varepsilon st(ag) \end{aligned}$$

$\qquad\qquad(1.36)$

where

$$o_i(st(ag_1), st(ag_2), .., st(ag_k)) = st(ag) \text{ and } o_i(x_1, x_2, .., x_k) = x.$$

Uncertainty relations express the agents knowledge about relationships among uncertainty coefficients of the agent ag and uncertainty coefficients of its children.

8. $Unc - rule(ag)$ is the set of *uncertainty rules* f_j where $f_j : [0,1]^k \longrightarrow [0,1]$ is a function which has the property that
   if $x_1 \in U(ag_1), x_2 \in U(ag_2), .., x_k \in U(ag_k)$ satisfy the condition

$$x_i \mu(ag_i)_{\varepsilon(ag_i)} st(ag_i)$$

   then $o_j(x_1, x_2, ..., x_k) \mu(ag)_{f_j(\varepsilon(ag_1), \varepsilon(ag_2), ..., \varepsilon(ag_k))} st(ag)$ where all parameters are as above.

9. $Dec - rule(ag)$ is a set of *decomposition rules* $dec - rule_i$ and

$$(\Phi(ag_1), \Phi(ag_2), .., \Phi(ag_k), \Phi(ag)) \in dec - rule_i$$

   where $\Phi(ag_1) \in L(ag_1), \Phi(ag_2) \in L(ag_2), .., \Phi(ag_k) \in L(ag_k)$,
   $\Phi(ag) \in L(ag)$ and $ag_1 ag_2 .. ag_k ag \in Link(ag))$ if there exists a collection of standards $st(ag_1), st(ag_2), ..., st(ag_k), st(ag)$ with the properties that $o_j(st(ag_1), st(ag_2), ..., st(ag_k)) = st(ag), st(ag_i)$ satisfies $\Phi(ag_i)$ for $i = 1, 2, .., k$ and $st(ag)$ satisfies $\Phi(ag)$. Decomposition rules are decomposition schemes in the sense that they describe the standard $st(ag)$ and the standards $st(ag_1)$,
   ..., $st(ag_k)$ from which the standard $st(ag)$ is assembled under o_i in terms of predicates which these standards satisfy.

Approximate synthesis of features in distributed information systems

The process of synthesis of a complex feature (e.g. a signal, an action) by the above defined scheme of agents consists of the two communication stages viz. the top - down communication/negotiation process and the bottom - up communication/assembling process. We outline the two stages here in the language of approximate formulae.

We assume for simplicity that the relation $ag' \leq ag$, which holds for agents $ag', ag \in Ag$ iff there exists a string $ag_1 ag_2 .. ag_k ag \in Link(ag)$ with $ag' = ag_i$ for some $i \leq k$, orders the set Ag into a tree.

We also assume that $O(ag) = \{o(ag)\}$ for $ag \in Ag$ i.e. each agent has a unique assembling operation.

We define a logic $L(Ag)$ in which we can express global properties of the synthesis process.

Elementary formulae of $L(Ag)$ are of the form $< st(ag), \Phi(ag), \varepsilon(ag) >$ where

$st(ag) \in St(ag), \Phi(ag) \in L(ag), \varepsilon(ag) \in [0,1]$ for any $ag \in Ag$. Formulae of $L(ag)$ form the smallest extension of the set of elementary formulae closed under propositional connectives $\lor$, $\land$, $\neg$ and under the modal operators $\Box$, $\Diamond$.

The meaning of a formula $\Phi(ag)$ is defined classically as the set $[\Phi(ag)] = \{u \in U(ag) : u$ has the property $\Phi(ag)\}$; we denote satisfaction by $u \models \Phi(ag)$. For $x \in U(ag)$, we say that x *satisfies* $< st(ag), \Phi(ag), \varepsilon(ag) >$, in symbols:

$$x \models < st(ag), \Phi(ag), \varepsilon(ag) >,$$

if

$$\begin{aligned}
&(i) \; st(ag) \models \Phi(ag) \\
&(ii) \; x\mu(ag)_{\varepsilon(ag)}st(ag)
\end{aligned} \tag{1.37}$$

We extend satisfaction over formulae by recursion as usual.

By a *selection* over Ag we mean a function sel which assigns to each agent ag an object $sel(ag) \in U(ag)$. For two selections sel, sel' we say that sel *induces* sel', in symbols

$$sel \rightarrow_{Ag} sel'$$

when

$$\begin{aligned}
&(i) \; sel(ag) = sel'(ag) \; for \; any \; ag \in Leaf(Ag) \\
&(ii) \; sel'(ag) = o(ag)(sel'(ag_1), sel'(ag_2), ..., sel'(ag_k)) for \; any \\
&\qquad ag_1 ag_2 ... ag_k ag \in Link
\end{aligned} \tag{1.38}$$

We extend the satisfiability predicate $\models$ to selections: for an elementary formula $< st(ag), \Phi(ag), \varepsilon(ag) >$, we let

$$\begin{aligned}
&sel \models < st(ag), \Phi(ag), \varepsilon(ag) > \; iff \\
&sel(ag) \models < st(ag), \Phi(ag), \varepsilon(ag) >
\end{aligned} \tag{1.39}$$

We now let $sel \models \Diamond < st(ag), \Phi(ag), \varepsilon(ag) >$ in case there exists a selection sel' satisfying the condition

$$\begin{aligned}
&(i) \; sel \rightarrow_{Ag} sel' \\
&(ii) \; sel' \models < st(ag), \Phi(ag), \varepsilon(ag) >
\end{aligned} \tag{1.40}$$

In terms of $L(Ag)$ it is possible to express the problem of synthesis of an approximate solution to the problem posed to Ag. We denote by $head(Ag)$ the root of the tree $(Ag, \leq)$ and by $Leaf(Ag)$ the set of leaf-agents in Ag. In the process of top − down communication, a requirement Ψ received by the scheme from an external source (which may be called a *customer*) is decomposed into approximate specifications of the form $< st(ag), \Phi(ag), \varepsilon(ag) >$ for any agent ag of the scheme. The decomposition process is initiated at the agent $head(Ag)$ and propagated down the tree. We now are able to formulate the synthesis problem.

The Synthesis Problem. *Given a formula*

$\alpha :< st(head(Ag)), \Phi(head(Ag)), \varepsilon(head(Ag)) >$

find a selection sel over the tree $(Ag, \leq)$ *with the property that* $sel \models \alpha$.

A solution to the Synthesis Problem with a given formula α is found by negotiations among the agents based on uncertainty rules and their successful result can be expressed by a top − down recursion in the tree $(Ag, \leq)$ as follows: for a team $ag_1 ag_2 ... ag_k ag$ with the formula $< st(ag), \Phi(ag), \varepsilon(ag) >$ already chosen, it is sufficient that each agent ag_i choose a standard $st(ag_i) \in U(ag_i)$, a formula $\Phi(ag_i) \in L(ag_i)$ and a coefficient $\varepsilon(ag_i) \in [0, 1]$ such that

$$(i)\ (\Phi(ag_1), \Phi(ag_2), ..., \Phi(ag_k), \Phi(ag)) \in Dec - rule(ag)$$
$$with\ standards\ st(ag), st(ag_1), ..., st(ag_k) \qquad (1.41)$$
$$(ii)\ f(\varepsilon(ag_1), .., \varepsilon(ag_k)) \geq \varepsilon(ag)$$

where f satisfies $unc - rule(ag)$ with $st(ag)$, $st(ag_1)$, ..., $st(ag_k)$ and $\varepsilon(ag_1)$, ..., $\varepsilon(ag_k)$, $\varepsilon(ag)$.

For a formula α, we call an α - *scheme* an assignment of a formula $\alpha(ag)$: $< st(ag), \Phi(ag), \varepsilon(ag) >$ to each $ag \in Ag$ in such manner that (i),(ii) in 1.41 above are satisfied and $\alpha(head(Ag))$ is

$$< st(head(Ag)), \Phi(head(Ag)), \varepsilon(head(Ag)) > .$$

We denote this scheme with the symbol

$$sch(< st(head(Ag)), \Phi(head(Ag)), \varepsilon(head(Ag)) >).$$

We say that a selection *sel* is *compatible* with a scheme

$$sch(< st(head(Ag)), \Phi(head(Ag)), \varepsilon(head(Ag)) >)$$

in case $sel(ag)\mu(ag)_{\varepsilon(ag)} st(ag)$ for each leaf agent $ag \in Ag$.

The goal of negotiations can be summarized now as follows.

Proposition 1.2. *(The Sufficiency Condition) Given a formula*

$$< st(head(Ag)), \Phi(head(Ag)), \varepsilon(head(Ag)) > \qquad (1.42)$$

if a selection sel is compatible with a scheme

$$sch(< st(head(Ag)), \Phi(head(Ag)), \varepsilon(head(Ag)) >) \qquad (1.43)$$

then

$$sel \models \Diamond < st(head(Ag)), \Phi(head(Ag)), \varepsilon(head(Ag)) > \qquad (1.44)$$

Rough mereological approach to complex information systems has other merits: the dominant one is the presence of the *class operator* of mereology (cf. [Polkowski–Skowron01]). The class operator allows for representation of a collection of objects as an individual object and – depending on the context – it may be applied as a clustering, granule–forming, or neighborhood–forming operator.

Due to class operator, rough mereology has been applied to problems of granulation of knowledge and computing with words (cf. [Polkowski–Skowron01], [Polkowski–Polkowski00]).

1.10 Generalized and approximate reducts

We have outlined above some approaches to decision rule induction based on application of similarity relations instead of equivalence relations. Under these approaches, reducts undergo changes: we may expect that new reducts will be subsets of classically defined reducts thus leading to shorter decision rules with better classification qualities.

Such motivations are behind the general problem of relevant feature selection to which the rough set community has devoted a considerable attention. We now present some more general approaches to reducts leading to new notions of reducts and a fortiori to new mechanisms of decision rule generation and to new classifiers based on them.

1.10.1 Frequency based reducts

In classical deterministic decision systems, reducts are minimal sets of attributes which preserve information about decision. This observation may be a starting point for various ramifications of the notion of a reduct: given a measure of information related to decision, one may search for various sets of attributes preserving or approximately preserving this information with respect to the assumed measure.

Some of these measures may be based on frequency characteristics of decision.

The reader will find a detailed discussion of frequency based measures and reducts in [Ślęzak00]; here, we give a few introductory examples.

In a general setting, one has to accept the fact that values of an attribute a may be complex i.e. they may be intervals, or subsets of the value set V_a; accordingly, the notion of an information set becomes more complex

$$Inf_B(u) = \prod_{a \in B} \{\langle a, a(u) \rangle\} \tag{1.45}$$

i.e. elements of the information sets are threads (i.e. elements) in the Cartesian product of values of $a \in B$ on u.

Elements of the set

$$V_B^U = \{Inf_B(u) : u \in U\} \tag{1.46}$$

of all supported information patterns on B correspond to equivalence classes of indiscernibility relation

$$IND_B = \{(u_1, u_2) \in U \times U : Inf_B(u_1) = Inf_B(u_2)\} \tag{1.47}$$

We denote by w_B an element of Inf_B.

Generalized decision reducts

Representing knowledge via decision rules requires attaching to each information pattern over considered conditional attributes a local decision information. As an example, let us consider the notion of a generalized decision function $\partial_{d/B} : V_B \longrightarrow 2^{V_d}$, which, for any subset $B \subseteq A$, labels information patterns with possible decision values

$$\partial_{d/B}(w_B) = \{d(u) : u \in Inf_B^{-1}(w_B)\} \tag{1.48}$$

For each $w_B \in V_B^U$, the subset $\partial_{d/B}(w_B) \subseteq V_d$ contains exactly these decision values which occur in the indiscernibility class of w_B. Thus the implication

$$(B, w_B) \longrightarrow \bigvee_{v_d \in \partial_{d/B}(w_B)} (d, v_d) \tag{1.49}$$

is in some sense optimal with respect to A_d. Namely, disjunction in its right side is the minimal one which keeps consistency of the rule with A_d: any implication with (B, w_B) as predecessor and the disjunction of a smaller number of atomic elements than $|\partial_{d/B}(w_B)|$ as the successor is not true in A_d.

The generalized decision function leads to a new notion of a decision reduct for inconsistent data.

For a decision table $A_d = (U, A, d)$, we say that a subset $B \subseteq A$ ∂–*defines* d if for any $w_A \in V_A^U$ the equality holds

$$\partial_{d/B}\left(w_A^{\downarrow B}\right) = \partial_{d/A}(w_A) \tag{1.50}$$

where $w_A^{\downarrow B}$ is the thread w_A restricted to B.

A subset $B \subseteq A$ which ∂-defines d is called a ∂–*decision reduct* if none of its proper subsets has this property.

Approximate discernibility reducts

Classical reducts were obtained by means of discernibility matrices; they discerned among all possible pairs of objects. Usually, exactly discerning reducts turn out to be much longer than ∂-decision ones, because they have to discern among many more object pairs.

As for consistent decision tables both types are equivalent to the notion of an exact decision reduct, their common extension which could lead to yet shorter (and possibly, better) decision rules might be based on a notion(s) of *approximate discernibility*.

To this end, one may introduce discernibility measures which assign to each $B \subseteq A$ a numerical value of a degree in which B discerns pairs of objects with different decision values.

For instance, let us consider discernibility measure $N_A(B) =$

$$\frac{|\{(u, u') \in U \times U : (d(u) \neq d(u')) \wedge (Inf_B(u) \neq Inf_B(u'))\}|}{|\{(u, u') \in U \times U : (d(u) \neq d(u')) \wedge (Inf_A(u) \neq Inf_A(u'))\}|} \tag{1.51}$$

labeling subsets $B \subseteq A$ with the ratio of object pairs discerned by B, which are:

1. pairs necessary to be discerned because their elements belong to different decision classes.

2. pairs possible to be discerned by the whole of A.

It is easy to notice that a given $B \subseteq A$ exactly discerns d if and only if equality $N_A(B) = 1$ holds.

Employing this discernibility measure 1.51, one may define the notion of an *ε–discerning decision reduct*.

Given a decision table A=(U, A, d) and an approximation threshold $\varepsilon \in [0, 1)$, we say that a subset $B \subseteq A$ *ε–discerns* d if and only if the inequality

$$N_A(B) \geq 1 - \varepsilon \tag{1.52}$$

holds.

A subset $B \subseteq A$ which ε-discerns d is called a *ε–discerning decision reduct* if it is a minimal set of attributes with this property.

Approximate reducts keep therefore a judicious balance between the (reduced) number of conditions and the loss of determinism with respect to the initial table.

By adaptive tuning of ε, one is likely to obtain interesting results via appropriate heuristics. The reader is referred to [Bazan00] for a study of a possible design of such heuristics aimed at extraction of approximate association rules from data.

Approximate reducts based on conditional frequency

The idea of finding reducts based on a global discernibility measure may be refined yet: one may search for reducts preserving a given frequency characteristic of decision. To have a glimpse into this venue of research, let us focus on conditional representation based on rough membership functions $\mu_{d/B} : V_d \times V_B^U \rightarrow [0,1]$, defined, for any fixed $B \subseteq A$, by the formula

$$\mu_{d/B}\left(v_d/w_B\right) = \frac{\left|Inf_B^{-1}\left(w_B\right) \cap d^{-1}\left(v_d\right)\right|}{\left|Inf_B^{-1}\left(w_B\right)\right|} \qquad (1.53)$$

where the number of objects with the pattern value w_B on B and the decision value v_d is divided by the number of objects with the pattern value w_B on B.

The value of $\mu_{d/B}\left(v_d/w_B\right)$ is, actually, the conditional frequency of occurrence of the given $v_d \in V_d$ under the condition $w_B \in V_B^U$ on $B \subseteq A$.

A rational approach is to choose, for any given $w_B \in V_B^U$, the decision value with the highest conditional frequency of occurrence, i.e., such $v_d \in V_d$ that

$$\mu_{d/B}\left(v_d/w_B\right) = \max_{k=1,..,|V_d|} \mu_{d/B}\left(v_k/w_B\right) \qquad (1.54)$$

which assigns to each pattern over $B \subseteq A$ a decision value which minimizes the risk of wrong classification.

We then define the *majority decision function* $m_{d/B} : V_B^U \rightarrow 2^{V_d}$ via

$$m_{d/B}\left(w_B\right) = \left\{ v_d \in V_d : \mu_{d/B}\left(v_d/w_B\right) = \max_k \mu_{d/B}\left(v_k/w_B\right) \right\} \qquad (1.55)$$

Now, one may define a new notion of a reduct preserving the information stored in the majority decision function.

Given a decision table A$=(U, A, d)$, we say that a subset $B \subseteq A$ *m–defines* d if

$$m_{d/B}\left(w_A^{\downarrow B}\right) = m_{d/A}\left(w_A\right) \qquad (1.56)$$

for any $w_A \in V_A^U$.

A subset B which m–defines d is called an *m–decision reduct* if no proper subset of it m–defines d.

This line of research is continued and in the search for optimal classifiers various measures of information are employed e.g. *entropy, distance measures* etc. As with classical reducts, problems of finding a reduct of any of these types is NP–complete, and the respective problem of finding an (optimal) reduct is NP–hard (cf. [Ślęzak00] for a discussion of complexity issues). The reader will find in [Bazan00] a discussion of relevant heuristics.

1.10.2   Local reducts

Another important type of reducts are local reducts. A *local reduct* $r(x_i) \subseteq A$ (or a *reduct relative to decision and object* $x_i \in U$) where x_i is called a *base object*, is a subset of A such that

$$(i)\ \forall x_j \in U(d(x_i) \neq d(x_j) \implies \exists a_k \in r(x_i)a_k(x_i) \neq a_k(x_j))$$
$$(ii)\ r(x_i)\ is\ minimal\ with\ respect\ to\ inclusion \qquad (1.57)$$

Local reducts may be better suited to the task of classification as they reflect the local information structure. In [Bazan00] a detailed discussion of local reducts and algorithms generating coverings of the universe with local reducts may be found.

1.10.3 Dynamic reducts

Methods based on calculation of reducts allow to compute, for a given decision table A_d, descriptions of all decision classes in the form of decision rules (see previous sections). In search for robust decision algorithms, the idea of dynamic reducts has turned out to be a happy one: as suggested by experiments, rules calculated by means of dynamic reducts are better predisposed to classify unseen cases, because these reducts are the most stable reducts in a process of random sampling of the original decision table.

For a decision system $A_d = (U, A, d)$, we call any system $B = (U', A, d)$ such that $U' \subseteq U$ a *sub-table* of A. By the symbol $P(A_d)$, we denote the set of all sub-tables of A_d. For $F \subseteq P(A_d)$, by $DR(A, F)$ we denote the set

$$RED(A_d) \cap \bigcap_{B \in F} RED(B).$$

Any element of $DR(A, F)$ is called an F-*dynamic reduct* of A.

From the definition of a dynamic reduct it follows that a relative reduct of A is dynamic if it is also a reduct of all sub-tables from a given family F. This notion can be sometimes too much restrictive and one may relax it to the notion of an (F, ε)-*dynamic reduct*, where $\varepsilon \in [0, 1]$. The set $DR_\varepsilon(A, F)$, of all (F, ε)-dynamic reducts is defined by
$DR_\varepsilon(A, F) = \{C \in RED(A_d):$

$$\frac{card(\{B \in F : C \in RED(B)\})}{card(F)} \geq 1 - \varepsilon\}$$

For $C \in RED(A_d)$, the number:

$$\frac{card(\{B \in F : C \in RED(B)\})}{card(F)}$$

is called the *stability coefficient* of the reduct C (relative to F).

1.10.4 Generalized dynamic reducts

From the definition of a dynamic reduct it follows that a relative reduct of any table from a given family F of sub-tables of A can be dynamic if it is also a reduct of the table A. This can be sometimes not convenient because we are interested in useful sets of attributes which are not necessarily reducts of the table A. Therefore we have to generalize the notion of a dynamic reduct.

Let $A_d = (U, A, d)$ be a decision table and $F \subseteq P(A_d)$. By the symbol $GDR(A_d, F)$, we denote the set

$$\bigcap_{B \in F} RED(B).$$

Elements of $GDR(A_d, F)$ are called F-*generalized dynamic reducts* of A.

From the above definition it follows that any subset of A is a generalized dynamic reduct if it is a reduct of all sub–tables from a given family F. As with dynamic reducts, one defines a more general notion of an (F, ε)-*generalized dynamic reduct*, where $\varepsilon \in [0, 1]$. The set $GDR_\varepsilon(A_d, F)$ of all (F, ε)-generalized dynamic reducts is defined by $GDR_\varepsilon(A_d, F) = \{C \subseteq A :$

$$\frac{card(\{B \in F : C \in RED(B)\})}{card(F)} \geq 1 - \varepsilon\}.$$

The number

$$\frac{card(\{B \in F : C \in RED(B)\})}{card(F)}$$

is called the *stability coefficient* of the generalized dynamic reduct C (relative to F).

Dynamic reducts are determined on the basis of a family of sub–tables, each of which may be regarded as a perturbed original table; thus one may expect that these reducts are less susceptible to noise. This expectation is borne out by experiments. In [Bazan00] the reader will find a detailed discussion of algorithmic issues related to dynamic reducts.

1.10.5   Genetic and hybrid algorithms in reduct computation

Recently, much attention is paid to hybrid algorithms for reduct generation, employing rough set algorithms in combination with other techniques e.g. evolutionary programming.

To exploit advantages of both genetic and heuristic algorithms, one can use a hybridization strategy. The general scheme of a *hybrid algorithm* is as follows:

1. Find a strategy (a heuristic algorithm) which gives an approximate result.

2. Modify (parameterize) the strategy using a control sequence, so that the result depends on this sequence (recipe).

3. Encode the control sequence as a chromosome.

4. Use a genetic algorithm to produce control sequences. Proceed with the heuristic algorithm controlled by the sequence. Evaluate an object generated by the algorithm and use its quality measure as the fitness of the control sequence.

5. A result of evolution is the best control sequence, i.e. the sequence producing the best object. Send this object to the output of the hybrid algorithm.

Hybrid algorithms proved to be useful and efficient in many areas, including NP-hard problems of combinatorics. The short reduct finding problem also can be solved efficiently by this class of algorithms.

Finding reducts using Genetic Algorithms

An order-based genetic algorithm is one of the most widely used components of various hybrid systems. In this type of genetic algorithm a chromosome is an n-element permutation σ, represented by a sequence of numbers: $\sigma(1)\ \sigma(2)\ \sigma(3)\ \dots\ \sigma(n)$. Mutation of an order-based individual means one random transposition of its genes (a transposition of random pair of genes). There are various methods of recombination (*crossing–over*) considered in literature like PMX (Partially Matched Crossover), CX (Cycle Crossover) and OX (Order Crossover) cf. [Goldberg89]. After crossing–over, fitness function of every individual is calculated. In the case of a hybrid algorithm a heuristics is launched under control of an individual; a fitness value depends on the result of heuristics. A new population is generated using the "roulette wheel" algorithm: the fitness value of every individual is normalized and treated as a probability distribution on the population;one chooses randomly M new individuals using this distribution. Then all steps are repeated.

In the hybrid algorithm a simple, deterministic method may be used for reduct generation:

Algorithm 8. *Finding a reduct basing on permutation.*

Input:
     1. *decision table* $\mathbf{A} = (U, \{a_1, ..., a_n\} \cup \{d\})$
     2. *permutation τ generated by genetic algorithm*

Output: *a reduct R generated basing on permutation τ*

Method:
     $R = \{a_1, ..., a_n\}$
     $(b_1 \dots b_n) = \tau\,(a_1 \dots a_n)$
     for $i = 1$ **to** n **do**
     begin
         $R = R - b_i$
         if not $Reduct(R, \mathbf{A})$ **then** $R = R \cup b_i$
     end
     return R □

□

The result of the algorithm will always be a reduct. Every reduct can be found using this algorithm, the result depends on the order of attributes. The genetic algorithm is used to generate the proper order. To calculate the function of fitness for a given permutation (order of attributes) one run of the deterministic algorithm is performed and the length of the found reduct is calculated. In the selection phase of the genetic algorithm a fitness function

$$F(\tau) = n - L_\tau + 1 \qquad (1.58)$$

where L_τ is the length of the reduct, may be used. Yet another approach is to select a reduct due to the number of rules it generates rather than to its length. Every reduct generates an indiscernibility relation on the universe and in most cases it identifies some pairs of objects. When a reduct generates a smaller number of rules it means that the rules are more general and they should better recognize new objects.

The hybrid algorithm described above can be used to find reducts generating the minimal number of rules. The only thing to change is the definition of the fitness function:

$$F(\tau) = m - R_\tau + \frac{n - L_\tau + 1}{n} \qquad (1.59)$$

where R_τ denotes the number of rules generated by the reduct. Now, the primary criterion of optimization is the number of rules, while the secondary is the reduct length. The results of experiments show, that the classification system based on the reducts optimized due to the number of rules performs better (or not worse) than the short reduct based one. Moreover, due to the rule set reduction, it occupies less memory and classifies new objects faster.

A discussion of this topic will be found in [Bazan00].

1.11 Template techniques

Feature extraction may be also regarded abstractly as concerned with developing methods of searching for *regular patterns* (regularities) hidden in data sets.

There are two two fundamental kinds of regularities: regularities defined by *templates* and regularities defined in a *relational language*. The process of extracting regularities should find templates of high quality. In the general case, we distinguish between numerical (orderable) value sets and symbolic (nominal) value sets. A unified treatment of both types of attributes may be provided by means of (generalized) descriptors.

1.11.1 Templates

Let A= (U, A) be an information system. Any clause of the form $D = (a \in W_a)$ is called the *descriptor* and the value set $W_a \subseteq V_a$ is called the *range* of D. In case a is a numeric attribute, one usually restricts the range of descriptors for a to real intervals, that means $W_a = [v_1, v_2] \subseteq V_a$. In case of symbolic attributes, W_a can be any non-empty subset $W_a \subseteq V_a$. The *volume* of a given descriptor $D = (a \in W_a)$ is defined by

$$Volume(D) = |W_a \cap a(U)| \qquad (1.60)$$

By $Prob(D)$, where $D = (a \in W_a)$, we denote the *hitting probability* of the set W_a i.e.

$$Prob(D) = \frac{|W_a|}{|V_a|} \qquad (1.61)$$

For a a numeric attribute,

$$Prob(D) = \frac{v_2 - v_1}{\max(V_a) - \min(V_a)} \qquad (1.62)$$

where $\max(V_a)$, $\min(V_a)$ denote the maximum and the minimum value in the value set V_a, respectively.

A formula $T : \bigwedge_{a \in B} (a \in W_a)$ is called *a template* of A. A template T is *simple*, if any descriptor of T has range of one element. Templates with descriptors consisting of more than one element are called *generalized*. For any $X \subseteq U$, the set $\{x \in X : \forall_{a \in B}\ a(x) \in W_a\}$ of objects from X satisfying T is denoted by $[T]_X$. One defines: $support_X(T) = |[T]_X|$.

Any template $T = D_1 \wedge ... \wedge D_k$, where $D_i = (a_i \in W_i)$, is characterized by the following parameters:

1. $length(T)$, which is the number of descriptors occurring in T.

2. $support(T) = support_U(T)$, which is the number of objects in U satisfying T.

3. $applength(T)$, which is $\sum_{1 \leq i \leq k} \frac{1}{Volume(D_i)}$ called the *approximated length* of the generalized template T.

Functions *applength* and *length* coincide for simple templates.

The complexity and algorithmic problems related to search for good templates with applications are given a throughout examination in [Nguyen Sinh Hoa00] to which the reader is invited for details.

In some applications we are interested in templates "well matching" some chosen decision classes. Such templates are called *decision templates*. The template associated with one decision class is called a *decision rule*. The precision of any decision template (e.g. $T \Longrightarrow (d \in V)$) is estimated by its *confidence ratio.* The *confidence* of a decision rule $T \implies (d, i)$ (associated with the decision class X_i) is defined as

$$confidence_{X_i}(T) = \frac{support_{X_i}(T)}{support(T)} \qquad (1.63)$$

1.11.2 Template goodness measures

Template quality should be evaluated taking into account the context of an individual application. In general, templates with high quality have the following properties:

1. They are supported by a *large* number of objects.

2. They are of high *specificity*, i.e. they should be described by a sufficiently *large* number of descriptors and the set of values W_a in any descriptor $(a \in W_a)$ should be *small*.

Moreover, decision templates (decision rules) used for classification tasks should be of high *predictive accuracy*. Decision templates (decision rules) of high quality should have the following properties

3. They should be supported by a *large* number of objects.

4. They should have high predictive accuracy.

For a given template T, its support is defined by the function $support(T)$ and its length by the function $applength(T)$. Any quality measure is a combination of these functions. One may use one of the following exemplary functions

$$
\begin{aligned}
(i)\ &quality(T) = support(T) + applength(T) \\
(ii)\ &quality(T) = support(T) \cdot applength(T)
\end{aligned} \qquad (1.64)
$$

In case of decision templates, one should take into consideration the confidence ratios. Let T be a decision template associated with the decision class X_i. The quality of T can be measured by one of the following functions

$$
\begin{aligned}
(i)\ &quality(T) = confidence(T) \\
(ii)\ &quality(T) = support_{C_i}(T) \cdot applength(T) \\
(iii)\ &quality(T) = support_{C_i}(T) \cdot confidence(T)
\end{aligned} \qquad (1.65)
$$

In search for good template–based classifiers one may applied various strategies. Here we comment in passing on the two basic methodologies.

One can start with the *empty descriptor set* and extend it by adding the most relevant descriptors from the original set. One can also begin with the original *full set* of descriptors and remove irrelevant descriptors from it. The former strategy is the *sequential forward* strategy and the latter is the *sequential backward* strategy. Every strategy can be classified as *deterministic* or *random* depending on a method used for descriptor selection. A deterministic strategy always chooses the descriptor with the *optimal fitness*, whereas the random selector chooses templates according to some probability distribution.

Sequential forward generation

This strategy begins with the *empty* template T. The template T is extended by adding one descriptor at a time that well fits the existing template. Let T_i be a temporary template obtained in the i-th step of construction. A new descriptor is selected according to the function $fitness_{T_i}$. For the temporary template T_i, the fitness of any descriptor D measured relatively to T_i reflects its potential ability to create a new template $T_{i+1} = T_i \wedge D$ of high quality. In the *Deterministic Sequential Forward Generation* (DSFG), the template T_i is extended by a descriptor with the maximum value of fitness. In general DSFG detects a template of high quality. The drawback of this method is that it can be stuck in a local extreme (the best template at the moment). To avoid this situation, a strategy *Random Sequential Forward Generation* (RSFG) is applied. In RSFG, descriptors are chosen randomly according to a probability distribution defined by fitness function.

Let P be a set of descriptors, which can be used to extend T_i. Let p_0 be a descriptor in P. In the simplest case, the distribution function *Prob* can be defined by $Prob(p_0) = \frac{fitness_{T_i}(p_0))}{\sum_{p \in P} fitness_{T_i}(p)}$

Sequential backward generation

The sequential backward method uses top–down strategy rather than bottom–up strategy used for DSFG. Starting from a *full template* T_{full}, (which is often defined by an information vector $Inf_A(x)$ for some object x), the algorithm finds irrelevant descriptors and removes one descriptor at a time. After a removal, the quality of a new template is estimated. The descriptor p is *irrelevant* if the template T without this descriptor is of better quality. Attributes are selected according to *fitness function*, similarly as in DSFG.

Assume the full template T_{full} is of the form $T_{full} = \bigwedge_{1 \le i \le k} (a_i \in W_i)$, where k is a number of attributes in a data table and W_i are fixed subsets of V_{a_i}.

Analogously to the forward strategy, a descriptor to be removed can also be chosen randomly. For $fitness_{T_i}(p_0)$ denoting the fitness of a descriptor p_0 according to the template T_i, the distribution function for irrelevant descriptor elimination can be defined as follows: $Prob(p_0) = 1 - \frac{fitness_{T_i}(p_0))}{\sum_{p \in T_i} fitness_{T_i}(p)}$.

For any algorithm based on sequential schemes, the following three parameters are fixed:

1. estimation of the descriptor fitness.

2. estimation of the template quality.

3. a method of searching for the best descriptor $(a \in W_a)$ of a given attribute a.

1.11.3   Searching for optimal descriptors

One can regard a descriptor $(a \in W_a)$ as of "good quality" if the template $T \wedge (a \in W_a)$ has large support, although the set W_a is small. Hence the fitness function is defined using the following two parameters

- the number of objects supporting the template $T \wedge (a \in W_a)$.

- the cardinality of the set W_a.

The fitness function should be proportional to the first parameter and inversely proportional to the second parameter. It can be defined by the following formula

$$fitness_T(a \in W_a) = support_{[T]_U}(a \in W_a) \cdot Prob^{-1}(W_a) \qquad (1.66)$$

The descriptor $a \in W_a$ is *optimal* if $fitness(a \in W_a)$ is maximal.

A searching heuristics for an optimal descriptor may be based on a fixed similarity relation.

Let $R_a \subseteq V_a \times V_a$ be a given similarity relation. The range of W_a may be computed by taking the similarity pre-class $[v, R_a]$ of the properly chosen *generator v*.

Searching heuristics for optimal similarity relations are investigated in detail in [Nuyen Sinh Hoa00].

1.12   Discretization

Complex attributes may have continuous values or the number of values may be large so they require value grouping etc. In such cases a pre-processing step is necessary which would return a data table with a sufficiently small number of attribute values. We focus here on discretization of continuous attributes as an example of pre-processing stage (cf. [Nguyen Hung Son98], [Nguyen–Nguyen 98^b]).

Discretization is a step that is not specific to the rough set approach but the majority of rule or tree induction algorithms require it in order to perform well. The search for appropriate cut–off points can be reduced to finding some minimal Boolean expressions cf. [Nguyen–Skowron99].

The reported results (op.cit., op.cit.) are showing that the discretization problems and symbolic value partition problems are of high computational complexity (i.e. NP-complete or NP-hard) which clearly justifies the importance of designing efficient heuristics.

To implement discretization, one can use the idea of *cuts*. These are pairs (a, c) where $c \in V_a$.

1.12.1 Value partition via cuts

Any cut defines a new conditional attribute with binary values. For example the attribute corresponding to the cut $(a, 1.2)$ takes 0 on x if $a(x) < 1.2$, otherwise it takes 1 on x.

Similarly, any set P of cuts defines a new conditional attribute a_P for any a. One should consider a partition of the value set of a by cuts from P and assign new names to elements of this partition. Let us take the set of cuts: $P = \{(a, 0.9), (a, 1.5), (b, 0.75), (b, 1.5)\}$. This set of cuts is gluing together all values of a less then 0.9, all values in the interval $[0.9, 1.5)$ and all values in the interval $[1.5, 4)$. Similarly for b.

The problem of finding a (minimal) set of cuts may be solved by means of boolean reasoning.

Let $A_d = (U, A, d)$ be a decision system where $U = \{x_1, x_2, \ldots, x_n\}$, $A = \{a_1, \ldots, a_k\}$, and $d : U \to \{1, \ldots, r\}$. We assume that $V_a = [l_a, r_a) \subseteq \mathbf{R}^1$ is an interval of real numbers for any $a \in A$ and A_d is a deterministic decision system. Any pair (a, c) where $a \in A$ and $c \in \mathbf{R}^1$ will be called a *cut on* V_a. Let P_a be a partition of V_a (for $a \in A$) into sub–intervals i.e. $P_a = \{[c_0^a, c_1^a), [c_1^a, c_2^a), \ldots, [c_{k_a}^a, c_{k_a+1}^a)\}$ for some integer k_a, where $l_a = c_0^a < c_1^a < c_2^a < \ldots < c_{k_a}^a < c_{k_a+1}^a = r_a$ and $V_a = [c_0^a, c_1^a) \cup [c_1^a, c_2^a) \cup \ldots \cup [c_{k_a}^a, c_{k_a+1}^a)$. Hence any partition P_a is uniquely defined and often identified as the set of cuts: $\{(a, c_1^a), (a, c_2^a), \ldots, (a, c_{k_a}^a)\} \subset A \times \mathbf{R}^1$.

Any set of cuts $P = \bigcup_{a \in A} P_a$ defines from A_d a new decision system $A_d{}^P = (U, A^P, d)$ called the P-*discretization of* A_d, where $A^P = \{a^P : a \in A\}$ and $a^P(x) = i \Leftrightarrow a(x) \in [c_i^a, c_{i+1}^a)$ for $x \in U$ and $i \in \{0, \ldots, k_a\}$.

Two sets of cuts $\mathbf{P'}, \mathbf{P}$ are *equivalent* if $A^P = A^{P'}$.

We say that the set of cuts P is A_d–*consistent* if $\partial_{A_d} = \partial_{A^P}$, where ∂_{A_d} and $\partial_{A_d^P}$ are generalized decisions of A_d and $A_d{}^P$, respectively.

The A_d-consistent set of cuts P^{irr} is A_d–*irreducible* if P is not A_d–consistent for any $P \subset P^{irr}$.

The A_d–consistent set of cuts P^{opt} is A_d–*optimal* if $card(P^{opt}) \leq card(P)$ for any A_d–consistent set of cuts P.

One can show that the decision problem of checking if for a given decision system A_d and an integer k there exists an irreducible set of cuts P such that $card(\mathbf{P}) < k$ is NP-complete.

The problem of searching for an optimal set of cuts P in a given decision system A_d is NP-hard.

1.12.2 Heuristics

One can construct efficient heuristics returning semi-minimal sets of cuts. A simple heuristics may be based on the Johnson strategy. Using this strategy one can look for a cut discerning the maximal number of object pairs, then one can eliminate all already discerned object pairs and repeat the procedure until all object pairs to be discerned are discerned. It is interesting to note that this can be realized by computing the minimal relative reduct of the corresponding decision system. Heuristics of this type are discussed in [Bazan00], [Nguyen Sinh Hoa00].

The $M(aximal)D(iscernibility)$ heuristics is based on searching for a cut with maximal number of object pairs discerned by this cut. The idea is analogous to the Johnson approximation algorithm and can be briefly formulated as follows.

First, a new decision system is constructed from the decision system $A_d = (U, A, d)$; for each $a \in A$, we denote by C_a the set

$$\{(a, \frac{c_1^a + c_2^a}{2}), ..., (a, \frac{c_{n_a-1}^a + c_{n_a}^a}{2})\}$$

of all possible cuts concerning the attribute a where

$$\{a(x) : x \in U\} = \{c_1^a, ..., c_{n_a}^a\}$$

and

$$C_A = \bigcup_{a \in A} C_a$$

denotes the set of all cuts on A.

The decision system $A_d^c = (U^c, A^c, d^c)$ is defined as follows

$$U^c = \{(x, y) \in U^2 : d(x) \neq d(y)\} \cup \{new\} \text{ where } new \notin U^2$$
$$A^c = \{b_{(a,c)} : (a, c) \in C_A\} \text{ where}$$
$$b_{(a,c)}(x) = \begin{array}{l} 0 \text{ in case } a(x) < c \\ 1 \quad otherwise \end{array} \tag{1.67}$$

$$d^c(u) = \begin{array}{l} 0 \text{ in case } u = new \\ 1 \text{ in case } u \neq new \end{array} \tag{1.68}$$

Now, the algorithms follows.

ALGORITHM MD-heuristic (Semi-optimal family of partitions)

Step 1 Construct the table $A_d{}^c$ from A_d and denote by B the table $A_d{}^c$ with the row *new* deleted. Let $D = \emptyset$.

Step 2 Choose a column c_{max} from B with the maximal number of occurrences of 1.

Step 3 Delete from B the column chosen in Step 2 and all rows marked in this column with 1. Insert c_{max} into D and remove it from C_A.

Step 4 **If** $B \neq \emptyset$ then go to Step 2 **else** Stop.

Let us observe that the new features in the considered case of discretization are of the form $a \in V$, where $V \subseteq V_a$ and V_a is the set of values of attribute a. One can extend the presented approach to the case of symbolic (nominal, qualitative) attributes as well as to the case when in a given decision system nominal and numeric attributes appear. The received heuristics are of very good quality. The exhaustive discussion of this topic will be found in [Bazan00], [Nguyen Sinh Hoa00].

Works quoted

[Bazan00] J. Bazan, H.S. Nguyen, S. H. Nguyen, P. Synak, and J. Wróblewski, *Rough set algorithms in classification problems*, in: [Polkowski–Tsumoto–Lin], pp. 49–88.

[Bazan98] J.G. Bazan, Nguyen Hung Son, Nguyen Tuan Trung, A. Skowron, and J. Stepaniuk , *Decision rules synthesis for object classification*, in: [Orłowska98], pp. 23–57.

[Goldberg89] D.E. Goldberg, *GA in Search, Optimisation, and Machine Learning*, Addison–Wesley, 1989.

[Greco99] S. Greco, B. Matarazzo, R. Słowiński, *Fuzzy dominance as basis for rough approximations*, in: Proceedings: the 4th Meeting of the EURO WG on Fuzzy Sets and 2nd Internat. Conf. on Soft and Intelligent Computing, (EUROFUSE-SIC'99), Budapest, Hungary, May 1999, pp. 273-278.

[Greco98] S. Greco, B. Matarazzo, and R. Słowiński, *On joint use of indiscernibility, similarity and dominance in rough approximation of decision classes*, in: Proceedings: the 5th International Conference of the Decision Sciences Institute, Athens, Greece, July 1999, pp. 1380–1382.

[Iwinski87] T. B. Iwiński, *Algebraic approach to rough sets*, Bull. Polish Acad. Ser. Sci. Math., 35 (1987), pp. 673–683.

[Mitchell98] T. Mitchell, *Machine Learning*, McGraw–Hill, Boston, 1998.

[Nguyen Hung Son98] Nguyen Hung Son, *From optimal hyperplanes to optimal decision trees*, Fundamenta Informaticae, 34(1-2) (1998), pp. 145–174.

[Nguyen–Nguyen98^a] Nguyen Sinh Hoa and Nguyen Hung Son, *Pattern extraction from data*, Fundamenta Informaticae, 34(1-2) (1998), pp. 129–144.

[Nguyen–Nguyen98^b] Nguyen Hung Son and Nguyen Sinh Hoa, *Discretization methods in Data Mining*, in: [Polkowski–Skowron98^a], pp. 451–482.

[Nguyen Sinh Hoa00] Nguyen Sinh Hoa, *Regularity analysis and its applications in Data Mining*, in: [Polkowski–Tsumoto–Lin], pp. 289–378.

[Nguyen–Skowron99] Nguyen Hung Son and A. Skowron, *Boolean reasoning scheme with some applications in Data Mining*, in: Proceedings: Principles of Data Mining and Knowledge Discovery PKDD'99, Prague, Czech Republic, September 1999, LNAI vol. 1704, Springer Verlag, Berlin, 1999, pp. 107–115.

[Novotný–Pawlak89] M. Novotný and Z. Pawlak, *Algebraic theory of independence in information systems*, Report 51, Institute of Mathematics of the Czechoslovak Academy of Sciences, 1989.

[Novotný–Pawlak88^a] M. Novotný and Z. Pawlak, *Partial dependency of attributes*, Bull. Polish Acad. Sci. Math., 36 (1989), pp. 453–458.

[Novotný–Pawlak85^a] M. Novotný and Z. Pawlak, *Characterization of rough top equalities and rough bottom equalities*, Bull. Polish Acad. Sci. Math., 33 (1985), pp. 91–97.

[Novotný–Pawlak85^b] M. Novotný and Z. Pawlak, *On rough equalities*, Bull. Polish Acad. Sci. Math., 33 (1985), pp. 99-104.

[Novotný–Pawlak85^c] M. Novotný and Z. Pawlak, *Black box analysis and rough top equality*, Bull. Polish Acad. Sci. Math., 33 (1985), pp. 105–113.

[Novotný–Pawlak85^d] M. Novotný and Z. Pawlak, *Independence of attributes*, Bull. Polish Acad. Sci. Tech., 33 (1985), pp. 459–465.

[Obtulowicz85] A. Obtułowicz, *Rough sets and Heyting algebra valued sets*, Bull. Polish Acad. Sci. Math., 33(1985), pp. 454–476.

[Orlowska90] E. Orłowska, *Kripke semantics for knowledge representation*, Studia Logica, 49 (1990), pp. 255–272.

[Orlowska89] E. Orłowska, *Logic for reasoning about knowledge*, Z. Math. Logik u. Grund. d. Math., 35(1989), pp. 559–572.

[Orlowska85] E. Orłowska, *Logic approach to information systems*, Fundamenta Informaticae, 8 (1985), pp. 359–378.

[Orlowska84] E. Orłowska, *Modal logics in the theory of information systems*, Z. Math. Logik u. Grund.d. Math., 30(1984), pp. 213–222.

[Orlowska–Pawlak84^a] E. Orłowska and Z. Pawlak, *Logical foundations of knowledge representation*, Reports of the Comp. Centre of the Polish Academy of Sciences, 537, 1984.

[Orlowska–Pawlak84^b] E. Orłowska and Z. Pawlak, *Representation of nondeterministic information*, Theor. Computer Science, 29 (1984), pp. 27–39.

[Pagliani98^a] P. Pagliani, *Rough set theory and logic–algebraic structures*, in: [Orlowska98], pp. 109–192.

[Pawlak01] Z. Pawlak, *Combining rough sets and Bayes' rule*, Computational Intelligence: An Intern. Journal, 17, 2001, pp. 401–408.

[Pawlak91] Z. Pawlak, *Rough Sets. Theoretical Aspects of Reasoning about Data*, Kluwer, Dordrecht, 1991.

[Pawlak87^a] Z. Pawlak, *Decision tables–a rough set approach*, Bull. EATCS, 33 (1987), pp. 85–96.

[Pawlak87^b] Z. Pawlak, *Rough logic*, Bull. Polish Acad. Sci. Tech., 35 (1987), pp. 253–258.

[Pawlak86] Z. Pawlak, *On decision tables*, Bull. Polish Acad. Sci. Tech., 34 (1986), pp. 553–572.

[Pawlak85^a] Z. Pawlak, *On rough dependency of attributes in information systems*, Bull. Polish Acad. Sci. Tech., 33 (1985), pp. 551–559.

[Pawlak85^b] Z. Pawlak, *Rough sets and decision tables*, LNCS vol. 208, Springer Verlag, Berlin, 1985, pp. 186–196.

[Pawlak82^a] Z. Pawlak, *Rough sets*, Intern. J. Comp. Inform. Sci., 11 (1982), pp. 341–356.

[Pawlak81^a] Z. Pawlak, *Information Systems–Theoretical Foundations* (in Polish), PWN–Polish Scientific Publishers, Warsaw, 1981.

[Pawlak81^b] Z. Pawlak, *Information systems–theoretical foundations*, Information Systems, 6 (1981), pp. 205–218.

[Pawlak–Rauszer85] Z. Pawlak and C. Rauszer, *Dependency of attributes in information systems*, Bull. Polish Acad. Sci. Math., 33 (1985), pp. 551–559.

[Pawlak–Skowron94] Z. Pawlak and A. Skowron, *Rough membership functions*, in: R.R. Yaeger, M. Fedrizzi, and J. Kacprzyk, eds., *Advances in the Dempster–Schafer Theory of Evidence*, Wiley, New York, 1994, pp. 251–271.

[Polkowski93^a] L. Polkowski, *Metric spaces of topological rough sets from countable knowledge bases*, Foundations of Computing and Decision Sciences, 18(1993), pp. 293–306.

[Polkowski93^b] L. Polkowski, *Mathematical morphology of rough sets*, Bull. Polish Acad. Sci. Math., 41(1993), pp. 241–273.

[Polkowski–Polkowski00] L. Polkowski and M. Semeniuk–Polkowska, *Towards usage of natural language in approximate computation: a granular semantics employing formal languages over mereological granules of knowledge*, Scheda Informaticae (Sci. Fasc. Jagiellonian University), 10 (2000), pp. 131–146.

[Polkowski–Skowron01] L. Polkowski and A. Skowron, *Rough mereological calculi of granules: a rough set approach to computation*, Computational Intelligence: An Intern. Journal, 17 (2001), pp. 472–492.

[Polkowski–Skowron98^a] L. Polkowski and A. Skowron, *Rough Sets in Knowledge Discovery 1. Methodology and Applications*, Physica Verlag, Heidelberg, 1998.

[Polkowski–Skowron98^b] L. Polkowski and A. Skowron (eds.), *Rough Sets in Knowledge Discovery 2. Applications, Case Studies and Software Systems*, Studies in Fuzziness and Soft Computing, vol. 19, Physica Verlag, Heidelberg, 1998.

[Polkowski–Tsumoto–Lin00] L. Polkowski, S. Tsumoto, and T. Y. Lin, eds., *Rough Set Methods and Applications. New Developments in Knowledge Discovery in Information Systems*, Studies in Fuzzines and Soft Computing vol. 56, Physica Verlag, Heidelberg, 2000.

[Pomykala88] J. Pomykała and J. A. Pomykała, *The Stone algebra of rough sets*, Bull. Polish Acad. Ser. Sci. Math., 36 (1988), 495–508.

[Rasiowa–Skowron86^a] H. Rasiowa and A. Skowron, *Rough concept logic*, LNCS vol. 208, Springer Verlag, Berlin, 1986, pp. 288–297.

[Rasiowa–Skowron86^b] H. Rasiowa and A. Skowron, *The first step towards an approximation logic*, J. Symbolic Logic, 51 (1986), p. 509.

[Rasiowa–Skowron86^c] H. Rasiowa and A. Skowron, *Approximation logic*, Proc. Conf. on Mathematical Methods of Specification and Synthesis of Software Systems, Akademie Verlag, Berlin, 1986, pp. 123–139.

[Rissanen83] J. Rissanen, *A universal prior for integers and estimation by minimum description length*, The Annals of Statistics, 11 (1983), pp. 416–431.

[Skowron89] A. Skowron, *The implementation of algorithms based on discernibility matrix*, manuscript, 1989.

[Skowron88] A. Skowron, *On topology in information systems*, Bull. Polish Acad. Sci. Math., 36 (1988), pp. 477–480.

[Skowron–Rauszer92] A. Skowron and C. Rauszer, *The discernibility matrices and functions in information systems*, in: R. Słowiński, ed., *Intelligent Decision Support. Handbook of Applications and Advances of the Rough Set Theory*, Kluwer, Dordrecht, 1992, pp. 311–362.

[Skowron–Stepaniuk96] A. Skowron and J. Stepaniuk, *Tolerance approximation spaces*, Fundamenta Informaticae, 27 (1996), pp. 245–253.

[Slowinski0x] R. Słowiński and D. Vanderpooten, *A generalized definition of rough approximations based on similarity*, IEEE Transactions on Data and Knowledge Engineering, to appear.

[Stepaniuk00] J. Stepaniuk, *Knowledge discovery by application of rough set model*, in: [Polkowski–Tsumoto–Lin00], pp. 137–234.

[Ślęzak00] D. Ślęzak, *Various approaches to reasoning with frequency based decision reducts: a survey*, in: [Polkowski–Tsumoto–Lin00], pp. 235–288.

[Vakarelov89] D. Vakarelov, *Modal logics for knowledge representation systems*, Lecture Notes in Computer Science, vol. 363 (1989), Springer Verlag, Berlin, pp. 257–277.

1.13   Selected bibliography on rough sets

The bibliography begins with a list [A] of books and conference proceedings dedicated to rough set theory and its applications. Then selected papers on rough sets are listed in [B].This bibliography is a sequel to the bibliography of 1077 papers in: L. Polkowski and A. Skowron (eds.), *Rough Sets in Knowledge Discovery 2. Applications, Case Studies and Software Systems*, Series Studies in Fuzziness in Soft Computing, vol. 19, Physica Verlag and it does extend and update the bibliography in : L. Polkowski, S. Tsumoto and T. Y. Lin, *Rough Set Methods and Applications. New Developments in Knowledge Discovery in Information Systems*, same Series, vol. 56.

[A] Books and conference proceedings

1. S. K. Pal, L. Polkowski and A. Skowron (eds.), *Rough–Neuro Computing: Techniques for Computing with Words*, Springer-Verlag, 2002, to appear.

2. L. Polkowski, S. Tsumoto and T. Y. Lin (eds.), *Rough Set Methods and Applications. New Developments in Knowledge Discovery in Information Systems*, this Series, vol. 56, Physica–Verlag, Heidelberg, 2001.

3. S. K. Pal and A. Skowron (eds.), *Rough Fuzzy Hybridization: A New Trend in Decision–Making*, Springer–Verlag, Singapore, 1999.

4. L. Polkowski and A. Skowron (eds.), *Rough Sets in Knowledge Discovery 1. Methodology and Applications*, this Series, vol. 18, Physica–Verlag, Heidelberg, 1998.

5. L. Polkowski and A. Skowron (eds.), *Rough Sets in Knowledge Discovery 2. Applications, Case Studies and Software Systems*, this Series, vol. 19, Physica–Verlag, Heidelberg, 1998.

6. T. Y. Lin and N. Cercone (eds.), *Rough Sets and Data Mining. Analysis of Imprecise Data*, Kluwer Academic Publishers, Dordrecht, 1997.

7. S. Hirano, M. Inuiguchi, and S. Tsumoto (eds.), *Proceedings of the Int. Workshop on Rough Set Theory and Granular Computing RSTGC'2001*, Bull. Intern. Rough Set Society, 5 (2001); also Lecture Notes in AI, vol. 2253, Springer-Verlag, Berlin, 2002.

8. W. Ziarko and Y. Y. Yao (eds.), *Proceedings of 2nd International Conference on Rough Sets and Current Trends in Computing RSCTC'2000*, Technical Report CS-2000-07, University of Regina, Regina, Canada, 2001; also Lecture Notes in AI, vol. 2205, Springer - Verlag, Berlin, 2001.

9. N. Zhong, A. Skowron, and S. Ohsuga (eds.), *New Directions in Rough Sets, Data Mining, and Granular–Soft Computing, Proceedings: the 7th International Workshop* (RSFDGrC'99), Ube–Yamaguchi, Japan, November 1999, LNAI 1711, Springer–Verlag, Berlin, 1999.

10. L. Polkowski and A. Skowron (eds.), *Rough Sets and Current Trends in Computing, Proceedings: the First International Conference on Rough Sets and Current Trends in Computing* (RSCTC'98), Warsaw, Poland, June 1998, LNAI 1424, Springer–Verlag, Berlin, 1998.

[B] **Journal and monograph articles, and conference papers**

1. P. Apostoli and A. Kanda, Parts of the Continuum: Towards a modern ontology of science, to appear in: Poznań Series on the Philosophy of Science and the Humanities.

2. G. Arora, F. Petry, and T. Beaubouef, New information measures for fuzzy sets, in: *Proceeings: the 7th IFSA World Congress*, Prague, the Czech Republic, June 1997.

3. G. Arora, F. Petry, and T. Beaubouef, Information measure of type β under similarity relations, in: *Proceedings: the 6th IEEE International Conference on Fuzzy Systems* (FUZZ-IEEE'97), Barcelona, Spain, July 1997.

4. C. Baizán, E. Menasalvas, J. Peña, A new approach to efficient calculation of reducts in large databases, in: *Proceedings: the 5th International Workshop on Rough Sets and Soft Computing* (RSSC'97) at *Proceedings: the 3rd Joint Conference on Information Sciences* (JCIS'97), Research Triangle Park NC, March 1997, pp. 340–344.

5. C. Baizán, E. Menasalvas, J. Peña, Using rough sets to mine socio-economic data, in: *Proceedings: SMC'97*, 1997, pp. 567–571.

6. C. Baizán, E. Menasalvas, J. Peña, S. Millán, and E. Mesa, Rough dependencies as a particular case of correlation: Application to the calculation of approximate reducts, in: *Proceedings: Principles of Data Mining and Knowledge Discovery* (PKDD'99), Prague, Czech Republic, September 1999, LNAI 1704, Springer-Verlag, Berlin, 1999, pp. 335–341.

7. C. Baizán, E. Menasalvas, J. Peña, and J. Pastrana, Integrating KDD algorithms and RDBMS code, in: *Proceedings: First International Conference on Rough Sets and Current Trends in Computing* (RSCTC'98), Warsaw, Poland, June 1998, LNAI 1424, Springer-Verlag, Berlin, 1998, pp. 210–214.

8. C. Baizán, E. Menasalvas, J. Peña, and J. Pastrana, RSDM system, *Bull. Intern. Rough Set Society*, 2, 1998, pp. 21–24.

9. C. Baizán, E. Menasalvas, J. Peña, C. P. Peréz and E. Santos, The lattices of generalizations in a KDD process, in: *Proceedings: Cybernetics and Systems'98*, 1998, pp. 181–184.

10. C. Baizán, E. Menasalvas, and A. Wasilewska, A model for RSDM Implementation, in: *Proceedings: the First international Conference on Rough Sets and Current Trends in Computing* (RSCTC'98), Warsaw, Poland, June 1998, LNAI 1424, Springer–Verlag, Berlin, 1998, pp. 186–196.

11. M. Banerjee, S. Mitra, and S.K. Pal, Rough Fuzzy MLP: Knowledge encoding and classification, *IEEE Trans. Neural Networks* 9(6), 1998, pp. 1203–1216.

12. M. Banerjee and S.K. Pal, Roughness of a fuzzy set, *Information Science* 93(3/4), 1996, pp. 235–246.

13. W. Bartol, X. Caicedo, and F. Rosselló, Syntactical content of finite approximations of partial algebras, in: *Proceedings: the First International Conference on Rough Sets and Current Trends in Computing* (RSCTC'98), Warsaw, Poland, June 1998, LNAI 1424, Springer-Verlag, Berlin, 1998, pp. 408–415.

14. J.G. Bazan, Approximate reasoning in decision rule synthesis, in: *Proceedings of the Workshop on Robotics, Intelligent Control and Decision Support Systems*, Polish–Japanese Institute of Information Technology, Warsaw, Poland, February 1999, pp. 10–15.

15. J.G. Bazan, Discovery of decision rules by matching new objects against data tables, in: *Proceedings: the First International Conference on Rough Sets and Current Trends in Computing* (RSCTC-98), Warsaw, Poland, June 1998, LNAI 1424, Springer–Verlag, Berlin, 1998, pp. 521–528.

16. J.G. Bazan, A comparison of dynamic and non–dynamic rough set methods for extracting laws from decision table, in: L. Polkowski, A. Skowron (eds.), *Rough Sets in Knowledge Discovery 1. Methodology and Applications*, Physica–Verlag, Heidelberg, 1998, pp. 321–365.

17. J. G. Bazan, Approximate reasoning methods for synthesis of decision algorithms (in Polish), Ph.D. Dissertation, supervisor A. Skowron, Warsaw University, 1998, pp. 1–179.

18. J.G. Bazan, Nguyen Hung Son, Nguyen Tuan Trung, A. Skowron, and J. Stepaniuk , Decision rules synthesis for object classification, in: E. Orłowska (ed.), *Incomplete Information: Rough Set Analysis*, Physica–Verlag, Heidelberg, 1998, pp. 23–57.

19. T. Beaubouef and R. Lang, Rough set techniques for uncertainty management in automated story generation, in: *the 36th Annual ACM Southeast Conference*, Marietta GA, April 1998.

20. T. Beaubouef, F. Petry, and G. Arora, Information measures for rough and fuzzy sets and application to uncertainty in relational databases, in: S. Pal and A. Skowron (eds.), *Rough-Fuzzy Hybridization: A New Trend in Decision-Making*, Springer-Verlag, Singapore, 1998, pp. 200–214.

21. T. Beaubouef, F. Petry, and G. Arora, Information–theoretic measures of uncertainty for rough sets and rough relational databases, *Information Sciences* 109(1–4),1998, pp. 185–195.

22. T. Beaubouef, F. Petry, and G. Arora, Information–theoretic measures of uncertainty for rough sets and rough relational databases, in: *Proceedings: the 5th International Workshop on Rough Sets and Soft Computing* (RSSC'97), Research Triangle Park NC, March 1997.

23. T. Beaubouef, F. Petry, and J. Breckenridge, Rough set based uncertainty management for spatial databases and Geographical Information Systems, in: *Proceedings: Fourth On–line World Conference on Soft Computing in Industrial Applications* (WSC4), September 1999, pp. 21–30.

24. Chien–Chung Chan, Distributed incremental Data Mining fron very large databases: a rough multi–set approach, in: *Proceedings SCI'2001: World Multiconference on Systemics, Cybernetics and Informatics*, Orlando, Fla., July 2001, vol. VII, pp. 517–522.

25. Chien-Chung Chan and J. W. Grzymala–Busse, On the lower boundaries in learning rules from examples, in: E. Orlowska (ed.), *Incomplete Information: Rough Set Analysis*, Physica–Verlag, Heidelberg, 1998, pp. 58–74.

26. I. V. Chikalov, On average time complexity of decision trees and branching programs, *Fundamenta Informaticae* 39(4), 1999, pp. 337–357.

27. I. V. Chikalov, On decision trees with minimal average depth, in: *Proceedings: the First International Conference on Rough Sets and Current Trends in Computing* (RSCTC'98), Warsaw, Poland, LNAI 1424, Springer–Verlag, Berlin, 1998, pp. 506–512.

28. I. V. Chikalov, Bounds on average weighted depth of decision trees depending only on entropy, in: *Proceedings: the 7th International Conference of Information Processing and Management of Uncertainty in Knowledge-Based Systems*, La Sorbonne, Paris, France, July 1998, pp. 1190–1194.

29. B. Chlebus and Nguyen Sinh Hoa, On finding optimal discretization on two attributes, in: *Proceedings: the First International Conference on Rough Sets and Current Trends in Computing* (RSCTC'98), Warsaw, Poland, June 1998, LNAI 1424, Springer–Verlag, Berlin, 1998, pp. 537–544.

30. Vhunnian Liu and Ning Zhong, Rough problem settings for ILP dealing with imperfect data, *Computational Intelligence: An Intern. Journal*, 17, 2001, pp. 446–459.

31. A. Czyżewski, Soft Processing of Audio Signals, in: L. Polkowski and A. Skowron (eds.), *Rough Sets in Knowledge Discovery 2. Applications, Case Studies and Software Systems*, Physica–Verlag, Heidelberg, 1998, pp. 147–165.

32. A. Czyżewski, Speaker – independent recognition of isolated words using rough sets, *J. Information Sciences* 104, 1998, pp. 3–14.

33. A. Czyżewski, Learning algorithms for audio signal enhancement. Part 2: Implementation of the rough set method for the removal of hiss, *J. Audio Eng. Soc.* 45(11), 1997, pp. 931-943.

34. A. Czyżewski, Speaker–independent recognition of digits – experiments with neural networks, fuzzy logic and rough sets, *J. Intelligent Automation and Soft Computing* 2(2), 1996, pp. 133-146.

35. A. Czyżewski and B. Kostek, Tuning the perceptual noise reduction algorithm using rough sets, in: *Proceedings: the First international Conference on Rough Sets and Current Trends in Computing* (RSCTC'98), Warsaw, Poland, June 1998, LNAI 1424, Springer–Verlag, Berlin, 1998, pp. 467–474.

36. A. Czyżewski and B. Kostek, Rough set–based filtration of sound applicable to hearing prostheses, In: *Proceedings: the 4th Intern. Workshop on Rough Sets, Fuzzy Sets, and Machine Discovery* (RSFD'96), Tokyo, Japan, November 1996, pp. 168–175.

37. A. Czyżewski and B. Kostek, Restoration of old records employing Artificial Intelligence methods, In: *Proceedings: the IASTED Internat. Conference – Artificial Intelligence, Expert Systems and Neural Networks*, Honolulu, Hawaii, 1996, pp.372–375.

38. A. Czyżewski, B. Kostek , H. Skarżyński, and R. Królikowski, Evaluation of some properties of the human auditory system using rough sets, In: *Proceedings: the 6th European Congress on Intelligent Techniques and Soft Computing* (EUFIT'98), Aachen, Germany, September 1998, Verlag Mainz, Aachen, 1998, pp. 965-969.

39. A. Czyżewski and R. Królikowski, Noise reduction in audio signals based on the perceptual coding approach, in: *Proceedings: the IEEE Workshop on Applications of Signal Processing to Audio and Acoustics*, New Paltz NY, October 1999, pp. 147–150.

40. A. Czyżewski and R. Królikowski, Noise reduction algorithms employing an intelligent inference engine for multimedia applications, in: *Proceedings: the IEEE 2nd Workshop on Multimedia Signal Processing*, Redondo Beach CA, December 1998, pp.125–130.

41. A. Czyżewski, R. Królikowski, S. K. Zieliński, and B. Kostek, Echo and noise reduction methods for multimedia communication systems, in: *Proceedings: the IEEE Signal Processing Society 1999 Workshop on Multimedia Signal Processing*, Copenhagen, Denmark, September 1999, pp. 239–244.

42. A. Czyżewski, H. Skarżyński, B. Kostek, and R. Królikowski, Rough set analysis of electro–stimulation test database for the prediction of post–operative profits in cochlear implanted patients, in: *Proceedings: the 7th Intern. Workshop on Rough Sets, Fuzzy Sets, Data Mining, and Granular-Soft Computing* (RSFDGrC'99), Ube–Yamaguchi, Japan, November 1999, LNAI 1711, Springer–Verlag, Berlin, 1999, pp. 109-117.

43. A. Czyżewski, H. Skarżyński, and B. Kostek, Multimedia databases in hearing and speech pathology, in: *Proceedings: the World Automation Congress* (WAC'98), Anchorage, Alaska, May 1998, pp. IFMIP-052.1–052.6.

44. R. Deja, Conflict analysis, in: *Proceedings: the 7th European Congress on Intelligent Techniques and Soft Computing* (EUFIT'99), Aachen, Germany, September 1999.

45. S. Demri, A class of decidable information logics, *Theoretical Computer Science*, 195(1), 1998, pp. 33–60.

46. S. Demri and B. Konikowska, Relative similarity logics are decidable: reduction to FO^2 with equality, in: *Proceedings: JELIA '98*, LNAI 1489, Springer-Verlag, Berlin, 1998, pp. 279–293.

47. S. Demri and E. Orłowska, Informational representability: Abstract models versus concrete models, in: D. Dubois and H. Prade (eds.), *Fuzzy sets, Logics and Reasoning about Knowledge*, Kluwer Academic Publishers, Dordrecht, 1999, pp. 301–314.

48. S. Demri and E. Orłowska, Informational representability of models for information logics, in: E. Orłowska (ed.), *Logic at Work. Essays Dedicated to the Memory of Helena Rasiowa*, Physica–Verlag, Heidelberg, 1998, pp. 383–409.

49. S. Demri, E. Orłowska, and D. Vakarelov, Indiscernibility and complementarity relations in Pawlak's information systems, in: *Liber Amicorum for Johan van Benthem's 50th Birthday*, 1999.

50. A.I. Dimitras, R. Słowiński, R. Susmaga, and C. Zopounidis: Business failure prediction using rough sets, *European Journal of Operational Research* 114, 1999, pp. 49–66.

51. Ju. V. Dudina and A. N. Knyazev, On complexity of language word recognition generated by context-free grammars with one non–terminal symbol (in Russian), *Bulletin of Nizhny Novgorod State University. Mathematical Simulation and Optimal Control* 19, 1998, pp. 214–223.

52. A. E. Eiben, T. J. Euverman, W. Kowalczyk, and F. Slisser, Modeling customer retention with statistical techniques, rough data models, and genetic programming, in: S. K. Pal and A. Skowron (eds.), *Rough Fuzzy Hybridization: A New Trend in Decision–Making*, Springer–Verlag, Singapore, 1999, pp. 330–348.

53. P. Ejdys and G. Góra, The More We Learn the Less We Know? - On Inductive Learning from Examples, *Proceedings: the 11th International Symposium on Methodologies for Intelligent Systems, Foundations of Intelligent Systems* (ISMIS'99), Warsaw, Poland, June 1999, LNAI, Springer–Verlag, Berlin, in print.

54. L. Goodwin, J. Prather, K. Schlitz, M. A. Iannacchione, M. Hage, W. E. Hammond Sr., and J. W. Grzymala-Busse, Data mining issues for improved birth outcomes, *Biomedical Sciences Instrumentation* 34, 1997, pp. 291–296.

55. S. Greco, B. Matarazzo, and R. Słowiński, Dominance–based rough set approach to rating analysis, *Gestion 2000 Magazine*, to appear.

56. S. Greco, B. Matarazzo, and R. Słowiński, Decision rules, in: *Encyclopedia of Management*, 4th edition, 2000, to appear.

57. S. Greco, B. Matarazzo, and R. Słowiński, Rough set processing of vague information using fuzzy similarity relation, in: C. Calude and G. Paun (eds.), *Finite versus Infinite – Contributions to an Eternal Dilemma*, Springer-Verlag, Berlin, to appear.

58. S. Greco, B. Matarazzo, and R. Słowiński, Rough approximation of a preference relation by dominance relations, *European Journal of Operational Research* 117, 1999, pp. 63–83.

59. S. Greco, B. Matarazzo, and R. Słowiński, The use of rough sets and fuzzy sets in MCDM, in: T. Gal, T. Stewart, and T. Hanne (eds.), *Advances in Multiple Criteria Decision Making* , Kluwer Academic Publishers, Boston, 1999, Chapter 14: pp. 14.1–14.59.

60. S. Greco, B. Matarazzo, R. Słowiński, Fuzzy dominance as basis for rough approximations, in: *Proceedings: the 4th Meeting of the EURO WG on Fuzzy Sets and 2nd Internat. Conf. on Soft and Intelligent Computing*, (EUROFUSE-SIC'99), Budapest, Hungary, May 1999, pp. 273-278.

61. S. Greco, B. Matarazzo, and R. Słowiński, Handling missing values in rough set analysis of multi–attribute and multi–criteria decision problems, in: *Proceedings: New Directions in Rough Sets, Data Mining and Granular-Soft Computing* (RSFSGrC'99), Ube–Yamaguchi, Japan, November 1999, LNAI 1711, Springer–Verlag, Berlin, 1999, pp. 146–157.

62. S. Greco, B. Matarazzo, and R. Słowiński, Fuzzy dominance as a basis for rough approximations, in: *Proceedings: Workshop Italiano sulla Logica Fuzzy* (Wilf'99), Genova, Italy, June 1999, pp.14–16.

63. S. Greco, B. Matarazzo, and R. Słowiński, Misurazione congiunta e incoerenze nelle preferenze, in: *Atti del Ventitreesimo Convegno A.M.A.S. E.S.*, Rende-Cosenza, Italy, September 1999, pp. 255–269.

64. S. Greco, B. Matarazzo, and R. Słowiński, L'approcio dei rough sets all'analisi del rating finanziario, in: *Atti del Ventitreesimo Convegno A. M. A. S. E. S.*, Rende-Cosenza, Italy, September 1999, pp. 271–286.

65. S. Greco, B. Matarazzo, and R. Słowiński, On joint use of indiscernibility, similarity and dominance in rough approximation of decision classes, in: *Proceedings: the 5th International Conference of the Decision Sciences Institute*, Athens, Greece, July 1999, pp. 1380–1382; also in: *Research Report RA–012/98*, Inst. Comp. Sci., Poznań Univ. Technology, 1998.

66. S. Greco, B. Matarazzo, and R.Słowiński, A new rough set approach to evaluation of bankruptcy risk, in: C. Zopounidis (ed.), *Operational Tools in the Management of Financial Risks*, Kluwer Academic Publishers, Dordrecht, 1998, pp. 121–136.

67. S. Greco, B. Matarazzo, and R. Słowiński, Fuzzy similarity relation as a basis for rough approximations, in: *Proceedings: Rough Sets and Current Trends in Computing* (RSCTC'98), Warsaw, Poland, June 1998, LNAI 1424, Springer–Verlag, Berlin, 1998, pp. 283–289.

68. S. Greco, B. Matarazzo, and R. Słowiński, A new rough set approach to multi–criteria and multi–attribute classification, in: *Proceedings: Rough Sets and Current Trends in Computing* (RSCTC'98), Warsaw, Poland, LNAI 1424, Springer–Verlag, Berlin, 1998, pp. 60–67.

69. S. Greco, B. Matarazzo, and R. Słowiński, Rough approximation of a preference relation in a pair–wise comparison table, in: L. Polkowski and A. Skowron (eds.), *Rough Sets in Knowledge Discovery 2. Applications, Case Studies and Software Systems*, Physica-Verlag, Heidelberg, 1998, pp. 13–36.

70. S. Greco, B. Matarazzo, and R. Słowiński, Rough set theory approach to decision analysis, in: *Proceedings: the 3rd European Workshop on Fuzzy Decision Analysis and Neural Networks for Management, Planning and Optimization* (EFDAN'98), Dortmund, Germany, June 1998, pp. 1–28.

71. S. Greco, B. Matarazzo, and R. Słowiński: Conjoint measurement, preference inconsistencies and decision rule model, in: *Proceedings: the 2nd International Workshop on Preferences and Decisions*, Trento, Italy, July 1998, pp. 49–53.

72. S. Greco, B. Matarazzo, and R. Słowiński, New developments in the rough set approach to multi–attribute decision analysis, in: *Tutorials and Research Reviews: 16th European Conference on Operational Research* (EURO XVI), Brussels, Belgium, July 1998, 37 pp.

73. S. Greco, B. Matarazzo, and R. Słowiński, Rough set handling of ambiguity, in: *Proceedings: the 6th European Congress on Intelligent Techniques and Soft Computing* (EUFIT'98), Aachen, Germany, September 1998, Verlag Mainz, Aachen, 1998, pp. 3–14.

74. S. Greco, B. Matarazzo, and R. Słowiński, Fuzzy measures as a technique for rough set analysis, in: *Proceedings: the 6th European Congress on Intelligent Techniques and Soft Computing* (EUFIT'98), Aachen, Germany, September 1998, Verlag Mainz, Aachen, 1998, pp. 99–103.

75. S. Greco, B. Matarazzo, and R. Słowiński, Un nuovo approccio dei rough sets alla classificazione multiattributo e multicriteriale, in: *Atti del Ventiduesimo Convegno A. M. A. S. E. S.*, Genova, Italy, September 1998, Bozzi Editore, Genova, 1998, pp. 249–260.

76. S. Greco, B. Matarazzo, and R. Słowiński, Modellizzazione delle preferenze per mezzo di regole di decisione, in: *Atti del Ventiduesimo Convegno A. M. A. S. E. S.*, Genova, Italy, September 1998, Bozzi Editore, Genova, 1998, pp. 233–247.

77. S. Greco, B. Matarazzo, and R. Słowiński, The rough set approach to decision support, in: *Proceedings: the Annual Conference of the Operational Research Society of Italy* (AIRO), Treviso, Italy, September 1998, pp. 561–564.

78. S. Greco, B. Matarazzo, R. Słowiński, and A. Tsoukias, Exploitation of a rough approximation of the outranking relation in multi-criteria choice and ranking, in: T.J. Stewart and R.C. van den Honert (eds.), *Trends in Multi-criteria Decision Making*, LNEMS 465, Springer–Verlag, Berlin, 1998, pp. 45–60.

79. S. Greco, B. Matarazzo, R. Słowiński, and S. Zanakis, Rough set analysis of information tables with missing values, in: *Proceedings: the 5th International Conference of the Decision Sciences Institute*, Athens, Greece, July 1999, pp. 1359–1362.

80. J. P. Grzymała –Busse, J. W. Grzymała–Busse, and Z. Hippe, Prediction of melanoma using rule induction based on rough sets, in: *Proceedings SCI'2001: World Multiconference on Systemics, Cybernetics and Informatics*, Orlando, Fla., July 2001, vol. VII, pp. 523–527.

81. J. W. Grzymala–Busse, Applications of the rule induction system LERS, in: L. Polkowski and A. Skowron (eds.), *Rough Sets in Knowledge Discovery 1. Methodology and Applications*, Physica–Verlag, Heidelberg, 1998, pp. 366–375.

82. J. W. Grzymala–Busse, LERS : A knowledge discovery system, in: L. Polkowski and A. Skowron (eds.),*Rough Sets in Knowledge Discovery 2. Applications, Case Studies and Software Systems*, Physica–Verlag, Heidelberg, 1998, pp. 562–565.

83. J. W. Grzymala–Busse, Rule induction system LERS, *Bull. of Intern. Rough Set Society* 2, 1998, pp. 18–20.

84. J. W. Grzymala–Busse, Classification of unseen examples under uncertainty, *Fundamenta Informaticae* 30, 1997, pp. 255–267.

85. J. W. Grzymala–Busse, A new version of the rule induction system LERS, *Fundamenta Informaticae* 31, 1997, pp. 27–39.

86. J. W. Grzymała–Busse, W. J. Grzymała–Busse, and L. K. Goodwin, Coping with missing attribute values based on closest fit in preterm birth data: a rough set approach, *Computational intelligence: An Intern. Journal*, 17, 2001, pp. 425–434.

87. J. W. Grzymala–Busse and L. K. Goodwin, Predicting pre–term birth risk using machine learning from data with missing values, *Bull. of Intern. Rough Set Society* 1, 1997, pp. 17–21.

88. J. W. Grzymala–Busse, L. K. Goodwin, and Xiaohui Zhang, Pre–term birth risk assessed by a new method of classification using selective partial matching, in: *Proceedings: the 11th International Symposium on Methodologies for Intelligent Systems* (ISMIS'99), Warsaw, Poland, June 1999, LNAI 1609, Springer Verlag, Berlin, 1999, pp. 612–620.

89. J. W. Grzymala–Busse, L. K. Goodwin, and Xiaohui Zhang, Increasing sensitivity of pre–term birth by changing rule strengths, in: *Proceedings: the 8th Workshop on Intelligent Information Systems* (IIS'99), Ustroń, Poland, June 1999, pp. 127–136.

90. J. W. Grzymała–Busse, P. Loupe, and S. Schroeder, Analysis of behavioral responsiveness of rats to GBR12909 using data mining system LERS, in: *Proceedings SCI'2001: World Multiconference on Systemics, Cybernetics and Informatics*, Orlando, Fla., July 2001, vol. VII, pp. 528–533.

91. J. W. Grzymala–Busse, W. J. Grzymala–Busse, and L. K. Goodwin, A closest fit approach to missing attribute values in pre–term birth data, in: *Proceedings: the 7th International Workshop on Rough Sets, Fuzzy Sets, Data Mining and Granular–Soft Computing* (RSFDGrC'99), Ube–Yamaguchi, Japan, November 1999, LNAI 1711, Springer–Verlag, Berlin, 1999, pp. 405–413.

92. J. W. Grzymala–Busse and L. J. Old, A machine learning experiment to determine part of speech from word-endings, in: *Proceedings: the 10th Intern. Symposium on Methodologies for Intelligent Systems* (IS-MIS'97), Charlotte NC, October 1997, LNAI 1325, Springer-Verlag, Berlin, 1997, pp. 497–506.

93. J. W. Grzymala–Busse, S. Y. Sedelow, and W. A. Sedelow Jr., Machine learning and knowledge acquisition, rough sets, and the English semantic code, in: T. Y. Lin and N. Cercone (eds.), *Rough Sets and Data Mining. Analysis of Imprecise Data*, Kluwer Academic Publishers,Dordrecht, 1997, pp. 91–107.

94. J. W. Grzymala–Busse and Soe Than, Inducing simpler rules from reduced data, in: *Proceedings: the Seventh Workshop on Intelligent Information Systems* (IIS'98), Malbork, Poland, June 1998, pp. 371–378.

95. J. W. Grzymala–Busse and J. Stefanowski, Two approaches to numerical attribute discretization for rule induction, in: *Proceedings: the 5th International Conference of the Decision Sciences Institute*, Athens, Greece, July 1999, pp. 1377–1379.

96. J. W. Grzymala–Busse and J. Stefanowski, Discretization of numerical attributes by direct use of the rule induction algorithm LEM2 with interval extension, in: *Proceedings: the Sixth Symposium on Intelligent Information Systems* (IIS'97), Zakopane, Poland, June 1997, pp. 149–158.

97. J. W. Grzymala–Busse and Ta-Yuan Hsiao, Dropping conditions in rules induced by ID3, in: *Proceedings: the 6th International Workshop*

on Rough Sets, Data Mining and Granular Computing (RSDMGrC'98) at *the 4th Joint Conference on Information Sciences* (JCIS'98), Research Triangle Park NC, October 1998, pp. 351–354.

98. J. W. Grzymala–Busse and A. Y. Wang, Modified algorithms LEM1 and LEM2 for rule induction from data with missing attribute values, in: *Proceedings: the 5th Intern. Workshop on Rough Sets* (RSSC'97) at *the 3rd Joint Conference on Information Sciences* (JCIS'97), Research Triangle Park NC, March 1997, pp. 69–72.

99. J. W. Grzymala–Busse and P. Werbrouck, On the best search method in the LEM1 and LEM2 algorithms, in :E. Orłowska (ed.), *Incomplete Information: Rough Set Analysis*, Physica–Verlag, Heidelberg, 1998, pp. 75–91.

100. J. W. Grzymala–Busse and Xihong Zou, Classification strategies using certain and possible rules, in: *Proceedings: the First International Conference on Rough Sets and Current Trends in Computing*(RSCTC'98), Warsaw, Poland, June 1998, LNAI 1424, Springer Verlag, Berlin, 1998, pp. 37–44.

101. Hoang Kiem and Do Phuc, A combined multi-dimensional Genetic Algorithm and Kohonen Neural Network for cluster discovery in Data Mining, in: *Proceedings: the 3rd International Conference on Data Mining* (PAKDD'99), Beijing, China, 1999.

102. Hoang Kiem and Do Phuc, A Rough Genetic Kohonen Neural Network for conceptual cluster discovery, in: *Proceedings: the 7th Work-Shop on Rough Set, Fuzzy Set, Granular Computing and Data Mining* (RSFDGrC'99), Ube–Yamaguchi, Japan, November 1999, LNAI 1711, Springer Verlag, Berlin, 1999, pp. 448–452.

103. Hoang Kiem and Do Phuc, On the association rules based extension of the dependency of attributes in rough set theory for classification problem, *Magazine of Science and Technology* 1, 1999, Vietnam National University.

104. Hoang Kiem and Do Phuc, Discovering the binary and fuzzy association rules from database, *Magazine of Science and Technology* 4, 1999, Vietnam National University.

105. V. Jog, W. Michałowski, R. Słowiński, and R. Susmaga, The rough set analysis and the neural networks classifier – a hybrid approach to predicting stocks' performance, in: *Proceedings: the 5th International Conference of the Decision Sciences Institute*, Athens, Greece, July 1999, pp. 1386–1388.

106. R. E. Kent, Soft concept analysis, in: S. K. Pal and A. Skowron (eds.), *Rough Fuzzy Hybridization: A New Trend in Decision–Making*, Springer–Verlag, Singapore, 1999, pp. 215–232.

107. A. N. Knyazev, On word recognition in language generated by 1-context-free grammar (in Russian), in: *Proceedings: 12th International Conference on Problems of Theoretical Cybernetics*, Nizhny Novgorod, Russia, 1999, Moscow State University Publishers, Moscow, Part 1 (1999), pp. 96.

108. A. N. Knyazev, On recognition of words from languages generated by linear grammars with one non–terminal symbol, in: *Proceedings: the First International Conference on Rough Sets and Current Trends in Computing* (RSCTC'98), Warsaw, Poland, 1998, LNAI 1424, Springer–Verlag, Berlin, 1998, pp. 111–114.

109. J. Komorowski, L. Polkowski, and A. Skowron, Rough Sets: A Tutorial, in: *Lecture Notes for ESSLLI'99: the 11th European Summer School in Language, Logic and Information*, Utrecht, Holland, August 1999, 111 pp.

110. J. Komorowski, Z. Pawlak, L. Polkowski, and A. Skowron, Rough sets: A tutorial, in: S. K. Pal and A. Skowron (eds.), *Rough Fuzzy Hybridization: A New Trend in Decision Making*, Springer Verlag, Singapore, 1999, pp. 3–98.

111. J. Komorowski, L. Polkowski, and A. Skowron, Towards a rough mereology–based logic for approximate solution synthesis, *Studia Logica* 58(1), 1997, pp. 143–184.

112. J. Komorowski, L. Polkowski, and A. Skowron, Rough sets for Data Mining and Knowledge Discovery (Tutorial–abstract), in: *Proceedings: the First European Symposium on Principles of Data Mining and Knowledge Discovery*, Trondheim, Norway, LNAI 1263, Springer–Verlag, Berlin, pp. 395–395.

113. B. Kostek, *Soft Computing in Acoustics, Applications of Neural Networks, Fuzzy Logic and Rough Sets to Musical Acoustics* in the Series: *Studies in Fuzziness and Soft Computing* (J. Kacprzyk (ed.)), vol. 31, Physica–Verlag, Heilderberg, 1999.

114. B. Kostek, Assessment of concert hall acoustics using rough set and fuzzy set approach, in: S. K. Pal and A. Skowron (eds.), : *Rough–Fuzzy Hybridization: A New Trend in Decision Making*, Springer-Verlag, Singapore, 1999, pp. 381–396.

115. B. Kostek, Rough–fuzzy method of subjective test result processing, in: *Proceedings: the 8th International Symposium on Sound Engineering*

and Mastering (ISSEM'99), Gdańsk, Poland, September 1999, pp. 11–18.

116. B. Kostek, Soft computing–based recognition of musical sounds, in: L. Polkowski and A. Skowron (eds.), *Rough Sets in Knowledge Discovery 2. Applications, Case Studies and Software Systems*, Physica–Verlag, Heidelberg, 1998, pp. 193–213.

117. B. Kostek, Computer–based recognition of musical phrases using the rough set approach, *J. Information Sciences* 104, 1998, pp. 15–30.

118. B. Kostek, Soft set approach to the subjective assessment of sound quality, in: *Proceedings: the Conference FUZZ-IEEE'98 at the World Congress on Computational Intelligence*(WCCI'98), Anchorage, Alaska, May 1998, pp. 669–674.

119. B. Kostek, Automatic recognition of sounds of musical instruments: An expert media application, in: *Proceedings: the World Automation Congress* (WAC'98), pp. IFMIP–053.

120. B. Kostek, Sound quality assessment based on the rough set classifier, in: *Proceedings: the 5th European Congress on Intelligent Techniques and Soft Computing* (EUFIT'97), Aachen, Germany, September 1997, Verlag Mainz, Aachen, 1997, pp. 193–195.

121. B. Kostek, Soft set approach to the subjective assessment of sound quality, in: *Proceedings: the 9th Intern. Conference on Systems Research Informatics and Cybernetics* (InterSymp'97), Baden–Baden, Germany, 1997.

122. B. Kostek, Rough set and fuzzy set methods applied to acoustical analyses, *J. Intelligent Automation and Soft Computing*, 2(2), 1996, pp. 147–160.

123. K. Krawiec, R. Słowiński, and D. Vanderpooten, Learning of decision rules from similarity based rough approximations, in: L. Polkowski, A. Skowron (eds.), *Rough Sets in Knowledge Discovery 2. Applications, Case Studies and Software Systems*, Physica–Verlag, Heidelberg, 1998, pp. 37–54.

124. R. Królikowski and A. Czyżewski, Noise reduction in telecommunication channels using rough sets and neural networks, in: *Proceedings: the 7th Intern. Workshop on Rough Sets, Fuzzy Sets, Data Mining, and Granular-Soft Computing* (RSFDGrC'99), Ube–Yamaguchi, Japan, November 1999, LNAI 1711, Springer–Verlag, Berlin, 1999, pp. 109–117.

125. R. Królikowski and A. Czyżewski, Applications of rough sets and neural nets to noisy audio enhancement, in: *proceedings: the 7th European Congress on Intelligent Techniques and Soft Computing* (EUFIT'99), Aachen, Germany, September 1999.

126. M. Lifantsev and A. Wasilewska, A decision procedure for rough sets equalities, in: *Proceedings: the 18th International Conference of the North American Fuzzy Information Processing Society* (NAFIPS'99), New York NY, June 1999, pp.786–791.

127. T. Y. Lin, Data Mining and machine oriented modeling: A granular computing approach, *Journal of Applied Intelligence*, in print.

128. T. Y. Lin, Theoretical sampling for Data Mining, in: *Proceedings: the 14th Annual International Symposium Aerospace/Defense Sensing, Simulation, and Controls* (SPIE) 4057, Orlando Fla., April 2000, to appear.

129. T. Y. Lin, Attribute transformations on numerical databases: Applications to stock market data, in: *Methodologies for Knowledge Discovery and Data Mining*, LNAI, Springer–Verlag, Berlin, 2000, to appear.

130. T. Y. Lin, Belief functions and probability of fuzzy sets, in: *Proceedings: the 8th IFSA World Congress* (IFSA'99), Taipei, Taiwan, August 1999, pp. 219–223.

131. T. Y. Lin, Discovering patterns in numerical sequences using rough set theory, in: *Proceedings: the 3rd World Multi–Conference on Systemics, Cybernetics and Informatics*, Orlando Fla., July 1999, 5, pp. 568–572.

132. T. Y. Lin, Measure theory on granular fuzzy sets, in : *Proceedings: the 18th International Conference of North America Fuzzy Information Processing Society*, June 1999, pp. 809–813.

133. T. Y. Lin, Data Mining: Granular computing approach, in: *Methodologies for Knowledge Discovery and Data Mining, the 3rd Pacific–Asia Conference*, Beijing, China, April 1999, LNAI 1574, Springer–Verlag, Berlin, 1999, pp. 24–33.

134. T. Y. Lin, Granular computing: Fuzzy logic and rough sets, in: L.A. Zadeh and J. Kacprzyk (eds), *Computing with Words in Information Intelligent Systems 1*, Physica–Verlag, Heidelberg, 1999, pp. 183–200.

135. T. Y. Lin, Granular computing on binary relations II: Rough set representations and belief functions, in: L. Polkowski and A. Skowron (eds), *Rough Sets In Knowledge Discovery 1.Methodology and Applications*, Physica–Verlag, Heidelberg, 1998, pp. 121–140.

136. T. Y. Lin, Granular computing on binary relations I: Data Mining and neighborhood systems, in: L. Polkowski and A. Skowron (eds.), *Rough Sets In Knowledge Discovery 1. Methodology and Applications*, Physica–Verlag, Heidelberg, 1998, pp. 107–121.

137. T. Y. Lin, Context free fuzzy sets and information tables, in: *Proceedings: the Sixth European Congress on Intelligent Techniques and Soft Computing* (EUFIT'98) , Aachen, Germany, September 1998, Verlag Mainz, Aachen, pp. 76–80.

138. T. Y. Lin, Granular fuzzy sets: Crisp representation of fuzzy sets, in: *Proceedings: the Sixth European Congress on Intelligent Techniques and Soft Computing* (EUFIT'98) , Aachen, Germany, September 1998, Verlag Mainz, Aachen, pp. 94–98.

139. T. Y. Lin, Fuzzy partitions : Rough set theory, in: *Proceedings: the Conference on Information Processing and Management of Uncertainty in Knowledge–Based Systems* (IPMU'98), La Sorbonne, Paris, France, July 1998, pp. 1167–1174.

140. T. Y. Lin, Sets with partial memberships: A Rough sets view of fuzzy sets, in: *Proceedings: the FUZZ-IEEE International Conference, 1998 IEEE World Congress on Computational Intelligence* (WCCI'98), Anchorage, Alaska, May 1998.

141. T. Y. Lin and Q. Liu, First–order rough logic revisited, in: *Proceedings: the 7th International Workshop on rough Sets, Fuzzy Sets, Data Mining and Granular–Soft computing* (RSFSGrC'99), Ube–Yamaguchi, Japan, November 1999, LNAI 1711, Springer–Verlag, Berlin, pp. 276–284.

142. T. Y. Lin and E. Louie, A Data Mining approach using machine oriented modeling: finding association rules using canonical names, in: *Proceedings: the 14th Annual International Symposium Aerospace/Defense Sensing, Simulation, and Controls* (SPIE) 4057, Orlando Fla., April 2000, to appear.

143. T. Y. Lin, Ning Zhong, J. J. Dong, and S. Ohsuga, An incremental, probabilistic rough set approach to rule discovery, in: *Proceedings: the FUZZ-IEEE International Conference, 1998 IEEE World Congress on Computational Intelligence* (WCCI'98), Anchorage, Alaska, May 1998.

144. T. Y. Lin, Ning Zhong, J. J. Dong, and S. Ohsuga, Frameworks for mining binary relations in data, in : *Proceedings: the First International Conference on Rough Sets and Current Trends in Computing* (RSCTC'98), Warsaw, Poland, June 1998, LNAI 1424, Springer–Verlag, Berlin, 1998, pp. 387–393.

145. T. Y. Lin and S. Tsumoto, Context-free fuzzy sets in Data Mining context, in: *Proceedings: the 7th International Workshop on Rough Sets, Fuzzy Sets, Data Mining and Granular–Soft computing* (RSFSGrC'99), Ube–Yamaguchi, Japan, November 1999, LNAI 1711, Springer–Verlag, Berlin, pp. 212–220.

146. P. Lingras and C. Davies, Applications of rough genetic algorithms, *Computational Intelligence: An Intern. journal*, 17, 2001, pp. 435–445.

147. B. Marszał–Paszek, Linking α–approximation with evidence theory, in: *Proceedings: the 6th International Conference Information Processing and Management of Uncertainty in Knowledge-Base System* (IPMU'96), Granada, Spain, 1996, pp. 1153–1158.

148. B. Marszał–Paszek and P. Paszek, Searching for attributes which well determinate decision in the decision table, in: *Proceedings: Intelligent Information Systems VIII*, Ustroń, Poland, 1999, pp. 146–148.

149. B. Marszał–Paszek and P. Paszek, Extracting strong relationships between data from decision table, in: *Proceedings: Intelligent Information Systems VII*, Malbork, Poland, 1998, pp. 396–399.

150. V. W. Marek and M. Truszczyński, Contributions to the theory of rough sets, *Fundamenta Informaticae* 39(4), 1999, pp. 389–409.

151. E. Martienne and M. Quafafou, Learning fuzzy relational descriptions = using the logical framework and rough set theory, in: *Proceedings: the 7th IEEE International Conference on Fuzzy Systems* (FUZZ-IEEE'98), IEEE Neural Networks Council, 1998.

152. E. Martienne and M. Quafafou, Learning logical descriptions for document understanding: a rough sets based approach, in: *Proceedings: the First International Conference on Rough Sets and Current Trends in Computing* (RSCTC'98), Warsaw, Poland, June 1998, LNAI 1424, Springer–Verlag, Berlin, 1998.

153. E. Martienne and M. Quafafou, Vagueness and data reduction in learning of logical descriptions, in: *Proceedings: the 13th European Conference on Artificial Intelligence* (ECAI'98), Brighton, UK, August 1998, John Wiley and Sons, Chichester, 1998.

154. P. Mitra, S. Mitra, and S.K. Pal, Staging of cervical cancer with Soft Computing, *IEEE Trans. Bio-Medical Engineering*, in print.

155. S. Mitra, M. Banerjee, and S.K. Pal, Rough Knowledge-based networks, fuzziness and classification, *Neural Computing and Applications* 7, 1998, pp. 17–25.

156. S. Mitra, P. Mitra, and S. K. Pal, Evolutionary design of modular Rough Fuzzy MLP, *Neurocomputing*, communicated.

157. S. Miyamoto and Kyung Soo Kim, Images of fuzzy multi–sets by one-variable functions and their applications (in Japanese), *Journal of Japan Society for Fuzzy Theory and Systems*10(1), 1998, pp. 150–157.

158. S. Miyamoto, Application of rough sets to information retrieval, *Journal of the American Society for Information Science* 47(3), 1998, pp. 195–205.

159. S. Miyamoto, Indexed rough approximations and generalized possibility theory, in: *Proceedings: FUZZ-IEEE'98*, May 4-9, 1998, Anchorage, Alaska, pp. 791-795.

160. S. Miyamoto, Fuzzy multi–sets and a rough approximation by multi-set–valued function, in: L. Polkowski and A. Skowron (eds.), *Rough Sets in Knowledge Discovery 1. Methodology and Applications*, Physica–Verlag, Heidelberg,1998, pp. 141–159.

161. H. Moradi, J. W. Grzymala–Busse, and J. A. Roberts, Entropy of English text: Experiments with humans and a machine learning system based on rough sets, *Information Sciences. An International Journal* 104, 1998, pp. 31–47.

162. M. Ju. Moshkov, Time complexity of decision trees (in Russian), in: *Proceedings: the 9th Workshop on Synthesis and Complexity of Control Systems*, Nizhny Novgorod, Russia, 1998, Moscow State University Publishers, Moscow, 1999, pp. 52–62.

163. M. Ju. Moshkov, Local approach to construction of decision trees, in: S.K.Pal and A. Skowron (eds.), *Rough Fuzzy Hybridization. A New Trend In Decision–Making*, Springer-Verlag, Singapore, 1999, pp. 163–176.

164. M. Ju. Moshkov, On complexity of deterministic and nondeterministic decision trees (in Russian), in: *Proceedings: the 12th International Conference on Problems of Theoretical Cybernetics*, Nizhny Novgorod, Russia, 1999, Moscow State University Publishers, Moscow, 1999, p. 164.

165. M. Ju. Moshkov, On the depth of decision trees (in Russian), *Doklady RAN*, 358(1), 1998, p. 26.

166. M. Ju. Moshkov, Some relationships between decision trees and decision rule systems, in: *Proceedings: the First International Conference on Rough Sets and Current Trends in Computing* (RSCTC'98), Warsaw, Poland, June 1998, LNAI 1424, Springer–Verlag, Berlin, 1998, pp. 499–505.

167. M. Ju. Moshkov, Three ways for construction and complexity estimation of decision trees, in: *Program: the 16th European Conference on Operational Research*(EURO XVI), Brussels, Belgium, July 1998, pp. 66-67.

168. M. Ju. Moshkov, Rough analysis of tree–program time complexity, in: *Proceedings: the 7th International Conference of Information Processing and Management of Uncertainty in Knowledge-Based Systems*, La Sorbonne, Paris, France, July 1998, pp. 1376–1380.

169. M. Ju. Moshkov, On time complexity of decision trees, in: L. Polkowski and A. Skowron (eds.), *Rough Sets in Knowledge Discovery 1. Methodology and Applications*, Physica–Verlag, Heidelberg, 1998, pp. 160–191.

170. M. Ju. Moshkov, On time complexity of decision trees (in Russian), in: *Proceedings: International Siberian Conference on Operational Research*, Novosibirsk, Russia, 1998, pp. 28–31.

171. M. Ju. Moshkov, On complexity of decision trees over infinite information systems, in: *Proceedings: the Third Joint Conference on Information Sciences* (JCIS'97), Duke University, USA, 1997, pp. 353–354.

172. M. Ju. Moshkov, Algorithms for constructing of decision trees, in: *Proceedings: the First European Symposium on Principles of Data Mining and Knowledge Discovery* (PKDD'97), Trondheim, Norway, 1997, LNAI 1263, Springer–Verlag, Berlin, 1997, pp. 335–342.

173. M. Ju. Moshkov, Unimprovable upper bounds on time complexity of decision trees, *Fundamenta Informaticae* 31(2), 1997, pp. 157–184.

174. M. Ju. Moshkov, Rough analysis of tree-programs, in: *Proceedings: the 5th European Congress on Intelligent Techniques and Soft Computing* (EUFIT' 97), Aachen, Germany, September 1997, Verlag Mainz, Aachen, pp. 231–235.

175. M. Ju. Moshkov, Complexity of deterministic and nondeterministic decision trees for regular language word recognition, in: *Proceedings : the 3rd International Conference on Developments in Language Theory*, Thessaloniki, Greece, 1997, pp. 343–349.

176. M. Ju. Moshkov, Comparative analysis of time complexity of deterministic and nondeterministic tree–programs (in Russian), in: *Actual Problems of Modern Mathematics* 3, Novosibirsk University Publishers, Novosibirsk, 1997, pp. 117–124.

177. M. Ju. Moshkov and I. V. Chikalov, On effective algorithms for conditional test construction (in Russian), in: *Proceedings: the 12th International Conference on Problems of Theoretical Cybernetics*, Nizhny

Novgorod, Russia, 1999, Moscow State University Publishers, Moscow, 1999, pp. 165.

178. M. Ju. Moshkov and I. V. Chikalov, Bounds on average depth of decision trees, in: *Proceedings: the Fifth European Congress on Intelligent Techniques and Soft Computing* (EUFIT'97), Aachen, Germany, September 1997, Verlag Mainz, Aachen, pp. 226–230.

179. M. Ju. Moshkov and I. V. Chikalov, Bounds on average weighted depth of decision trees, *Fundamenta Informaticae* 31(2), 1997, pp. 145–156.

180. M. Ju. Moshkov and A. Moshkova, Optimal bases for some closed classes of Boolean functions, in: *Proceedings: the 5th European Congress on Intelligent Techniques and Soft Computing* (EUFIT 97), Aachen, Germany, September 1997, Verlag Mainz, Aachen, pp. 1643–1647.

181. A. M. Moshkova, On complexity of "retaining" fault diagnosis in circuits (in Russian), in: *Proceedings: the 12th International Conference on Problems of Theoretical Cybernetics*, Nizhny Novgorod, Russia, 1999, Moscow State University Publishers, Moscow, 1999, p. 166.

182. A. M. Moshkova, On diagnosis of retaining faults in circuits, in: *Proceedings: the First International Conference on Rough Sets and Current Trends in Computing* (RSCTC'98), Warsaw, Poland, June 1998, LNAI 1424, Springer-Verlag, Berlin, 1998, pp. 513–516.

183. A. M. Moshkova, Diagnosis of "retaining" faults in circuits (in Russian), *Bulletin of Nizhny Novgorod State University. Mathematical Simulation and Optimal Control* 19, 1998, pp. 204–213.

184. A. Nakamura, Conflict logic with degrees, in: S. K. Pal and A. Skowron (eds.), *Rough Fuzzy Hybridization: A New Trend in Decision–Making*, Springer–Verlag, Singapore, 1999, pp. 136–150.

185. Nguyen Hung Son, From optimal hyperplanes to optimal decision trees, *Fundamenta Informaticae* 34(1-2), 1998, pp. 145–174.

186. Nguyen Hung Son, Discretization problems for rough set methods, in: *Proceedings: the First International Conference on Rough Sets and Current Trend in Computing* (RSCTC'98), Warsaw, Poland, June 1998, LNAI 1424, Springer–Verlag, Berlin, 1998, pp. 545–552.

187. Nguyen Hung Son, Discretization of real value attributes. Boolean reasoning approach, Ph.D. Dissertation, supervisor A. Skowron, Warsaw University, Warsaw, 1997, pp. 1–90.

188. Nguyen Hung Son, Rule induction from continuous data, in: *Proceedings: the 5th International Workshop on Rough Sets and Soft Computing* (RSSC'97) at *the 3rd Annual Joint Conference on Information Sciences* (JCIS'97), Durham NC, March 1997, pp. 81–84.

189. Nguyen Hung Son and Nguyen Sinh Hoa, An application of discretization methods in control, in: *Proceedings: the Workshop on Robotics, Intelligent Control and Decision Support Systems*, Polish-Japanese Institute of Information Technology, Warsaw, Poland, February 1999, pp. 47–52.

190. Nguyen Hung Son and Nguyen Sinh Hoa, Discretization methods in Data Mining, in: L. Polkowski and A. Skowron (eds.): *Rough Sets in Knowledge Discovery 1. Methodology and Applications*, Physica–Verlag, Heidelberg, 1998, pp. 451–482.

191. Nguyen Hung Son and Nguyen Sinh Hoa, Discretization methods with back–tracking, in: *Proceedings: the 5th European Congress on Intelligent Techniques and Soft Computing* (EUFIT'97), Aachen, Germany, September 1997, Verlag Mainz, Aachen, 1997, pp. 201–205.

192. Nguyen Hung Son, Nguyen Sinh Hoa, and A. Skowron, Decomposition of task specifications, in: *Proceedings: the 11th International Symposium on Methodologies for Intelligent Systems, Foundations of Intelligent Systems* (ISMIS'99), Warsaw, Poland, June 8–11, LNAI 1609, Springer–Verlag, Berlin, pp. 310–318.

193. Nguyen Hung Son and A. Skowron, Boolean reasoning scheme with some applications in Data Mining, in: *Proceedings: Principles of Data Mining and Knowledge Discovery* (PKDD'99), Prague, Czech Republic, September 1999, LNAI 1704, Springer–Verlag, Berlin, 1999, pp. 107–115.

194. Nguyen Hung Son and A. Skowron, Task decomposition problem in multi–agent system, in: *Proceedings: the Workshop on Concurrency, Specification and Programming*, Berlin, Germany, September 1998, Informatik Bericht 110, Humboldt–Universität zu Berlin, pp. 221–235.

195. Nguyen Hung Son and A. Skowron, Boolean reasoning for feature extraction problems, in: *Proceedings: the 10th International Symposium on Methodologies for Intelligent Systems, Foundations of Intelligent Systems* (ISMIS'97), Charlotte NC, October 1997, LNAI 1325, Springer–Verlag, Berlin, 1997, pp. 117–126.

196. Nguyen Hung Son and A. Skowron, Quantization of real value attributes: Rough set and boolean reasoning approach, *Bulletin of International Rough Set Society* 1(1), 1997, pp. 5–16.

197. Nguyen Hung Son, A. Skowron, and J. Stepaniuk, Granular computing: a rough set approach, *Computational Intelligence: An Intern. Journal*, 17, 2001, pp. 514–544.

198. Nguyen Hung Son, M. Szczuka, and D. Ślęzak, Neural network design: Rough set approach to continuous data, in: *Proceedings: the First European Symposium on Principles of Data Mining and Knowledge Discovery* (PKDD'97), Trondheim, Norway,June 1997, LNAI 1263, Springer–Verlag, Berlin, 1997, pp. 359–366.

199. Nguyen Hung Son and D. Ślęzak, Approximate reducts and association rules: correspondence and complexity results, in: *Proceedings: the 7th International Workshop on New Directions in Rough Sets, Data Mining, and Granular–Soft Computing* (RSFDGrC'99), Ube–Yamaguchi, Japan, November 1999, LNAI 1711, Springer–Verlag, Berlin, 1999, pp. 137–145.

200. Nguyen Sinh Hoa, Discovery of generalized patterns, in: *Proceedings: the 11th International Symposium on Methodologies for Intelligent Systems, Foundations of Intelligent Systems* (ISMIS'99), Warsaw, Poland, June 1999, LNAI 1609, Springer–Verlag, Berlin, in print.

201. Nguyen Sinh Hoa, Data regularity analysis and applications in data mining, Ph.D. Dissertation, supervisor B. Chlebus, Warsaw University, Warsaw, Poland, 1999.

202. Nguyen Sinh Hoa and Nguyen Hung Son, Pattern extraction from data, in: *Proceedings: the Conference of Information Processing and Management of Uncertainty in Knowledge-Based Systems* (IPMU'98), La Sorbonne, Paris, France, July 1998, pp. 1346–1353.

203. Nguyen Sinh Hoa and Nguyen Hung Son, Pattern extraction from data, *Fundamenta Informaticae* 34(1-2), 1998, pp. 129–144.

204. Nguyen Sinh Hoa, Nguyen Tuan Trung, L. Polkowski, A. Skowron, P. Synak, and J. Wróblewski, Decision rules for large data tables, in: *Proceedings: CESA'96 IMACS Multiconference: Computational Engineering in Systems Applications* (CESA'96), Lille, France, July 1996, pp. 942–947.

205. Nguyen Sinh Hoa and A. Skowron, Searching for relational patterns in data, in: *Proceedings: the First European Symposium on Principles of Data Mining and Knowledge Discovery* (PKDD'97), Trondheim, Norway, June 1997, LNAI 1263, Springer–Verlag, Berlin, 1997, pp. 265–276.

206. Nguyen Sinh Hoa, A. Skowron, and P. Synak, Discovery of data patterns with applications to decomposition and classification problems, in: L. Polkowski and A. Skowron (eds.), *Rough Sets in Knowledge Discovery 2. Applications, Case Studies and Software Systems*, Physica–Verlag, Heidelberg, 1998, pp. 55–97.

207. Nguyen Sinh Hoa, A. Skowron, P. Synak, and J. Wróblewski, Knowl-
     edge discovery in data bases: Rough set approach, in: *Proceedings: the
     7th International Fuzzy Systems Association World Congress* (IFSA'97),
     Prague, the Czech Republic, June 1997, Academia, Prague, 1997, pp.
     204–209.

208. A. Øhrn, J. Komorowski, A. Skowron, and P. Synak, The design and
     implementation of a knowledge discovery toolkit based on rough se-
     ts– The ROSETTA system, in: L. Polkowski and A. Skowron (eds.),
     Rough Sets in Knowledge Discovery 1. Methodology and Applications,
     Physica–Verlag, Heidelberg, 1998, pp. 376–399.

209. A. Øhrn, J. Komorowski, A. Skowron, and P. Synak, The ROSETTA
     software system, In: L. Polkowski and A. Skowron (eds.), *Rough Sets
     in Knowledge Discovery 2. Applications, Case Studies and Software
     Systems*, Physica–Verlag, Heidelberg, 1998, pp. 572–576.

210. A. Øhrn, J. Komorowski, A. Skowron, and P. Synak, A software system
     for rough data analysis, *Bulletin of the International Rough Set Society*
     1(2), 1997, pp. 58–59.

211. P. Paszek and A. Wakulicz–Deja, Optimalization diagnose in progres-
     sive encephalopathy applying the rough set theory, in: *Intelligent In-
     formation Systems V*, Dęblin, Poland, 1996, pp. 142–151.

212. G. Paun, L. Polkowski, and A. Skowron, Rough set approximations of
     languages, *Fundamenta Informaticae* 32(2), 1997, pp. 149–162.

213. G. Paun, L. Polkowski, and A. Skowron, Parallel communicating gram-
     mar systems with negotiations, *Fundamenta Informaticae* 28(3-4), 1996,
     pp. 315–330.

214. G. Paun, L. Polkowski, and A. Skowron, Rough–set–like approxima-
     tions of context–free and regular languages, in: *Proceedings: Informa-
     tion Processing and Management of Uncertainty in Knowledge Based
     Systems* (IPMU-96), Granada, Spain, July 1996, pp. 891–895.

215. Z. Pawlak, Combining rough sets and Bayes' rule, *Computational In-
     telligence: An Intern. Journal*, 17, 2001, pp. 401–408.

216. Z. Pawlak, Granularity of knowledge, indiscernibility, and rough sets,
     in: *Proceedings: IEEE Conference on Evolutionary Computation*, An-
     chorage, Alaska, May 5-9, 1998, pp. 106–110; also in: *IEEE Transac-
     tions on Automatic Control* 20, 1999, pp. 100–103.

217. Z. Pawlak, Rough set theory for intelligent industrial applications, in:
     *Proceedings: the 2nd International Conference on Intelligent Processing
     and Manufacturing of Materials*, Honolulu, Hawaii, 1999, pp. 37–44.

218. Z. Pawlak, Data Mining - a rough set perspective, in: *Proceedings: Methodologies for Knowledge Discovery and Data Mining. The 3rd Pacific–Asia Conference*, Beijing, China, Springer– Verlag, Berlin, 1999, pp. 3–11.

219. Z. Pawlak, Rough sets, rough functions and rough calculus, in: S.K. Pal, A. Skowron (eds.), *Rough Fuzzy Hybridization: A New Trend in Decision Making*, Springer-Verlag, Singapore, 1999, pp. 99–109.

220. Z. Pawlak, Logic, Probability and Rough Sets, in: J. Karhumaki, H. Maurer, G. Paun, and G. Rozenberg (eds.), *Jewels are Forever. Contributions to Theoretical Computer Science in Honor of Arto Salomaa*, Springer–Verlag, Berlin, 1999, pp. 364–373.

221. Z. Pawlak, Decision rules, Bayes' rule, and rough sets, in: *Proceedings: the 7th International Workshop on rough Sets, Fuzzy Sets, Data Mining and Granular–Soft computing* (RSFSGrC'99), Ube–Yamaguchi, Japan, November 1999, LNAI 1711, Springer–Verlag, Berlin, pp. 1–9.

222. Z. Pawlak, An inquiry into anatomy of conflicts, *Journal of Information Sciences* 109, 1998, pp. 65–78.

223. Z. Pawlak, Sets, fuzzy sets, and rough sets, in: *Proceedings: Fuzzy – Neuro Systems – Computational Intelligence*, Muenchen, Germany, March 18-20, 1998, pp. 1–9.

224. Z. Pawlak, Reasoning about data–a rough set perspective, in: *Proceedings: First International Conference on Rough Sets and Current Trends in Computing* (RSCTC'98), LNAI 1424, Springer–Verlag, Berlin, 1998, pp. 25–34.

225. Z. Pawlak, Rough sets theory and its applications to data analysis, *Cybernetics and Systems* 29, 1998, pp. 661–688.

226. Z. Pawlak, Rough set elements, in: L. Polkowski and A. Skowron (eds.), *Rough Sets in Knowledge Discovery 1. Methods and Applications*, Physica–Verlag, Heidelberg, 1998, pp. 10–30.

227. Z. Pawlak, Rough Modus Ponens, in: *Proceedings: the 7th Conference on Information Processing and Management of Uncertainty in Knowledge Based Systems* (IPMU'98), La Sorbonne, Paris, France, July 1998, pp. 1162–1165.

228. Z. Pawlak, Rough set approach to knowledge-based decision support, *European Journal of Operational Research* 29(3), 1997, pp. 1–10.

229. Z. Pawlak, Rough sets and Data Mining, in: *Proceedings: the International Conference on Intelligent Processing and Manufacturing Materials*, Gold Coast, Australia, 1997, pp. 1–5.

230. Z. Pawlak, Rough sets, in: T.Y. Lin, N. Cercone (eds.), *Rough Sets and Data Mining. Analysis of Imprecise Data*, Kluwer Academic Publishers, Dordrecht, 1997, pp. 3–8.

231. Z. Pawlak, Rough real functions and rough controllers, in: T.Y. Lin, N. Cercone (eds.), *Rough Sets and Data Mining. Analysis of Imprecise Data*, Kluwer Academic Publishers, Dordrecht, 1997, pp. 139–147.

232. Z. Pawlak, Conflict analysis, in: *Proceedings: the 5th European Congress on Intelligent Techniques and Soft Computing* (EUFIT'97), Aachen, Germany, September 9-11, Verlag Mainz , Aachen, 1997, pp. 1589–1591.

233. Z. Pawlak, Rough sets and their applications, *Proceedings: Fuzzy Sets 97*, Dortmund, Germany, 1997.

234. Z. Pawlak, Vagueness–a rough set view, in: *Structures in Logic and Computer Science*, LNCS 1261, Springer–Verlag, Berlin, 1997, pp. 106–117.

235. Z. Pawlak, Data analysis with rough sets, in: *Proceedings: CODATA'96*, Tsukuba, Japan, October 1996.

236. Z. Pawlak, Rough sets, rough relations and rough functions, in: *Fundamenta Informaticae* 27(2-3), 1996, pp. 103–108.

237. Z. Pawlak, Data versus Logic: A rough set view, in: *Proceedings: the 4th International Workshop on Rough Sets, Fuzzy Sets, and Machine Discovery* (RSFD'96), Tokyo, November 1996, pp. 1–8.

238. Z. Pawlak, Rough sets: Present state and perspectives, in: *Proceedings: the Sixth International Conference, Information Processing and Management of Uncertainty in Knowledge-Based Systems* (IPMU'96), Granada, Spain, July 1996, pp. 1137–1145.

239. Z. Pawlak, Some remarks on explanation of data and specification of processes, *Bulletin of International Rough Set Society* 1(1), 1996, pp. 1–4.

240. Z. Pawlak, Why rough sets?, in: *Proceedings: the 5th IEEE International Conference on Fuzzy Systems* (FUZZ-IEEE'96), New Orleans, Louisiana, September 1996, pp. 738–743.

241. Z. Pawlak, Rough Sets and Data Analysis, in: *Proceedings: the 1996 Asian Fuzzy Systems Symposium - Soft Computing in Intelligent Systems and Information Processing*, Kenting, Taiwan ROC, December 1996, pp. 1–6.

242. Z. Pawlak, On some Issues Connected with Indiscernibility, in: G. Paun (ed.), *Mathematical Linguistics and Related Topics*, Editura Academiei Romane, Bucureşti, 1995, pp. 279–283.

243. Z. Pawlak, Rough sets, in: *Proceedings of ACM : Computer Science Conference*, Nashville TN, February 28–March 2, 1995, pp. 262–264.

244. Z. Pawlak, Rough real functions and rough controllers, in: *Proceedings: the Workshop on Rough Sets and Data Mining at 23rd Annual Computer Science Conference*, Nashville TN, March 1995, pp. 58–64.

245. Z. Pawlak, Vagueness and uncertainty: A Rough set perspective, *Computational Intelligence: An International Journal* 11(2), 1995 (a special issue: W. Ziarko (ed.)), pp. 227–232.

246. Z. Pawlak, Rough set approach to knowledge-based decision support, in: *Towards Intelligent Decision Support. Semi–Plenary Papers: the 14th European Conference of Operations Research – 20th Anniversary of EURO*, Jerusalem, Israel, July 1995.

247. Z. Pawlak, Rough set theory, in: *Proceedings: the 2nd Annual Joint Conference on Information Sciences* (JCIS'95), Wrightsville Beach NC, September 28–October 1, 1995, pp. 312–314.

248. Z. Pawlak, Rough sets: Present state and further prospects, in: T. Y. Lin and A. M. Wildberger (eds.), *Soft Computing: Rough Sets, Fuzzy Logic, Neural Networks, Uncertainty Management, Knowledge Discovery*, Simulation Councils Inc., San Diego CA, 1995, pp. 78–85.

249. Z. Pawlak, Hard and soft sets, in: W. Ziarko (ed.), *Rough Sets, Fuzzy Sets and Knowledge Discovery* (RSKD'93), Workshops in Computing, Springer–Verlag and British Computer Society, Berlin and London, 1994, pp. 130–135.

250. Z. Pawlak, Knowledge and uncertainty - A rough sets approach, in: *Proceedings: Incompleteness and Uncertainty in Information Systems; SOFTEKS Workshop on Incompleteness and Uncertainty in Information Systems*, Concordia Univ., Montreal, Canada,1993; also in: W. Ziarko (ed.), *Rough Sets, Fuzzy Sets and Knowledge Discovery* (RSKD' 93), Workshops in Computing, Springer–Verlag and British Computer Society, Berlin and London, 1994, pp. 34–42.

251. Z. Pawlak, An inquiry into vagueness and uncertainty, in: *Proceedings: the 3rd International Workshop on Intelligent Information Systems*, Wigry, Poland, June 1994, Institute of Computer Science, Polish Academy of Sciences, Warsaw, 1994, pp. 338–359.

252. Z. Pawlak, Rough sets: Present state and further prospects, in: *Proceedings: the 3rd International Workshop on Rough Sets and Soft Computing* (RSSC94), San Jose, California, November 10-12, pp. 3–5.

253. Z. Pawlak, Rough sets and their applications, *Microcomputer Applications* 13(2), 1994, pp. 71–75.

254. Z. Pawlak, Anatomy of conflict, *Bull. of the European Association for Theoretical Computer Science* 50, 1993, pp. 234–247.

255. Z. Pawlak, Rough sets. Present state and the future, in: *Proceedings: the First International Workshop on Rough Sets: State of the Art and Perspectives*, Kiekrz – Poznań, Poland, September 1992, pp. 51–53.

256. Z. Pawlak, E. Czogała, and A. Mrózek, Application of a rough fuzzy controller to the stabilization of an inverted pendulum, in: *Proceedings: the 2nd European Congress on Intelligent Techniques and Soft Computing* (EUFIT'94), Aachen, Germany, Verlag Mainz, Aachen, pp. 1403–1406.

257. Z. Pawlak, E. Czogała and A. Mrózek, The idea of a rough fuzzy controller and its applications to the stabilization of a pendulum-car system, *Fuzzy Sets and Systems* 72, 1995, pp. 61–73.

258. Z. Pawlak, J.W. Grzymala–Busse, W. Ziarko, and R. Słowiński, Rough sets, *Communications of the ACM* 38/11, 1995, pp. 88–95.

259. Z. Pawlak, A.G. Jackson, and S.R. LeClair, Rough sets and the discovery of new materials, *Journal of Alloys and Compounds*, 1997, pp. 1-28.

260. Z. Pawlak and T. Munakata, Rough control: Application of rough set theory to control, in: *Proceedings: 4th European Congress on Intelligent Techniques and Soft Computing* (EUFIT'96), Aachen, Germany, Verlag Mainz, Aachen, 1996, pp. 209–218.

261. Z. Pawlak and A. Skowron, Helena Rasiowa and Cecylia Rauszer's research on logical foundations of Computer Science, in: A. Skowron (ed.), *Logic, Algebra and Computer Science. Helena Rasiowa and Cecylia Rauszer in Memoriam, Bulletin of the Section of Logic* 25(3-4), 1996 (a special issue), pp. 174–184.

262. Z. Pawlak and A. Skowron, Rough membership functions, in: R.R. Yaeger, M. Fedrizzi, and J. Kacprzyk (eds.), *Advances in the Dempster–Shafer Theory of Evidence*, John Wiley and Sons Inc., New York, 1994, pp. 251–271.

263. Z. Pawlak and A. Skowron, Rough membership functions: A tool for reasoning with uncertainty, in: C. Rauszer (ed.), *Algebraic Methods in Logic and Computer Science*, Banach Center Publications 28, Polish Academy of Sciences, Warsaw, 1993, pp. 135–150.

264. Z. Pawlak and A. Skowron, A rough set approach for decision rules generation, in: *Proceedings: the Workshop W12: The Management of Uncertainty in AI* at the 13th IJCAI, Chambery Savoie, France, August 30, 1993.

265. Z. Pawlak and R. Słowiński, Decision analysis using rough sets, *International Transactions in Operational Research* 1(1), 1994, pp. 107–114.

266. Z. Pawlak and R. Słowiński, Rough set approach to multi-attribute decision analysis, *European Journal of Operational Research* 72, 1994, pp. 443–45.

267. W. Pedrycz, Shadowed sets : bridging fuzzy and rough sets, in: S. K. Pal and A. Skowron (eds.), *Rough Fuzzy Hybridization: A New Trend in Decision–Making*, Springer–Verlag, Singapore, 1999, pp. 179–199.

268. J. F. Peters, L. Han, and S. Ramanna, Rough neural computing in signal processing, *Computational Intelligence: An Intern. Journal*, 17, 2001, pp. 493–513.

269. J. E. Peters, W. Pedrycz, S. Ramanna, A. Skowron, and Z. Suraj, Approximate real – time decision making: Concepts and rough Petri net models, *Intern. Journal Intelligent Systems*, 14(8), 1999, pp. 805–840.

270. J. E. Peters and S. Ramanna, A rough set approach to assessing software quality: concepts and rough Petri net models, in: S. K. Pal and A. Skowron (eds.), *Rough Fuzzy Hybridization: A New Trend in Decision–Making*, Springer–Verlag, Singapore, 1999, pp. 349–380.

271. J. E. Peters, S. Ramanna, A. Skowron, and Z. Suraj, Graded transitions in rough Petri nets, in: *Proceedings: the 7th European Congress on Intelligent Techniques and Soft Computing* (EUFIT'99), Aachen, Germany, September 1999.

272. J. F. Peters, A. Skowron, and Z. Suraj, An application of rough set methods in control design, *Fundamenta Informaticae*, 43, 2000, pp. 269–290.

273. J. E. Peters, A. Skowron, and Z. Suraj, An application of rough set methods in control design, in : *Proceedings: the Workshop on Concurrency, Specification and Programming* (CS&P'99), Warsaw, Poland, September 1999, pp. 214–235.

274. J. E. Peters, A. Skowron, Z. Suraj, S. Ramanna, and A. Paryzek, Modeling real–time decision–making systems with rough fuzzy Petri nets, in: *Proceedings: the Sixth European Congress on Intelligent Techniques and Soft Computing* (EUFIT'98) , Aachen, Germany, September 1998, Verlag Mainz, Aachen, pp. 985–989.

275. L. Polkowski, On connection synthesis via rough mereology, *Fundamenta Informaticae*, 46, 2001, pp. 83–96.

276. L. Polkowski, Approximate mathematical morphology. Rough set approach, in: S. K. Pal and A. Skowron (eds.), *Rough Fuzzy Hybridization: A New Trend in Decision-Making*, Springer-Verlag, Singapore, 1999, pp. 151–162.

277. L. Polkowski, Rough set approach to mathematical morphology: Approximate compression of data, in: *Proceedings: the 7th International Conference on Information Processing and Management of Uncertainty in Knowledge – Based Systems* (IPMU'98), La Sorbonne, Paris, France, July, pp. 1183–1189.

278. L. Polkowski and M. Semeniuk–Polkowska, Towards usage of natural language in approximate computation: A granular semantics employing formal languages over mereological granules of knowledge, *Scheda Informaticae (Fasc. Jagiellonian University)*, 10, 2000, pp. 131–145.

279. L. Polkowski and M. Semeniuk–Polkowska, Concerning the Zadeh idea of computing with words: Towards a formalization, in: *Proceedings: Workshop on Robotics, Intelligent Control and Decision Support Systems*, Polish–Japanese Institute of Information Technology, Warsaw, Poland, February 1999, pp. 62–67.

280. L. Polkowski and A. Skowron, Rough mereology in information systems with applications to qualitative spatial reasoning, *Fundamenta Informaticae*, 43, 2000, pp.291–320.

281. L. Polkowski and A. Skowron, Towards adaptive calculus of granules, in: L.A. Zadeh and J. Kacprzyk (eds.), *Computing with Words in Information/Intelligent Systems*, Physica–Verlag, Heidelberg, 1999, pp. 201–228.

282. L. Polkowski and A. Skowron, Grammar systems for distributed synthesis of approximate solutions extracted from experience, in: Gh. Paun, A.Salomaa (eds.), *Grammar Systems for Multi–agent Systems*, Gordon and Breach Science Publishers, Amsterdam, 1999, pp. 316–333.

283. L. Polkowski and A. Skowron, Rough mereology and analytical morphology, in: E. Orlowska (ed.), *Incomplete Information: Rough Set Analysis*, Physica–Verlag, Heidelberg, 1998, pp. 399–437.

284. L. Polkowski and A. Skowron, Rough sets: A perspective, in: L. Polkowski and A. Skowron (eds.), *Rough Sets in Knowledge Discovery 1. Methodology and Applications*, Physica–Verlag, Heidelberg, 1998, pp. 31–56.

285. L. Polkowski and A. Skowron, Introducing the book, in: L. Polkowski and A. Skowron (eds.), *Rough Sets in Knowledge Discovery 1. Methodology and Applications* , Physica–Verlag, Heidelberg, 1998, pp. 3–9.

286. L. Polkowski and A. Skowron, Introducing the book, in: L. Polkowski and A. Skowron (eds.), *Rough Sets in Knowledge Discovery 2. Applications, Case Studies and Software Systems*, Physica–Verlag, Heidelberg, 1998, pp. 1–9.

287. L. Polkowski and A. Skowron, Rough mereological foundations for design, analysis, synthesis, and control in distributed systems, *Information Sciences. An International Journal* 104(1-2), Elsevier Science, New York, 1998, pp. 129–156.

288. L. Polkowski and A. Skowron, Rough mereological approach - A survey, *Bulletin of International Rough Set Society* 2(1), 1998, pp. 1–13.

289. L. Polkowski and A. Skowron, Rough mereological formalization, in: W. Pedrycz and J. F. Peters III (eds.), *Computational Intelligence and Software Engineering*, World Scientific, Singapore, 1998, pp. 237–267.

290. L. Polkowski and A. Skowron, Towards adaptive calculus of granules, in: *Proceedings: the FUZZ-IEEE International Conference, 1998 IEEE World Congress on Computational Intelligence* (WCCI'98), Anchorage, Alaska, May 1998, pp. 111–116.

291. L. Polkowski and A. Skowron, Synthesis of complex objects: Rough mereological approach, in: *Proceedings: Workshop W8 on Synthesis of Intelligent Agents from Experimental Data* (at ECAI'98), Brighton, UK, August 1998, pp. 1–10 .

292. L. Polkowski and A. Skowron, Calculi of granules for adaptive distributed synthesis of intelligent agents founded on rough mereology, in: *Proceedings: the 6th European Congress on Intelligent Techniques and Soft Computing* (EUFIT'98), Aachen, Germany, Verlag Mainz, Aachen, 1998, pp. 90–93.

293. L. Polkowski and A. Skowron, Towards information granule calculus, in: *Proceedings: the Workshop on Concurrency, Specification and Programming* (CS&P'98), Berlin, Germany, September 1998, Humboldt University Berlin, Informatik Berichte 110, pp. 176–194.

294. L. Polkowski and A. Skowron, Synthesis of decision systems from data tables, in: T.Y. Lin, N. Cercone (eds.), *Rough sets and data mining: Analysis of imprecise data*, Kluwer Academic Publishers, Dordrecht, 1997, pp. 259–299.

295. L. Polkowski and A. Skowron, Decision algorithms: A survey of rough set theoretic methods, *Fundamenta Informaticae* 30(3-4), 1997, pp. 345–358.

296. L. Polkowski and A. Skowron, Approximate reasoning in distributed systems, in: *Proceedings of the Fifth European Congress on Intelligent Techniques and Soft Computing* (EUFIT'97), Aachen, Germany, Verlag Mainz, Aachen, 1997, pp. 1630–1633.

297. L. Polkowski and A. Skowron, Mereological foundations for approximate reasoning in distributed systems (plenary lecture), in: *Proceedings of the Second Polish Conference on Evolutionary Algorithms and Global Optimization*, Rytro, September 1997, Warsaw University of Technology Press, 1997, pp. 229–236.

298. L. Polkowski and A. Skowron, Adaptive decision–making by systems of cooperative intelligent agents organized on rough mereological principles, *Intelligent Automation and Soft Computing, An International Journal* 2(2), 1996, pp. 121–132.

299. L. Polkowski and A. Skowron, Rough mereology: A new paradigm for approximate reasoning, *Intern. Journal Approx. Reasoning* 15(4), 1996, pp. 333–365.

300. L. Polkowski and A. Skowron, Analytical morphology: Mathematical morphology of decision tables, *Fundamenta Informaticae* 27(2-3), 1996, pp. 255–271.

301. L. Polkowski and A. Skowron, Rough mereological controller, in: *Proceedings of The Fourth European Congress on Intelligent Techniques and Soft Computing* (EUFIT'96), Aachen, Germany, September 1996, Verlag Mainz, Aachen, 1996, pp. 223–227.

302. L. Polkowski and A. Skowron, Learning synthesis scheme in intelligent systems, in: *Proceedings: the 3rd International Workshop on Multistrategy Learning* (MSL-96), Harpers Ferry, West Virginia, May 1996, George Mason University and AAAI Press 1996, pp. 57–68.

303. L. Polkowski and A. Skowron, Implementing fuzzy containment via rough rough inclusions: rough mereological approach to distributed problem solving, in: *Proceedings: the 4th IEEE International Conference on Fuzzy Systems* (FUZZ-IEEE'96), New Orlean LA, September 1996, pp. 1147–1153.

304. L. Polkowski, A. Skowron, and J. Komorowski, Approximate case-based reasoning: A rough mereological approach, in: *Proceedings: the 4th German Workshop on Case-Based Reasoning. System Development and Evaluation*, Berlin, Germany, April 1996, Informatik Berichte 55, Humboldt University, Berlin, pp. 144–151.

305. B. Prędki, R. Słowiński, J. Stefanowski, R. Susmaga, and S. Wilk, ROSE – software implementation of the rough set theory, in: *Proceedings: the First International Conference on Rough Sets and Current Trends in Computing* (RSCTC'98),Warsaw , Poland, June 1998, LNAI 1424, Springer-Verlag, Berlin, 1998, pp. 605–608.

306. M. Quafafou, α–RST : A generalization of Rough Set Theory, *Information Systems*, 1999, to appear.

307. M. Quafafou, Learning flexible concepts from uncertain data, in: *Proceedings: the 10th International Symposium on Methodologies for Intelligent Systems* (ISMIS'97), Charlotte NC, 1997.

308. M. Quafafou and M. Boussouf, Generalized rough sets based feature selection, *Intelligent Data Analysis Journal* 4(1), 1999.

309. M. Quafafou and M. Boussouf, Induction of strong feature subsets, in: *Proceedings: the First European Symposium on Principles of Data Mining and Knowledge Discovery*, Trondheim, Norway, June 1997, LNAI 1263, Springer–Verlag, Berlin, 1997.

310. S. Radev, Argumentation systems, *Fundamenta Informaticae* 28, 1996, pp. 331–346.

311. Z. W. Ras, Discovering rules in information trees, in: *Proceedings: Principles of Data Mining and Knowledge Discovery* (PKDD'99), Prague, Czech Republic, September 1999, LNAI 1704, Springer–Verlag, Berlin, 1999, pp. 518–523.

312. Z. W. Ras, Intelligent query answering in DAKS, in: O. Pons, M. A. Vila, and J. Kacprzyk (eds.), *Knowledge Management in Fuzzy Databases*, Physica–Verlag, Heidelberg, 1999, pp. 159–170.

313. Z. W. Ras, Answering non-standard queries in distributed knowledge-based systems, in: L. Polkowski and A. Skowron (eds.), *Rough Sets in Knowledge Discovery 2. Applications, Case Studies and Software Systems*, Physica Verlag, Heidelberg, 1998, pp. 98–108.

314. Z. W. Ras, Handling queries in incomplete CKBS through knowledge discovery, in: *Proceedings: the First International Conference on Rough Sets and Current Trends in Computing* (RSCTC'98), Warsaw, Poland, LNAI 1424, Springer–Verlag, Berlin, 1998, pp. 194–201.

315. Z. W. Ras, Knowledge discovery objects and queries in distributed knowledge systems, in: *Proceedings: Artificial Intelligence and Symbolic Computation*(AISC'98), LNAI 1476, Springer–Verlag, Berlin, 1998, pp. 259–269.

316. Z. W. Ras and A. Bergmann, Maintaining soundness of rules in distributed knowledge systems, in: *Proceedings: the Workshop on Intelligent Information Systems* (IIS'98), Malbork, Poland, June 1998, Polish Academy of Sciences, Warsaw, 1998, pp. 29–38.

317. Z. W. Ras and J. M. Żytkow, Mining for attribute definitions in a distributed two-layered DB system, *Journal of Intelligent Information Systems* 14(2-3), 2000, in print.

318. Z. W. Ras and J. M. Żytkow, Mining distributed databases for attribute definitions, in: *Proceedings: SPIE. Data Mining and Knowledge Discovery: Theory, Tools, and Technology*, Orlando, Florida, April 1999, pp. 171–178.

319. Z. W. Ras and J. M. Żytkow, Discovery of equations and the shared operational semantics in distributed autonomous databases, in: *Proceedings: Methodologies for Knowledge Discovery and Data Mining* (PAKKD'99), Beijing, China, 1999, LNAI 1574, Springer-Verlag, Berlin, 1999, pp. 453–463.

320. L. Rossi, R. Słowiński, and R. Susmaga, Rough set approach to evaluation of storm water pollution, *International Journal of Environment and Pollution*, to appear.

321. L. Rossi, R. Słowiński, and R. Susmaga, Application of the rough set approach to evaluate storm water pollution, in: *Proceedings: the 8th International Conference on Urban Storm Drainage* (8th ICUSD), Sydney, Australia, 1999, vol. 3, pp. 1192–1200.

322. H.Sakai, Some issues on non–deterministic knowledge bases with incomplete and selective information, in: *Proceedings: the First International Conference on Rough Sets and Current Trends in Computing* (RSCTC'98), LNAI 1424, Springer–Verlag, Berlin, 1998, pp. 424–431.

323. H. Sakai, Another fuzzy Prolog, in: *Proceedings: The Fourth International Workshop on Rough Sets, Fuzzy Sets, and Machine Discovery* (RSFD'96), Tokyo, November 1996, pp. 261–268.

324. H. Sakai and A. Okuma, An algorithm for finding equivalence relations from tables with non–deterministic information, in: *Proceedings: the 7th International Conference on Rough Sets, Fuzzy Sets, Data Mining and Granular-Soft Computing* (RSFDGrC99), LNAI 1711, Springer–Verlag, Berlin, 1999, pp. 64–72.

325. M. Semeniuk–Polkowska, On Applications of Rough Set Theory in Humane Sciences (in Polish), Warsaw University Press, Warsaw, 2000.

326. M. Semeniuk–Polkowska, On Rough Set Theory in Library Sciences (in Polish), Warsaw University Press, Warsaw, 1996.

327. V. I. Shevtchenko, On the depth of decision trees for diagnosis of non–elementary faults in circuits (in Russian), in: *Proceedings: the 9th Workshop on Synthesis and Complexity of Control Systems*, Nizhny Novgorod, Russia, 1998, Moscow State University Publishers, Moscow, 1999, pp. 94–98.

328. V. I. Shevtchenko, On complexity of non–elementary fault detection in circuits (in Russian), in: *Proceedings: the 12th International Conference on Problems of Theoretical Cybernetics*, Nizhny Novgorod, Russia, 1999, Moscow State University Publishers, Moscow, 1999, p. 254.

329. V. I. Shevtchenko, On complexity of confused connection diagnosis in circuits (in Russian), in: *Proceedings: the 9th All–Russian Conference on Mathematical Methods of Pattern Recognition*, Moscow, Russia, 1999, pp. 129–131.

330. V. I. Shevtchenko, On the depth of decision trees for diagnosing of nonelementary faults in circuits, in: *Proceedings: the First International Conference on Rough Sets and Current Trends in Computing* (RSCTC'98), Warsaw, Poland, June 1998, LNAI 1424, Springer–Verlag, Berlin, 1998, pp. 517–520.

331. V. I. Shevtchenko, On complexity of "OR" ("AND")–closing detection in circuits (in Russian), in: *Proceedings: the International Conference on Discrete Models in Theory of Control Systems*, Moscow, Russia, 1997 pp. 61–62.

332. V. I. Shevtchenko, On complexity of "OR" ("AND")–closing diagnosis in circuits (in Russian), in: *Proceedings: the 8th All-Russian Conference on Mathematical Methods of Pattern Recognition*, Moscow, Russia, 1997, pp. 125–126.

333. A. Skowron and J. Stepaniuk, Information granules: Towards foundations for spatial and temporal reasoning, *Journal of Indian Science Academy*, in print.

334. A. Skowron and J. Stepaniuk, Information granules in distributed systems, in: *Proceedings: the 7th International Workshop on Rough Sets, Fuzzy Sets, Data Mining and Granular-Soft Computing* (RSFDGrC'99), Ube – Yamaguchi, Japan, November 1999, Lecture Notes in Artificial Intelligence 1711, Springer–Verlag, Berlin, 1999, pp. 357–365.

335. A. Skowron and J. Stepaniuk, Towards discovery of information granules, in: *Proceedings: Principles of Data Mining and Knowledge Discovery* (PKDD'99), Prague, the Czech Republic, September 1999, LNAI 1704, Springer–Verlag, Berlin, 1999, pp. 542–547.

336. A. Skowron and J. Stepaniuk, Information granules and approximation spaces, in: *Proceedings: the 7th International Conference on Information Processing and Management of Uncertainty in Knowledge – Based Systems* (IPMU'98), La Sorbonne, Paris, France, July 1998, pp. 1354–1361.

337. A. Skowron and J. Stepaniuk, Information Reduction Based on Constructive Neighborhood Systems, in: *Proceedings: the 5th International Workshop on Rough Sets Soft Computing* (RSSC'97) at the 3rd Annual Joint Conference on Information Sciences (JCIS'97), Durham NC, October 1997, pp. 158–160.

338. A. Skowron and J. Stepaniuk, Constructive information granules, in: *Proceedings: the 15th IMACS World Congress on Scientific Computation, Modeling and Applied Mathematics*, Berlin, Germany, August 1997; also in: *Artificial Intelligence and Computer Science* 4, pp. 625–630.

339. A. Skowron and J. Stepaniuk, Tolerance approximation spaces, *Fundamenta Informaticae* 27(2-3), 1996, pp. 245–253.

340. A. Skowron , J. Stepaniuk, and S. Tsumoto, Information granules for spatial reasoning, *Bulletin Intern. Rough Set Society* 3(4), 1999, pp. 147–154.

341. A. Skowron and Z. Suraj, A parallel algorithm for real-time decision making: A rough set approach, *Journal of Intelligent Information Systems* 7, 1996, pp. 5–28.

342. K. Słowiński and J. Stefanowski, Medical information systems – problems with analysis and ways of solutions, in: S. K. Pal and A. Skowron (eds.), *Rough Fuzzy Hybridization: A New Trend in Decision–Making*, Springer–Verlag, Singapore, 1999, pp. 301.

343. R. Słowiński, Rough set data analysis – a new way of solving some decision problems in transportation, *Proceedings: Modeling and Management in Transportation* (MMT'99), Kraków – Poznań, October 1999, pp. 63–66.

344. R. Słowiński, Multi–criterial decision support based on rules induced by rough sets (in Polish), in: T.Trzaskalik (ed.),*Metody i Zastosowania Badań Operacyjnych*, Part 2, Wydawnictwo Akademii Ekonomicznej w Katowicach, Katowice, 1998, pp. 19–39.

345. R. Słowiński and J. Stefanowski, Handling inconsistency of information with rough sets and decision rules, in: *Proceedings: Intern. Conference on Intelligent Techniques in Robotics, Control and Decision Making*, Polish–Japanese Institute of Information Technology, Warsaw, February 1999, pp. 74–81.

346. R. Słowiński and J. Stefanowski, Rough family – software implementation of the rough set theory, in: L. Polkowski, A. Skowron (eds.), *Rough Sets in Knowledge Discovery 2. Applications, Case Studies and Software Systems*, Physica-Verlag, Heidelberg, 1998, pp. 581–586.

347. R. Słowiński, J. Stefanowski, S. Greco, and B. Matarazzo, Rough sets processing of inconsistent information in decision analysis, *Control and Cybernetics*, to appear.

348. R. Słowiński and D. Vanderpooten, A generalized definition of rough approximations based on similarity, *IEEE Transactions on Data and Knowledge Engineering*, to appear.

349. R. Słowiński, C. Zopounidis, A. I. Dimitras, and R. Susmaga, Rough set predictor of business failure, in: R. A. Ribeiro, H.-J. Zimmermann, R. R. Yager, and J. Kacprzyk (eds.), *Soft Computing in Financial Engineering*, Physica–Verlag, Heidelberg, 1999, pp. 402–424.

350. J. Stefanowski and A. Tsoukias, Incomplete information tables and rough classification, *Computational Intelligence: An Intern. Journal*, 17, 2001, pp. 545–566.

351. J. Stepaniuk, Rough set based data mining in diabetes mellitus data table, in: *Proceedings: the 6th European Congress on Intelligent Techniques and Soft Computing* (EUFIT'98), Aachen, Germany, September 1998, pp. 980–984; for extended version see: : *Proceedings: the 11th International Symposium on Methodologies for Intelligent Systems, Foundations of Intelligent Systems* (ISMIS'99), Warsaw, Poland, June 1999, LNAI 1609, Springer–Verlag, Berlin, 1999.

352. J. Stepaniuk, Optimizations of rough set model, *Fundamenta Informaticae* 36(2-3), 1998, pp. 265–283.

353. J. Stepaniuk, Rough relations and logics, in: L. Polkowski, A. Skowron (eds.), *Rough Sets in Knowledge Discovery 1. Methodology and Applications*, Physica–Verlag, Heidelberg, 1998, pp. 248–260.

354. J. Stepaniuk, Approximation spaces, reducts and representatives, in: L. Polkowski, A. Skowron (eds.), *Rough Sets in Knowledge Discovery 2. Applications, Case Studies and Software Systems*, Physica–Verlag, Heidelberg, 1998, pp. 109–126.

355. J. Stepaniuk, Approximation spaces in extensions of rough set theory, in: *Proceedings: the First International Conference on Rough Sets and Current Trends in Computing* (RSCTC'98), Warsaw, Poland, June 1998, LNAI 1424, Springer–Verlag, Berlin, 1998, pp. 290–297.

356. J. Stepaniuk, Rough sets similarity based learning, in: *Proceedings: the 5th European Congress on Intelligent Techniques and Soft Computing* (EUFIT' 97), Aachen, Germany, September 1997, Verlag Mainz, Aachen, 1997, pp. 1634–1639.

357. J. Stepaniuk, Similarity relations and rough set model, in: *Proceedings: the International Conference MENDEL97*, Brno, the Czech Republic, June 1997.

358. J. Stepaniuk, Attribute discovery and rough sets, in: *Proceedings:the First European Symposium on Principles of Data Mining and Knowledge Discovery* (PKDD'97), Trondheim, Norway, June 1997, LNAI 1263, Springer–Verlag, Berlin, 1997, pp. 145–155.

359. J. Stepaniuk, Searching for optimal approximation spaces, in: *Proceedings: the 6th International Workshop on Intelligent Information Systems* (ISMIS' 97), Zakopane, Poland, June 1997, Publ. Institute of Computer Science, Polish Academy of Sciences, pp. 86–95.

360. Z. Suraj, An application of rough sets and Petri nets to controller design, in: *Workshop on Robotics, Intelligent Control and Decision Support Systems*, Polish-Japanese Institute of Information Technology, Warsaw, Poland, February 1999, pp. 86–96.

361. Z. Suraj, The synthesis problem of concurrent systems specified by dynamic information systems, in: L. Polkowski and A. Skowron (eds.), *Rough Sets in Knowledge Discovery 2. Applications, Case Studies and Software Systems*, Physica–Verlag, Heidelberg, 1998, pp. 418–448.

362. Z. Suraj, Reconstruction of cooperative information systems under cost constraints: A rough set approach, *Journal of Information Sciences* 111, 1998, pp. 273–291.

363. Z. Suraj, Reconstruction of cooperative information systems under Cost constraints: A rough set approach, in: *Proceedings: the First International Workshop on Rough Sets and Soft Computing* (RSSC'97), Durham NC, March 1997, pp. 364–371.

364. Z. Suraj, Discovery of concurrent data models from experimental tables, *Fundamenta Informaticae* 28(3-4), 1996, pp. 353–376.

365. R. Susmaga, R. Słowiński, S. Greco, and B. Matarazzo, Computation of reducts for multi–attribute and multi–criteria classification, in: *Proceedings: the 7th Workshop on Intelligent Information Systems* (IIS'99), Ustroń, Poland, June 1999, pp. 154–163.

366. M. Szczuka, Refining decision classes with neural networks, in: *Proceedings: the 7th International Conference on Information Processing and Management of Uncertainty in Knowledge–Based Systems* (IPMU'98), La Sorbonne, Paris, France, July 1998, pp. 1370–1375.

367. M. Szczuka, Rough Sets and Artificial Neural Networks, in: L. Polkowski and A. Skowron (eds.), *Rough Sets in Knowledge Discovery 2. Applications, Case Studies and Software Systems*, Physica–Verlag, Heidelberg, 1998, pp. 449–470.

368. M. Szczuka, Rough set methods for constructing neural networks, in: *Proceedings: the 3rd Biennial Joint Conference On Engineering Systems Design Analysis, Session on Expert Systems*, Montpellier, France, 1996, pp. 9–14.

369. M. Szczuka, D. Ślęzak, and S. Tsumoto, An application of reduct networks to medicine – chaining decision rules, in: *Proceedings: the 5th International Workshop on Rough Sets and Soft Computing* (RSSC'97) at *the 3rd Annual Joint Conference on Information Sciences* (JCIS'97), Duke University, Durham NC, USA, 1997, pp. 395–398.

370. P. Synak, Adaptation of decomposition tree to extended data, in: *Proceedings SCI'2001: World Multiconference on Systemics, Cybernetics and Informatics*, Orlando, Fla., July 2001, vol. VII, pp. 552–556.

371. D. Ślęzak, Foundations of entropy based bayesian networks: Theoretical results & rough set based extraction from data, in : *Proceedings: the 8th International Conference on Information Processing and Management of Uncertainty in Knowledge–Based Systems* (IPMU'00), Madrid, Spain, July 2000, in print.

372. D. Ślęzak, Normalized decision functions and measures for inconsistent decision tables analysis, *Fundamenta Informaticae*, in print.

373. D. Ślęzak, Decomposition and synthesis of decision tables with respect to generalized decision functions, in: S. Pal and A. Skowron (eds.), *Rough Fuzzy Hybridization: A New Trend in Decision–Making*, Springer–Verlag, Singapore, 1999, pp. 110–135.

374. D. Ślęzak, Decision information functions for inconsistent decision tables analysis, in: *Proceedings: the 7th European Congress on Intelligent Techniques & Soft Computing* (EUFIT'99), Aachen, Germany, p. 127.

375. D. Ślęzak, Searching for Dynamic Reducts in Inconsistent Decision Tables, in: *Proceedings: the 7th International Conference on Information Processing and Management of Uncertainty in Knowledge–Based Systems* (IPMU'98), La Sorbonne, Paris, France, July 1998, pp. 1362–1369.

376. D. Ślęzak, Searching for frequential reducts in decision tables with uncertain objects, in: *Proceedings: the First International Conference on Rough Sets and Current Trends in Computing* (RSCTC'98), Warsaw, Poland, LNAI 1424, Springer-Verlag, Berlin, 1998, pp. 52–59.

377. D. Ślęzak, Rough set reduct networks, in: *Proceedings: the 5th Intern. Workshop on Rough Sets and Soft Computing* (RSSC'97) at *the 3rd Annual Joint Conference on Information Sciences* (JCIS'97), Durham NC, 1997, pp. 77–81.

378. D. Ślęzak, Attribute set decomposition of decision tables, in: *Proceedings: the 5th European Congress on Intelligent Techniques and Soft Computing* (EUFIT'97), Aachen, Germany, Verlag Mainz, Aachen, 1997, pp. 236–240.

379. D. Ślęzak, Decision value oriented decomposition of data tables, in: *Proceedings: the 10th International Symposium on Methodologies for Intelligent Systems, Foundations of Intelligent Systems* (ISMIS'97), Charlotte NC, October 1997, LNAI 1325, Springer–Verlag, Berlin, 1997, pp. 487–496.

380. D. Ślęzak, Approximate reducts in decision tables, in: *Proceedings: the 6th International Conference, Information Processing and Management of Uncertainty in Knowledge-Based Systems* (IPMU'96), Granada, Spain, July 1996, pp. 1159–1164.

381. D. Ślęzak, Tolerance dependency model for decision rules generation, in: *Proceedings: the 4th International Workshop on Rough Sets, Fuzzy Sets, and Machine Discovery* (RSFD'96), Tokyo, Japan, November 1996, pp. 131–138.

382. D. Ślęzak and M. Szczuka, Hyperplane–based neural networks for real-valued decision tables, in: *Proceedings: the 5th International Workshop on Rough Sets Soft Computing* (RSSC'97) at the 3rd Annual Joint Conference on Information Sciences (JCIS'97), Durham NC, 1997, pp. 265–268.

383. D. Ślęzak and J. Wróblewski, Classification algorithms based on linear combinations of features, in: *Proceedings: Principles of Data Mining and Knowledge Discovery* (PKDD'99), Prague, Czech Republic, September 1999, LNAI 1704, Springer-Verlag, Berlin, 1999, pp. 548–553.

384. I. Tentush, On minimal absorbents and closure properties of rough inclusions: new results in rough set theory, Ph.D. Dissertation, supervisor L. Polkowski, Institute of Fundamentals of Computer Science, Polish Academy of Sciences, Warsaw, Poland, 1997.

385. S. Tsumoto, Induction of expert decision rules using rough sets and set–inclusion, in: S.K. Pal and A. Skowron (eds.), *Rough Fuzzy Hybridization: A New Trend in Decision–Making*, Springer–Verlag, Singapore, 1999, pp. 316–329.

386. S. Tsumoto, Discovery of rules about complications, in: *Proceedings: the 7th International Workshop on Rough Sets, Fuzzy Sets, Data Mining and Granular–Soft Computing* (RSFDGrC'99), Ube–Yamaguchi, Japan, November 1999, Lecture Notes in AI 1711, Springer–Verlag, Berlin, 1999, pp. 29–37.

387. S. Tsumoto, Extraction of expert's decision rules from clinical databases using rough set model, *J. Intelligent data Analysis* 2(3), 1998.

388. S. Tsumoto, Automated induction of medical expert system rules from clinical databases based on rough set theory, *Information Sciences* 112, 1998, pp. 67–84.

389. V. Uma Maheswari, A. Siromoney, K. M. Mehata, and K. Inoue, The variable precision rough set inductive logic programming model and strings, *Computational Intelligence: An Intern. Journal*, 17, 2001, pp. 460–471.

390. A. Wakulicz–Deja, M. Boryczka, and P. Paszek, Discretization of continuous attributes on decision system in mitochondrial encephalomyopathies, in: *Proceedings: the First International Conference on Rough Sets and Current Trends in Computing* (RSCTC'98), Warsaw, Poland, June 1998, LNAI 1424, 1998, pp. 483–490.

391. A. Wakulicz–Deja, B. Marszał–Paszek, P. Paszek, and E. Emich–Widera, Applying rough sets to diagnose in children's neurology, in: *Proceedings: the 6th International Conference Information Processing and Management of Uncertainty in Knowledge-Base Systems* (IPMU'96), Granada, Spain, 1996, pp. 1463–1468.

392. A. Wakulicz–Deja and P. Paszek, Optimalization of decision problems on medical knowledge bases, in: *Proceedings: Intelligent Information Systems VI*, Zakopane, Poland, 1997, pp. 204–210.

393. A. Wakulicz–Deja and P. Paszek, Optimalization of decision problems on medical knowledge bases, in: *Proceedings: the 5th European Congress on Intelligent Techniques and Soft Computing* (EUFIT'97), Aachen, Germany, September 1997, Verlag Mainz, Aachen, 1997, pp. 1607–1610.

394. A. Wakulicz–Deja and P. Paszek, Diagnose progressive encephalopathy applying the rough set theory, *International Journal of Medical Informatics* 46, 1997, pp. 119–127.

395. A. Wakulicz–Deja and P. Paszek, Optimalization of diagnose in progressive encephalopathy applying the rough set theory, in: *Proceedings: the 4th European Congress on Intelligent Techniques and Soft Computing* (EUFIT'96), Aachen, Germany, Verlag Mainz, Aachen, 1996, pp. 192–196.

396. A. Wakulicz–Deja, P. Paszek, and B. Marszał–Paszek, Optymalizacja procesu podejmowania decyzji (diagnozy) w medycznych bazach wiedzy (in Polish), in: *Proceedings: II Krajowa Konferencja Techniki Informatyczne w Medycynie*, Jaszowiec, Poland, 1997, pp. 279–286.

397. H. Wang and Nguyen Hung Son, Text classification using Lattice Machine, in: *Proceedings: the 11th International Symposium on Methodologies for Intelligent Systems, Foundations of Intelligent Systems* (ISMIS'99), Warsaw, Poland, June 1999 , LNAI 1609, Springer–Verlag, Berlin, 1999.

398. A. Wasilewska, Topological rough algebras, in: T. Y. Lin and N. Cercone (eds.), *Rough Sets and Data Mining. Analysis of Imprecise Data*, Kluwer Academic Publishers, Dordrecht, 1997, pp. 411–425.

399. A. Wasilewska, E. Menasalvas, and M. Hadjimichael, A generalization model for implementing a Data Mining system, in: *Proceedings: IFSA'99*, Taipei, Taiwan, August 1999, pp. 245–251.

400. A. Wasilewska and L. Vigneron, Rough algebras and automated deduction, in: L. Polkowski and A. Skowron (eds.), *Rough Sets in Knowledge Discovery 1. Methodology and Applications*, Physica–Verlag, Heidelberg, 1998, pp. 261–275.

401. A. Wasilewska and L. Vigneron, On Generalized rough sets, in: *Proceedings: the 5th Workshop on Rough Sets and Soft Computing* (RSSC'97) at the 3rd Joint Conference on Information Sciences (JCIS'97), Research Triangle Park NC, March 1997.

402. P. Wojdyłło, Wavelets, rough sets and artificial neural networks in EEG analysis, in: *Proceedings: First International Conference on Rough Sets and Current Trends in Computing* (RSCTC'98), Warsaw, Poland, LNAI 1424, Springer–Verlag, Berlin, pp. 444–449.

403. J. Wróblewski, Genetic algorithms in decomposition and classification problem, in: L. Polkowski and A. Skowron (eds.), *Rough Sets in Knowledge Discovery 2. Applications, Case Studies and Software Systems*, Physica–Verlag, Heidelberg, 1998, pp. 471–487.

404. J. Wróblewski, Covering with reducts – a fast algorithm for rule generation, in: *Proceedings: the First International Conference on Rough Sets and Current Trends in Computing* (RSCTC'98), Warsaw, Poland, June 1998, LNAI 1424, Springer-Verlag, Berlin, 1998, pp. 402–407.

405. J. Wróblewski, A Parallel Algorithm for Knowledge Discovery System, in: *Proceedings: the International Conference on Parallel Computing in Electrical Engineering* (PARELEC'98), Białystok, Poland, September 1998, The Press Syndicate of the Technical University of Białystok, 1998, pp. 228–230.

406. J. Wróblewski, Theoretical Foundations of Order-Based Genetic Algorithms, *Fundamenta Informaticae* 28(3-4), 1996, pp. 423–430.

407. L. Vigneron, Automated deduction techniques for studying rough algebras, *Fundamenta Informaticae* 33(1), 1998, pp. 85–103.

408. L. Vigneron and A. Wasilewska, Rough sets based proofs visualisation, in: *Proceedings: the 18th International Conference of the North American Fuzzy Information Processing Society* (NAFIPS'99)(invited session on Granular Computing and Rough Sets), New York NY, 1999, pp. 805–808.

409. L. Vigneron and A. Wasilewska, Rough sets congruences and diagrams, in: *Proceedings: the 16th European Conference on Operational Research* (EURO XVI), Brussels, Belgium, July 1998.

410. L. Vigneron and A. Wasilewska, Rough diagrams, in: *Proceedings: the sixth Workshop on Rough Sets, Data Mining and Granular Computing* (RSDMGrC'98) at the 4th Joint Conference on Information Sciences (JCIS'98), Research Triangle Park NC, October 1998.

411. L. Vigneron and A. Wasilewska, Rough and modal algebras, in: *Proceedings: the International Multi–conference (Computational Engineering in Systems Applications), Symposium on Modelling, Analysis and Simulation* (IMACS/IEEE CESA'96), Lille, France, July 1996, pp. 1107–1112.

412. Zhang Qi and Han Zhenxiang, Rough sets : theory and applications, *Control Theory and Applications* 16(2), 1999, pp. 153–157, S. China Univ. Technology Press, Guangzhou, China.

413. Zhang Qi and Han Zhenxiang, A new method for alarm processing in power systems using rough set theory, *Electric Power* 31(4), 1998, pp. 32–38, China Electric Power Press, Beijing.

414. Zhang Qi, Han Zhenxiang, and Wen Fushuan, Analysis of Rogers ratio table for transformer fault diagnosis using rough set theory, in: *Proceedings: CUS-EPSA'98*, Harbin, China, 1998, pp. 386–391.

415. Zhang Qi, Han Zhenxiang, and Wen Fushuan, A new approach for fault diagnosis in power systems based on rough set theory, in: *Proceedings: APSCOM'97*, Hong Kong, 1997, pp. 597–602.

416. W. Ziarko, Probabilistic decision tables in the variable precision rough set model, *Computational Intelligence: An Intern. Journal*, 17, 2001, pp. 593–603.

417. W. Ziarko, Decision making with probabilistic decision tables, in: : *Proceedings: the 7th International Workshop on Rough Sets, Fuzzy Sets, Data Mining and Granular-Soft Computing* (RSFDGrC'99), Ube–Yamaguchi, Japan, November 1999, Lecture Notes in Artificial Intelligence 1711, Springer–Verlag, Berlin, 1999, pp. 463–471.

418. W. Ziarko, Rough sets as a methodology for data mining, in: L. Polkowski and A. Skowron (eds.), *Rough Sets in Knowledge Discovery 1. Methodology and Applications*, Physica–Verlag, Heidelberg, 1998, pp. 554–576.

419. W. Ziarko, KDD–R: rough sets based data mining system, in: L. Polkowski and A. Skowron (eds.), *Rough Sets in Knowledge Discovery 2. Applications, Case Studies and Software Systems*, Physica–Verlag, Heidelberg, 1998, pp. 598–601.

420. W. Ziarko, Approximation region–based decision tables, in : *Proceedings: the First International Conference on Rough Sets and Current Trends in Computing* (RSCTC'98), Warsaw, June 1998, Lecture Notes in Artificial Intelligence 1424, Springer–Verlag, Berlin, 1998, pp. 178–185.

421. C. Zopounidis, R. Słowiński, M. Doumpos, A.I. Dimitras, and R. Susmaga, Business failure prediction using rough sets – a comparison with multivariate analysis techniques, *Fuzzy Economic Review* 4, 1999(1), pp. 3–33.

PART 2:
PREREQUISITES

Chapter 2

The Sentential Calculus

By the very act of arguing, you awake the patient's reason; and once it is awake, who can foresee the result?

C.S. Lewis, *The Screwtape Letters*, I

2.1 Introduction

We begin with a discussion of the basic tool of mathematical reasoning: the sentential calculus (or, *calculus of propositions, propositional calculus, propositional logic*). By a *proposition, sentence* we mean any statement about reality of interest to us whose truth value can be established with certainty. By a *truth value* of a proposition, we understand either of two possible values: *truth* (T or 1), *falsity* (F or 0). A proposition p may be therefore either *true* or *false* and only one of the two possibilities actually holds for p. For instance, the statement *"if today is Monday then tomorrow is Tuesday"* is according to our best knowledge true while the statement of ordinary arithmetic *" 2+2 = 3"* is false. In the sequel, we denote truth values with the symbols $0, 1$.

The sentential calculus deals with propositions considered in their symbolic form: there is no reason to consider all possible propositions (apart from the real possibility of listing all of them); instead, the aim of the theory of sentential calculus is to reveal all properties of this system which hold regardless of the nature of statements considered i.e. which depend only on truth values of these statements (one says also that such properties are *truth functional*). Thus, we consider instead of actual propositions *propositional variables* $p, q, r, s, t, \ldots$ ranging over propositions and we aim at establishing general properties of the sentential calculus which are preserved when we substitute for any of variables $p, q, r, \ldots$ a specific proposition.

As with any language, in propositional calculus we may construct complex expressions from simpler ones. Passing from simpler to more complex propositions is effected by means of *proposition–forming functors* (or, *functors*, shortly). In propositional calculus we have four basic functors with which to form more complex expressions. Let us look at them now.

2.2   Functors

The functor of *negation* denoted N or $\neg$ is a unary functor acting on single propositions: for a proposition p, the expression Np (or, $\neg p$) is a proposition. To describe fully the action of N on propositions, we must describe its semantic impact, thus we have to give truth values of Np for respective values of p. So, we let $Np = 0$ when $p = 1$ and $Np = 1$ when $p = 0$. Thus, Np is true if and only if p is false. The proposition Np is read: *"it is not true that p"* or, shortly, *"not p"*. To give a concise description of the semantic content of the functor N, we use the following conditions

$$N0 = 1$$
$$N1 = 0$$
$$\text{(2.1)}$$

The binary functor of *alternation* denoted OR (or, $\vee$) acts on pairs of propositions p, q to produce the proposition $ORpq$ (denoted also $pORq$, $p \vee q$; we here prefer whenever possible the *Polish notation* of Jan Łukasiewicz in which the operator symbol precedes the symbols of arguments). The proposition $ORpq$ is read: *either p or q*. Truth functional description of OR is given by the following conditions:

$$OR00 = 0$$
$$OR01 = 1$$
$$OR10 = 1$$
$$OR11 = 1$$
$$\text{(2.2)}$$

It follows that $ORpq$ is true if and only if at least one of p, q is a true proposition.

The binary functor AND of *conjunction* (denoted also $\wedge$) acts on pairs p, q to produce the expression $ANDpq$ (denoted also $p \wedge q$). The following conditions determine the semantics of AND:

$$AND00 = 0$$
$$AND01 = 0$$
$$AND10 = 0$$
$$AND11 = 1$$
$$\text{(2.3)}$$

Hence, the proposition $ANDpq$ is true if and only if each of propositions p, q is true.

The binary functor of *implication* C (denoted also $\Rightarrow$) acts on pairs p, q of propositions to form the proposition Cpq (denoted also $p \Rightarrow q$) defined semantically as follows:

$$
\begin{aligned}
C00 &= 1 \\
C01 &= 1 \\
C10 &= 0 \\
C11 &= 1
\end{aligned}
\qquad (2.4)
$$

2.3 Meaningful expressions

We may now introduce the notion of a *meaningful expression(a formula)* of propositional calculus by stating that:
1. Each propositional variable is a meaningful expression,
2. $N\alpha$ is a meaningful expression whenever α is a meaningful expression,
3. Expressions $OR\alpha\beta$, $AND\alpha\beta$, $C\alpha\beta$ are meaningful when α, β are meaningful expressions.

Thus, we may – invoking the notion of a set which is yet to be discussed – say that the meaningful expressions are the least set containing all propositional variables and closed under propositional functors N, C, OR, AND.

We will be interested from now on only in meaningful expressions formed in accordance with the rules given in this paragraph. For a meaningful expression α, we will sometimes write down α in the form $\alpha(p_1, p_2, ..., p_k)$ to denote the fact that propositional variables occurring in α are exactly $p_1, p_2, ..., p_k$. Clearly, any meaningful expression $\alpha(p_1, p_2, ..., p_k)$ takes a truth value either 0 or 1 on assuming that each of p_i $(i = 1, 2, ..., k)$ takes a truth value 0 or 1 (a simple proof may be done by induction on number of connectives in α).

For instance, the meaningful expression $CNpq$ (which may be also written down as $\neg p \Rightarrow q$) takes the value 0 for $p = 0 = q$, the value 1 for $p = 0, q = 1$, the value 1 for $p = 1, q = 0$ and the value 1 for $p = 1, q = 1$.

We will say that two meaningful expressions α, β are *equivalent* when they take the same truth value for each substitution of the same truth values for equiform propositional variables occurring in either of them. To give an important example, we observe that

Proposition 2.1. *1. The meaningful expressions $ORpq$, $CNpq$ are equivalent*

2. The meaningful expressions $ANDpq$, $NCpNq$ are equivalent.

Proof. It is sufficient to compare truth values of both expressions for each substitution of truth values for p, q. In case 1:

$$OR00 = 0;\ CN00 = 0$$
$$OR01 = 1;\ CN01 = 1$$
$$OR10 = 1;\ CN10 = 1$$
$$OR11 = 1;\ CN11 = 1$$

Thus, expressions $ORpq$ and $CNpq$ are equivalent.

In case 2, for expressions $ANDpq$ and $NCpNq$ (which may be written down also respectively as $p \wedge q$, $\neg[p \Rightarrow \neg q]$), we have:

$$AND00 = 0;\ NC0N0 = NC01 = N1 = 0$$
$$AND01 = 0;\ NC0N1 = NC00 = N1 = 0$$
$$AND10 = 0;\ NC1N0 = NC11 = N1 = 0$$
$$AND11 = 1;\ NC1N1 = NC10 = N0 = 1$$

In both cases, substituting same truth values for occurrences of p, q in either expression yields the same truth value of these expressions. Thus, they are equivalent.   □

We will call a *theorem* of the sentential calculus any meaningful expression $\alpha(p_1, ..., p_k)$ with the property that the truth value of α is 1 for every substitution of truth values for $p_1, ..., p_k$.

We will write $\alpha = \beta$ in case α, β are equivalent meaningful expressions. Let us derive two important observations from the above proof.

1. For any meaningful expression α, the truth value of α for any substitution of truth values of propositional variables in α may be checked in a finite number of steps. In particular, it may be checked in a finite number of steps whether α is a *theorem* of propositional calculus i.e. the truth value of α is 1 for any substitution of truth values for propositional variables in α.

2. Connectives OR and AND may be defined by means of connectives C, N: in any meaningful expression, $ORpq$ may be replaced with $CNpq$ and $ANDpq$ may be replaced with $NCpNq$ without any effect on the truth value of that expression. It follows that in forming meaningful expressions of propositional calculus, we may apply connectives C, N only.

2.4   The sentential calculus

When discussing meaningful expressions of propositional calculus, we are interested above all in *theorems* of this calculus i.e. as defined above, in expressions which are true regardless of truth values of propositional variables that occur in these expressions. The reason for this lies primarily in the fact that theorems of propositional calculus may be regarded as *laws of thought* leading always to valid (true) conclusions regardless of truth values (i.e. propositional variables or propositions substituted for them). The notion of a theorem is a *semantic* one, relying on our checking of truth values. On the other hand, we have defined a *syntactic mechanism* for generating meaningful expressions. The importance of relations between the syntactic and semantic aspects of propositional calculus had been fully recognized first by Gottlob Frege [Frege874].

It had been his idea to distinguish between meaningful expressions and *derivation rules* i.e. rules for forming new meaningful expressions from given ones. The Frege program assumed the existence of certain apriorical expressions called *axioms* and the existence of certain rules by means of which meaningful expressions may be derived from axioms. Accepted axioms and derivation rules constitute a *deduction system* for propositional calculus. Meaningful expressions derived from axioms by means of specified rules are called *theses* of the system.

Clearly, in our choice of axioms and derivation rules, we are guided by certain constraints. Above all, we would like to invent axioms and rules of derivation which would give us as theses of the system all theorems of propositional calculus; also, we would like to have the situation when all theorems of propositional calculus are theses of our system. When both constraints are met, we have the case when syntactic and semantic aspects of propositional calculus are mutually expressible: theses of the deductive system = theorems of propositional calculus.

Many axiom systems for sentential calculus were proposed by various authors (cf. Historical remarks). We present here the deductive system for the sentential calculus proposed by Jan Łukasiewicz [Lukasiewicz63].

First, we should discuss derivation (inference) rules applied in this system. We will say that two meaningful expressions are *equiform* when their reading symbol by symbol from left to right produces at each step the same (equiform) symbol of a functor or a propositional variable, or of an auxiliary symbol. With this notion, we may state the first derivation rule of *substitution*.

Substitution *For any thesis α of the system, the formula obtained from α by substituting meaningful expressions for propositional variables, in such a way that equiform propositional variables are replaced by equiform meaningful expressions, is a thesis of the system.*

The second derivation rule *detachment* (or, *modus ponens*) consists in the following.

Detachment *For any pair of meaningful expressions* α, β, *if* α *and* $C\alpha\beta$ *are theses of the system then* β *is a thesis of the system.*

The axioms of propositional calculus proposed by Jan Łukasiewicz are the following:

$$\text{(T1) } CCpqCCqrCpr \text{ (or, } (p \Rightarrow q) \Rightarrow [(q \Rightarrow r) \Rightarrow (p \Rightarrow r)])$$
$$\text{(T2) } CCNppp \text{ (or, } (\neg p \Rightarrow p) \Rightarrow p)$$
$$\text{(T3) } CpCNpq \text{ (or, } p \Rightarrow (\neg p \Rightarrow q))$$

Given axioms and derivation rules, one may derive from axioms new theses of the system. We will give a formal definition of a *derivation* (a *proof*) of a thesis.

For a meaningful expression α, a *derivation* of α from axioms is a sequence

$$\alpha_0, \alpha_1, .., \alpha_n$$

of meaningful expressions such that

1. α_0 is an axiom,
2. α_n is equiform with α,
3. Each α_i for $i = 1, 2, ..., n$ is obtained either from some α_j with $j < i$ by means of substitution or from some α_j, α_k with $j, k < i$ by means of detachment.

A meaningful expression having a derivation from axioms is called *a thesis* of the axiomatic system. To denote the fact that α is a thesis, we use the acceptance symbol $\vdash$ so $\vdash \alpha$ reads "α is a thesis".

We denote with the symbol (L) the deductive system consisting of axioms (T1)–(T3) and rules of substitution and detachment.

Checking whether a given meaningful expression is a thesis requires a trial–and–error procedure. It is customary to apply the deductive mechanism of sentential calculus to derive a number of theses on basis of which one may establish fundamental properties of this system. We give – following closely derivations in [Łukasiewicz63] – derivations of some basic theses which come useful later. Other important theses will be found and their derivations hinted at in exercises to this Chapter.

We denote by the symbol p/α the operation of substitution of α for p; by detachment of α from an expression β we mean inferring γ such that β is $C\alpha\gamma$.

2.5   Exemplary derivations in (Ł)

I. We begin with (T1) and substitute p/Cpq, $q/CCqrCpr$, r/s. The result is

$$CCCpqCCqrCprCCCCqrCprsCCpqs \qquad (2.5)$$

and detachment of (T1) from it yields

(T4) $CCCCqrCprsCCpqs$

II. In (T4), we substitute p/s, $s/CpCsr$; the result is

(1) $CCCCqrCsrCpCsrCCsqCpCsr$.

In (T4), we substitute q/Cqr, r/Csr, $s/CCsqCpCsr$. The result is:

(2) $CCCCCqrCsrCpCsrCCsqCpCsrCCpCqrCCsqCpCsr$

and detachment of (1)from (2) gives

(T5) $CCpCqrCCsqCpCsr$

III. In (T4), we substitute $s/CCCprsCCqrs$; the result is

(3) $CCCCqrCprCCCprsCCqrsCCpqCCCprsCCqrs$.

In (T1), we substitute p/Cqr, q/Cpr, r/s; the result is

(4) $CCCqrCprCCCprsCCqrs$

and detaching (4) from (3) we get

(T6) $CCpqCCCprsCCqrs$

IV. In (T5), we substitute p/Cpq, $q/CCprs$, $r/CCqrs$, s/t. The result is

(5) $CCCpqCCCprsCCqrsCCtCCprsCCpqCtCCqrs$.

Detachment of (T6) from (5) yields

(T7) $CCtCCprsCCpqCtCCqrs$

V. In (T1), we substitute $q/CNpq$ to get

(6) $CCpCNpqCCCNpqrCpr$.

Detaching (T3) from (6) gives us

(T8) $CCCNpqrCpr$

VI. In (T8), substituting $r/CCCNpppCCqpp$ yields

(7) $CCCNpqCCCNpppCCqppCpCCCNpppCCqpp.$

Substitution in (T6) of p/Np, r/p, s/p yields

(8) $CCNpqCCCNpppCCqpp$

and detachment of (8) from (7) results in

(T9) $CpCCCNpppCCqpp$

VII. In (T9), we substitute $p/CCNppp$ to get

(9) $CCCNpppCCCNCCNpppCCNpppCCNpppCCqCCNpppCCNppp.$

Detachment of (T2) yields

(10) $CCCNCCNpppCCNpppCCNpppCCqCCNpppCCNppp.$

Substituting in (T2) $p/CCNppp$ yields

(11) $CCNCCNpppCCNpppCCNppp$

and on detaching (11) from (10) we get

(T10) $CCqCCNpppCCNppp$

VIII. In (T8), we substitute p/t, $q/CCNppp$, $r/CCNppp$ and the result is

(12) $CCCNtCCNpppCCNpppCtCCNppp.$

Substituting q/Nt in (T10), gives

(13) $CCNtCCNpppCCNppp$

and on detachment of (13) from (12) we get

(T11) $CtCCNppp$

IX. Substituting in (T7) p/Np, r/p, s/p yields

(14) $CCtCCNpppCCNpqCtCCqpp.$

Detachment of (T11) from (14) results in

(T12) $CCNpqCtCCqpp$.

Derivations included above may serve as examples. More derivations of theses that allow us to exhibit certain basic features of the system may be found in Exercises 1–38 to which we refer in what follows. First, we generalize slightly the notion of a derivation. For a meaningful expression γ, we will say that a meaningful expression β has a derivation from γ, in symbols $\gamma \vdash \beta$, if there exists a sequence $\alpha_1, \alpha_2, ..., \alpha_k$ with α_1 equiform with either an axiom or γ, α_k equiform with β, and any α_i for $i = 2, .., k$ is either equiform with γ or obtained by substitution from some α_j with $j < i$, or obtained by detachment from some α_j, α_m with $j, m < i$.

This notion extends in the obvious sense to the case when β has a derivation from a finite set $\gamma_1, ..., \gamma_n$, in symbols $\gamma_1, ..., \gamma_n \vdash \beta$, (simply, in the above definition we may substitute in place of γ any of γ_i's).

This notion of derivation is instrumental in a basic property of the system we present here expressed by the *Herbrand deduction theorem*. This result is essentially the following

Proposition 2.2. *(Herbrand) If* $\gamma \vdash \beta$ *then* $\vdash C\gamma\beta$.

Let us observe that this proposition extends easily to a more general one viz. if $\gamma_1, .., \gamma_k \vdash \beta$ then $\gamma_1, ..., \gamma_{k-1} \vdash C\gamma_k\beta$. Simply please check that the proof below holds with minor changes in this more general case. Our proof below rests on some theses of the system (L) which are proposed to be proved in Exercise Section; nevertheless, as they are provided with detailed hints, their proofs will not cause the reader any difficulty and thus the proof of the Herbrand theorem will be complete.

Proof of the deduction theorem

Consider a derivation $\alpha_1, ..., \alpha_k$ of β from γ. The proof is by induction on i of the statement $\vdash C\gamma\alpha_i$. For $i = 1$, either α_1 is an axiom or α_1 is γ.

In the former case as (T16): $CqCpq$ is a thesis (cf. Exercise 4), we have after an adequate substitution to (T16) that $\vdash C\alpha_1 C\gamma\alpha_1$ and from this we infer by means of detachment that $\vdash C\gamma\alpha_1$.

In the latter case, by the thesis (T49) (cf. Exercise 37):Cpp after the substitution p/γ we have $\vdash C\gamma\gamma$ i.e. $\vdash C\gamma\alpha_1$. Assuming that $\vdash C\gamma\alpha_i$ for $i < n \leq k$, we consider α_n in the three cases:

(i) α_n is an axiom

(ii) α_n is equiform with γ

(iii) α_n follows by means of detachment from some α_r, α_s with $r, s < n$ i.e. α_r is equiform with $C\alpha_s\alpha_n$.

Clearly, only the case (iii) is new, since cases (i) and (ii) may be done with similarly to the case when $i = 1$. For (iii), we thus have (iv) $\vdash C\gamma\alpha_s$,

(v) $\vdash C\gamma C\alpha_s\alpha_n$. As (T32) (cf. Exercise 20):

$$CCpCqrCCpqCqr$$

is a thesis, substituting in (T32) $p/\gamma, q/\alpha_s, r/\alpha_n$ gives
$\vdash CC\gamma C\alpha_s\alpha_n CC\gamma\alpha_s C\gamma\alpha_n$ from which by applying twice detachment with
(iv), (v), we arrive at $\vdash C\gamma\alpha_n$. Thus inductive step is completed and the
deduction theorem follows.

The deduction theorem allows for derivations of new theses but for us its
principal merit is in the fact that it leads quickly to some other very important properties of the system of propositional calculus. Let us also notice that

Remark
     In case $\gamma_1, ..., \gamma_k \vdash \beta$, it follows by applying the deduction theorem con-
secutively k number of times that $\vdash C\gamma_1 C\gamma_2...C\gamma_k\beta$.

Let us observe that

Proposition 2.3. *Each thesis of our system is a theorem of propositional
calculus.*

     Indeed, we check directly that each of axioms (T1)–(T3) is a theorem and
clearly any substitution preserves this property. This leaves only detachment
but in this case if β follows from α and $C\alpha\beta$ by an application of detachment
and both $\alpha, C\alpha\beta$ are theorems then β is also a theorem as $C1x = 1$ implies
$x = 1$ in the table of truth values for the implication functor C.

This brings forth the question whether the converse holds i.e. whether every
theorem of the sentential calculus is its thesis. This property is known as
the *completeness property* of the system. It turns out that our system has
this property i.e. it is *complete*. We include an elegant proof of this fact due
to Laszlo Kalmár. Before we embark on this proof, we have to state some
auxiliary facts.

2.6   Completeness of (Ł)

     Let us consider a meaningful expression $\alpha(p_1, ..., p_n)$ (recall that this nota-
tion tells us that propositional variables occurring in α are $p_1, ..., p_n$). Assume
that we have fixed truth values of those variables and hence the truth value
of α is also determined.
     We introduce new meaningful expressions

$$p_1', ..., p_n', \alpha'$$

as follows:

1. If the chosen truth value for p_i is 1 then we let p_i' to be equiform with p_i; otherwise, we let p_i' to be equiform with Np_i,

2. If the truth value of α is 1 then we let α' to be equiform with α; otherwise, we let α' to be equiform with $N\alpha$.

Let us marginally mention that expressions of the form p' defined in point 1 above are called *literals*.

We may now state the auxiliary

Proposition 2.4. *For* $p_1', ..., p_n', \alpha'$ *defined above, we have* $p_1', ..., p_n' \vdash \alpha'$

Proof. The proof is by induction on the number m of propositional functors C, N in α. In case $m = 0$, α is equiform with a propositional variable p and the claim follows from (T49): Cpp or its variant obtained by the substitution $p/Np : CNpNp$ depending on the assigned truth value of p.

Assume now that the claim has been proved for the number of connectives in a meaningful expression less than m and consider α. We may distinguish between the two cases.

In the first case, α is equiform with $N\beta$ and so α' is equiform with $NN\beta$. There are two subcases:
    (i) the truth value of β is 1
    (ii) the truth value of β is 0.
    In case (i), we have that α' is equiform with $N\alpha$ and by inductive assumption $p_1', ..., p_n' \vdash \beta$. As (T25) (cf. Exercise 13):

$$CCCpqqCCqpp$$

is a thesis of our system, we have after a substitution $p/Np, q/p$, detaching (T2), and detaching (T8) with substituted $q/NNp, r/NNp$ that $C\beta NN\beta$ is a thesis and thus an application of detachment yields $p_1', ..., p_n' \vdash NN\beta$ i.e. $p_1', ..., p_n' \vdash \alpha'$.

In case (ii), we have that β' is equiform with $N\beta$ and α' is equiform with α. By inductive assumption, $p_1', ..., p_n' \vdash \beta'$ i.e. $p_1', ..., p_n' \vdash \alpha'$.

Now, for the second case when α is equiform with $C\gamma\beta$ for some γ, β. There are three cases to consider viz.
    (i) the truth value of γ is 0 so α is true
    (ii) the truth value of β is 1 so α is true
    (iii) the truth value of γ is 1 and the truth value of β is 0 so α is false.

In case(i), α' is equiform with α and γ' is equiform with $N\gamma$. By inductive assumption, $p_1', ..., p_n' \models N\gamma$, so by the thesis (T33) (cf. Exercise 21):

$$CNpCpq$$

with substitutions $p/\gamma, q/\beta : CN\gamma C\gamma\beta$, it follows by an application of detachment that $p'_1, ..., p'_n \vdash C\gamma\beta$ i.e. $p'_1, ..., p'_n \vdash \alpha$. Cases (ii), (iii) are dealt with in a similar manner: for (ii), we have γ' equiform with γ and $p'_1, ..., p'_n \vdash \gamma$ so by the thesis (T16) (cf. Exercise 4):

$$CqCpq$$

with an appropriate substitution, $p'_1, ..., p'_n \vdash \alpha$ follows. For (iii), we have β' equiform with β, γ' equiform with $N\gamma$, $p'_1, ..., p'_n \vdash \beta$, and $p'_1, ..., p'_n \vdash N\gamma$. As (T48) (cf. Exercise 36):

$$CpCNqNCpq$$

is a thesis, after an adequate substitution and two–fold application of detachment we get $p'_1, ..., p'_n \vdash \alpha'$. The proof is concluded.   □

The completeness theorem for our system follows now

Proposition 2.5. *(the Completeness Theorem) Every theorem of propositional calculus is a thesis of the system* (L) *and the converse holds as well.*

Proof. In the notation of Proposition 2.4, for a theorem α we have that α' is equiform with α regardless of truth values of propositional variables $p_1, .., p_n$. Hence, by this proposition, $p'_1, ..., p'_n \vdash \alpha$ holds regardless of the form of literals $p'_1, ..., p'_n$. In particular, we have $p'_1, ..., p_n \vdash \alpha$ as well as $p'_1, ..., Np_n \vdash \alpha$. By the thesis (T46) (cf. Exercise 34) :

$$CCpqCCNpqq,$$

in which we substitute p/p_n, q/α, and an application of detachment, we arrive at

$$p'_1, ..., p'_{n-1} \vdash \alpha.$$

Thus, we have reduced the number of literals by one. Repeating this argument $n - 1$ number of times gives us $\vdash \alpha$ i.e. our claim that α is a thesis of our system follows which concludes the proof.   □

The Completeness Theorem allows now for checking in an algorithmic manner whether a given meaningful expression is a thesis of our system: it is sufficient to this end to determine the truth value of this expression for each possible selection of literals and to check whether in each case the truth value of the expression is 1. If so, the expression, being a theorem, is a thesis; otherwise, the answer is in the negative. As the checking requires only at most 2^k steps for the number k of propositional variables, this procedure is finite. One says that propositional calculus is *decidable*.

2.7 The sequent approach

We outline here yet another approach to formalization of the sentential calculus which goes back to Gentzen [Gentzen34]. This approach is based on the idea of a *sequent*. We will refer to this formalization in what follows especially in our discussions involving the intuitionistic calculus (cf. Chapter 9, Chapter 12).

To define a sequent, we introduce a new symbol of the *Gentzen implication* $\longrightarrow$ and we consider finite sequences Γ, Δ of meaningful expressions of the sentential calculus. A sequent is an expression of the form $\Gamma \longrightarrow \Delta$ in which Γ is the *antecedent* and Δ is the *succedent*. To convey to us the intended meaning of this expression let us quote from [Gentzen34] p. 181:

"*Die Sequenz* $A_1, ..., A_n \longrightarrow B_1, ..., B_m$ *bedeutet inhaltlich genau dasselbe wie die Formel* $A_1 \wedge ... \wedge A_n \supset B_1 \vee ... \vee B_m$...*ist das Antezedens leer, so ist die Formel* $B_1 \vee ... \vee B_m$ *gemeint. Ist das Sukzedens leer, so bedeutet die Sequenz dasselbe wie die Formel* $\neg A_1 \wedge ... \wedge A_n$ *oder* $A_1 \wedge ... \wedge A_n \supset \perp$ *(Falsch). Sind beide leer, so bedeutet die Sequenz wie* $\perp$, *also eine falsche Aussage.*

Thus, informally, a sequent $\Gamma \longrightarrow \Delta$ is regarded as true when each expression in Γ is true and some expression in Δ is true.

In presenting the Gentzen–type formalization of the sentential calculus, we make use of the variant due to Kanger [Kanger57]. The reader will undoubtedly notice the intended meaning of presented rules.

We present first the corresponding *sequent calculus* formalized as with the sentential calculus in the Łukasiewicz version discussed above by means of *axioms* and *inference (derivation) rules*.

An inference rule is of either of the following forms:

$$\frac{S_1}{S}, \quad \frac{S_1; S_2}{S}.$$

In the former case, we say that the sequent S is *directly inferrable* from the sequent S_1 and in the latter case the sequent S is *directly inferrable* from sequents S_1, S_2.

2.7.1 Axioms and inference rules for the sequent calculus of the sentential calculus

We begin with an axiom for the Gentzen implication. We use capital letters $A, B, C, ...$ to denote meaningful expressions.

1.(the axiom) $\Gamma', A, \Gamma \longrightarrow \Theta', A, \Theta$

We now present inference rules which split into rules governing sentential connectives $C, \wedge, \vee, N$ and each case splits further into two subcases governing the occurrence of connectives in antecedents or succedents of sequents.

Inference rules for C

2. $(C \longrightarrow)$ $\dfrac{\Gamma'',\Gamma',A,\Gamma \longrightarrow \Theta'',\Theta',B,\Theta}{\Gamma'',\Gamma',\Gamma \longrightarrow \Theta'',CAB,\Theta',\Theta}$

3. $(C \longleftarrow)$ $\dfrac{\Gamma'',\Gamma',\Gamma \longrightarrow \Theta'',\Theta',A,\Theta;\Gamma'',\Gamma',B,\Gamma \longrightarrow \Theta'',\Theta',\Theta}{\Gamma'',\Gamma',CAB,\Gamma \longrightarrow \Theta'',\Theta',\Theta}$

Inference rules for $\wedge$

4. $(\wedge \longrightarrow)$ $\dfrac{\Gamma \longrightarrow \Theta'',\Theta',A,\Theta;\Gamma \longrightarrow \Theta'',\Theta',B,\Theta}{\Gamma \longrightarrow \Theta'',\Theta',A \wedge B,\Theta}$

5. $(\wedge \longleftarrow)$ $\dfrac{\Gamma'',\Gamma',A,B,\Gamma \longrightarrow \Theta}{\Gamma'',\Gamma',A \wedge B,\Gamma \longrightarrow \Theta}$

Inference rules for $\vee$

6. $(\vee \longrightarrow)$ $\dfrac{\Gamma \longrightarrow \Theta'',\Theta',A,B,\Theta}{\Gamma \longrightarrow \Theta'',\Theta',A \vee B,\Theta}$

7. $(\vee \longleftarrow)$ $\dfrac{\Gamma'',\Gamma',A,\Gamma \longrightarrow \Theta;\Gamma'',\Gamma',B,\Gamma \longrightarrow \Theta}{\Gamma'',\Gamma',A \vee B,\Gamma \longrightarrow \Theta}$

Inference rules for N

8. $(N \longrightarrow)$ $\dfrac{\Gamma'',\Gamma',A,\Gamma \longrightarrow \Theta'',\Theta}{\Gamma'',\Gamma',\Gamma \longrightarrow \Theta'',NA,\Theta}$

9. $(N \longleftarrow)$ $\dfrac{\Gamma'',\Gamma \longrightarrow \Theta'',\Theta',A,\Theta}{\Gamma'',NA,\Gamma \longrightarrow \Theta'',\Theta',\Theta}$

The idea of a proof is the same as with the Łukasiewicz formalization. A sequent S is *provable* from a set $\mathcal{S}$ of sequents if there exists a sequence $S_1, S_2, .., S_k, ...$ of sequents such that

(i) S_1 is S

(ii) each S_i is either an instance of the axiom 1 or it is a member of $\mathcal{S}$, or it is directly inferrable from either S_k, S_j or S_k (depending on the rule applied) with $k, j > i$.

In case the sequence $S_1, S_2, ...$ is finite and the set $\mathcal{S}$ is empty, we say that S has a *proof* $S_1, S_2, ...$; in this case, S is a *thesis* of the sequent calculus.

We include here some exemplary proofs.

Example. $\dfrac{\longrightarrow Cpp}{p \longrightarrow p}$ is the proof of the thesis $\longrightarrow Cpp$ from an instance $p \longrightarrow p$ of the axiom 1, in which the rule $C \longrightarrow$ has been applied.

Example. $\dfrac{\dfrac{\longrightarrow p \vee Np}{\longrightarrow p, Np}}{p \longrightarrow p}$ is the proof of the thesis $\longrightarrow p \vee Np$ making use of inference rules $N \longrightarrow$ and $\vee \longrightarrow$.

We augment the syntactic mechanism defined above with a semantic ingredient; by a *valuation* v we understand a function which assigns to each propositional variable p the truth value $v(p) \in \{0,1\}$. Given a valuation v, we define inductively the induced valuation $Tr(v, A)$ on meaningful expressions in the following way

$$
\begin{aligned}
&(i) \; Tr(v, p) = v(p) \\
&(ii) \; Tr(v, NA) = 1 \; iff \; Tr(v, A) = 0 \\
&(iii) \; Tr(v, CAB) = 1 \; iff \; Tr(v, A) = 0 \; or \; Tr(v, B) = 1 \\
&(iv) \; Tr(v, A \wedge B) = 1 \; iff \; Tr(v, A) = 1 \; and \; Tr(v, B) = 1 \qquad (2.6) \\
&(v) \; Tr(v, A \vee B) = 1 \; iff \; Tr(v, A) = 1 \; or \; Tr(v, B) = 1 \\
&(vi) \; Tr(v, \Gamma \longrightarrow \Delta) = 0 \; iff \; Tr(v, A) = 1 \; for \; each \; A \in \Gamma \; and \\
&\qquad\qquad Tr(v, B) = 0 \; for \; each \; B \in \Delta
\end{aligned}
$$

A sequent S is *true* if and only if $Tr(v, S) = 1$ for every valuation v. Let us observe that for a meaningful expression A, the sequent $\longrightarrow A$ is true if and only if A is true which by completeness of the sentential calculus yields

Proposition 2.6. *A meaningful expression A is a thesis of the sentential calculus if and only if the sequent $\longrightarrow A$ is true.*

As with the sentential calculus, the question of completeness of the sequent calculus arises. To this end, let us observe that by induction on complexity of sequents one verifies easily that

Proposition 2.7. *Each thesis of the sequent calculus is true.*

Indeed, each instance of the axiom 1 is true, and each inference rule 2–9 gives true consequence from true premise(s).

The converse statement follows from the following, easy to be verified, statements. We observe that a derivation of a meaningful expression A involves sequents forming a tree (a derivation tree). Each branch of this tree constitutes a partial derivation in which an economy of proof i.e. removing redundant applications of rules leads to a sequence of sequents such that any occurrence of an expression A with the last applied in it connective X is a result of applying the corresponding inference rule $X \longrightarrow$ or $X \longleftarrow$ (such a branch is called *explicit*). Let us observe further that if an explicit branch contains no instance of the axiom 1, then no expression A may occur in an antecedent of one sequent in the branch and in the succedent of another sequent in the branch: to see this it is sufficient to scrutinize the inference rules. From the definition of provability it follows that if a sequent S is not provable, then some explicit deduction branch contains no instance of the axiom 1 (cf. [Kanger57]). We sum up these findings.

Proposition 2.8. *1. If an explicit branch Π in the derivation of a sequent S contains no instance of the axiom 1, then there is no expression B such that B occurs in an antecedent of some sequent in Π and B occurs in the succedent of some other sequent in Π*
2. If S is not provable then there exists an explicit deduction branch for S without any instance of the axiom 1.

Assume now that a sequent S has an explicit deduction branch Π without any occurrence of an instance of the axiom 1. We define a *canonical valuation* v^c as follows (cf. [Kanger57]): $v^c(p) = 1$ if p occurs in an antecedent of a sequent in Π and $v^c(p) = 0$ if p occurs in a succedent of a sequent in Π. By Proposition 2.8(1) no conflict of values is possible here and the definition of v^c is correct. By induction on complexity of expressions, we may prove that the defining property extends over expressions i.e. $Tr(v^c, A) = 1$ if A occurs in an antecedent of a sequent from Π and $Tr(v^c, A) = 0$ if A occurs in the succedent of a sequent in Π. It thus follows that

Proposition 2.9. *If there exists an explicit deduction branch for a sequent S without any instance of the axiom 1, then $Tr(v^c, S) = 0$.*

Indeed, by definition of $Tr(v^c, .)$, and (2.6(vi)), above we have $Tr(v^c, S) = 0$.

Combining the last Proposition 2.9 with Proposition 2.8 (ii), we obtain (cf. [Kanger57])

Proposition 2.10. *1. If a sequent S is not provable, then S is not true*
2. If S is true then S is a thesis of the sequent calculus
3. a sequent S is true if and only if it is a thesis of the sequent calculus.

Indeed, (2) is a paraphrase of (1) and (3) follows from (2) and Proposition 2.7.

It follows that the sequent calculus is complete. Combining Proposition 13.31 with Proposition 2.6 we finally arrive at

Proposition 2.11. *A meaningful expression A of the sentential calculus is a thesis of the sentential calculus if and only if the sequent $\longrightarrow A$ is a thesis of the sequent calculus.*

Historic remarks

The sentential calculus i.e. logic of propositions is the second in the line of systems of logic after the syllogistics of Aristotle. The Stoic school (Zeno of Kithion (ca. 336–ca. 264 BC), Chrysippus (ca.280– ca. 204 BC)) is credited with its creation (cf. [Lukasiewicz70]) even if their system was formulated in terms of inference rules rather than theses. All four functors defined above were known to Stoics. The distinction between inference rules and theses as well as axiomatization of the logic of propositions are due to Gottlob Frege [Frege874]. In the Frege system, axioms were the following

1. $CqCpq$

2. $CCpCqrCqCpr$

3. $CCpCqrCCpqCpr$

4. $CNNpp$

5. $CpNNp$

6. $CCpqCNqNp$.

Completeness of the sentential calculus was proved by Emil Post [Post21]. Ideas for the proof of completeness presented here are due to Kalmár [Kalmár 34]. The sequent calculus was introduced in [Gentzen34]; the Gentzen–type calculus presented here is due to [Kanger57].

Exercises

In 1–37, we include some basic theses instrumental in proving completeness etc. along the lines chosen above. The reader may find here detailed suggestions about proofs of those theses. We would like to stress that hints to those exercises are closely modeled on derivations in [Łukasiewicz63, II.5]

1. Prove that (T13) is a thesis of the sentential calculus. [Hint: substitution $p/CNpq$, $q/CtCCqpp$ in (T1) gives (10):

$$CCCNpqCtCCqppCCCtCCqpprCCNpqr.$$

Detaching (T12) from (10) results in (T13)]

(T13) $CCCtCCqpprCCNpqr$

2. Prove that (T14) is a thesis of the sentential calculus. [Hint: in (T13), substitute $t/NCCqpp$, $r/CCqpp$ to get (11):

$$CCCNCCqppCCqppCCqppCCNpqCCqpp$$

and in (T2) substitute $p/CCqpp$ which results in (12):

$$CCNCCqppCCqppCCqpp.$$

Detachment of (12) from (11) results in (T14)]

(T14) $CCNpqCCqpp$

3. Show that (T15) is a thesis of the sentential calculus. [Hint: in (T8), substitution $r/CCqpp$ results in (13): $CCCNpqCCqppCpCCqpp$ and detachment of (T14) from (13) gives (T15)]

(T15) $CpCCqpp$

4. Prove that (T16) is a thesis of the sentential calculus. [Hint: in (T5), substitute p/q, $q/CNpq$, r/q, s/p which gives (14):

$$CCqCCNpqqCCpCNpqCqCpq;$$

in (T15), substitute p/q, q/Np to get (15): $CqCCNpqq$.Detachment of (15) from (14) yields (T16)]

(T16) $CqCpq$

5. Prove that (T17) is a thesis of the sentential calculus. [Hint: in (T1), substitute p/q, q/Cpq. From the result $CCqCpqCCCpqrCqr$, by detachment of (T16), obtain (T17)]

(T17) $CCCpqrCqr$

6. Show that (T18) is a thesis of the sentential calculus. [Hint: in (T17), substitute p/Nq, q/p, $r/CCpqq$ to obtain (16):

$$CCCNqpCCpqqCpCCpqq.$$

In (T14), substitute p/q, q/p to the result (17): $CCNqpCCpqq$. Detachment of (17) from (16) yields (T18)]

(T18) $CpCCpqq$

7. Demonstrate that (T19) is a thesis of the sentential calculus. [Hint: in (T5), substitute p/q, q/Cqr, s/p to obtain (18):

$$CCqCCqrrCCpCqrCqCpr.$$

In (T18), substitute p/q, q/r to obtain (19): $CqCCqrr$. Detachment of (19) from (18) yields (T19)]

(T19) $CCpCqrCqCpr$

8. Check that (T20) is a thesis of the sentential calculus. [Hint: in (T19), substitute p/Cpq, q/Cqr, r/Cpr to result in (20):

$$CCCpqCCqrCprCCqrCCpqCpr.$$

Detachment of (T1) from (20) yields (T20)]

(T20) $CCqrCCpqCpr$

9. Verify that (T21) is a thesis of the sentential calculus. [Hint: in (T1), substitute $p/CpCqr$, $q/CqCpr$, r/s to get (21):

$$CCCpCqrCqCprCCCqCprsCCpCqrs.$$

Detachment of (T19) from (21) yields (T21)]

(T21) $CCCqCprsCCpCqrs$

10. Show that (T22) is a thesis of the sentential calculus. [Hint: in (T21), substitute q/Np, r/q, $s/CCCpqpp$ which results in (22):

$$CCCNpCpqCCCpqppCCpCNpqCCCpqpp.$$

Substitution q/Cpq in (T14) results in (23): $CCNpCpqCCCpqpp$. Detachment of (23) from (22) results in (24): $CCpCNpqCCCpqpp$ and detachment of (T3) from (24) yields (T22)]

(T22) $CCCpqpp$

11. Prove that (T23) is a thesis of the sentential calculus. [Hint: in (T19), substitute p/Cpq, $q/CCprs$, $r/CCqrs$ to obtain (25):

$$CCCpqCCCprsCCqrsCCCprsCCpqCCqrs.$$

Detachment of (T6) from (25) yields (T23)]

(T23) $CCCprsCCpqCCqrs$

12. Prove that (T24) is a thesis of the sentential calculus. [Hint: in (T23), substitute p/Cpq, r/p, s/p, q/r to the result (26):

$$CCCCpqppCCCpqrCCrpp.$$

Detachment of (T22) from (26) yields (T24)]

(T24) $CCCpqrCCrpp$

13. Prove that (T25) is a thesis of the sentential calculus. [Hint: substitute r/q in (T24)]

(T25) $CCCpqqCCqpp$

14. Show that (T26) is a thesis of the sentential calculus. [Hint: in (T1), substitute $p/CCpqr$, $q/CCrpp$, r/s to obtain (27):

$$CCCCpqrCCrppCCCCrppsCCCpqrs.$$

Detachment of (T24) from (27) yields (T26)]

(T26): $CCCCrppsCCCpqrs$

15. Verify that (T27) is a thesis of the sentential calculus. [Hint: in (T26), substitute $s/CCprr$ to get (28):

$$CCCCrppCCprrCCCpqrCCprr.$$

In (T25), substitute p/r, q/p with the result (29): $CCCrppCCprr$. Detachment of (29) from (28) yields (T27)]

(T27) $CCCpqrCCprr$

16. Check that (T28) is a thesis of the sentential calculus. [Hint: in (T7), substitute $t/CCpqr$, s/r, q/s to get (30):

$$CCCCpqrCCprrCCpsCCCpqrCCsrr.$$

Detachment of (T27) from (30) yields (T28)]

(T28) $CCpsCCCpqrCCsrr$

17. Verify that (T29) is a thesis of the sentential calculus. [Hint: in (T19), substitute p/Cps, $q/CCpqr$, $r/CCsrr$ to get (31):

$$CCCpsCCCpqrCCsrrCCCpqrCCpsCCsrr.$$

Detachment of (T28) from (31) results in (T29)]

(T29) $CCCpqrCCpsCCsrr$

18. Prove that (T30) is a thesis of the sentential calculus. [Hint: in (T29), substitute q/r, $r/CqCpr$ to the result (32):

$$CCCprCqCprCCpsCCsCqCprCqCpr.$$

In (T16), substitute q/Cpr, p/q to get (33): $CCprCqCpr$. Detachment of (33) from (32) yields (T30)]

(T30) $CCpsCCsCqCprCqCpr$

19. Prove that (T31) is a thesis of the sentential calculus. [Hint: in (T19), substitute p/Cps, $q/CsCqCpr$, $r/CqCpr$ which results in (34):

$$CCCpsCCsCqCprCqCprCCsCqCprCCpsCqCpr.$$

Detach (T30) from (33)]

(T31) $CCsCqCprCCpsCqCpr$

20. Verify that (T32) is a thesis of the sentential calculus. [Hint: in (T31), substitute $s/Cqr, q/Cpq$ to get (35):

$$CCCqrCCpqCprCCpCqrCCpqCpr.$$

Detach (T20) from (35)]

(T32) $CCpCqrCCpqCpr$

21. Show that (T33) is a thesis of the sentential calculus. [Hint: substitute $q/Np, r/q$ in (T19), to get (36): $CCpCNpqCNpCpq$. Detach (T3) from (36)]

(T33) $CNpCpq$

22. Verify that (T34) is a thesis of the sentential calculus. [Hint: in (T1), substitute $p/Np, q/Cpq$ to get (37):

$$CCNpCpqCCCpqrCNpr.$$

Detach (T33) from (37)]

(T34) $CCCpqrCNpr$

23. Prove that (T35) is a thesis of the sentential calculus. [Hint: in (T34), substitute $p/Np, q/p, r/p$ to get (38) $CCCNpppCNNpp$. Detach (T2) from (38)]

(T35) $CNNpp$

24. Prove that (T36) is a thesis of the sentential calculus. [Hint: in (T1), substitute $p/NNp, q/p, r/q$ to get (39): $CCNNppCCpqCNNpq$. Detach (T35) from (39)]

(T36) $CCpqCNNpq$

25. Verify that (T37) is a thesis of the sentential calculus. [Hint: in (T1), substitute $p/Cpq, q/CNNpq$ with the result (40):

$$CCCpqCNNpqCCCNNpqrCCpqr.$$

Detach (T36) from (40)]

(T37) $CCCNNpqrCCpqr$

26. Prove that (T38) is a thesis of the sentential calculus. [Hint: in (T37), substitute $r/CCqNpNp$ for (41):

$$CCCNNpqCCqNpNpCCpqCCqNpNp.$$

In (T14), substitute p/Np with the result (42): $CCNNpqCCqNpNp$. Detach (42) from (41)]

(T38) $CCpqCCqNpNp$

27. Check that (T39) is a thesis of the sentential calculus. [Hint: in (T5), substitute $p/Cpq, q/CqNp, r/Np$ to produce (43):

$$CCCpqCCqNpNpCCsCqNpCCpqCsNp.$$

Detach (T38) from (43)]

(T39) $CCsCqNpCCpqCsNp$

28. Verify that (T40) is a thesis of the sentential calculus. [Hint: in (T39), substitute s/Nq to get (44): $CCNqCqNpCCpqCNqNp$. In (T33), substitute $p/q, q/Np$ which gives (45): $CNqCqNp$. Detach (45) from (44)]

(T40) $CCpqCNqNp$

29. Prove that (T41) is a thesis of the sentential calculus. [Hint: in (T5), substitute $p/CNpq, q/Cqp, r/p$ to get (46):

$$CCCNpqCCqppCCsCqpCCNpqCsp.$$

Detach (T14) from (46)]

(T41) $CCsCqpCCNpqCsp$

30. Verify that (T42) is a thesis of the sentential calculus. [Hint: in (T41), substitute $s/q, q/Nq$ to get (47): $CCqCNqpCCNpNqCqp$. In (T3), substitute $p/q, q/p$ to get (48): $CqCNqp$. Detach (48) from (47)]

(T42) $CCNpNqCqp$

31. Show that (T43) is a thesis of the sentential calculus. [Hint: in (T41), substitute s/Nq to get (49): $CCNqCqpCCNpqCNqp$. In (T33), substitute $p/q, q/p$ to get (50): $CNqCqp$. Detach (50) from (49)]

(T43) $CCNpqCNqp$

32. Demonstrate that (T44) is a thesis of the sentential calculus. [Hint: in (T1), substitute $p/CNpq, q/CNqp$ to get (51):

$$CCCNpqCNqpCCCNqprCCNpqr.$$

Detach (T43) from (51)]

(T44) $CCCNqprCCNpqr$

33. Verify that (T45) is a thesis of the sentential calculus. [Hint: in (T44), substitute $r/CCpqq$ to get (52):

$$CCCNqpCCpqqCCNpqCCpqq.$$

In (T14), substitute $p/q, q/p$ to get (53): $CCNqpCCpqq$. Detach (53) from (52)]

(T45) $CCNpqCCpqq$

34. Prove that (T46) is a thesis of the sentential calculus. [Hint: in (T19), substitute $p/CNpq, q/Cpq, r/q$ to get (54):

$$CCCNpqCCpqqCCpqCCNpqq.$$

Detach (T45) from (54)]

(T46) $CCpqCCNpqq$

35. Show that (T47) is a thesis of the sentential calculus. [Hint: in (T20), substitute $q/Cqr, r/CNrNq$ to get (55):

$$CCCqrCNrNqCCpCqrCpCNrNq.$$

In (T40), substitute $p/q, q/r$, to get (56): $CCqrCNrNq$. Detach (56) from (55)]

(T47) $CCpCqrCpCNrNq$

36. Verify that (T48) is a thesis of the sentential calculus. [Hint: in (T47), substitute $q/Cpq, r/q$ to get (57): $CCpCCpqqCpCNqNCpq$ and detach (T18) from (57)]

(T48) $CpCNqNCpq$

37. Prove that (T49) is a thesis of the sentential calculus. [Hint: in (T8), substitute $q/p, r/p$ to get (58): $CCCNpppCpp$. Detach (T2) from (58)]

(T49) Cpp

In exercises 38–41 we address consistency and independence issues for the sentential calculus. A system of sentential calculus is *consistent* when there is no meaningful expression α with the property that both $\alpha, N\alpha$ are theses. Independence issues concern axioms of the system. An axiom α is *independent* when there is no proof of it from the remaining axioms. A method of independence proof of α due to Bernays [Bernays26] (cf. also [Lukasiewicz63, p. 73]) consists in constructing a matrix of truth values for propositional functors in such a way that all theses derived from axioms distinct from α take on the value 1 at any substitution of truth values for propositional variables while the axiom α takes on a value other than 1 at some substitution.

38. Prove consistency of the system (L). [Hint: Apply the completeness theorem]

39. [Lukasiewicz63] Prove that the assignment of values $0, 1, 2$ to functors C, N in the following way

   1. $C00 = 1, \ C01 = 1, \ C02 = 1$

   2. $C10 = 0, \ C11 = 1, \ C12 = 0$

   3. $C20 = 1, \ C21 = 1, \ C22 = 0$

   4. $N0 = 1, \ N1 = 0, \ N2 = 2$

verifies independence of (T1) from (T2) and (T3) i.e. (T2), (T3) and all theses derived from them take the value 1 at any substitution of truth values $0, 1, 2$ for propositional variables while (T1) does not have this property.

40. [Lukasiewicz63] Prove that the assignment of values $0, 1$ to functors C, N in the following way

   1. $C00 = 1, \ C01 = 1$

   2. $C10 = 0, \ C11 = 1$

   3. $N0 = 0, \ N1 = 0$

witnesses independence of (T2) from (T1) and (T3).

41. [Lukasiewicz63] Prove that the assignment of values $0, 1$ to functors C, N in the following way

   1. $C00 = 1, \ C01 = 1$

   2. $C10 = 0, \ C11 = 1$

   3. $N0 = 1, \ N1 = 1$

witnesses independence of (T3) from (T1), (T2).

In the third section of exercises, we will discuss relations of meaningful expressions of the sentential calculus to boolean functions. A *boolean function* of n variables is a function $f : \{0, 1\}^n \to \{0, 1\}$ which assigns a value 0 or 1 to any sequence of 0's and 1's of length n. Clearly, any meaningful expression α of the sentential calculus induces a boolean function f_α: assigning truth values $x_1, x_2, ..., x_n$ to propositional variables $p_1, p_2, ..., p_n$ occurring in α, we define the value of $f_\alpha(x_1, ..., x_n)$ as the truth value of α under this assignment. It turns out that the converse is also true.

42. Prove that to any boolean function $f(x_1, ..., x_n)$ there exists a meaningful expression α of the sentential calculus such that $f = f_\alpha$. [Hint: consider propositional variables $p_1, ..., p_n$, for any choice σ of $x_1, ..., x_n$ such that $f(x_1, ..., x_n) = 1$ let $p'_i = p_i$ when $x_i = 1$ and $p'_i = Np_i$ when $x_i = 0$ and let α_σ to be $p'_1 \wedge ... \wedge p'_n$. Then let α to be $\vee_\sigma \alpha_\sigma$. In case f is identically 0 let α to be $p_1 \wedge Np_1$] (The property demonstrated here is called the *functional completeness*).

In the following exercise, we address the sequent calculus. We propose to verify the Gentzen *Hauptsatz* (cf. [Gentzen34], p. 196) in the formulation due to [Kanger57].

43. Prove, making use of the completeness of the sequent calculus, that: a sequent $\Gamma \longrightarrow \Delta$ is a thesis of the sequent calculus if and only if each of sequents $\Gamma, A \longrightarrow \Delta$, $\Gamma \longrightarrow \Delta, A$ is a thesis of the sequent calculus.

Works quoted

[Bernays26] P. Bernays, *Axiomatische Untersuchung den Aussagenkalküls der "Principia Mathematica"*, Mathematische Zeitschrift, 25 (1926).

[Frege874] G. Frege, *Begriffsschrift, eine der mathematischen nachgebildete Formelsprache des Reinen Denkens*, Halle, 1874.

[Gentzen34] G. Gentzen, *Untersuchungen über das logische Schliessen. I, II.*, Mathematische Zeitschrift 39 (1934–5), pp. 176–210, 405–431.

[Herbrand30] J. Herbrand, *Recherches sur la theórie de la démonstration*, Travaux de la Soc.Sci. Lettr. de Varsovie, III, 33 (1930), pp. 33–160.

[Kalmár34] L. Kalmár, *Über die Axiomatisierbarkeit des Aussagenkalküls*, Acta Scientiarum Mathematicarum, 7 (1934–5), pp.222–243.

[Kanger57] S. Kanger, *Provability in Logic*, Acta Universitatis Stockholmiensis, Stockholm Studies in Philosophy I, 1957.

[Łukasiewicz70] J. Łukasiewicz, *On the history of the logic of propositions*,

in: Jan Łukasiewicz. Selected Works, L. Borkowski (ed.), North Holland –
Polish Scientific Publishers, Amsterdam – Warsaw, 1970, pp. 197–217].

[Lukasiewicz63] J. Łukasiewicz, *Elements of Mathematical Logic*, Pergamon
Press – Polish Scientific Publishers, Oxford – Warsaw, 1963.

[Post21] E. Post, *Introduction to a general theory of elementary propositions*,
Amer. J. Math., 43(1921), 163–185.

Chapter 3

Logical Theory of Approximations

To find out what is natural, we must study specimens which retain their nature and not those which have been corrupted

Aristotle, *Politics*, I, V

3.1   Introduction

It may seem surprising that the first formalized system aimed at capturing basic laws of thought and rules for reasoning had been conceived before the propositional calculus was recognized as a legitimate subject of independent study by the Stoic school. But such had been the case with the system of logic created by Aristotle of Stagira and known as *The Syllogistic*. One of reasons seems to be that propositional calculus was used by Aristotle intuitively, as a tool not in need of formalization, in his reasoning about syllogisms and the necessity for its more formal study had not been felt before the Stoic school undertook such a study.

As evident from our introduction to rough set theory, operations of containment and intersection which are the subject of Syllogistic are basic ingredients in approximation theory of general concepts by means of exact ones, and thus it seems imperative to include an account of this theory here.

3.2   Figures of syllogisms

Aristotle's syllogisms i.e. meaningful expressions of Syllogistic, are of the form of an implication:

$$\text{(S)   If } p \text{ and } q \text{ then } r$$

where p, q, r are of one of four possible forms each of which expresses either a universal or an existential relation between general terms. These forms are

1. All a is b

2. Some a is b

3. Some a is not b

4. No a is b

Following traditional usage, we will use the notation:

Aab in case 1: *all a is b* (universal affirmative qualification)

Iab in case 2: *some a is b* (existential affirmative qualification)

Oab in case 3: *some a is not b* (existential negative qualification)

Eab in case 4: *no a is b* (universal negative qualification)

In the following Figures, the reader will find diagrams (known as Euler' or Venn' diagrams) which illustrate the qualifications A, O, E, I. The reader is advised to illustrate with such diagrams the figures of syllogisms which will be produced later on. On the basis of those diagrams and basic set theory, one may attempt a proof of completeness of this logic (cf. [Słupecki 49] mentioned below).

In syllogisms of the form (S), either of four types of statements may appear for any of p, q, r. This uniformity had been one of reasons which prompted Aristotle to exclude singular, individual terms from his Syllogistic (in spite of some examples in which individual terms do appear) cf. [Łukasiewicz57]. Also, we have to presume that all terms we consider are non–empty (*loc. cit.*). Looked at from contemporary point of view, syllogistic may be regarded as an attempt at deriving in a systematic way all rules of arguing with non–empty non–singleton sets (i.e. collective properties) based on containment (i.e. being a subset) and on non–empty intersection.

Aristotle himself had been giving proofs of his syllogisms by means of naive (i.e. not formalized) propositional calculus and its self–evident rules, by falsifying not accepted syllogisms by means of substitutions of specific terms in place of general variables and by using some auxiliary rules (e.g. laws of conversion cf. below). A historic merit of Aristotle's system is also first usage of variables.

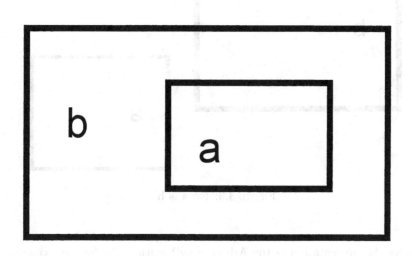

Figure 3.1: All a is b

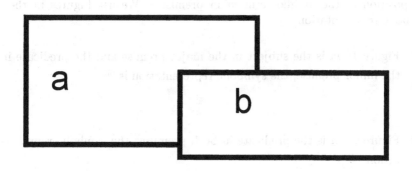

Figure 3.2: Some a is b, Some a is not b

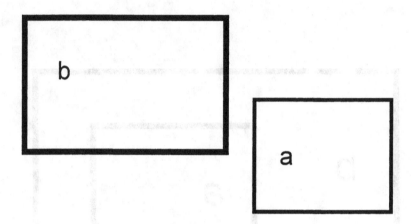

Figure 3.3: No a is b

We give the presentation of the Aristotle Syllogistic in its formal axiomatic version due to Jan Łukasiewicz [Łukasiewicz39, 57, 63]. First, we give some terminology related to syllogisms. In the statement Xab, a is the *subject* and b is the *predicate* where $X = A, I, O, E$. For a syllogistic figure of the form (S), statements p, q are *premises* of which p is the *major* and q is the *minor* and r is the *conclusion*. The predicate in the conclusion is the *major* term, the subject in the conclusion is the *minor* term, and the remaining term is the *middle* term.

All valid syllogisms are traditionally divided in four *Figures* depending on the position of the middle term m in premises. We list Figures in their symbolic representation.

1. Figure 1: m is the subject in the major premise and the predicate in the minor premise; the symbolic representation is $\frac{mb}{\frac{am}{ab}}$

2. Figure 2: m is the predicate in both premises; the symbolic representation: $\frac{bm}{\frac{am}{ab}}$

3. Figure 3: m is the subject in both premises; symbolically, $\frac{mb}{\frac{ma}{ab}}$

4. Figure 4: m is the predicate in the major premise and the subject in the minor premise; symbolically, $\frac{bm}{\frac{ma}{ab}}$

An easy count shows that each Figure may consist of at most 4^3 distinct syllogisms, one for each of 4^3 ways to select three out of the four functors A, I, O, E. Thus the total number of syllogisms does not exceed $4 \times 4^3 = 256$. It turns out that only 24 of them are valid, as the remaining ones may be falsified by examples.

Syllogistic was given a form of a deductive system by Jan Łukasiewicz [Łukasiewicz39] and we now proceed with a description of this system of Syllogistic following closely [Łukasiewicz39, 57, 63]. Following our practice with propositional expressions, we use the implication symbol C in the prefix notation in expressions of the form *if ... then* and we use the symbol $\wedge$ for the *and* connective.

3.3 Syllogistic as a deductive system

First, we adopt propositional calculus as presented in the preceding Chapter. Having now the Completeness Theorem at our disposal, we will perform the semantic check whether an expression is a thesis by calculating truth values. All expressions of propositional calculus used in what follows are theses of this calculus. Thus, among axioms of Syllogistic, we have axioms (T1)-(T3) of propositional calculus.

We adopt, following Łukasiewicz (op.cit.) the following expressions as axioms
(S1) Aaa (*all a is a*)
(S2) Iaa (*some a is a*)
(S3) $CAmb \wedge Aam\ Aab$ (or, $Amb \wedge Aam \Rightarrow Aab$)
(S4) $CAmb \wedge Ima\ Iab$ (or, $Amb \wedge Ima \Rightarrow Iab$)

Axioms (S1), (S2) are clearly not genuine syllogisms but their inclusion is motivated by historic reasons as well as by technical reasons (cf. a discussion in [Łukasiewicz57]). Axioms (S3), (S4) are genuine syllogisms, which may be regarded as self–evident and thus not requiring any proof. Traditionally, syllogisms are known by their names in which functors appearing in them are listed. Thus (S3) has been called Barbara and (S4) has been called Datisi.

Rules of derivation are the following:

Detachment (or, modus ponens) Once $\alpha, C\alpha\beta$ are derived, β is regarded as derived.

Substitution In theses of propositional calculus, we may substitute for propositional variables meaningful expressions containing constants of propositional logic as well as of Aristotle's Syllogistic.[0.25cm] **Replacement by equivalents** We may replace expressions on right–hand sides of the following definitions with their left–hand sides equivalents:
D1 $Oab = NAab$
D2 $Eab = NIab$

Now, we may repeat the derivation process familiar from our discussion of the sentential calculus. The first group of theses are those of auxiliary character from which genuine syllogisms will follow in the further steps. Let us again stress that our presentation here follows strictly the exposition by Jan Łukasiewicz. We denote the deductive system expounded above with the symbol $(A\text{Ł})$.

I. In Cpp, we substitute $p/NAab$ to get $CNAabNAab$ from which replacement via D1 yields

(S5) $COabNAab$ (*if some a is not b then it is not true that all a is b*)

II. In Cpp we substitute $p/NAab$ to get $CNAabNAab$ from which replacement via D2 yields

(S6) $CNAabOab$ (*if it is not true that all a is b then some a is not b*)

III. In $CCpNqCqNp$, we substitute $p/Oab, q/Aab$ to get

$$CCOabNAabCAabNOab$$

from which detachment of (S5) yields

(S7) $CAabNOab$ (*if all a is b then it is not true that some a is not b*)

IV. In $CCNpqCNqp$, we substitute $p/Aab, q/Oab$ to get

$$CCNAabOabCNOabAab$$

from which detachment of (S6) yields

(S8) $CNOabAab$ (*if it is not true that some a is not b then all a is b*)

V. In Cpp, we substitute $p/NIab$ to get $CNIabNIab$ from which replacement via D2 yields

(S9) $CEabNIab$ (*if no a is b then it is not true that some a is b*)

VI. In Cpp we substitute $p/NIab$ to get $CNIabNIab$ from which replacement via D2 yields

(S10) $CNIabEab$ (*if it is not true that some a is b then no a is b*)

VII. In $CCpNqCqNp$, we substitute $p/Eab, q/Iab$ to get

$$CCEabNIabCIabNEab$$

from which detachment of (S9) yields

(S11) $CIabNEab$ (*if some a is b then it is not true that no a is b*)

VIII. In $CCNpqCNqp$, we substitute $p/Iab, q/Eab$ to get

$$CCNIabEabCNEabIab$$

from which detachment of (S10) yields

(S12) $CNEabIab$ (*if it is not true that no a is b then some a is b*)

We may conclude that the above laws hold by the virtue of laws of propositional calculus and definitions D1, D2 establishing basic *laws of contradiction*.

IX. In $CCp \wedge qrCqCpr$ we substitute $p/Aab, q/Iaa$ to get

$$CCAab \wedge IaarCIaaCAabr$$

in which we substitute r/Iab to get $CCAab \wedge IaaIabCIaaCAabIab$ from which we obtain by detachment of $CAab \wedge IaaIab$ (being a case of (S4) with m/a) the result $CIaaCAabIab$. Detachment of (S2) yields

(S13) $CAabIab$ (*if all a is b then some a is b*)

X. In $CCpqCNqNp$, we substitute $p/Aab, q/Iab$ to get

$$CCAabIabCNIabNAab$$

from which by detachment of (S13) we obtain

(S14) $CNIabNAab$ (*if it is not true that some a is b then it is not true that all a is b*)

XI. In (S14), replacements via D1, D2 yield

(S15) $CEabOab$ (*if no a is b then some a is not b*)

XII. In $CCpqCNqNp$, we substitute $p/Eab, q/Oab$ to get

$$CCEabOabCNOabNEab$$

from which detachment of (S15) yields

(S16) $CNOabNEab$

These laws express *relations of sub – alternation* among various functors.

XIII. From (S14) by replacement via D1 it follows

(S17) $CEabNAab$

XIV. In $CCpNqCqNp$, we substitute $p/Eab, q/Aab$ to get

$$CCEabNAabCAabNEab$$

from which by detachment of (S17) we obtain

(S18) $CAabNEab$

XV. From (S14) via replacement with D1 it follows that

(S19) $CNIabOab$

XVI. In $CCNpqCNqp$, we substitute $p/Iab, q/Oab$ to get

$$CCNIabOabCNOabIab$$

from which via detachment of (S19) we get

(S20) $CNOabIab$

An important group of theses of Syllogistic consists of *laws of conversion*, rules of thought which were used by Aristotle in his proofs of valid syllogisms.

XVII. In $CCp \wedge qrCpCqr$ we perform consecutively: substitution p/Aaa, $q/Iab, r/Iba$, detachment of (S4) in which substitution $m/a, a/b, b/a$ was performed and detachment of (S1) to get

(S21) $CIabIba$ (*if some a is b then some b is a*)

XVIII. In $CCpqCCqrCCpr$ we substitute $p/Aab, q/Iab, r/Iba$ and perform detachment twice, first of (S13) then of (S21) to get

(S22) $CAabIba$ (*if all a is b then some b is a*)

XIX. In $CCpqCNqNp$ we substitute $p/Iba, q/Iab$ and then detach (S21) to get

(S23) $CNIabNIba$ (*if it is not true that some a is b then it is not true that some b is a*)

XX. Replacement via D2 in (S23) yields

(S24) $CEabNIba$ (*if no a is b then it is not true that some a is b*)

XXI. Replacement in (S24) via D2 produces

(S25) $CEabEba$ (*if no a is b then no b is a*)

3.4 Selected syllogisms

Now, we present derivations of syllogisms following closely derivations in Łukasiewicz [Lukasiewicz57]. We give here derivations of syllogisms of the first Figure i.e. with the arrangement $mb - am - ab$ of terms. Other syllogisms will be found in Exercise section.

XXII. (Barbari) In $CC(p \wedge q)rCCsqC(p \wedge s)r$ we substitute
$p/Amb, q/Ima, r/Iab, s/Aam$ and then we detach (S4) to get (1):

$$CCAamImaCAmb \wedge AamIab.$$

Now we detach from (1) the expression (S22) where we substitute b/m to produce

(S26) $CAmb \wedge AamIab$ (*if all m is b and all a is m then some a is b*)

XXIII. (Celarent) In $CC(p \wedge q)rC(p \wedge Nr)Nq$ we substitute
$p/Amb, q/Ima, r/Iab$ and then detach (S4) to get (2):

$$CAmb \wedge NIabNIma.$$

In $CC(p \wedge q)rCCsqC(p \wedge s)r$ we substitute $p/Amb, q/NIab, r/NIma$, s/Eba and detach from the resulting expression the expression (2) to get (3):

$$CCEbaNIabNImaCCEbaNIabCAmb \wedge EbaNIma.$$

Detachment from (3) of (S24) where we have substituted $a/b, b/a$ yields (4):
$CAmb \wedge EbaNIma$.

In $CC(p \wedge q)rCCrsC(q \wedge p)s$ we substitute $p/Aam, q/Emb, r/NIab$, s/Eab and then detach (4) where substitutions $m/a, b/m, a/b$ have been made to get (5):

$$CCNIabEabCEmb \wedge AamEab.$$

Detachment of (S10) from (5) gives

(S27) $CEmb \wedge AamEab$ (*if no m is b and all a is m then no a is b*)

XXIV. (Celaront) In $CCpqCCqrCCpr$ we substitute $p/Emb \wedge Aam$, $q/Eab, r/Oab$ and perform detachment twice: first of (S27), next of (S15) to obtain

(S28) $CEmb \wedge AamOab$ (*if no m is b and all a is m then some a is not b*)

XXV. (Darii) In $CC(p \wedge q)rCCsqC(p \wedge q)r$ we substitute $p/Amb, q/Ima$, $r/Iab, s/Iam$ and perform detachment twice: first of (S4) then of (S21) where we have substituted b/m to get

(S29) $CAmb \wedge IamIab$ (*if all m is b and some a is m then some a is b*)

XXVI. (Ferio) Substituting in $CCp \wedge qrCNr \wedge qNp : p/Amb, q/Ima$, r/Iab and then detaching of (S4) yields (6):

$$CNIab \wedge ImaNAmb.$$

Substitution $a/m, m/a$ in (6) and then replacement twice: first via D2 where substitution a/m has been performed and next with D1 results in

(S30) $CEmb \wedge IamOab$ (*if no m is b and some a is b then some a is not b*)

We refer the reader for further syllogisms and their derivations to Exercise section (cf. Exercises 1–18). It had been proved by Słupecki [Słupecki49] that the system (AL) is a complete logical theory of the relations of inclusion and intersection of non–empty classes. A fortiori, the relations listed in Exercise 18 (i)–(xxvi) provide a complete theory of lower and upper approximations.

Historic remarks

Syllogistic was created by Aristotle of Stagira (384–322 BC), with its principles being exposed in *Prior Analytics, Bks. I, II*. A formalization as a deductive system of Syllogistic was proposed by Jan Łukasiewicz [Łukasiewicz39] and exposed in [Łukasiewicz57, 63]. A historic account is given in Bocheński [Bocheński61].

Exercises

Following the pattern of derivations presented above, in Exercises 1–18 we propose to practice derivations of other valid syllogisms. Hints to exercises are closely patterned on derivations in [Łukasiewicz39, 57]. In Exercises 1–6, syllogisms of Figure 2 are discussed.

1. Prove the syllogism (**Cesare**) (S31). [Hint: in the thesis

$$CC(p \wedge q)rCCspC(s \wedge q)r$$

of propositional calculus, substitute $p/Emb, q/Aam, r/Eab, s/Ebm$ and then detach first (S27) and then (S25) where substitution $a/b, b/m$ had been done]

(S31) $Ebm \wedge AamEab$ (*if no b is m and all a is m then no a is b*)

2. Prove the syllogism (**Cesaro**) (S32). [Hint: in the thesis

$$CCpqCCqrCCpr$$

of propositional calculus, substitute $p/Ebm \wedge Aam, q/Eab, r/Oab$ and then detach first (S31) then (S15)]

(S32) $CEbm \wedge AamOab$ (*if no b is m and all a is m then some a is not b*)

3. Verify that (**Camestres**) (S33) is a thesis of the system (AŁ). [Hint: in the thesis
$$CCp \wedge qrCCrsCq \wedge ps$$
of propositional calculus, substitute $p/Eam, q/Abm, r/Eba, s/Eab$ and apply detachment twice: first of (S31) where substitution $b/a, a/b$ had been made and then of (S25) where substitution $a/b, b/a$ had been made]

(S33) $CAbm \wedge EamEab$ (*if all b is m and no a is m then no a is b*)

4. Check that (**Camestrop**) (S34) is a thesis of the system (AŁ). [Hint: in the thesis
$$CCpqCCqrCCpr$$
of the sentential calculus, substitute $p/Abm \wedge Eam, q/Eab, r/Oab$ and then detach first (S33) then (S15)]

(S34) $CAbm \wedge EamOab$ (*if all b is m and no a is m then some a is not b*)

5. Prove the syllogism (**Festino**) (S35). [Hint: in the thesis

$$CCp \wedge qrCCspCs \wedge qr$$

of the sentential calculus, substitute $p/Emb, q/Iam, r/Oab, s/Ebm$ and then detach first (S30) then (S25) where substitution $a/b, b/m$ had been made]

(S35) $CEbm \wedge IamOab$ (*if no b is m and some a is m then some a is not b*)

6. Prove the syllogism (**Baroco**) (S36). [Hint: in the thesis

$$CC(p \wedge q)rC(Nr \wedge q)Np$$

of the sentential calculus, substitute $p/Amb, q/Aam, r/Aab$ and then detach (S3) to result in (i) $CAmb \wedge NAabNAam$. Substitute $m/b, b/m$ in (i) and apply replacement via D1]

(S36) $CAbm \wedge OamOab$ (*if all b is m and some a is not m then some a is not b*)

Exercises 7–11 are concerned with syllogisms of Figure 3.

7. Verify the syllogism (**Darapti**) (S37). [Hint: in the thesis

$$CCp \wedge qrCCsqCp \wedge sr$$

of the sentential calculus, substitute $p/Amb, q/Ima, r/Iab, s/Ama$ and detach first (S4) next (S13)]

(S37) $CAmb \wedge AmaIab$ (*if all m is b and all m is a then some a is b*)

8. Prove the syllogism (**Felapton**) (S38). [Hint: in the thesis

$$CCp \wedge qrCCsqCp \wedge sr$$

of the sentential calculus, substitute $p/Emb, q/Iam, r/Oab, s/Ama$ and detach first (S30) then (S22) where substitution $a/m, b/a$ had been made]

(S38) $CEmb \wedge AmaOab$ (*if no m is b and all m is a then some a is not b*)

9. Prove the syllogism (**Disamis**) (S39). [Hint: in the thesis

$$CCp \wedge qrCp \wedge NrNq$$

of the sentential calculus, substitute $p/Ama, q/Imb, r/Iba, s/Iab$ and detach (S4) where substitution $b/a, a/b$ had been made and next (S21) where substitution $b/a, a/b$ had been made]

(S39) $CImb \land AmaIab$ (*if some m is b and all m is a then some a is b*)

10. Prove the syllogism (**Bocardo**) (S40). [Hint: in the thesis

$$CCp \land qrCNr \land qNp$$

of the sentential calculus, substitute $p/Amb, q/Aam, r/Aab$ and detach (S3) to get (ii) $CNAab \land AamNAab$. In (ii), substitute $a/m, m/a$ and apply twice replacement via D1]

(S40) $COmb \land AmaOab$ (*if some m is not b and all m is a then some a is not b*)

11. Prove the syllogism (**Ferison**) (S41). [Hint: in the thesis

$$CCp \land qrCCsqCp \land sr$$

of the sentential calculus, substitute $p/Emb, q/Iam, r/Oab, s/Ima$ and then detach (S30). The result is (iii): $CCImaIamCEmb \land ImaOab$ from which detach (S21)]

(S41) $CEmb \land ImaOab$ (*if no m is b and some m is b then some a is not b*)

This concludes syllogisms of Figure 3 (as (S4) (Datisi) is one of them) and we pass to syllogisms of Figure 4 in Exercises 12–17.

12. Check that (**Bamalip**) is a thesis of the system (AL). [Hint: in the thesis

$$CC(p \land q)rCCspC(s \land q)r$$

of the sentential calculus, substitute $p/Imb, q/Ama, r/Iab, s/Abm$ and detach first (S39) then (S22) where substitution $a/b, b/m$ had been made]

(S42) $CAbm \land AmaIab$ (*if all b is m and all m is a then some a is b*)

13. Prove the syllogism (**Calemes**) (S43). [Hint: in the thesis

$$CC(p \land q)rCCsqC(p \land s)r$$

of the sentential calculus, substitute $p/Abm, q/Eam, r/Eab, s/Ema$ and from the result detach (S33) and then (S25)]

(S43) $CAbm \land EmaEab$ (*if all b is m and no m is a then no a is b*)

14. Verify that (**Calemop**) (S44) is a thesis of the system (AL). [Hint: substitute $p/Abm \land Ema, q/Eab, r/Oab$ in the thesis

$$CCpqCCqrCCpr$$

of the sentential calculus and detach (S43) and then (S15)]

(S44) $CAbm \land EmaOab$ (*if all b is m and no m is a then some a is not b*)

15. Prove the syllogism (**Dimatis**) (S45). [Hint: in the thesis

$$CC(p \land q)rCCspC(s \land q)r$$

of the sentential calculus, substitute $p/Imb, q/Ama, r/Iab, s/Ibm$ and detach first (S39) and next (S21) where substitution $a/b, b/m$ had been made]

(S45) $CIbm \land AmaIab$ (*if some b is m and all m is a then some a is b*)

16. Prove the syllogism (**Fesapo**) (S46). [Hint: in the thesis

$$CC(p \land q)rCCspC(s \land q)r$$

of the sentential calculus, substitute $p/Emb, q/Ama, r/Oab, s/Ebm$ and detach (S38) first followed by detachment of (S25) where substitution $a/b, b/m$ had been made

(S46) $CEbm \land AmaOab$ (*if no b is m and all m is a then some a is not b*)

17. Prove the syllogism (**Fresison**) (S47). [Hint: in the thesis

$$CC(p \land q)rCCspC(s \land q)r$$

of the sentential calculus, substitute $p/Emb, q/Ima, r/Oab, s/Ebm$ and detach (S41) and after that (S25) where substitution $a/b, b/m$ had been made]

(S47) $CEbm \land ImaOab$ (*if no b is m and some m is a then some a is not b*)

We have completed the list of all valid syllogisms. One may check that all remaining candidates for syllogisms are not valid being falsified by appropriate substitutions into terms and that no further derivations are possible.

We propose to interpret syllogisms in yet another way. We introduce the following terminology and notation.

1. *a in $\underline{b}$* is equivalent to *all a is b*

2. *a in $\bar{b}$* is equivalent to *some a is b*

By definitions D1, D2, we have: Oab is equivalent to *not(a in $\underline{b}$)* and Eab is equivalent to *not(a in $\bar{b}$)*.

18. Verify that syllogisms derived above lead to the following transcripts in terms of $\underline{a}$, $\bar{a}$.

(i) a in $\underline{a}$

(ii) a in $\bar{a}$

(iii) (Barbara) if m in $\underline{b}$ and a in $\underline{m}$ then a in $\underline{b}$

(iv) (Barbari) if m in $\underline{b}$ and a in $\underline{m}$ then a in $\underline{b}$

(v) (Celarent) if a in $\underline{m}$ and m not in $\bar{b}$ then not a in $\bar{b}$

(vi) (Celaront) if not m in $\bar{b}$ and a in $\underline{m}$ then not a in $\underline{b}$

(vii) (Darii) if m in $\underline{b}$ and a in $\overline{m}$ then a in $\bar{b}$

(viii) (Ferio) if not m in $\bar{b}$ and a in $\underline{m}$ then not a in $\underline{b}$

(ix) (Cesare) if not b in $\overline{m}$ and a in $\underline{m}$ then not a in $\bar{b}$

 noindent (x) (Cesaro) if not b in $\overline{m}$ and a in $\underline{m}$ then not a in $\underline{b}$

(xi) (Camestres) if b in $\underline{m}$ and not a in $\overline{m}$ then not a in $\bar{b}$

(xii) (Camestrop) if b in $\underline{m}$ and not a in $\overline{m}$ then not a in $\underline{b}$

(xiii) (Festino) if not b in $\overline{m}$ and a in $\overline{m}$ then not a in $\underline{b}$

(xiv) (Baroco) if b in $\underline{m}$ and not a in $\underline{m}$ then not a in $\underline{b}$

(xv) (Datisi) if m in $\underline{b}$ and m in $\bar{a}$ then a in $\bar{b}$

(xvi) (Darapti) if m in $\underline{b}$ and m in $\bar{a}$ then a in $\bar{b}$

(xvii) (Felapton) if not m in $\bar{b}$ and m in $\underline{a}$ then not a in $\underline{b}$

(xviii) (Disamis) if m in $\bar{b}$ and m in $\underline{a}$ then a in $\bar{b}$

(xix) (Bocardo) if not m in $\underline{b}$ and m in $\bar{a}$ then not a in $\underline{b}$

(xx) (Ferison) if not m in $\bar{b}$ and m in $\bar{a}$ then not a in $\underline{b}$

(xxi) (Bamalip) if b in $\underline{m}$ and m in $\underline{a}$ then a in $\bar{b}$

(xxii) (Calemes) if b in $\underline{m}$ and not m in $\bar{a}$ then not a in $\bar{b}$

(xxiii) (Calemop) if b in $\underline{m}$ and not m in $\bar{a}$ then not a in $\underline{b}$

(xxiv) (Dimatis) if b in $\overline{m}$ and m in $\underline{a}$ then a in $\bar{b}$

(xxv) (Fesapo) if not b in $\overline{m}$ and m in $\underline{a}$ then not a in $\underline{b}$

(xxvi) (Fresison) if not b in $\overline{m}$ and m in $\bar{a}$ then not a in $\underline{b}$

Works quoted

[Bocheński61] I. M. Bocheński, *A History of Formal Logic*, Notre Dame Univ. Press, 1961.

[Lukasiewicz63] Jan Lukasiewicz, *Elements of Mathematical Logic*, Oxford – Warsaw, 1963

[Lukasiewicz57] Jan Lukasiewicz, *Aristotle's Syllogistic from the Standpoint of Modern Formal Logic*, 2nd ed., Oxford, 1957.

[Lukasiewicz39] Jan Lukasiewicz, *On Aristotle's Syllogistic* (in Polish), Compt. Rend. Acad. Polon. Lettr., Cracovie, 44 (1939).

[Słupecki49] J. Słupecki, *On Aristotle's Syllogistic*, Studia Philosophica (Poznań), 4(1949–50), pp. 275–300.

Chapter 4

Many–Valued Sentential Calculi

Nothing which implies contradiction falls under omnipotence of God

Thomas Aquinas, *Summa Theol.*

4.1 Introduction

It had been assumed by schools and individuals developing logical systems up to the beginning of the 20th century that propositions of logic should be restricted to be either true or false and the twain should exclude themselves (cf. in this respect [Lukasiewicz70]). Those systems are therefore called *two–valued*. On the other hand, from the very beginning of logic, as witnessed by the legacy of Aristotle, there was conviction that there are statements which cannot be true or false e.g. statements about future and that such statements may be interpreted in *modal* terms as e.g. *possible*. It is therefore understandable that first systems of logic in which propositions were not necessarily bi–valent were constructed in attempts to build a formal logic of modal expressions.

Jan Łukasiewicz was the first to consider a many–valued logic (actually, a 3–valued one) cf. [Lukasiewicz18, 20, 30^a] in which, in addition to already familiar truth values 1 (true) and 0 (false), other truth values, representing various partial *degrees* of truth were present (in particular, the value $\frac{1}{2}$ in 3–valued logic). The idea of such a logic was based on the truth matrix of propositional calculus

in which functors C, N are represented.

Looking at the part referring to C i.e. at the sub–matrix composed of columns

C	0	1	N
0	1	1	1
1	0	1	0

Figure 4.1: Truth table of the sentential calculus

C	0	1	2	N
0	1	1	1	1
1	0	1	2	0
2	2	1	1	2

Figure 4.2: Truth table of the 3–valued calculus

1 and 2, we may observe that the logical value $v(Cpq)$ depends on logical values $v(p), v(q)$ e.g. according to the formula

$$v(Cpq) = min\{1, 1 - v(p) + v(q)\} \qquad (4.1)$$

If we admit an intermediate logical value $\frac{1}{2}$ then in the case when $v(p), v(q)$ take values $0, 1, \frac{1}{2}$, also $v(Cpq)$ does take one of this values. Taking into account a formula defining negation in propositional calculus:

$$v(Np) = 1 - v(p) \qquad (4.2)$$

we may construct the matrix for this new 3–valued logic as the following one (in which for typing convenience the value of $\frac{1}{2}$ has been replaced by the value of 2).

We denote this 3–valued logic with the symbol L$_3$. Let us remark here that the formulae for C, N presented above allow for n–valued logics (with truth values $0, 1, \frac{k}{n-1}$ where $k = 1, 2, ..., n - 2$) as well as for *real valued logics* (with real truth values) all of which were proposed by Jan Łukasiewicz and investigated in the 20's and 30's cf. [Lukasiewicz30^b]. Independently, n–valued logics were proposed by Emil Post [Post21].

4.2   A formal development

As with propositional logic, we will call a *theorem* of this 3–valued logic any meaningful expression whose truth value is 1 for any choice of truth values of propositional variables that occur in it.

The problem of axiomatization of the logic L$_3$ was solved by M. Wajsberg [Wajsberg31] who proposed the following axioms:

(W1) $CqCpq$

(W2) $CCpqCCqrCpr$

(W3) $CCCpNppp$

(W4) $CCNqNpCpq$

along with already familiar to us derivation rules of substitution and detachment. We denote this deductive system with the symbol (W).

As proved by Wajsberg (op. cit.), the system (W) is complete i.e. a meaningful expression of this system is a thesis of this system if and only if it is a theorem of the 3-valued logic L_3.

Let us observe that – as one should expect – some axioms of propositional calculus are no longer theorems of the 3-valued logic of Łukasiewicz, witness for this is e.g. $CCNppp$: its truth value in case $v(p) = 2$ is 2.

The original proof by Wajsberg of the completeness of his system may be found in [Wajsberg31]. We include here for the sake of completeness of exposition a proof proposed by Goldberg, Leblanc and Weaver [Goldberg–Leblanc–Weaver74]. The proof exploits basic properties of syntactic consequence and thus it presents a new approach to the completeness proving.

We recall (cf. Chapter 2) that we use the symbol $\Gamma \vdash \alpha$ to denote the fact that a meaningful expression α has a proof from a set Γ of meaningful expressions (we recall that in particular $\vdash \alpha$ denotes the fact that α is a thesis).

We say that a set Γ is *syntactically consistent* if and only if there exists no α with $\Gamma \vdash \alpha$ and $\Gamma \vdash N\alpha$. A set Γ is a *maximal syntactically consistent set* if and only if Γ is syntactically consistent and for each meaningful expression α if $\Gamma \cup \{\alpha\}$ is syntactically consistent then $\Gamma \vdash \alpha$. When Γ is not syntactically consistent, we will say that it is *syntactically inconsistent*.

As with propositional logic (Chapter 2), we need a good deal of information about syntactic properties of (W) in order to attempt a proof of completeness. We list here the basic facts needed in the proof of completeness. These facts are proposed to be proved in Exercise section below (in case of (I)–(XIII)) and in Exercise Section of Chapter 12 in case of (XIV)–(XX) where proofs exploit the relationships between the 3-logic L_3 and Łukasiewicz algebras).

(I) $p \vdash Cqp$
(II) $\vdash CNpCpq$
(III) $Cpq \vdash CCqrCpr$
(IV) $Cpq, Cqr \vdash Cpr$
(V) $\vdash CNNpCqp$
(VI) $\vdash CCCpqqCNpq$
(VII) $\vdash CNNpp$

(VIII) $\vdash CpNNp$

(IX) $\vdash Cpp$

(X) $\vdash CCpqCNqNp$

(XI) $\vdash CCpqCNNpq$

(XII) $\vdash CCNNpqCNNpNNq$

(XIII) $\vdash CCpqCNNpNNq$

(XIV) $\vdash CCpCpNpCpNp$

(XV) $\vdash CCCpNpNCpNpp$

(XVI) $\vdash CNCpqp$

(XVII) $\vdash CNCpqNq$

(XVIII) $\vdash CpCNqNCpq$

(XIX) $\vdash CCpNpCCNqNNqCpq$

(XX) $\vdash CCpCpCqrCCpCpqCpCpr$

(XXI) $NCpp$ is false for each substitution of truth values for p

4.3   Consistent sets of meaningful expressions

The proof begins with the following list of basic properties of consistent sets.

1. $\Gamma \vdash \alpha$, $\Gamma \subseteq \Gamma'$ imply $\Gamma' \vdash \alpha$

2. $\Gamma \vdash \alpha$ implies the existence of a finite $\Gamma' \subseteq \Gamma$ with the property that $\Gamma' \vdash \alpha$

3. $\Gamma \vdash \alpha$, $\Gamma \vdash C\alpha\beta$ imply $\Gamma \vdash \beta$

4. $\Gamma \cup \{\alpha\} \vdash \beta$ implies $\Gamma \vdash C\alpha C\alpha\beta$

5. Γ is not syntactically consistent if and only if $\Gamma \vdash f$ where f denotes any proposition such that Nf is a thesis

6. If $\Gamma \cup \{\alpha\}$ is syntactically inconsistent then $\Gamma \vdash C\alpha N\alpha$

7. If $\Gamma \cup \{C\alpha N\alpha\}$ is syntactically inconsistent then $\Gamma \vdash \alpha$

Indeed, 1, 2, 3 follow by the definition of the relation $\vdash$; to prove 4, assume that $\Gamma \cup \{\alpha\} \vdash \beta$ and choose a proof $\alpha_1, \alpha_2, ..., \alpha_n$ of β from $\Gamma \cup \{\alpha\}$. By induction on j, we prove the following claim

(c_j) $\Gamma \vdash C\alpha C\alpha\alpha_j$

We assume (c_j) is valid for $j < i$. There are some cases to be considered.

Case 1. α_i is α. In this case (c_i) follows by (W1) with $p, q/\alpha$.

Case 2. α_i is an instance of an axiom or $\alpha_i \in \Gamma$. As $\Gamma \vdash \alpha_i$, applying (I) twice yields $\Gamma \vdash CaC\alpha\alpha_i$.

Case 3. α_i follows by detachment from α_k and $C\alpha_k\alpha_i$ with $k < i$. In this case we have by the inductive assumption

(i) $\Gamma \vdash CaC\alpha\alpha_k$

(ii) $\Gamma \vdash CaCaCaC\alpha_k\alpha_i$

It now suffices to apply (XX) along with the property 3 and (i), (ii). This concludes the proof of the property 4.

To begin with the proof of the property 5, assume that Γ is not syntactically consistent i.e. $\Gamma \vdash \alpha$ and $\Gamma \vdash N\alpha$ for some meaningful expression α. Then we have $\vdash CN\alpha Caf$ by (I) and detachment applied twice yields $\Gamma \vdash f$.

Conversely, assuming that $\Gamma \vdash f$ for any false f, we have in particular $\Gamma \vdash NCpp$ by (XXI); as it follows from (IX) that $\Gamma \vdash Cpp$, we conclude that Γ is syntactically inconsistent.

Concerning the property 6, assume that $\Gamma \cup \{\alpha\}$ is not syntactically consistent. From (I), as in the proof of property 5, we infer that $\Gamma \cup \{\alpha\} \vdash \beta$ for any meaningful expression β in particular $\Gamma \cup \{\alpha\} \vdash N\alpha$ hence $\Gamma \vdash CaCaN\alpha$ by the property 4 and it follows from (XIV) by detachment that $\Gamma \vdash CaN\alpha$.

The converse property 7 follows from (XV) and the property 6: by the property 6, from the assumption that $\Gamma \cup \{CaN\alpha\}$ is syntactically inconsistent it follows that $\Gamma \vdash CCaN\alpha NcaN\alpha$ so (XV) along with the property 3 implies $\Gamma \vdash \alpha$.

4.4 Completeness

This concludes the proof of basic properties of the relation $\vdash$ and we may pass to the proof of the completeness theorem. We recall our setting: we have a (countable) set of *propositional variables* $p_0, p_1, ..., p_n, ...$ from which meaningful expressions are formed in same way as in the sentential calculus. A *valuation* is an assignment of truth values to propositional variables; clearly, any valuation v induces a valuation on meaningful expressions according to the semantic rules governing functors C, N. We denote by the symbol v^* the induced valuation on meaningful expressions. A set Γ of meaningful expressions is said to be *semantically consistent* if there exists a valuation v with the property that $v^*(\alpha) = 1$ for each $\alpha \in \Gamma$; we will use the shortcut $v^*(\Gamma) = 1$ in this case.

The notion of *semantic consequence* follows: a meaningful expression α is a semantic consequence of a set Γ of meaningful expressions if and only if for each valuation v with $v * (\Gamma) = 1$ we have $v * (\alpha) = 1$. The fact that α is a semantic consequence of Γ is denoted with the symbol $\Gamma \models \alpha$.

Following [Goldberg–Leblanc–Weaver74], we state

Proposition 4.1. *(Strong Completeness Theorem for L_3) For each meaningful expression α and each set Γ of meaningful expressions, $\Gamma \vdash \alpha$ if and only if $\Gamma \models \alpha$.*

Proof. One way the proof goes simply: if $\Gamma \vdash \alpha$ then we may choose a proof $\alpha_1, ..., \alpha_n$ of α from Γ and a valuation v with $v^*(\Gamma) = 1$ and check directly by induction that $v^*(\alpha_i) = 1$ for $i = 1, 2, ..., n$.

The proof in the reverse direction requires some work. It follows from the following Claim.

Claim. *If Γ is syntactically consistent that Γ is semantically consistent.*

The proof of the Claim relies on the notion of a maximal consistent set. In this proof we use some intuitively understandable results of Set Theory. First, we enumerate all propositional variables into a countable set $V = \{p_0, p_1, ...\}$ and then we form the set M of meaningful expressions, enumerating its elements: $M = \{\alpha_0, \alpha_1, ...\}$; this is possible as the set M is countable since each meaningful expression uses only a finite number of propositional variables.

This done, we look at Γ and we construct a sequence $\Gamma_0 = \Gamma, \Gamma_1, ...$ of sets as follows. Given Γ_i we let

$$\Gamma_{i+1} = \Gamma_i \cup \{\alpha_{i+1}\}$$

in case $\Gamma_i \cup \{\alpha_{i+1}\}$ is syntactically consistent and

$$\Gamma_{i+1} = \Gamma_i,$$

otherwise. Having the sequence $(\Gamma_i)_i$, we let

$$\Gamma^* = \bigcup_i \Gamma_i.$$

Properties of Γ^* of importance to us are collected in the following sub–Claims.

Sub–Claim 1. Γ^* *is syntactically consistent.*

Indeed, assume to the contrary that $\Gamma^* \vdash \alpha$ and $\Gamma^* \vdash N\alpha$ for some α. Then by the property 2, there exists a finite $\Delta \subseteq \Gamma^*$ with the property that $\Delta \vdash \alpha$, $\Delta \vdash N\alpha$. By construction of Γ^*, there exists Γ_i with $\Delta \subseteq \Gamma_i$ and thus the property 1 implies that $\Gamma_i \vdash \alpha$, $\Gamma_i \vdash N\alpha$, a contradiction. It follows that Γ^*

is syntactically consistent.

Sub–Claim 2. Γ^* *is a maximal syntactically consistent set.*

Assume that it is not true that $\Gamma^* \vdash \alpha$ for some meaningful expression α; as α is α_i for some i, we have that $\Gamma_{i-1} \cup \{\alpha\}$ is not syntactically consistent hence by the property 5, $\Gamma_{i-1} \cup \{\alpha\} \vdash f$ and the property 1 implies that $\Gamma^* \cup \{\alpha\} \vdash f$ so by the property 5 applied in the reverse direction we have that $\Gamma^* \cup \{\alpha\}$ is not syntactically consistent. It thus follows that if $\Gamma^* \cup \{\alpha\}$ is syntactically consistent then $\Gamma^* \vdash \alpha$ which means that Γ^* is a maximal consistent set.

We now exploit Γ^* to introduce a certain canonical valuation on the set V of propositional variables. To this end, we define a valuation v_0 as follows

$$v_0(p) = 1 \ in \ case \ \Gamma^* \vdash p$$
$$v_0(p) = 0 \ in \ case \ \Gamma^* \vdash Np \qquad (4.3)$$
$$v_0(p) = 2 \quad otherwise$$

Now, we establish the vital property of the induced valuation v_0^* on the set M expressed in the following

Sub–Claim 3.

$$v_0^*(\alpha) = 1 \ in \ case \ \Gamma^* \vdash \alpha$$
$$v_0(\alpha) = 0 \ in \ case \ \Gamma^* \vdash N\alpha \qquad (4.4)$$
$$v_0(\alpha) = 2 \ otherwise$$

The proof of Sub–claim 3 goes by structural induction.

Case 1. α is of the form $N\beta$ and Sub–Claim 3 holds for β.
When $\Gamma^* \vdash N\beta$ we have $v_0^*(\beta) = 0$ hence $v_0^*(N\beta) = 1$ so the claim holds for α.
When $\Gamma^* \vdash NN\beta$ we have $v_0^*(N\beta) = 0$ so the claim again holds with α.
When finally neither of two previous cases hold we conclude by (VII) that it is not true that $\Gamma^* \vdash \beta$ so $v_0^*(\beta) = 2$ hence $v_0^*(\alpha) = 2$. The claim holds with α.

Case 2. α is of the form $C\beta\gamma$ and Sub–Claim 3 holds for β, γ.
We have to examine the three possible cases again.
When $\Gamma^* \vdash C\beta\gamma$, we need to check that

$$v_0^*(C\beta\gamma) = 1$$

which depends on valuations of β, γ and we have some sub–cases to consider.

When $v_0^*(\gamma) = 1$ which happens when $\Gamma^* \vdash \gamma$, we have $v_0^*(\alpha) = 1$; when $v_0^*(\beta) = 0$ we have again $v_0^*(\alpha) = 1$.

The sub–case $v_0^*(\gamma) = 0$ i.e. $\Gamma^* \vdash N\gamma$ remains but then by (X) we have $\Gamma^* \vdash N\beta$ already settled in the positive.

So finally we have the sub–case when neither of $\beta, N\beta, \gamma, N\gamma$ is derivable from Γ^* and then $v_0^*(\beta) = 2 = v_0^*(\gamma)$ a fortiori $v_0^*(\alpha) = 1$.

The next case is when

$$\Gamma^* \vdash NC\beta\gamma.$$

By (XVI) and (XVII) we obtain $\Gamma^* \vdash \beta$ and $\Gamma^* \vdash N\gamma$ so $v_0^*(\beta) = 1$, $v_0^*(\gamma) = 0$ and it follows that $v_0^*(\alpha) = 0$.

Finally, we are left with the case when neither $\Gamma^* \vdash C\beta\gamma$ nor $\Gamma^* \vdash NC\beta\gamma$. It follows from (II) that it is not true that $\Gamma^* \vdash N\beta$ and similarly (W1) implies that it is not true that $\Gamma^* \vdash \gamma$. Thus $v^*(\beta) \neq 0$ and $v^*(\gamma) \neq 1$. We are left with the following sub–cases.

We have $v^*(\beta) = 1$ i.e. $\Gamma^* \vdash \beta$ and then by (XVIII) we have that it is not true that $\Gamma^* \vdash N\gamma$ so $v^*(\gamma) \neq 0$ and hence $v^*(\gamma) = 2$ and in consequence $v^*(\alpha) = 2$.

We have $v^*(\beta) = 2$ i.e. it is not true that $\Gamma^* \vdash \beta$. It is finally here that we exploit maximality of Γ^*.

It follows by maximality of Γ^* that $\Gamma^* \cup \{\beta\}$ is not syntactically consistent so by the property 6 we have $\Gamma^* \vdash C\beta N\beta$; was $v^*(\gamma) = 2$ we would have by the same argument that

$$\Gamma^* \vdash CN\gamma NN\gamma$$

and then we would have by (XIX) that $\Gamma^* \vdash C\beta\gamma$, a contradiction.

Thus $v^*(\gamma) = 0$ and $v^*(\alpha) = 2$. It follows that Sub–Claim 3 holds for α which completes the inductive step and concludes the proof of Sub–Claim 3.

As $\Gamma \subseteq \Gamma^*$, we have $v^*(\alpha) = 1$ whenever $\alpha \in \Gamma$ i.e. Γ is semantically consistent. This concludes the proof of Claim.

Now, we return to the Strong Completeness Theorem; we assume that $\Gamma \models \alpha$. Thus $v^*(\Gamma) = 1$ implies $v^*(\alpha) = 1$ and hence $v^*(C\alpha N\alpha) = 0$ i.e.

$$\Gamma \cup \{C\alpha N\alpha\}$$

is semantically inconsistent hence by Claim this set is syntactically inconsistent and by the property 7 we have $\Gamma \vdash \alpha$. This concludes the proof of the Strong Completeness Theorem. $\square$

A specialization of the Strong Completeness Theorem (Proposition 4.1) to the case when Γ is the set of axioms (W1)–(W4) yields

Proposition 4.2. *(the completeness theorem for the system (W)) A meaningful expression α is a thesis of the system (W) if and only if it is a theorem of the logic (L_3).*

4.5   n–valued logics

We have seen above the proposition of Jan Łukasiewicz for n–valued logic. In his logic functors C, N were defined via functions on truth values. In general, we may observe that statements of many–valued logic are of the form $F_i(p_1, ..., p_k)$ where F_i is a functor (e.g. C, N) and $p_1, .., p_k$ are propositional variables. Assuming that any of p_i may be assigned any of truth values $1, 2, ..., n$, the functor F_i may be described by a matrix of values $f_i(v_1, ..., v_k)$ where each v_i runs over truth values $1, 2, .., n$ which may be assigned to p_i. Any meaningful expression $\alpha(p_1, .., p_m)$ is then assigned a matrix of truth values $g(v_1, .., v_m)$ where $v_1, .., v_m$ run over truth values $1, .., n$ of propositional variables $p_1, ..., p_m$ occurring in α and g is the truth function of α composed of truth functions f_j of functors F_j occurring in α.

The idea of a true or false expression has to be generalized to the idea of acceptance or rejection of a meaningful expression which can be realized in an n–valued logic by a selection of a threshold value $1 \leq s < n$ and declaring a meaningful expression $\alpha(p_1, .., p_m)$ accepted when $g(v_1, ..., v_m) \leq s$ for any choice of truth values $1, 2, ..., m$ for $v_1, .., v_m$ where g is the truth function of α. Otherwise, α is rejected. We may observe that formulae with truth function taking the value 1 only are accepted regardless of choice of s (we may say that they are accepted with certainty) while formulae whose truth function takes on the value n only are rejected regardless of choice of s (we may say that they are rejected with certainty). Other formulae may be accepted or rejected depending on the chosen value of s so their status is less precise and they may be termed to be *accepted in a degree*.

We have seen in case of 3–valued logic of Łukasiewicz that truth matrices have been described in a compressed form by functions: $f_1(v_1, v_2) = min\{1, v_2 - v_1 + 1\}$ for the functor C and $f_2(v) = 1 - v$ for N. In the setting of an n–valued logic as proposed above, these formulae should be dualized to ensure that the greater truth value does correspond to a lesser acceptance degree. Thus the Łukasiewicz formulae in the case of n–valued logic are modified to the form

$$f_1(v_1, v_2) = max\{1, v_2 - v_1 + 1\}, f_2(v) = 1 - v + n \qquad (4.5)$$

as proposed by Rosser and Turquette [Rosser – Turquette58].

We now follow the scheme proposed by Rosser and Turquette [Rosser – Turquette58]. The functors in this scheme are C, N familiar to us already as well as unary functors J_k for $k = 1, 2, ..., n$ construed as to satisfy the condition that the truth function j_k does satisfy the requirement that $j_k(v) = 1$ in case $v = k$ and $j_k(v) = n$ otherwise, for $k = 1, 2, ..., n$. Let us outline construction of J'_ks. First, we let

$$H_1(p) = N(p)$$
$$H_{i+1} = C(p, H_i(p)) \tag{4.6}$$

We denote by the symbol h_i the truth function of H_i. Then we have

$$h_i(v) = max\{1, n - (v - 1) \cdot i\} \ for \ each \ i \tag{4.7}$$

Indeed, inducting on i, we have $h_1(v) = f_2(v) = n - v + 1 = n - (v - 1) = max\{1, n - (v - 1)\}$ and $h_{i+1}(v) = f_1(v, h_i(v)) = max\{1, max\{1, n - (v - 1) \cdot i - (v - 1)\}\} = max\{1, max\{1, n - (v - 1) \cdot (i + 1)\}\} = max\{1, n - (v - 1) \cdot (i + 1)\}$.

Now it is possible to introduce functors $J_1, ..., J_n$; we resort to recursion. To begin with, we let

$$J_1(p) = N(H_{n-1}(p)) \tag{4.8}$$

Then the truth function $j_1(v)$ has the required property viz. $j_1(1) = 1, j_1(v > 1) = n$. Indeed, as $h_{n-1}(1) = n$ and $h_{n-1}(v) = 1$ for $v > 1$, it follows that $j_1(1) = n - n + 1 = 1$ and $j_1(v) = n - 1 + 1 = n$ for $v > 1$.

To define remaining functors J_k, one has to introduce auxiliary parameter κ; given $1 < k \leq m$, we let $\kappa = \lceil \frac{m-k}{k-1} \rceil$. Clearly, κ is a decreasing function of k. Other properties of κ necessary in what follows are summed up below; the proofs are elementary and hence left to the reader.

$$\kappa < \frac{m - 1}{k - 1} \tag{4.9}$$

$$h_\kappa(k) = n - (k - 1) \cdot \kappa \tag{4.10}$$

$$1 < h_\kappa(k) \leq k \tag{4.11}$$

By property (4.11), either $k = h_\kappa(k)$ or $k > h_\kappa(k)$. Definition of J_k splits into two distinct formulae depending on which is the case.

For $1 < k < n - 1$, one defines J_k via

$$J_k(p) = J_1(C(H_\kappa(p) \vee p)(H_\kappa(p) \wedge p))$$

in case $k = h_\kappa(k)$ and

$$J_k(p) = J_{\kappa(k)}(H_{\kappa(k)}(p))$$

in case $k > h_\kappa(k)$.

Then the truth function j_k of J_k does satisfy the requirement that $j_k(k) = 1$ and $j_k(v \neq k) = n$.

Indeed, in case $k = h_\kappa(k)$, we have

$$j_k(k) = j_1(max\{1, max\{h_{\kappa(k)}, k\} - min\{h_{\kappa(k)}, k\} + 1\}) = j_1(1) = 1 \quad (4.12)$$

For $v < k$, we have

$$h_\kappa(v) > h_\kappa(k) = k > v \qquad\qquad (4.13)$$

so

$$j_k(v) = j_1(h_\kappa(v) - v + 1) = n \qquad\qquad (4.14)$$

In case $v > k$, we have $h_\kappa(v) < k < v$ so

$$j_k(v) = j_1(v - h_\kappa(v) + 1) = n \qquad\qquad (4.15)$$

In case $k > h_\kappa(k)$, we have

$$\begin{aligned} j_k(k) &= j_{h_\kappa(k)}(h_\kappa(k)) = 1 \\ j_k(v \neq k) &= n \ when \ h_\kappa(v) \neq h_\kappa(k) \end{aligned} \qquad (4.16)$$

The remaining functor J_n will be defined via the formula

$$J_n(p) = J_1(Np).$$

Then we have $j_n(n) = j_1(n - n + 1) = 1$ and for $v < n$ it follows that $j_n(v) = j_1(n - v + 1) = n$.

Thus, functors $J_1, ..., J_n$ are defined and their truth functions $j_1, ..., j_n$ satisfy the requirement that $j_k(k) = 1$ and $j_k(v) = n$ whenever $v \neq k$.

With the help of these functors, one may define new functors which would play some of the roles played by two–valued functors C, N in two–valued calculus of propositions. Notably, if one would like to preserve in an n–valued logic the property of two–valued negation that it does convert accepted statements into rejected ones and conversely then one should modify the functor N: as defined in the Łukasiewicz n–valued calculus by means of the truth function $f_2(v) = n - v + 1$ it does convert statements accepted with certainty into statements rejected with certainty while statements accepted or rejected in a degree are not necessarily converted into the opposite kind.

To restore this property, one may define a functor $--$ via

$$-- (p) = J_{s+1}(p) \vee ... \vee J_n(p)$$

where $\vee$ (denoted also A) is the functor of alternation (cf. Exercise 11). Then α is accepted if and only if $--\alpha$ is rejected and conversely. Let us mention that the truth function $n(v)$ of this negation is given by the formula

$$n(v) = max\{j_{s+1}(v), ..., j_n(v)\}$$

as here the truth functions of alternation A and conjunction K are adopted from the Łukasiewicz logic. (cf. Exercise 11, again).

Similarly, we may observe that the functor C defined above does not ensure the validity of detachment in the form:

if statements α and $C\alpha\beta$ are accepted then the statement β is accepted as well.

To remedy this, we may invoke theses $CCpq(Np \vee q)$ and $C(Np \vee q)Cpq$ of two–valued propositional logic and we may define a new functor $\sqsubset$ via

$$\sqsubset pq \text{ is } --p \vee q.$$

Then detachment rule is valid in the form:

if statements α and $\sqsubset \alpha\beta$ are accepted then the statement β is accepted

n–valued logics described above are completely axiomatizable; axiom schemes and a proof of completeness will be found in Rosser and Turquette [Rosser – Turquette58].

4.6   4-valued logic: modalities

The matrix method employed in constructing a 3–valued logic may be extended to higher dimensions; a simple method to construct higher–dimensional matrices adequate to propositional calculus is to carry out propositional calculus coordinate–wise. This was the idea of Jan Łukasiewicz who applied it to extend his 3–valued logic in order to accomodate *modal expressions*. Modal expressions were already familiar to Aristotle and the peripatetic school and analysis done then resulted in identifying two fundamental modal functors viz. the functor L of *necessity* and the functor M of *possibility*. Accordingly, we admit that given a meaningful expression p of the sentential calculus, expressions Lp (read *it is necessary that* p) and Mp (read *it is possible that* p) are also meaningful. As proposed by peripatetics, functors L, M are related to each other by means of equivalences: meaningful expressions Lp and $NMNp$ are equivalent and meaningful expressions Mp and $NLNp$ are equivalent (i.e. we accept statements *it is necessary that* p and *it is not possible that not* p as equivalent and similarly, we regard as equivalent statements *it is possible that* p and *it is not necessary that not* p).

What emerged as *basic modal logic* (the term is due to Łukasiewicz [Lukasiewicz53, 57]) from studies by Aristotle and other members of the peripatetic school is a system of postulates about L, M:

1. $\vdash CMpNLNp$

2. $\vdash CNLNpMp$

3. $\vdash CLpNMNp$

4. $\vdash CNMNpLp$

5. $\vdash CLpp$

6. $\vdash CpMp$

7. $\dashv CMpp$

8. $\dashv CpLp$

9. $\dashv Mp$

10. $\dashv NLp$

This list requires some comments. The idea of acceptance and rejection familiar to us from a discussion of n–valued logic is expressed here with the help of Frege symbols $\vdash$ of assertion and $\dashv$ of rejection (cf. [Lukasiewicz57]). Meaningful expressions following the acceptance symbol are accepted and they should be theses (a fortiori theorems) of any system of modal logic while meaningful expressions following the rejection symbol should not be theses nor theorems of any system of modal logic.

Statements 1–4 express the equivalence of functors L with NMN and M with NLN already mentioned in the introductory paragraph. Statement 5 does express the conviction that from the necessity of a proposition the proposition itself should be asserted and Statement 6 does express the same for propositions p and Mp. Statements 7, 8 does express rejection of expressions $CMpp$, $CpLp$ and Statements 9, 10 express respectively that Mp is not

C	1	3	2	0	N
1	1	3	2	0	0
3	1	1	2	2	0
2	1	3	1	3	3
0	1	1	1	1	1

Figure 4.3: Truth table of the 4-valued logic

always true and that Lp is not always false. All together these Statements express a certain interpretation of modal utterances *it is necessary that...*, *it is possible that.....*

We now present a system of 4–valued logic in which modal functors L, M are constructed. This system is due to Jan Łukasiewicz [Lukasiewicz53, 57]. As already mentioned the idea for constructing such a system is to consider as meaningful expressions of a new system pairs (p, q) of meaningful expressions of the sentential calculus and to apply sentential functors coordinate–wise.

To be more precise, we declare, following [Łukasiewicz opera cit.], as meaningful expressions in this case pairs (p, q) of meaningful expressions of the sentential calculus and we extend functors C, N by letting

$$C(p, q)(r, s) = (C(p, r), C(q, s))$$

and

$$N(p, q) = (Np, Nq).$$

The truth matrix for this calculus is the following (cf. Fig. 4.3, we adopt substitutions $(1, 1) = 1$, $(1, 0) = 3$, $(0, 1) = 2$, $(0, 0) = 0$).
    Clearly theses of propositional calculus are those meaningful expressions whose truth value is 1 at any substitution of truth values for propositional variables i.e. the above matrix is *adequate* for the sentential calculus.

Now, we may observe that in addition to N, we may have three more logical functions of one variable viz. V, F, S defined as follows

1. $V(0) = 1, V(1) = 1$

2. $F(0) = 0, F(1) = 0$

3. $S(0) = 0, S(1) = 1$

    It was the idea of Jan Łukasiewicz to define functors of two propositional variables by means of the formula

$$P(p, q) = (X(p), Y(q)) \qquad (4.17)$$

p	1	3	2	0
M	1	1	2	2

Figure 4.4: Truth table of the functor M

p	1	3	2	0
L	3	3	0	0

Figure 4.5: Truth table of the functor L

where X, Y are among V, N, S, F. In particular he defined the functor $M(p,q) = (Vp, Sq) = (Cpp, q)$ whose truth value matrix is shown in Fig. 4.4.

4.7 Modalities

If the above introduced functor M is interpreted as the possibility functor then the functor L of necessity is defined according to the equivalence of L with NMN by means of the truth matrix Fig. 4.5.

Derivation rules in this system are detachment and substitution for accepted statements and detachment and substitution for rejected statements, the latter in the following form

1. (⊣ MP) If $C\alpha\beta$ is accepted and β is rejected than α is rejected

2. (⊣ S) If β results from α via substitution and β is rejected than α is rejected

Theorems of this system are those meaningful expressions whose truth value at any substitution of truth values for propositional variables is 1; they are accepted and all other meaningful expressions are rejected. The system proposed by Lukasiewicz is complete in the sense : any meaningful expression is either accepted or rejected (cf. [Lukasiewicz53, 57]) We denote this system with the symbol (L_4).

We may look at some modal statements which are accepted in this system. Asserted meaningful expressions in the list of 1–10 above are accepted and rejected meaningful expressions in this list are rejected; we propose to verify this in Exercises that follow.

In the next Chapter we will present some systems of modal propositional logic, traditionally denoted $K, T, S4$. They are defined by means of some axiom schemes. We will check that these axiom schemes are satisfied in the 4–valued logic of Lukasiewicz .

The axiom scheme (K) is the following

(K) $CLCpqCLpLq$

We will try to find a valuation falsifying (K) i.e. such that for this valuation (K) has the truth value other than 1. As $LCpq$ may take truth values $3, 0$ only it follows that we need to check only the case when the truth value of $LCpq$ is 3 as the matrix for C shows that implication whose premise has the truth value 0 is true. In this case the truth value of Cpq is $1, 3$. To falsify (K), we should have the case when $CLpLq$ takes the truth value from $0, 2$. The only case when $CLpLq$ takes the truth value 0 is $Lp = 1, Lq = 0$ which is impossible. Thus the case when $CLpLq$ takes the truth value 2 remains in which case we are left with the following truth values for the pair (p, q): $(1, 2), (1, 0), (3, 2), (3, 0)$; but in all these cases Cpq takes the truth value $0, 2$ and then (K) has the truth value 1.

It follows that (K) is accepted by the system ($Ł_4$).

Next we consider the axiom scheme (T).

(T) $CLpp$

This is Statement 5 cf. Exercise 21.

The axiom scheme (S4) is as follows

(S4) $CLpLLp$

As truth values of LLp are identical with truth values of Lp (cf. the matrix of truth values for L) it follows that (S4) is a theorem of ($Ł_4$).

Finally, we may consider the axiom scheme

(S5) $CMpLMp$

We verify this proposition in its equivalent dual form $CMLpLp$ (cf. Exercise 37 and Chapter 5, p. 195). It turns out that assertions 1–8 above are not sufficient to derive from them (S5). Indeed, in case $p = 1$, we find that Lp takes the truth value 3 and MLp has the truth value 1 so $CMLpLp$ has the truth value 3 i.e. it is rejected.

Thus (S5) is not a theorem of the basic modal logic. We will discuss this axiom scheme in the next Chapter, giving it a semantic interpretation.

Historic remarks

First time many–valued logic was mentioned in print by Łukasiewicz [Łukasie-wicz18]. The first account of 3–valued logic of Łukasiewicz was given in [Łukasiewicz20]. A distinct system of n–valued logic was proposed by Post[Po-st21]. The formulae for the Łukasiewicz implication and negation were given in [Łukasiewicz30^a] and then repeated in [Łukasiewicz30^b]. The axiom schemes for 3–valued logic of Łukasiewicz were given by Wajsberg [Wajsberg31] along with the proof of completeness. The proof presented here comes from [Gold-berg–Leblanc–Weaver74] cf. also [Boicescu91]. Properties of the Wajsberg system were expounded in [Wajsberg31] and in Becchio [Becchio72, 78]. Functional completeness of many–valued logics augmented with the func-tor T was proved by Słupecki[Słupecki36]. A complete axiomatization for n–valued propositional logic was presented by Rosser and Turquette [Rosser – Turquette58].

Modal propositions were studied by Aristotle in *Prior Analytics* and *De In-terpretatione* and by his commentators cf. [Łukasiewicz57]. The basic modal system was proposed by Łukasiewicz [Łukasiewicz53, 57].

Exercises

In the first part, Exercises 1–22, we propose to prove facts about the sys-tem (W) listed as (I)–(XIII) along with some auxiliary facts (i)–(ix).

1. Prove (I) $p \vdash Cqp$ [Hint: apply (W1)]

2. Prove (III) $Cpq \vdash CCqrCpr$ [Hint: apply (W2)]

3. Prove (IV) $Cpq, Cqr \vdash Cpr$ [Hint: apply detachment to (III)]

4. Prove (II) $\vdash CNpCpq$ [Hint: substitute $p/Np, q/Nq$ in (W1) and $p/q, q/p$ in (W3) and apply (IV)]

5. Prove (V) $\vdash CNNpCqp$ [Hint: substitute $p/Np, q/Nq$ in (II) and ap-ply (IV) with the result and (W4)]

6. Prove (VI) $\vdash CCCpqqCNpq$ [Hint: substitute $p/Np, q/Cpq, r/q$ in (III) and use (II)]

7. Prove (VII) $\vdash CNNpp$ [Hint: substitute $q/CpNp$ in (V) and apply the result and (W3) with (IV)]

8. Prove (VIII) $\vdash CpNNp$ [Hint: substitute p/Np in (VII) and $p/NNp, q/p$

in (W4) and apply detachment]

9. Prove (IX) ⊢ Cpp [Hint: apply (VII), (VIII), and (IV)]

10. Prove (XI) ⊢ $CCpqCNNpq$ [Hint: substitute in (III) $p/NNp, q/p$, r/q and invoke (VII)]

11. Prove (i) p ⊢ $CCpqq$ [Hint: substitute in (I): $q/CqNq$ and then in (I): $p/Cpq, q/CCqNqp$. In (W3), substitute p/q and apply (IV)]

12. Prove (ii) $p, CqCpr$ ⊢ Cqr [Hint: in (i), substitute q/r and apply (IV) with substitutions $p/q, q/Cpr$]

13. Prove (iii) ⊢ $CCpqCCCprsCCqrs$ [Hint: in (W2), substitute p/Cqr, $q/Cpr, r/s$ and apply (IV)]

14. Prove (iv) ⊢ $CpCCpqq$ [Hint: in (iii), substitute $p/Cpp, q/p, r/q$, s/q, in (IV), substitute $q/CCppp, r/CCCCppqqCCpqq$. In (i), substitute p/Cpp and apply detachment in virtue of (IX). Then substitute in (ii) $p/CCCppqq, q/p, r/CCpqq$ and apply detachment]

15. Prove (v) ⊢ $CCCCqrrCprCqCpr$ [Hint: in (iv), substitute p/q, q/r and in (III) substitute $p/q, q/CCqrr, r/Cpr$ then apply detachment]

16. Prove (vi) ⊢ $CCpCqrCqCpr$ [Hint: in (W2), substitute q/Cqr and apply the result along with (v) via (IV)]

17. Prove (vii) $CpCqr$ ⊢ $CqCpr$ [Hint: apply detachment to (vi)]

18. Prove (viii) ⊢ $CCqrCCpqCpr$ [Hint: in (vi), substitute p/Cpq, $q/Cqr, r/Cpr$ and apply (W2) and detachment]

19. Prove (ix) Cqr ⊢ $CCpqCpr$ [Hint: apply detachment to (viii)]

20. Prove (XII) ⊢ $CCNNpqCNNpNNq$ [Hint: in (VIII), substitute p/q and in (ix), substitute $p/NNp, q/NNq$ then apply detachment]

21. Prove (XIII) ⊢ $CCpqCNNpNNq$ [Hint: apply (IV) to (XI) and (XII)]

22. Prove (X) ⊢ $CCpqCNqNp$ [Hint: in (W4), substitute $p/Np, q/Nq$, and apply (IV) to the result and (XIII)]

In Exercises 23–26, we are concerned with n–valued logics. In Exercise 23, we refer to the logic (L_3) and we recall that F_1pq denotes Cpq and F_2p

denotes Np with truth functions of F_1, F_2 equal respectively to $f_1(p,q) = min\{1, 1 - p + q\}, f_2(p) = 1 - p$. In Exercises 24–26, we refer to n–valued logics in the Rosser–Turquette formalization and with truth functions defined there as dual to f_1, f_2.

23. (Łukasiewicz) In n–valued logics of Łukasiewicz, functors of alternation and conjunctions, denoted A, K respectively are defined via

1. Apq is $CCpqq$

2. Kpq is $NANpNq$

(i) Prove that truth functions a, k of A, K, respectively, are given by the following formulae

1. $a(p,q) = max\{p,q\}$

2. $k(p,q) = min\{p,q\}$

[Hint: verify the formula $a(p,q) = min\{1, 1 - min\{1, 1 - p + q\} + q\}$ and check that $a(p,q) = q$ when $q \geq p$ and $a(p,q) = p$ when $q \leq p$; similarly for $k(p,q)$]

(ii) Prove that in n–valued logic as formalized according to Rosser–Turquette, truth functions of A, K are expressed as dual to a, k above i.e.

1. $a(p,q) = min\{p,q\}$

2. $k(p,q) = max\{p,q\}$

24. [Rosser – Turquette58] Prove that the negation functor $--(p) = J_{s+1}(p) \vee ... \vee J_n(p)$ defined above has indeed the property stated there viz. it does convert accepted expressions into rejected ones and vice versa.

25. [Rosser – Turquette58] Give an example for the thesis that the functor C does not guarantee the validity of detachment in the following form: if meaningful expressions α and $C\alpha\beta$ are accepted then the meaningful expression β is accepted.

26. [Rosser – Turquette58] Prove that the detachment rule holds in the form: if meaningful expressions α and $\sqsubset \alpha\beta$ are accepted then the meaningful expression β is accepted.

In Exercises 27–28, we sketch the argument for functional incompleteness of n–valued logics and we outline a remedy as proposed by Słupecki [Słupecki36].

27. Justify the thesis: n–valued logics are functionally incomplete i.e. truth functions of basic functors F_1, F_2 (i.e. of C, N) are insufficient in order to

define via them all possible truth functions on truth values $1, 2, ..., n$. [Hint: observe that any function composed of F_1, F_2 takes on either the value 1 or the value n in case arguments i.e. truth values of propositional variables are restricted to $1, n$ (e.g. in case (L_3), we cannot produce the constant function 2)]

28. [Słupecki36] Prove that n–valued logic becomes functionally complete on adding to F_1, F_2 the functor T whose truth function t is constant with the value 2. [Hint: a. Define functors $G_j p$ for $j = 1, 2, ..., n$ via

1. $G_1 p = F_1 pp$

2. $G_j p = F_1 (H_j (Tp) F_2 (Tp))$ for $j = 2, ..., n - 1$

3. $G_n p = F_2 (F_1 pp)$

b. Prove $g_j(p) = j$ for $j = 1, 2, ..., n$ where g_j is the truth function of G_j (this is obvious for $j = 1, n$ and for other j calculate $g_j(p) = f_1(h_j(2), f_2(2)) = f_1(max\{1, n - j\}, n - 1) = j$).

c. Consider an arbitrary truth function $g(p)$ of one variable and let $E_g(p) = G_{g(1)}(p) \wedge J_1(p) \vee ... \vee G_{g(n)}(p) \wedge J_n(p)$; prove that $e_g(p) = g(p)$ where e_g is the truth function of E_g. To this end, observe that $g(i) = j$ implies $e_g(i) = min\{max\{g_{g(1)}(i), j_1(i)\}, ..., max\{g_{g(n)}(i), j_n(i)\}\} = min\{n, j\} = j = g(i)$.

d. Consider an arbitrary truth function g of $m + 1$ variables: apply induction and assume that any function of m variables is expressible via F_1, F_2, T. For $i = 1, 2, ..., n$, consider the function $g_i = g(i, p_2, ..., p_{m+1})$ of m variables along with its realization $Q_i(p_2, ..., p_{m+1})$ in terms of F_1, F_2, T having truth function $q_i = g_i$. Let $Q(p_1, p_2, ..., p_{m+1}) = J_1(p_1) \wedge Q_i(p_2, ..., p_{m+1}) \vee ... \vee J_n(p_1) \wedge Q_n(p_2, ..., p_{m+1})$, denote with q the truth function of Q and prove along the lines of c. that $q(i, p_2, ..., p_{m+1}) = g(i, p_2, ..., p_{m+1})$]

In Exercises 29–37, we refer to the modal propositional logic (L_4). We propose to verify that statements 1–10 are valid in this logic (doing this, use the completeness theorem and check the semantic value of each expression) and also we propose to prove the dual form of ($S5$).

29. Prove that $CMpNLNp$ is accepted in the logic (L_4).

30. Prove that $CNLNpMp$ is accepted in the logic (L_4).

31. Prove that $CLpNMNp$ is accepted in the logic (L_4).

32. Prove that $CNMNpLp$ is accepted in the logic (L_4).

32. Prove that $CLpp$ is accepted in the logic ($\mathbf{L}_4$).

32. Prove that $CpMp$ is accepted in the logic ($\mathbf{L}_4$).

33. Prove that $CMpp$ is rejected in the logic ($\mathbf{L}_4$).

34. Prove that $CpLp$ is rejected in the logic ($\mathbf{L}_4$).

35. Prove that Mp is rejected in the logic ($\mathbf{L}_4$).

36. Prove that NLp is rejected in the logic ($\mathbf{L}_4$).

37. Verify the equivalence of $CMpLMp$ and $CMLpLp$. [Hint: use theses of the sentential calculus along with Statements 1–4, p. 173]

Works quoted

[Becchio78] D. Becchio, *Logique trivalente de Lukasiewicz*, Ann. Sci. Univ. Clermont–Ferrand, 16(1978), pp. 38–89.

[Becchio72] D. Becchio, *Nouvelle démonstration de la complétude du systeme de Wajsberg axiomatisant la logique trivalente de Lukasiewicz*, C.R. Acad. Sci. Paris, 275(1972), pp. 679–681.

[Boicescu91] V. Boicescu, A. Filipoiu, G. Georgescu, and S. Rudeanu, *Lukasiewicz–Moisil Algebras*, North Holland, Amsterdam, 1991.

[Borkowski70] *Jan Łukasiewicz. Selected Works*, L. Borkowski (ed.), North Holland–Polish Scientific Publishers, Amsterdam–Warsaw, 1970.

[Goldberg–Leblanc–Weaver74] H. Goldberg, H. Leblanc, and G. Weaver, *A strong completeness theorem for 3–valued logic*, Notre Dame J. Formal Logic, 15(1974), 325–332.

[Lukasiewicz57] J. Łukasiewicz, *Aristotle's Syllogistic from the Standpoint of Modern Formal Logic*, Oxford, 1957.

[Lukasiewicz53] J. Łukasiewicz, *A system of modal logic*, The Journal of Computing Systems, 1(1953), 111–149.

[Lukasiewicz30^a] J. Łukasiewicz, *Philosophische Bemerkungen zu mehrwertige Systemen des Aussagenkalküls*, C. R. Soc. Sci. Lettr. Varsovie, 23(1930), 51–77 [English translation in [Borkowski70], pp. 153–178].

[Łukasiewicz30^b] J. Łukasiewicz and A. Tarski, *Untersuchungen ueber den Aussagenkalkül*, C. R. Soc. Sci. Lettr. Varsovie, 23(1930), 39–50 [English translation in [Borkowski70], pp. 130–152].

[Łukasiewicz20] J. Łukasiewicz, *On three–valued logic* (in Polish), Ruch Filozoficzny 5(1920), 170–171 [English translation in [Borkowski70], pp. 87–88].

[Łukasiewicz18] J. Łukasiewicz, *Farewell Lecture by Professor Jan Lkasiewicz* (delivered in the Warsaw University Lecture Hall on March 7, 1918) [English translation in [Borkowski70], pp. 84–86].

[Post21] E. Post, *Introduction to a general theory of elementary propositions*, Amer. J. Math., 43(1921), 163–185.

[Rosser – Turquette58] J. B. Rosser and A. R. Turquette, *Many–valued Logics*, North Holland, Amsterdam, 1958.

[Słupecki36] J. Słupecki, *Der volle dreiwertige Aussagenkalkul*, C. R. Soc. Sci. Lettr. Varsovie, 29(1936), 9–11.

[Wajsberg31] M. Wajsberg, *Axiomatization of the three–valued sentential calculus* (in Polish, Summary in German), C. R. Soc. Sci. Lettr. Varsovie, 24(1931), 126–148.

Chapter 5

Propositional Modal Logic

This brings us to the Spindle–destiny, spun according to the ancient by the Fates. [...]; the Fates, with Necessity, Mother of the Fates, manipulate it and spin at the birth of every being so that all comes into existence through Necessity

Plotinus, *the Enneads*, II.3.9

5.1   Introduction

We witnessed in the previous Chapter a system of basic modal logic constructed in the framework of a 4–valued logic. In addition to propositional connectives, studied by us in Chapter 2 on propositional logic, some other statements common in everyday speech and reasoning were studied there. These statements called also *modal statements* encompass such phrases as *it is necessary that.., it is possible that....* These statements were interpreted semantically by means of truth functions of modal functors L (necessity) and M (possibility). We have seen that basic modal system have accomodated some basic postulates about modalities while rejecting some less natural postulates e.g. $(S5)$: $CMpLMp$ (*what is possible is necessarily possible*). It is desirable to look for other systems with eventually distinct semantics which would accept some of expressions rejected by the basic modal system.

We now present a survey of systems of modal propositional logic along with an approach to their semantics called *many world semantics* or *Kripke semantics*.

5.2   The system K

Modal logics of propositions are constructed by adding to calculus of propositions of new expressions involving symbols for modalities: L for necessity and M for possibility. Symbols of this calculus will be then symbols $p, q, r, ...$ for propositional variables, symbols C, N for propositional functors of implication, respectively negation, and symbols L, M along with various kinds of parentheses, etc. We admit some definitions i.e.

1. (D1) $p \vee q = CNpq$ for defining alternation

2. (D2) $p \wedge q = NCpNq$ for defining conjunction

3. (D3) $Mp = NLNp$ for expressing possibility via necessity

4. (D4) $Lp = NMNp$ for expressing necessity via possibility

As we deal with propositions we recall that we already have at our disposal facts about this calculus presented in Chapter 2. In particular, we do not check that propositions we invoke are theorems i.e. they are derivable in propositional calculus (it may be checked in each case that they are).

We begin with the modal propositional system K whose axioms are axioms

(T1) $CCpqCCqrCpr$ (or, $(p \Rightarrow q) \Rightarrow [(p \Rightarrow r) \Rightarrow (q \Rightarrow r)]$)

(T2) $CCNppp$ (or, $(\neg p \Rightarrow p) \Rightarrow p$)

(T3) $CpCNpq$ (or, $p \Rightarrow (\neg p \Rightarrow q)$)

of propositional calculus and the axiom

(T4) $CLCpqCLpLq$

The axiom (T4) is usually denoted (K) and it reflects a basic postulate about necessity: if one admits the necessity of the implication Cpq then upon admitting necessity of p one has also to admit necessity of q.

Derivation rules applied in this calculus are the usual rules of detachment (modus ponens) and substitution known from propositional calculus and a new rule of *necessitation*:

Necessitation For any meaningful expression α, if α is a thesis of the system then $L\alpha$ is also a thesis of the system.

Given axioms and derivation rules, one may derive from axioms new theses of the system. We will give a formal definition of a derivation (a proof) of a thesis. For a meaningful expression α, a *derivation* of α from axioms is a sequence $\alpha_0, \alpha_1, .., \alpha_n$ of meaningful expressions such that α_0 is an axiom, α_n is equiform with α, and each α_i for $i = 1, 2, ..., n$ is obtained either from some α_j with $j < i$ by means of substitution or necessitation, or from some α_j, α_k with $j, k < i$ by means of detachment.

A meaningful expression having a derivation from axioms is called *a thesis* of the axiomatic system. To denote the fact that α is a thesis, we use the acceptance symbol $\vdash$ so $\vdash \alpha$ reads "α is a thesis".

Let us observe that the deduction theorem formulated and proved for propositional calculus is valid in this modal propositional system. Let us notice the following properties of this system.

Proposition 5.1. *The following statements about the system K are valid*

1. *if* $\vdash Cpq$ *then* $\vdash CLpLq$

2. $\vdash CL(p \vee q)(Lp \vee Lq)$

3. $\vdash CL(p \wedge q)(Lp \wedge Lq)$

4. $\vdash C(Lp \wedge Lq)L(p \wedge q)$

Proof. Statement 1 follows by necessitation, detachment and the axiom (K): given $\vdash Cpq$, it follows by necessitation that $\vdash LCpq$ and from (K) it follows via detachment that $\vdash CLpLq$. In case of Statement 2, it follows from theses $Cp(p \vee q), Cq(p \vee q)$ by necessitation that $\vdash LCpL(p \vee q), LCqL(p \vee q)$ and from these two statements we obtain by the thesis $CCutCCrtC(u \vee r)t$, in which we substitute $u/Lp, t/L(p \vee q), r/Lq$, and detachment applied twice that $\vdash CL(p \vee q)(Lp \vee Lq)$.

For Statements 3,4, we first observe that from theses $C(p \wedge q)p$, $C(p \wedge q)q$ it follows that

$$\vdash CL(p \wedge q)(Lp \wedge Lq) \tag{5.1}$$

We use the symbol $\Leftrightarrow$ of the equivalence functor (*if and only if*) with which Statements 3, 4 may be jointly written down as $\vdash Lp \wedge Lq \Leftrightarrow L(p \wedge q)$.

Applying to a tautology $CpCq(p \wedge q)$ already proven Statement 1, we get $\vdash CLpLCq(p \wedge q)$. Substituting in (K) $p/q, q/p \wedge q$, we get $CLCq(p \wedge q)CLqL(p \wedge q)$. In the thesis $CCutCCtrCur$, we substitute

$$u/Lp, t/LCq(p \wedge q), r/CLqL(p \wedge q),$$

to obtain

$$CCLpLCq(p \wedge q)CCLCq(p \wedge q)CLqL(p \wedge q)CLpCLqL(p \wedge q) \tag{5.2}$$

Detaching twice from the last expression two previous expressions we get $\vdash CLpCLqL(p \wedge q)$. By the thesis $CCuCtrC(u \wedge t)r$, where we substitute $u/Lp, t/Lq, r/L(p \wedge q)$ from which we detach $CLpCLqL(p \wedge q)$, we get

$$\vdash CLp \wedge LqL(p \wedge q) \tag{5.3}$$

From equations 5.1, 5.3, by the thesis $CCpq \wedge Cqpp \Leftrightarrow q$ we arrive at Statements 3,4:

$$\vdash Lp \wedge Lq \Leftrightarrow L(p \wedge q).$$

$\square$

The system K is *consistent*: it is not possible to have α and $N\alpha$ as theses of the system for any meaningful expression α. To see this it is best to resort to the idea of *collapse* (proposed by S. Leśniewski cf. [Łukasiewicz57]): to every meaningful expression α of the system K, we assign a meaningful expression $\gamma(\alpha)$ of the sentential calculus obtained by deleting in α all occurrences of modal connectives L, M.

Then clearly $\gamma(N\alpha)$ is $N\gamma(\alpha)$ and $\gamma(C\alpha\beta)$ is $C\gamma(\alpha)\gamma(\beta)$. In particular the axiom (K) becomes the thesis Cpp. It follows that if α is a thesis of the system K then $\gamma(\alpha)$ is a thesis of the sentential calculus. From this observation the consistency claim follows: were $\alpha, N\alpha$ theses of the system K, then $\gamma(\alpha), N\gamma(\alpha)$ would be theses of the sentential calculus which would contradict its consistency.

Now, about semantics for the system K; we introduce the *Kripke* semantics. This semantics is distinct from the 4–valued semantics of Łukasiewicz in that it values modalities in a relational system. We may recall that (K) is a thesis of the basic modal system. A systematic discussion of relations will be found in Chapter 6 on Set Theory.

We will call an *interpretation frame* for the system K a triple $\mathcal{M} = (W, R, v)$ in which

- W is a set of objects called *possible worlds*

- $R \subseteq W$ is a relation on W

- v is a function which assigns to any pair (w, p), where $w \in W$ and p is a propositional variable, a value $v(w, p) \in \{0, 1\}$

The function v is a *valuation* for a given M and in case wRw' we say that w' is *accessible* from w.

Once an interpretation frame M is set, we may define inductively the meaning $[\alpha]_w^M$ of a meaningful expression α of the system (K) in M relative to a world w. To this end, we let

$$[p]_w^M = v(w, p) \tag{5.4}$$

$$[N\alpha]_w^M = 1 \; iff \; [\alpha]_w^M = 0 \tag{5.5}$$

$$[C\alpha\beta]_w^M = 1 \; iff \; [\alpha]_w^M = 0 \; or \; [\beta]_w^M = 1 \tag{5.6}$$

$$[L\alpha]_w^M = 1 \; iff \; [\alpha]_{w^*}^M = 1 \; for \; every \; w^* \in W \; such \; that \; wRw^* \tag{5.7}$$

Let us observe that the first three conditions realize in every world the boolean semantics for the sentential calculus while the fourth condition is different in character: in order that $L\alpha$ be true in a world w it is necessary that α be true in every world w^* accessible from w.

We say that a meaningful expression α is *true* in a given M, in symbols $\models_M \alpha$, if and only if $[\alpha]_w^M = 1$ for every $w \in W$. A meaningful expression α is *true*, in symbols $\models \alpha$, if and only if $\models_M \alpha$ for every interpretation frame M.

Proposition 5.2. *For any meaningful expression α of the system K, if α is a thesis then α is true.*

Proof. As with propositional calculus, it suffices to check that all axioms are true and all derivation rules yield true expressions when applied to true expressions. As axioms of propositional calculus are true in every world by default, they are true. Consider the axiom (K); for $M = (W, R, v)$ and $w \in W$, assume that $[LCpq]_w^M = 1$ and $[Lp]_w^M = 1$ so $[Cpq]_{w^*}^M = 1$ and $[p]_{w^*}^M = 1$ for each w^* with wRw^*. Then $[q]_{w^*}^M = 1$, each w^* with wRw^*, and thus $[Lq]_w^M = 1$ hence $[CLCpqCLpLq]_w^M = 1$. It follows that (K) is true. Clearly, detachment preserves truth so only necessitation needs a checking. If α is true then given M, w as above, we have $[\alpha]_{w^*}^M = 1$ any w^* hence $[L\alpha]_w^M = 1$. Thus $L\alpha$ is true. It follows that any thesis of the system (K) is true. $\square$

The converse is known as with propositional calculus, as the *completeness* problem. We will prove that the system K is complete i.e. every true meaningful expression is a thesis. To this end, we apply the technique of *canonical models* due to Lemmon and Scott [Lemmon–Scott63]. Informally speaking, these are interpretation frames in which worlds are construed as maximal consistent sets of meaningful expressions.

Let us recall that a set Γ of meaningful expressions is *consistent* if and only if there is no meaningful expression α with $\Gamma \vdash \alpha$ and $\Gamma \vdash N\alpha$. In this case we use a meta–predicate Con and $Con(\Gamma)$ would mean that Γ is consistent.

We will say that Γ is *maximal consistent* if and only if $Con(\Gamma)$ and for every consistent set Δ if $\Gamma \subseteq \Delta$ then $\Gamma = \Delta$. We use a meta–predicate Con_{max} to denote this fact so $Con_{max}(\Gamma)$ would mean that Γ is maximal consistent. We list some basic properties of maximal consistent sets.

Proposition 5.3. *For Γ with $Con_{max}(\Gamma)$, the following hold*

1. *for every α: $\alpha \in \Gamma$ if and only if $N\alpha \notin \Gamma$*

2. *$\alpha \vee \beta \in \Gamma$ if and only if $\alpha \in \Gamma$ or $\beta \in \Gamma$*

3. *$\alpha \wedge \beta \in \Gamma$ if and only if $\alpha \in \Gamma$ and $\beta \in \Gamma$*

4. *if $\vdash \alpha$ then $\alpha \in \Gamma$*

5. *if $\alpha \in \Gamma$ and $C\alpha\beta \in \Gamma$ then $\beta \in \Gamma$*

Proof. In case 1, assume that $\alpha \notin \Gamma$ for a meaningful expression α. This means that $\Gamma \cup \{\alpha\}$ is not consistent, a fortiori for a finite subset $\Lambda \subseteq \Gamma$ we have $\Lambda, \alpha \vdash \delta$ and $\Lambda, \alpha \vdash N\delta$. By the deduction theorem, we have $\Lambda \vdash C\alpha\delta$, $\Lambda \vdash C\alpha N\delta$. By the thesis $CCpqCCpNqp$ in which we substitute $p/\alpha, q/\delta$ we have applying detachment twice that $\Lambda \vdash N\alpha$.

It follows that $\Gamma \cup N\alpha$ is consistent hence $N\alpha \in \Gamma$.

For 2, assume to the contrary $\alpha \vee \beta \in \Gamma$, $\alpha \notin \Gamma$ and $\beta \notin \Gamma$. Then – by already proven 1 – $N\alpha, N\beta \in \Gamma$ and by the thesis $CpCq(p \wedge q)$ with substitutions $p/N\alpha, q/N\beta$ we obtain applying detachment twice that $\Gamma \vdash N\alpha \wedge N\beta$ i.e. $\Gamma \vdash N(\alpha \vee \beta)$ hence Γ is not consistent, a contradiction.

The proof of 3 is on the same lines. For 4, it is manifest that $\vdash \alpha$ implies $\Gamma \cup \{\alpha\}$ consistent hence by maximality $\alpha \in \Gamma$. The case 5 follows from 1 as $C\alpha\beta$ is equivalent to $N\alpha \vee \beta$ so $\alpha \in \Gamma$ implies $\beta \in \Gamma$.   □

Let us also observe that a set Γ of meaningful expressions is consistent if and only if for no finite subset $\Lambda \subseteq \Gamma$ we have $\vdash N \bigwedge \Lambda$. Indeed, to see necessity of this condition, assume Γ consistent and $\vdash N \bigwedge \Lambda$ for some finite subset $\Lambda \subseteq \Gamma$. By the thesis $CpCq(p \wedge q)$ applied $|\Lambda| - -1$ times, we get $\Gamma \vdash \bigwedge \Lambda$ hence $\Gamma \vdash \bigwedge \Lambda, N \bigwedge \Lambda$ contradicting the consistency of Γ.

To see sufficiency, assume that Γ is not consistent i.e. $\Gamma \vdash \beta, N\beta$ for some meaningful expression β. By compactness, $\Lambda \vdash \beta, N\beta$ for some finite $\Lambda \subseteq \Gamma$. Let $\Lambda = \{\gamma_1, ..., \gamma_k\}$; by the thesis $Cp_1 \wedge ... \wedge p_k p_i$ for $i = 1, 2, ..., k$, it follows that $\bigwedge \Lambda \vdash \beta, N\beta$ and then the thesis $CCpq \wedge CpNqNp$ implies that $\vdash N \bigwedge \Lambda$. Thus Γ is not consistent.

We now define a canonical interpretation frame $M^c = (W^c, R^c, v^c)$ by letting

- $W^c = \{\Gamma : Con_{max}(\Gamma)\}$ i.e. worlds are maximal consistent sets of meaningful expressions

- $\Gamma R^c \Gamma'$ if and only if $L\alpha \in \Gamma$ implies $\alpha \in \Gamma'$ for every meaningful expression α

- $v^c(\Gamma, p) = 1$ if and only if $p \in \Gamma$

We then have

Proposition 5.4. *For each meaningful expression α of the system K, we have $[\alpha]_\Gamma = 1$ if and only if $\alpha \in \Gamma$.*

Proof. We will induct on structural complexity of α. We will assume that α, β already satisfy our claim. For $N\alpha$, we have $[N\alpha]_\Gamma = 1$ if and only if $[\alpha]_\Gamma = 0$ i.e. $\alpha \notin \Gamma$ which by Proposition 5.3 (1) is equivalent to $N\alpha \in \Gamma$.

For the expression $C\alpha\beta$, we have $[C\alpha\beta]_\Gamma = 1$ if and only if $[\alpha]_\Gamma = 0$ or $[\beta]_\Gamma = 1$ i.e. $\alpha \notin \Gamma$ or $\beta \in \Gamma$ i.e. $N\alpha \vee \beta \in \Gamma$ by Proposition 5.3 (3) i.e. $C\alpha\beta \in \Gamma$.

Finally, we consider the expression $L\alpha$; there are two cases (i) $L\alpha \in \Gamma$ (ii) $L\alpha \notin \Gamma$. In case (i), for every Γ' with $\Gamma R^c \Gamma'$ we have $\alpha \in \Gamma'$ i.e. $[\alpha]_{\Gamma'} = 1$ from which $[L\alpha]_\Gamma = 1$ follows. In case (ii), when $L\alpha \notin \Gamma$, we have $NL\alpha \in \Gamma$. We check that the following holds

Claim $Con(\{\delta : L\delta \in \Gamma\} \cup \{N\alpha\})$.

Indeed, was the contrary true, we would have a finite set $\{\gamma_1, ..., \gamma_k\}$ with $\{L\gamma_1, .., L\gamma_k\} \subseteq \Gamma$ such that $\vdash N(\gamma_1 \wedge ... \wedge \gamma_k \wedge N\alpha)$ i.e. $\vdash C\gamma_1 \wedge ... \wedge \gamma_k \alpha$ hence $\vdash LC\gamma_1 \wedge ... \wedge \gamma_k \alpha$ by necessitation and thus it follows by the axiom (K) that $\vdash CL\gamma_1 \wedge ... \wedge L\gamma_k L\alpha$ implying that $\vdash N(L\gamma_1 \wedge ... \wedge L\gamma_k \wedge NL\alpha)$ contrary to the facts that $Con(\Gamma)$ and $\{L\gamma_1, ..., L\gamma_k, NL\alpha\} \subseteq \Gamma$. Our claim is proved.

Now, as $\{\delta : L\delta \in \Gamma\} \cup \{N\alpha\}$ is consistent it extends to a maximal consistent set Γ'; clearly, $\Gamma R^c \Gamma'$; as $N\alpha \in \Gamma'$ we have $[\alpha]_{\Gamma'} = 0$ and finally $[L\alpha]_\Gamma = 0$. The proposition is proved. □

We may now state the main result of this discussion.

Proposition 5.5. *(The Completeness Theorem)* $\models \alpha$ *implies* $\vdash \alpha$ *for every meaningful expression α.*

Proof. Implication: *if* $\models \alpha$ *then* $\vdash \alpha$ is equivalent to the implication (i) *if non* $\vdash \alpha$ *then non* $\models \alpha$ or, equivalently, *if non* $\vdash \alpha$ *then there exists* $M = (W, R, v)$ *with* $[\alpha]_w = 0$ *for some* $w \in W$. As the condition *non* $\vdash N\alpha$ is equivalent to the condition $Con(\{\alpha\})$ one may formulate a condition equivalent to (i) viz.

(ii) *if $Con(\alpha)$ then there exists $M = (W, R, v)$ with $[\alpha]_w = 1$ for some $w \in W$.*
But $Con(\{\alpha\})$ implies the existence of Γ with $Con_{max}(\Gamma)$ and $\alpha \in \Gamma$ hence
in the canonical interpretation frame M^c we have $[\alpha]_\Gamma = 1$ thus proving (ii)
hence (i) and the completeness theorem as well.   □

We apply the completeness theorem to verify that the meaningful expression $CLpp$ is not a thesis of the system K.

Example 5.1. *It is sufficient to consider an interpretation frame $M = (W,
R, v)$ in which there are worlds $w \neq w^* \in W$ with $W = \{w, w^*\}$, wRw^*,
non wRw, $[p]_w = 0$, and $[p]_{w^*} = 1$. Then $\models_M Lp$ and non $\models_M p$ thus non
$\models_M CLpp$.*

5.3   The system T

Adding to the axioms of the system K a new axiom

(T) $CLpp$

would lead to a new system denoted T. We may remember that (T) is one
of axioms of the basic modal system of Łukasiewicz. The collapse method
shows that T is consistent. Let us observe that the axiom (T) is true in every
interpretation frame $M = (W, R, v)$ in which the relation R is *reflexive* (i.e.
wRw for each $w \in W$): if $[Lp]_w = 1$ for a $w \in W$ then $[p]_{w^*} = 1$ for each w^*
with wRw^* hence $[p]_w = 1$ and thus $[CLpp]_w = 1$ for each $w \in W$. Example
5.1 shows that the converse is also true: if (T) is true in an interpretation
frame $M = (W, R, v)$ then R is reflexive.

Thus we may formulate a completeness theorem for the system T; denot-
ing by the symbol $\mathcal{M}_r$ the set of all interpretation frames with the relation R
reflexive we have the completeness theorem whose proof parallels the general
case.

Proposition 5.6. *(The Completeness Theorem for T) For each meaningful
expression α: α is a thesis of T if and only if α is true in every interpretation
frame in $\mathcal{M}_r$.*

5.4   The system S4

We may look at the meaningful expression (S4): $CLpLLp$. This meaningful expression is also a thesis of the basic modal system. It is easy to see
that (S4) is true in an interpretation frame $M = (W, R, v)$ if and only if R
is *transitive* (i.e. wRw^* and w^*Rw^{**} imply wRw^{**} for each triple w, w^*, w^{**}
$\in W$).

Yet another method of expressing ways in which sets are related one to another is to compare them with respect to *inclusion*.

We say that a set X is a *subset* of a set Y which is denoted $X \subseteq Y$ when the following condition is fulfilled

$$\forall x (x \in X \Rightarrow x \in Y) \tag{6.25}$$

Thus being a subset of a set means consisting of possibly not all elements of this set.

Being a subset does satisfy also certain laws of which we may mention the following easy to establish ones

$$X \subseteq Y \wedge Y \subseteq Z \Rightarrow X \subseteq Z \tag{6.26}$$

$$X \subseteq X \tag{6.27}$$

$$\emptyset \subseteq X \tag{6.28}$$

$$X \subseteq Y \wedge Y \subseteq X \Rightarrow X = Y \tag{6.29}$$

(6.26) does express *transitivity* of inclusion, (6.29) states the *weak symmetry* of it. (6.27) says that any set is included in itself and (6.28) states that the empty set is included in any set (by the fact that implication from false to true has the truth value 1).

In many considerations involving sets, we restrict ourselves to sets which are subsets of a given fixed set U called in such a case the *universe*. In this case we may have some shortcut notation notably the difference $U \setminus X$ for $X \subseteq U$ is denoted X^c and it is called the *complement* of X.

The reader will undoubtedly establish many more properties of operations on sets in case of need; we will meet some in the sequel. We introduce a new notion of a *field of sets*.

By a *field of sets*, we will understand a non–empty set B of subsets of a certain universe U, which is such that results of operations $\triangle, \cap$ performed on its elements are again its elements and U is its element. More formally, B does satisfy the following requirements

1. $\forall x (x \in B \Rightarrow x \subseteq U)$

2. $U \in B$

3. $\forall x, y(x, y \in \mathcal{B} \Rightarrow x \triangle y \in \mathcal{B})$

4. $\forall x, y(x, y \in \mathcal{B} \Rightarrow x \cap y \in \mathcal{B})$

Let us observe that as $\mathcal{B}$ contains at least one element e.g. U it does contain $U \triangle U = \emptyset$. Also the identity

$$x \cup y = (x \triangle y) \triangle x \cap y$$

following directly by 6.24 shows that

5. $\forall x, y(x, y \in \mathcal{B} \Rightarrow x \cup y \in \mathcal{B})$

Thus any field of sets is closed with respect to the union of its elements. Similarly, by the identity

$$x^c = U \triangle x$$

we have

6. $\forall x(x \in \mathcal{B} \Rightarrow x^c \in \mathcal{B})$

This means that any field of sets is closed with respect to complements of its elements. We get a characteristics of field of sets. Clearly, any set of sets closed on union, intersection, and complement is closed on symmetric difference.

Proposition 6.1. *A set $\mathcal{B}$ of subsets of a set U is a field of sets if and only if it is non–empty and it is closed with respect to operations of the union, the intersection and the complement.*

A simple example of a field of sets is the *power set of U* denoted 2^U consisting of all subsets of U i.e.

$$x \in 2^U \Leftrightarrow x \subseteq U.$$

6.3   A formal approach

As observed already, we may perform some operations on sets once we are given them but it has turned out in the historic process of development of set theory that naive, intuitive approach led to some serious problems and contradictions. This motivated attempts at laying strict foundations for set theory similar to those we witnessed with propositional calculi. Studies in this respect led to some axiomatic systems of set theory e.g. the *Zermelo–Fraenkel* system. We will not need the full power of this system so we first restrict ourselves to a subsystem which guarantees us the existence of some basic means for new set construction.

We state below some axiomatic statements about sets which guarantee the existence of some sets as well as the possibility of performing some operations on them.

(A1) (*there exists the empty set*): $\exists X (\forall x (x \in X \Rightarrow x \neq x))$

This axiom guarantees that there exists at least one set

(A2) (*the extensionality axiom*) $\forall x (x \in X \Leftrightarrow x \in Y) \Rightarrow X = Y$

This axiom does express our intuition that sets consisting of the same elements should be identical to each other. A fortiori it does guarantee the uniqueness of other sets implied by the axioms.

(A3) (*the unordered pair axiom*) *Given any two objects* a, b *the following holds*:
$$\exists X (\forall x (x \in X \Leftrightarrow x = a \lor x = b))$$

This axiom guarantees that for any two objects a, b there exists a set whose only elements are a and b. We denote this set by the symbol $\{a, b\}$ and we call it the *unordered pair* of elements a, b. In case $a = b$, we call this set a *singleton* and denote it by the symbol $\{a\}$.

From this axiom, the existence of the set $\{\{a\}, \{a, b\}\}$ follows immediately. We denote this set by the symbol $< a, b >$ and call it the *ordered pair* of elements a, b.

The reader will check the following basic property of the ordered pair.

Proposition 6.2. *Given elements* a, b, c, d *we have* $< a, b > = < c, d >$ *if and only if* $a = c, b = d$.

For the proof we hint that $< a, b > = < c, d >$ i.e. $\{\{a\}, \{a, b\}\} = \{\{c\}, \{c, d\}\}$ implies $\{a\} = \{c\}$ or $\{a, b\} = \{c\}$ so $a = c$. Similarly, $\{c, d\} = \{b\}$ or $\{c, d\} = \{a, b\}$ hence $b = c = d$ or $b = d$. So either $a = b = c = d$ or $a = b, c = d$ and the thesis follows.

The procedure of generating ordered pairs may be repeated: given objects $a_1, ..., a_k$, we may define inductively the *ordered* k–*tuple*

$$< a_1, ..., a_k >$$

by letting
$$< a_1, ..., a_k > = << a_1, ..., a_{k-1} >, a_k > .$$

Clearly, inducting on k, we get the following extension of the last proposition: $< a_1, ..., a_k > = < b_1, ..., b_k >$ *if and only if* $a_i = b_i$ *for* $i = 1, 2, ..., k$.

(A4) (*the union of sets axiom*)

$$\forall X(\exists Z\forall x(x \in Z \Leftrightarrow \exists Y(x \in y \wedge Y \in Z)))$$

This axiom guarantees that for any set X there exists a set (uniquely defined by extensionality) Z whose elements are those sets which are elements in some element of X. The set Z is denoted $\bigcup X$ and it is called the *union* of X. This setting may imply that X is a set whose elements are sets; for this reason, to avoid a phrase like *a set of sets*, X is said to be a *family of sets* so Z is the union of this family.

We may consider the case when X is looked on as a family of subsets of the universe $\bigcup X$ hence the set

$$\bigcap X = \bigcup X \setminus \bigcup \{\bigcup X \setminus A : A \in X)\} \tag{6.30}$$

is well–defined; we call it the *intersection* of the family X of sets. Clearly

$$x \in \bigcap X \Leftrightarrow \forall A(A \in X \Rightarrow x \in A) \tag{6.31}$$

Particular applications for this axiom we may find in case of the two sets X, Y: clearly, in this case we have $\bigcup\{X, Y\} = X \cup Y$. Another application is in the formation of finite unordered tuples: given objects $a_1, a_2, ..., a_k$, we may form the union of singletons $\{a_i\}$ for $i = 1, 2, ..., k$, by induction as follows: $\{a_1, ..., a_k\} = \{a_1, ..., a_{k-1}\} \cup \{a_k\}$. The set $\{a_1, ..., a_k\}$ is called the *unordered k–tuple* of elements $a_1, a_2, ..., a_k$.

To get more specialized sets, we need new axioms. In particular, given a set X and a predicate $\Psi(x)$ expressing a certain property of objects, we may want to consider the new set whose elements are those elements of X which after being substituted for x satisfy $\Psi(x)$. We need an axiom which would guarantee the existence of such a set.

(A5) (*the axiom schema of separation*) *Given a set X and a predicate $\Psi(x)$, the following holds:*

$$\exists Y\forall x(x \in Y \Leftrightarrow \Psi(x) \wedge x \in X).$$

The (unique) set Y is denoted often by the symbol $\{x \in X : \Psi(x)\}$. Let us observe for example, that given in addition a set Z, the difference $X \setminus Z$ may be obtained via (A6) as the set $\{x \in X : \neg(x \in Z)\}$. Similarly, the intersection $X \cap Z$, is the set $\{x \in X : x \in Z\}$.

We may also want for any set X to form a new set whose elements are all subsets of X i.e. the familiar to us field of sets 2^X. For this also we need a new axiom which however may be regarded from foundational point of view as too strong.

(A6) (*the axiom of the power set*) *Given a set* X *the following holds*:

$$\forall X(\exists Y \forall x(x \in Y \Leftrightarrow x \subseteq X))$$

The set Y is denoted with the symbol 2^X (*the exponential set of* X, *the power set of* X). It is unique by extensionality.

Finally, we may want to form a new set from a given one by a transformation via a predicate of functional character (in a sense, we want to *map* a given set onto a new one). For this we also need a new axiom.

(A7) (*the axiom schema of replacement*) *Consider a binary predicate* $\Psi(x, y)$ *with the following property* (i):

$$\forall x, y, z(\Psi(x, y) \wedge \Psi(x, z) \Rightarrow y = z)$$

Then we have : *if* $\Psi(x, y)$ *does satisfy* (i) *then*

$$\forall X \exists Y(y \in Y \Leftrightarrow \exists x(x \in X \wedge \Psi(x, y)))$$

The set Y may be called the *image* of the set X by Ψ and we may denote it by the symbol $\Psi(X)$.

The axiom (A7) is stronger than (A5): given $\Psi(x)$ as in the formulation of (A6), we may define a binary predicate Φ as follows: $\Phi(x, y)$ is $x = y$ when $\Psi(x)$ is satisfied and $\Phi(x, y)$ is $x = a$ where a is a fixed element not in the set X when $\Psi(x)$ is not satisfied. Then the set $\Phi(X)$ is identical to the set $\{x \in X : \Psi(x)\}$.

6.4 Relations and functions

The notion of an ordered pair allows for a formalization of the important notions of a *relation* as well as of a *function*.

Informally, a relation (say, *binary*) is a constraint on ordered pairs $< x, y >$ where $x \in X, y \in Y$; thus, a formalization of a relation is done via a choice of a set of pairs which satisfy this constraint.

First, given sets X, Y, we form the set $X \times Y$ of all ordered pairs of the form $< x, y >$ where $x \in X, y \in Y$. We thus let

$$X \times Y = \{< x, y > : x \in X, y \in Y\} \tag{6.32}$$

The existence of this set may be justified on the basis of axioms: we first may form the set $2^{X \cup Y}$ of all subsets of the set $X \cup Y$ and then the set $2^{2^{X \cup Y}}$, both by the axiom (A7). Letting $\Psi(A)$ to be satisfied by A if and only if $\exists x \exists y (x \in X \wedge y \in Y \wedge A = \{\{x\}, \{x, y\}\})$, we obtain $X \times Y$ as the set $\{A \in 2^{2^{X \cup Y}} : \Psi(A)\}$ by (A6).

The set $X \times Y$ is called the *Cartesian product* of the set X and the set Y.

A *binary relation* R on $X \times Y$ is a subset R of the set $X \times Y$. For x, y such that $< x, y > \in R$ we say that x, y *are in the relation* R and we write xRy to denote this fact. We let

$$domR = \{x \in X : \exists y (< x, y > \in R)\} \tag{6.33}$$

and we call $domX$ the *domain of R*; similarly, we let

$$codomR = \{y \in Y : \exists x (< x, y > \in R)\} \tag{6.34}$$

and we call $codomR$ the codomain of R.

As relations are sets, usual set operations may be performed on relations: for relations R, S on the Cartesian product $X \times Y$, the union $R \cup S$, the intersection $R \cap S$, and the difference $R \setminus S$ are defined in the usual way. There is however more to relations: due to the structure of their elements as ordered pairs, we may have more operations on relations than on ordinary sets and these operations constitute what is called the *Algebra of relations*.

6.4.1   Algebra of relations

First, for any relation $R \subseteq X \times Y$, we may introduce the *inverse* of R which is the relation $R^{-1} \subseteq Y \times X$ defined as follows:

$$yR^{-1}x \Leftrightarrow xRy \tag{6.35}$$

This operation does satisfy the following

Proposition 6.3. *1.* $(R \cup S)^{-1} = R^{-1} \cup S^{-1}$
2. $(R \cap S)^{-1} = R^{-1} \cap S^{-1}$
3. $(R \setminus S)^{-1} = R^{-1} \setminus S^{-1}$

Indeed, for instance in case 1,

$$< y, x > \in (R \cup S)^{-1} \Leftrightarrow < x, y > \in R \cup S \Leftrightarrow$$

$$< x, y > \in R \vee < x, y > \in S \Leftrightarrow < y, x > \in R^{-1} \vee < y, x > \in S^{-1} \Leftrightarrow$$

$$< y, x > \in R^{-1} \cup S^{-1}.$$

The proof in remaining cases goes along similar lines.

Relations may also be *composed*; for relations $R \subseteq X \times Y$, $S \subseteq Y \times Z$, the *composition* (or, *superposition*) $R \circ S \subseteq X \times Z$ is defined as follows

$$< x, z > \in R \circ S \Leftrightarrow \exists y (y \in Y \wedge < x, y > \in R \wedge < y, z > \in S) \qquad (6.36)$$

The operation of relation composition does satisfy easy to be checked properties

Proposition 6.4. *1.* $(R \cup S) \circ T = (R \circ T) \cup (S \circ T)$
2. $(R \cap S) \circ T = (R \circ T) \cap (S \circ T)$
3. $(R \circ S)^{-1} = S^{-1} \circ R^{-1}$
4. $(R \circ S) \circ T = R \circ (S \circ T)$

These properties are proved similarly to those of the preceding proposition.

For a set X, the relation $id_X = \{< x, x >: x \in X\}$ is called the *identity relation* on X. Given a relation R on $X \times Y$, we have

Proposition 6.5. *1.* $id_{domR} \subseteq R \circ R^{-1}$
2. $id_{co-domR} \subseteq R^{-1} \circ R$
3. $id_{domR} \circ R = R$
4. $R \circ id_{co-domR} = R$

Indeed, in case 1, for $x \in domR$, there is $y \in Y$ with $< x, y > \in R$ hence $< y, x > \in R^{-1}$ so $< x, x > \in R \circ R^{-1}$. The proof in case 2 is on the same lines. Cases 3 and 4 are obvious.

One more operation one may perform on relations is *restriction*. Given a relation $R \subseteq X \times Y$, and a subset $A \subseteq X$, the restriction $R|A$ is $R \cap A \times Y$.

Functions are a special class of relations singled out by the uniqueness property: a relation $R \subseteq X \times Y$ is a function if and only if the following property is observed:

$$\forall x \forall y \forall z (xRy \wedge xRz \Rightarrow y = z) \qquad (6.37)$$

In this case R is said to be a *function from the set X into the set Y* and the notation $R : X \to Y$ is in use. To denote functions, usually small letters $f, g, h, k, ...$ are used and in place of the formula xfy we write $y = f(x)$ calling y *the value of f at the argument x*.

A function $f : X \to Y$ is said to be *injective* in case the following is satisfied:

$$\forall x \forall x' (f(x) = f(x') \Rightarrow x = x') \qquad (6.38)$$

For a subset $A \subseteq X$, the *image* fA of A by a function $f : X \to Y$ is the set

$$\{y \in Y : \exists x \in A(y = f(x))\} \tag{6.39}$$

Similarly, for $B \subseteq Y$, the *inverse image* $f^{-1}B$ is the set

$$\{x \in X : \exists y \in B(y = f(x))\} \tag{6.40}$$

These operations do satisfy the following rules

Proposition 6.6. *1.* $f(A \cup B) = fA \cup fB$
2. $f(A \cap B) \subseteq fA \cap fB$
3. $f^{-1}(A \cup B) = f^{-1}A \cup f^{-1}B$
4. $f^{-1}(A \cap B) = f^{-1}A \cap f^{-1}B$
5. $f^{-1}(A \setminus B) = f^{-1}A \setminus f^{-1}B$

These facts are easy to establish; let us note the difference between 2 and 4: in case of 2 we have only inclusion as for some f even disjoint sets may be mapped onto the same image, while in 4 we have identity by the uniqueness of the value of a function.

A function $f : X \to Y$ maps X *onto* Y (f *is a surjection*) if and only if $fX = Y$. A function f which is injective and onto is said to be a *bijection* between X and Y. For a bijection f, the relation f^{-1} is again a function $f^{-1} : Y \to X$ called the *inverse* of f. Let us observe that the function f^{-1} is a bijection between Y and X. Clearly, for a bijection f, it may not happen that disjoint sets map to the same value so in 2 above we have the identity: $f(A \cap B) = fA \cap fB$.

6.5   Orderings

Relations on a Cartesian product $X \times X$ with $X = dom_R$ are called *relations on the set X*; they are classified according to some properties:

1. (*reflexivity*) $\forall x (xRx)$
2. (*symmetry*) $\forall x \forall y (xRy \Rightarrow yRx)$
3. (*weak anti–symmetry*) $\forall x \forall y (xRy \wedge yRx \Rightarrow x = y)$
4. (*linearity*) $\forall x \forall y (xRy \vee x = y \vee yRx)$
5. (*transitivity*) $\forall x \forall y \forall z (xRy \wedge yRz \Rightarrow xRz)$

A relation R on a set X is an *ordering* on X if and only if R is reflexive, weak anti–symmetric and transitive. In this case we write $x \leq_R y$ in place of xRy.

An ordering R on the set X is *linear* in case R is linear; we say also that X

is *linearly ordered* by R.

Somewhat more generally, we say that a subset $Y \subseteq X$ of a set X ordered by an ordering R is a *chain* in X when the restriction $R|Y \times Y$ is linear.

An ordering does stratify elements of its domain; $x, y \in X$ are *comparable* if $x \leq_R y \vee y \leq_R x$. In case $x \leq_R y$ we say that x is *less or equal to* y.

An element x is an *upper bound* (respectively a *lower bound*) of a set $A \subseteq X$ in case $a \leq_R x$ for each a in A (respectively, $x \leq_R a$ for each $a \in A$); an upper bound x of A is the *least upper bound* (*the supremum of A*) in case $x \leq_R y$ for each upper bound y of A and we denote x by the symbol *supA*.

Similarly, we call a lower bound x of A the *greatest lower bound* (*the infimum of A*) when $y \leq_R x$ for each lower bound y of A. We denote an infimum of A with the symbol *infA*. In case infimum respectively supremum are admitted i.e. *infA* $\in A$ respectively *supA* $\in A$ we call *infA* the *least element in A* respectively we call *supA the greatest element in A*.

We denote also by A^+ the set of all upper bounds of the set A and by A^- the set of all lower bounds of the set A. Thus, *supA = infA⁺* and *infA = supA⁻* whenever *supA*, respectively *infA* exists.

A set $Y \subseteq X$ is *bounded from above* (respectively, *bounded from below*) when there exists an upper bound for Y (respectively, a lower bound for Y).

In search of archetypical orderings, we may turn to inclusion $\subseteq$ on the power set of a given set X. It is clearly an ordering on any family of subsets of the set X. It is however not linear. To find suprema of families of sets with respect to inclusion, we have to turn to the sum axiom: given a family F of subsets of X we have $supF = \bigcup F$. Similarly, $infF = \bigcap F$. A family F is *closed with respect to unions* when $supF' \in F$ for any $F' \subseteq F$; by analogy, F is *closed with respect to intersections* when $infF' \in F$ for any $F' \subseteq F$.

A set X is *completely ordered* by an ordering R on X when for any subset $A \subseteq X$ there exist *supA* and *infA*. The ordering R is said to be *complete* in this case. An example of a complete ordering is inclusion on the power set 2^X for any set X and for any family of sets $F \subseteq 2^X$ its supremum is $supF = \bigcup F$ and its infimum is $infF = \bigcap F$.

Relations on various sets may be compared to each other by means of functions satisfying adequate conditions. In the general case of a relation R on the product $X \times X$ and a relation $S \subseteq Y \times Y$, we say that a function $f : X \rightarrow Y$ *agrees with R, S* when

$$\forall x \forall y (x R y \Rightarrow f(x) S f(y)) \qquad (6.41)$$

When f when is an injection, we say f is an *embedding* of R into S and when f is a bijection, then we say f is an *isomorphism* between R and S.

In the particular case when R, S are orderings on respectively sets X, Y, we call a function $f : X \to Y$ which agrees with R, S an *isotone* function. Thus we have for the isotone function f

$$x \leq_R y \to f(x) \leq_S f(y) \tag{6.42}$$

In general there is little one may say about isotone functions; however in the case when $R = S$ is a complete ordering, isotone functions between R and S have an important property of having a *fixed point* which means that $f(x) = x$ for some $x \in X$ called a *fixed point of f.*

Proposition 6.7. *(Knaster–Tarski) If $f : X \to X$ is an isotone function on a completely ordered set X then f has a fixed point.*

Proof. Clearly, the set $A = \{x \in X : x \leq_R f(x)\}$ has a supremum $sup A = a$. As for $x \in A$ we have $x \leq_R a$, also $x \leq_R f(x) \leq_R f(a)$ hence $a \leq_R f(a)$. It follows that $f(a) \leq_R f(f(a))$ so $f(a) \in A$ and hence $f(a) \leq a$ implying finally that $a = f(a)$ so a is a fixed point of f.   □

Given an ordered set X with an ordering $\leq$ one may construct a canonical isotone embedding of X into a completely ordered by inclusion family of sets. To this end, we make use of sets A^-, A^+ defined above for any set $A \subseteq X$. We check that the following properties of operations $(.)^{+,-}$ are fulfilled.

Proposition 6.8. *1. $A \subseteq B \Rightarrow B^- \subseteq A^-, B^+ \subseteq A^+$*
2. $A \subseteq (A^+)^-$
3. $B = (A^+)^-$ satisfies $B = (B^+)^-$

Proof. 1 is obvious: anything less (greater) than all elements of B is also less (greater) than all elements of A. 2 is also evident: for $x \in A$ we have $x \leq a$ for any $a \in A^+$ hence $x \in (A^+)^-$. For 3, we first notice that as $A \subseteq B$ by the property 2, we have by the property 1 that $B^+ \subseteq A^+$. But also $A^+ \subseteq B^+$ by transitivity of $\leq$. Since $A^+ = B^+$, we have $B = (A^+)^- = (B^+)^-$.   □

Consider now for $x \in X$ the set $\{x\}^-$ denoted also $(\leftarrow, x]$ and called the *left interval* of x. We have $((\leftarrow, x]^+)^- = (\leftarrow, x]$.

We may consider now the function f which assigns to $x \in X$ its left interval $(\leftarrow, x]$. We may consider f as a function from X into the set $I(X) = \{A \subseteq X : A = (A^+)^-\}$ ordered by inclusion. We have

Proposition 6.9. *(Dedekind, MacNeille) The function f is an isotone function from X into completely ordered by inclusion set $I(X)$. Moreover f preserves suprema and infima.*

Proof. Clearly, $x \leq y$ implies $(\leftarrow, x] \subseteq (\leftarrow, y]$. For any $A \subseteq I(X)$, we have $inf A = (\bigcap A^+)^-$, $sup A = (\bigcup A^+)^-$ so $I(X)$ is completely ordered. Finally, given $A \subseteq X$, with $a = sup A$, we have easily that

$$(\leftarrow, a] = (\bigcup \{(\leftarrow, x] : x \in A\}^+)^-$$

i.e. $f(sup A) = sup f(A)$ and in case of infima we have for $b = inf A$ that $(\leftarrow, b] = \bigcap \{f(x) : x \in A\}$ i.e. $f(inf A) = inf f(A)$. □

We may by this result embed any ordered set into a completely ordered by inclusion family of sets. Let us observe that the Dedekind–MacNeille embedding is the least one: it embeds X into any completely ordered set into which X may be embedded isotonically.

6.6   Lattices and Boolean algebras

An ordered set $(X, \leq)$ in which for each pair x, y there exist the supremum $sup\{x, y\}$ as well as the infimum $inf\{x, y\}$ is called a *lattice*. Usually, $sup\{x, y\}$ is denoted $x \cup y$ and it is called the *join* of x, y while $inf\{x, y\}$ is denoted $x \cap y$ and it is called the *meet* of x, y.

When additionally X is endowed with a unary operation $-$ of *complementation* and the join and the meet satisfy the following conditions

1. $x \cup y = y \cup x, x \cap y = y \cap x$

2. $x \cup (y \cup z) = (x \cup y) \cup z, x \cap (y \cap z) = (x \cap y) \cap z$

3. $(x \cap y) \cup y = y, (x \cup y) \cap y = y$

4. $x \cap (y \cup z) = (x \cap y) \cup (x \cap z), x \cup (y \cap z) = (x \cup y) \cap (x \cup z)$

5. $(x \cap -x) \cup y = y, (x \cup -x) \cap y = y$

then the lattice X is called a *Boolean algebra*. We may observe duality here: replacing $\cup/\cap, \cap/\cup$ we obtain a dual condition; we may also not confuse the symbols for the join and the meet used in general with the particular symbols of set union and intersection. The latter clearly satisfy 1–4. Let us observe that any field of sets is a Boolean algebra with operations of set union, intersection and complementation.

We will not dwell here on lattices as they will be studied systematically in one of the following Chapters. Boolean algebras will come also in the next Chapter in the context of their topological representation as *Stone spaces*.

6.7   Infinite sets

We have witnessed a good deal of standard set theory having basically only finite sets (in an informal sense of the word) at our disposal. To introduce infinite sets, we need a new axiom.

(A9) (*the infinity axiom*) *There exists a set* X *such that* (i) $\emptyset \in X$ (ii) *if* $A \in X$ *then* $A \cup \{A\} \in X$.

Taking the intersection of the family of all sets satisfying (A10), we obtain the set N which contains elements $0 = \emptyset$, $1 = \emptyset \cup \{\emptyset\}$, $2 = 1 \cup \{1\}$,..., $n + 1 = n \cup \{n\}$,... and only those elements. N is the set of *natural numbers*. Having N we may attempt at defining precisely *finite* and *infinite sets*. A standard way of arguing with natural numbers employs the *principle of mathematical induction*.

Proposition 6.10. *Assume* $A \subseteq N$ *is such that (i)* $0 \in A$ *(ii)* $\forall n(n \in A \Rightarrow n + 1 \in A)$. *Then* $A = N$.

Indeed, by (i), (ii), A does satisfy requirements of (A9) hence $N \subseteq A$ and finally $A = N$.

The principle of mathematical induction will serve us as a tool in proving basic facts about natural numbers.

Proposition 6.11. *For all natural numbers* n, m *the following statements are true:*

1. $n \in m \Leftrightarrow n \subseteq m \land n \neq m$
2. $n \in m \lor n = m \lor m \in n$
3. $n \subseteq N \land n \neq N$
4. $n \notin n$

Proof. For 1: we consider $A = \{m \in N : \forall n(n \in m \Rightarrow n \subseteq m)\}$. Clearly, $0 \in A$ by default. Assuming $m \in A$, consider $m + 1$ along with $n \in m + 1$. Then either $n \in m$ or $n = m$; in the former case $n \subseteq m$ hence $n \subseteq m + 1$ and in the latter case again $n \subseteq m$ whence $n \subseteq m + 1$. Thus $m + 1 \in A$. By the principle of mathematical induction, $A = N$. This proves the part (i) $n \in m \Rightarrow n \subseteq m$.

We now prove 4; clearly, $0 \notin 0$. Assume $n \notin n$ and $n + 1 \in n + 1$. Then either $n + 1 \in n$ or $n + 1 = n$. In the former case by (i) $n + 1 \subseteq n$ hence $n \in n$, a contradiction. So only $n + 1 = n$ may hold but then again $n \in n$, a contradiction. It follows that we have $n + 1 \notin n + 1$ and by the principle of mathematical induction 4 holds.

Returning to 1, we infer from 4 that (ii) $n \in m \Rightarrow n \subseteq m \wedge n \neq m$. It remains to prove the converse. Again we apply induction. We use $\subset$ instead of $\subseteq \wedge \neq$. So we are going to check that $n \subset m$ implies $n \in m$. For $m = 0$ this is true. Assume this is true with m and consider $n \subset m + 1$. Was $m \in n$ then $m \subset n$ and $m + 1 \subseteq n$, a contradiction with $n \subset m + 1$. Hence $m \notin n$ and thus $n \subseteq m$. We have either $n \subset m$ in which case by assumption $n \in m$ and a fortiori $n \in m+1$ or $n = m$ in which case $n \in m+1$ so 1 is proved.

3 follows easily: by 4 we have $n \neq N$, and by induction $n \subset N$ follows.

It remains to prove trichotomy 2. We apply induction proving that for each $n \in N$ the set $A_n = \{m \in N : n \in m \vee n = m \vee m \in n\}$ is N. Clearly, $A_0 = N$ as $0 \in m$ for each m. For A_n, it is clear that $0 \in A_n$. Assume that $m \in A_n$ and consider $m+1$. By assumption either $m \in n$ or $m = n$ or $n \in m$. In the first case $m \subset n$ by 1 hence either $m + 1 \subset n$ and thus $m + 1 \in n$ or $m+1 = n$. In both cases $m+1 \in A_n$. In the second case when $m = n$ clearly $n \in m + 1$ and again $m + 1 \in A_n$. Finally, when $n \in m$ a fortiori $n \in m + 1$ so $m + 1 \in A_n$. It follows by the induction principle that $A_n = N$ for each $n \in N$ so 2 is proved.   □

It follows that N is ordered linearly by the membership relation $\in$; using the symbol $<$ in place of $\in$, we have $0 < 1 < 2 < 3 < \ldots\ldots < n < n+1 < \ldots$. We have used here the symbol $<$ of a *strict ordering* related to the ordering $\leq$ by the identity $\leq = < \cup =$ i.e. $n \leq m$ if and only if $n < m$ or $n = m$. The relation $<$ has one more important property viz.

Proposition 6.12. *In any non–empty subset $A \subseteq N$ there exists the least element with respect to $<$.*

Indeed, there is $n \in A$ and clearly it suffices to consider the set $A|n = A \cap (\leftarrow, n]$. If $0 \in A|n$ then 0 is the least element in A; otherwise we may check whether $1 \in A|n$ etc. Proceeding in this way, after at most $n - 1$ steps, we find k such that $\{0, 1, ..., k-1\} \cap A = \emptyset$ and $k \in A$, Thus k is the least element in A.

A relation R of linear ordering on a set X with the property expressed in the last proposition is said to *well–order* the set X and it is called a *well–ordering* of X. Thus $\in$ well–orders N.

A question may arise at this point whether a well–ordering may exist for an arbitrary set X. We do not have by now any tool which would enable us to prove such a result. We need a new axiom which would guarantee us the existence of a well–ordering on any set.

6.8   Well–ordered sets

The axiomatic statement which would guarantee the existence of a well–ordering on each set is:

(A10) (the *axiom of well–ordering*) For each set X these exists a relation R on X which well–orders X

We may be aware of the fact that this axiom is non–effective and it does not give us any procedure for constructing a well–ordering on a given set; its value is ontological, allowing us to explore consequences of its content. We state first one of this consequences known as the *axiom of choice* of Zermelo.

Proposition 6.13. *(A10) implies the following: for any family $\mathcal{F}$ of non–empty sets there exists a function $f : \mathcal{F} \to \bigcup \mathcal{F}$ with the property that $f(X) \in X$ for each $X \in \mathcal{F}$.*

Indeed, by (A10), the set $\bigcup \mathcal{F}$ may be well–ordered by a relation $<$. To construct f with the desired property, it is sufficient to define $f(X)$ to be the least element in X with respect to $<$.

A function f with the property stated in the last proposition is called a *choice function* for the family $\mathcal{F}$ (or, a *selector* for $\mathcal{F}$). From the axiom of choice (hence from the axiom of well–ordering) one may derive some very important consequences. We now present some of them. Recall that a subset A of an ordered by $\leq$ set X is a *chain* if it is linearly ordered by the restriction $\leq |A| =\leq \cap A \times A$. An element c of X is *maximal* (respectively *minimal*) in X in case there is in X no element $d \neq c$ with $c \leq d$ (respectively $d \leq c$).

We may give a formulation of the *maximum principle* known also as the *Zorn lemma* due independently to Hausdorff, Kuratowski and Zorn.

Proposition 6.14. *(the Maximum Principle) Assume that a set X ordered by a relation $\leq$ has the property that any chain in X is bounded from above. Then in X there exists a maximal element.*

Proof. ([Balcar–Štěpánek86]). Let f be a selector on the set $2^X \setminus \{\emptyset\}$ of all non–empty subsets of X. Assume, to the contrary, that in X there is no maximal element i.e. for any $x \in X$ the set $M(x) = \{y \in X : x < y\}$ is non–empty and thus $s(x) = f(M(x))$ exists.

We will start with an element a in X and the set $S(a) = [a, \to)$, and we observe properties of $Y \subseteq S(a)$ that

$$(i)\ a \in Y$$
$$(ii)\ y \in Y \to s(y) \in Y \qquad\qquad (6.43)$$
$$(iii)\ if\ L\ a\ chain\ in\ Y\ then\ \sup L \in Y$$

For the family $\mathcal{F}(a)$ of all $Y \subseteq S(a)$ which satisfy (6.43), we form the intersection $\bigcap \mathcal{F}(a) = L$. Clearly, L satisfies (6.43) being a minimal set with this property. We would like to show that L is a chain.

To this end, first, we exploit minimality of L by looking at the set

$$K = \{y \in L : \forall x (x \in L \wedge x < y \Rightarrow s(x) \le y)\}.$$

Claim 1. *For $x \in K, y \in L$: $x \le y \vee y \le x$.*
To check this claim, we take $x \in K$ and we consider the set

$$K_x = \{y \in L : x \le y \vee y \le x\}.$$

We check that K_x does satisfy (6.43). Clearly, $a \in K_x$ so (i) holds. Assume $y \in K_x$. In case $x \le y$ we have $x < s(y)$; in case $x > y$, by definition of K, we have $s(y) \le x$. In either case, $s(y) \in K_x$ so (ii) is satisfied. For a chain C in K_x, either $x < c$ for some $c \in C$ hence $x < supC$ or $c \le x$ for every $c \in C$ hence $supC \le x$. In either case $supC \in K_x$ witnessing (iii). By minimality of L we must have $K_x = L$ so Claim 1 is verified.

Claim 2. *For $x \in K, y \in L$: $x < y \Rightarrow s(x) \le y$.*
We apply the same technique: given $x \in K$, we look at the set

$$K^x = \{y \in L : x < y \Rightarrow s(x) \le y\}.$$

Again (i) holds obviously, for (ii) we look at $y \in K^x$. By Claim 1, either $x \le y$ or $y \le x$. The latter case impossible, we are left with $x \le y$ so either $x = y$ hence $s(x) = s(y)$ or $x < y$ hence $s(x) \le s(y)$ as $y \in K^x$. Thus in all cases $s(y) \in K^x$ proving (ii). In case (iii), for a chain C in K^x, if $x < supC$ then $x < c$ for some $c \in C$ hence $s(x) \le c$ and thus $s(x) \le supC$. It follows that $supC \in K^x$ and (iii) holds. By minimality of L, we have $K^x = L$ which proves Claim 2.

Claim 3. *$K = L$.*
We check that K satisfies (6.43) so Claim 3 would follow by minimality of L. Clearly, $a \in K$ so (i) holds. For $x \in K$, let $y < s(x)$. We have by Claim 1 that $x \le y$ or $y \le x$. But $x \le y$ cannot hold by Claim 2 so only $y < x$ remains. Then $s(y) \le x$ hence $s(y) < s(x)$ witnessing $s(y) \in K$ so (ii) holds. Finally, consider a chain C in K with $y < supC$ for some $y \in L$. Then $y < c$ for some $c \in C$ hence $s(y) \le c$ and $s(y) \le supC$ implying that $supC \in K$ so (iii) holds. By minimality of L, $K = L$ and it follows that L is a chain. By (iii), $supL \in L$ and by (ii) $s(supL) \in L$ which is a contradiction with $supL < s(supL)$. Thus in X there exists a maximal element. $\square$

It is much easier to show that the axiom of well–ordering follows from the maximum principle.

Proposition 6.15. *The maximum principle implies the axiom of well–ordering.*

Proof. Consider a set X assuming the maximum principle. Now, consider the set $\mathcal{F}$ of all pairs of the form (a subset Y of X, a relation R well–ordering Y) i.e.

$$(Y, R) \in \mathcal{F} \Leftrightarrow Y \subseteq X \wedge R \; well--orders \; Y.$$

We introduce an ordering $\sqsubseteq$ on $\mathcal{F}$ by letting $(Y, R) \sqsubseteq (Z, S)$ if and only if $Y \subseteq Z$ and $S|Y = R$. Let us observe that any chain C in $\mathcal{F}$ with respect to $\sqsubseteq$ has a supremum viz. $(\bigcup F, \bigcup \{R : (Y, R) \in C\})$.

By the maximum principle (Proposition 6.14) there exists a maximal element (Y_0, R_0) in $\mathcal{F}$. If it was $Y_0 \neq X$ we could pick an element $a \in X \setminus Y_0$ and extend Y_0 to $Y_1 = Y_0 \cup \{a\}$ extending also R_0 to a new ordering R_1 by declaring a greater than all elements in Y_0 i.e. $R_1 = R_0 \cup \{(y, a) : y \in Y_0\}$. Then R_1 well–orders Y_1 and $(Y_0, R_0) \sqsubset (Y_1, R_1)$, a contradiction. Hence $Y_0 = X$ and R_0 well–orders X. $\qquad\qquad\qquad\qquad\qquad\qquad\qquad\qquad\qquad\quad$ □

It follows that the axiom of choice, the axiom of well–ordering and the maximum principle are equivalent. There are a number of other statements which turn out to be equivalent to those three. For instance the following are often used in various branches of mathematics.

1. (Birkhoff) In any ordered set X, any chain is contained in a chain maximal with respect to inclusion

2. (Vaught) Any family $\mathcal{F}$ of sets contains a maximal with respect to inclusion sub–family consisting of pair–wise disjoint sets

Indeed, the Birkhoff principle follows from the maximum principle by the fact that the union of a chain $\mathcal{L}$ of chains is a chain being the supremum of L hence in a set of all chains over X ordered by inclusion there exists a maximal chain. By analogy, the Vaught principle follows from the maximum principle by the fact that the union of a linearly ordered by inclusion family $\mathcal{L}$ of sub–families of $\mathcal{F}$ consisting of pair–wise disjoint sets consists of pair–wise sets and constitutes the supremum of $\mathcal{L}$.

6.9   Finite versus infinite sets

We now are in position to define formally the notion of *finiteness* and to contrast it with the notion of *being infinite*.

For a set X, we say that X is a *finite set* (in the Dedekind sense) in case there exist: a natural number $n \in N$ and a bijection $f : X \to n$. In this case we say that the *cardinality* of X is n, in symbols $|X| = n$. Let us observe that given a singleton k, we have $|X| = |X \times \{k\}|$. From this observation we may derive

Proposition 6.16.
1. *Any subset of a finite set is finite*
2. *The union $\bigcup_i^m X_i$ of finite sets $X_1, X_2, ..., X_m$ is a finite set.*

Indeed, if $f : X \to n$ witnesses $|X| = n$ and $Y \subseteq X$ then $f(Y) \subseteq n$ hence either $f(Y) = n$ and thus $|Y| = n$ or $f(Y) \subset n$ hence $|Y| = m < n$ by Proposition 6.11. This proves (i).

Given $X_1, X_2, ..., X_m$ with $|X_i| = m_i$ for $i = 1, 2, ..., m$ we may observe that (a) $\bigcup_i^m X_i$ embeds as a subset into $\oplus_i X_i = \bigcup_i X_i \times \{i\}$ and as the latter is the union of pair–wise disjoint sets we have $|\oplus_i X_i| = \sum_i m_i$ hence it is a finite set. By (i) $\bigcup_i^m X_i$ is finite.

Apparently distinct is the *finiteness in the Tarski sense* viz. a set X is finite in the sense of Tarski when the following *Tarski property* is observed :

in any non–empty subset $A \subseteq 2^X$ there exists a set maximal in A with respect to inclusion

One may try to relate the two notions of finiteness; it turns out that they are equivalent under the axiom of choice (or, for that matter under anyone of its equivalent statements). To this end we prove first that

Proposition 6.17. *Every natural number n is finite in the Tarski sense hence each set finite in the Dedekind sense is finite in the Tarski sense.*

Proof. We introduce a predicate *Tfin(k)* read k *is finite in the Tarski sense.* We check that the set $A = \{n \in N : Tfin(n)\}$ satisfies premises of the principle of mathematical induction.

Clearly, $0 \in A$. Assuming that $n \in A$, we prove that $n + 1 \in A$. We have $2^{n+1} = 2^n \cup \{X \cup \{n+1\} : X \in 2^n\}$. Now, for a set $Y \subseteq 2^{n+1}$, either $Y \subseteq 2^n$ in which case in Y there exists a maximal element or it is not true that $Y \subseteq 2^n$ in which case in the set $Y_1 = \{B \in 2^n : B \cup \{n+1\} \in Y\}$ there exists a maximal set B_0. Then $B_0 \cup \{n+1\}$ is a maximal set in Y. It follows that $n+1 \in A$ and $A = N$ by the principle of mathematical induction (Proposition 6.10) so each natural number is finite in the Tarski sense.

Now, if X is finite in the Dedekind sense, $|X| = n$ being witnessed by $f : X \to n$, the sets $2^X, 2^n$ are in bijective correspondence via $F : Y \to f(Y)$, hence 2^X has the Tarski property. $\qquad\square$

The converse fact that any set finite in the Tarski sense is finite in the Dedekind sense requires the use of axiom of choice. First, we have to say what an infinite set is. We say that X is an *infinite set* in either sense when it is not finite in this sense. We begin with

Proposition 6.18. *Under axiom of choice: if a set X is not finite in the sense of Dedekind then X contains a copy of the set N of natural numbers.*

Proof. Assume that X is not Dedekind–finite so no function $f : n \to X$ can be onto (otherwise, we could pick, by the axiom of choice, a point c_x from any non–empty set $f^{-1}(x)$ and then the function $g : X \to n$ defined via the formula $g(x) = c_x$ would be an embedding of X into n so $|X| = m < n$ by Proposition 6.11 , a contradiction to infinity of X. Let s be a selector on the set $2^X \setminus \{\emptyset\}$ of all non–empty subsets of X. We define a function $h : N \to X$ defining the value $h(n)$ by induction on n. We assign $h(0) = x_0$ where x_0 is an arbitrarily chosen element in X. Assume that we have defined values $h(0), h(1), ..., h(n)$ in such a way that $h(i) \neq h(j)$ whenever $i \neq j$. Then we let $h(n + 1) = s(X \setminus \{h(0), ..., h(n)\})$ so $h(n + 1)$ is distinct from any of $h(0), .., h(n)$. By the principle of mathematical induction (Proposition 6.10), a function $h : N \to X$ is defined with $h(i) \neq h(j)$ when $i \neq j$ i.e. h embeds N into X and $h(N)$ is a copy of the set N of natural numbers.   □

A proposition follows stating that any Tarski–finite set is Dedekind–finite.

Proposition 6.19. *Under axiom of choice, each set finite in the Tarski sense is finite in the Dedekind sense.*

Proof. We prove this fact by showing that any set which is Dedekind–infinite is also Tarski–infinite. Assume thus that X is Dedekind infinite. By Proposition 6.18, X contains a copy Y of the set N of natural numbers so $Y = \{x_1, x_2, ..., x_n, ...\}$. Consider the family $Y_n = \{\{x_1, x_2, ..., x_n\} : n \in N\}$ of subsets of X. Clearly, this family has no maximal element hence X is no Tarski–finite set.   □

Proposition 6.20. *Under axiom of choice both kinds of finiteness coincide.*

We will from now on simply discuss *finite sets* without referring to the kind of finiteness. It also follows from our discussion that any infinite set has to contain a copy of the set of natural numbers. As a particular case, the set N itself is infinite.

6.10   Equipotency

We have defined finite sets as sets that map bijectively onto a natural number. This notion may be extended viz. for two sets X, Y we say that X and Y are *equipotent* whenever there exists a bijection $f : X \to Y$. We are justified in such statement as clearly the relation of equipotency is symmetrical: if $f : X \to Y$ is a bijection of X onto Y then $f^{-1} : Y \to X$ is a bijection of Y onto X. It is also evident that any set X is equipotent with itself, and that the relation of equipotency is *transitive*: if X, Y are equipotent and Y, Z are equipotent then X, Z are equipotent because of the evident fact that if $f : X \to Y$, $g : Y \to Z$ are bijections then the composition $g \circ f : X \to Z$ is also a bijection. In case X, Y are equipotent, we say that their *cardinalities* coincide, in symbols $|X| = |Y|$.

We may attempt at ordering a given set of cardinalities: we may say that $|X| \leq |Y|$ if there exists an injection $f : X \to Y$. Then clearly, $|X| \leq |X|$, and from $|X| \leq |Y|, |Y| \leq |Z|$ it follows that $|X| \leq |Z|$. So, the only question that remains if we would like to claim that the relation $\leq$ is an ordering on any family of sets is that whether this relation is weakly symmetrical. It turns out to be such and this fact is by no means trivial and requires a proof.

Proposition 6.21. *(the Cantor–Bernstein theorem) If* $|X| \leq |Y|$ *and* $|Y| \leq |X|$ *then* $|X| = |Y|$.

Proof. We apply in the proof the Knaster –Tarski fixed point theorem (Proposition 6.7). Assume that $|X| \leq |Y|$ and $|Y| \leq |X|$ so injections $f : X \to Y$ and $g : Y \to X$ exist.

Consider the function $h : 2^X \to 2^X$ defined as follows: for $A \subseteq X$

$$h(A) = g[Y \setminus f(X \setminus A)] \qquad (6.44)$$

It is easy to check that h is isotone: if $A \subseteq B$ then $X \setminus B \subseteq X \setminus A$ hence $f(X \setminus B) \subseteq f(X \setminus A)$ so

$$Y \setminus f(X \setminus A) \subseteq Y \setminus f(X \setminus B)$$

and hence

$$g[Y \setminus f(X \setminus A)] \subseteq g[Y \setminus f(X \setminus B)].$$

As the power set 2^X is completely ordered by inclusion, by the Knaster–Tarski theorem it follows that for some $C \subseteq X$ we have $h(C) = C$ i.e.

$$g[Y \setminus f(X \setminus C)] = C.$$

This means that f maps $X \setminus C$ onto the set B such that $g(Y \setminus B) = C$.

We may define then a function $k : X \to Y$ by the recipe

$$\begin{aligned} k(x) &= f(x) \text{ when } x \in X \setminus C \\ k(x) &= g^{-1}(x) \text{ when } x \in C \end{aligned} \qquad (6.45)$$

the inverse g^{-1} defined for the restriction $g|B$ of the function g to the set B. As both f, g are injections, it is clear that k is a bijection of X onto Y so X, Y are equipotent i.e. $|X| = |Y|$. $\qquad \square$

We now have proved that the relation $\leq$ is an ordering on any set of cardinalities.

We denote in particular the cardinality $|N|$ of the set of natural numbers N by the symbol ω. One may ask whether there exist sets whose cardinalities exceed that of ω i.e. about *degrees of infinity*. The classical result which has told us that there are more potent sets that N and at the same time has posed very difficult questions about cardinal numbers is the *Cantor diagonal theorem*.

Proposition 6.22. *(Cantor) For no set X there exists a function $f : X \rightarrow 2^X$ from X onto 2^X.*

Indeed, it is sufficient to consider the subset $Y = \{x \in X : x \notin f(x)\}$. Assume that $Y = f(y)$ for some $y \in X$. Then: if $y \in Y$ then $y \notin f(y)$ i.e. $y \notin Y$ and vice versa, if $y \notin Y$ then $y \in f(y)$ i.e. $y \in Y$. Finally we get: $y \in Y$ if and only if $y \notin Y$, a contradiction. Thus, Y cannot be the value of f and f cannot map X onto 2^X.

The upshot of this result is that no set X can be equipotent with its power set 2^X. In particular we have $|2^N| > \omega$ as clearly N embeds into 2^N via the function $f(n) = \{n\}$. We may form higher and higher cardinalities by looking at sets $2^N, 2^{2^N}, ...$ For our purposes, it will be enough to stay with finite sets or sets equipotent with ω which all together are called *countable sets*.

6.11   Countable sets

To make the realm of countable sets more familiar to us, we should give some examples of countable sets besides already known to us finite sets and N itself. In particular, a question may arise about the cardinality of Cartesian products $N \times N$, $N \times N \times N$, and in general, about the cardinality of powers $N^k = N_1 \times N_2 \times ... \times N_k$ where $N_i = N$ for $i = 1, 2, .., k$. It turns out that all these sets are countable being equipotent with N.

To demonstrate this one should exhibit some bijections among these sets and N. We begin with the square N^2.

Proposition 6.23. *The function $f(m,n) = \frac{(m+n)\cdot(m+n+1)}{2} + m$ when $(m,n) \neq (0,0)$ and $f(0,0) = 0$ is a bijection from N^2 onto N.*

Proof. The function f is injective: consider points of the form (p,q) where $p, q \in N$ in the Euclidean plane: each (p,q) lies in the line $x + y = p + q$ and each line $x + y = c$ contains $c + 1$ such integer points, viz., from $(c, 0)$ to $(0, c)$. It follows that $(p+q) \cdot (p+q+1)/2 + p$ is the number of segments connecting neighboring integer points on lines of the form $x + y = c$ where $c = 0, 1, 2, ...$ necessary to reach the point (p,q) from $(0,0)$. So if $f(m,n) = f(p,q)$ then necessarily $m = p, n = q$.

The function f is onto: by the above interpretation, for any k, going k segments the way as described above brings us to some (p,q). One may give an explicit formula for the inverse of f: for a natural number r, we have $f^{-1}(r) = (p,q)$ where p, q follow uniquely from the two equations:

$$p + q = \lfloor (\lfloor (8r + 1)^{\frac{1}{2}} \rfloor + 1)/2 \rfloor - 1$$
$$3p + q = 2r - (\lfloor (\lfloor (8r + 1)^{\frac{1}{2}} \rfloor + 1)/2 \rfloor - 1)^2$$
(6.46)

To get these equations, it suffices to let $r = \frac{(p+q) \cdot (p+q+1)}{2}$ and observe that $8r + 1 = (2p + 2q + 1)^2 + 8p$ from which it follows that $2p + 2q + 1 \leq (8r + 1)^{\frac{1}{2}} \leq 2p + 2q + 3$ implying 1 and 2.

It follows that f does establish the equipotence of N^2 with N. We may write down $f^{-1}(r) = (k(r), l(r))$ where k, l are functions which satisfy $k(r) = p, l(r) = q$ in the above equations. $\square$

We may iterate the above bijection: given the set N^k of ordered k–tuples of natural numbers we may define inductively a bijection f_k from N^k onto N viz. assuming $f_1, ..., f_k$ defined, we define

$$f_{k+1}(n_1, n_2, .., n_{k+1}) = f(f_k(n_1, n_2, .., n_k), n_{k+1})$$

where f is the bijection of Proposition 6.23. One may prove by induction that all f_k are bijections. Thus, for any $k \in N$, the set N^k of all k–tuples of natural numbers is countable.

This result may be carried yet further: we may consider the set

$$N^* = \bigcup \{N^k : k \in N\}$$

of all finite tuples of natural numbers (sometimes called from the grammatical point of view *strings*, or *words*). As for each k the function $f_k : N^k \to N$ is a bijection, and sets N^k are pair–wise disjoint, we may construct a bijection h of N^* onto N by first defining a function $g : N^* \to N \times N$, by letting $g(n_1, ..., n_k) = (f_k(n_1, ..., n_k), k)$ for each $k \in N$ and each string $(n_1, ..., n_k) \in N^k$ and then defining $h(n_1, ..., n_k) = f(g(n_1, ..., n_k))$.

Thus, the set of all finite tuples (strings) of natural numbers is countable.

A step further in this direction would be the ultimate result that a union of countably many countable sets is countable. It is indeed so, but the proof will require the axiom of choice.

Proposition 6.24. *Under axiom of choice, the union of a countable set X whose elements are countable sets is countable.*

Proof. As X is countable, we may enumerate its elements as $\{X_n : n \in N\}$ (where in case X be finite, $X_n = X_{n+1} = X_{n+2} = ...$ for some n). For each n, the set B_n of all injections from X_n onto N is non–empty so by the axiom of choice we may select a function $k_n \in B_n$ for each n. For $S = \bigcup X$, we look for each $x \in S$ at the least $n(x)$ such that $x \in X_{n(x)}$. We define a function $h : S \to N^2$ by letting $h(x) = (k_{n(x)}(x), n(x))$; clearly, h is an injection so the composition $f \circ h$ embeds S into N. $\square$

6.12   Filters and ideals

In ordered sets, one may single out certain subsets regular with respect to the ordering relation. Among them of primary interest are *ideals* and *filters*. Here we restrict ourselves to ideals and filters in fields of sets ordered by inclusion. Our primary field of sets will be that of 2^X for a non–empty set X. The notions of a *filter* and of *ideal* are dual to each other in the sense that if a family $\mathcal{F}$ is a filter then the family $\mathcal{I}_{\mathcal{F}} = \{X \setminus A : A \in \mathcal{F}\}$ is an ideal (and vice versa). Hence, we restrict ourselves to filters as the exposition of ideals follows by duality.

We consider a field of sets $\mathcal{B} \subseteq 2^X$. A family $\mathcal{F} \subseteq \mathcal{B}$ is a filter when
(F1) $A, B \in \mathcal{F} \Rightarrow A \cap B \in \mathcal{F}$
(F2) $A \subseteq B \wedge A \in \mathcal{F} \Rightarrow B \in \mathcal{F}$
(F3) $\emptyset \notin \mathcal{F}$

A filter is then a family which is closed with respect to the intersection and taking a superset and is distinct from $\mathcal{B}$ as it cannot have $\emptyset$ as an element. In particular, $X \in \mathcal{F}$ by (F2). Thus, by duality, an ideal is a family of sets in $\mathcal{B}$ closed with respect to unions and taking subsets so any ideal contains $\emptyset$ as an element and does not contain X.

As an example of a filter, we may take (in case $\mathcal{B} = 2^X$) for any $x \in X$ the set $\mathcal{F}_x = \{A \in 2^X : x \in A\}$ which clearly satisfies (F1), (F2), and (F3) called the *principal filter induced by* x. By duality, *the principal ideal induced by* x is the ideal $\mathcal{I}_x = \{A \in \mathcal{B} : x \notin A\}$.

We may notice an important property of a principal filter (respective of a principal ideal): given any filter $\mathcal{F}$ with $\mathcal{F}_x \subseteq \mathcal{F}$ (respectively any ideal $\mathcal{I}$ with $\mathcal{I}_x \subseteq \mathcal{I}$) we have $\mathcal{F}_x = \mathcal{F}$ (respectively $\mathcal{I}_x = \mathcal{I}$).

Indeed, assuming that $\mathcal{F}_x \subset \mathcal{F}$ we may pick a set $A \in \mathcal{F} \setminus \mathcal{F}_x$ hence $x \notin A$ but then $\{x\} \cap A = \emptyset$ contrary to (F3). Thus, $\mathcal{F}_x$ is a maximal with respect to inclusion filter; maximal filters are called also *ultrafilters*. Similarly, the ideal $\mathcal{I}_x$ is a *maximal ideal*. Principal filters and ideals are the only examples of maximal filters respectively ideals which we may construct effectively. In all other cases, to justify the existence of an ultrafilter or a maximal ideal we have to resort to the maximum principle.

Proposition 6.25. *By the maximum principle, any filter $\mathcal{F}$ may be extended to an ultrafilter $\mathcal{F}^* \supseteq \mathcal{F}$.*

Proof. Consider the set $\mathbf{F}$, of all filters on a field of sets $\mathbf{B}$ containing the filter $\mathcal{F}$ as a subset, ordered by inclusion. It remains to check that the union $\mathcal{S} = \bigcup \mathbf{L}$ of any chain $\mathbf{L}$ of filters in $\mathbf{F}$ is again a filter. To this end, consider $A, B \in \mathcal{S}$: there are filters $F, G \in \mathbf{L}$ such that $A \in F, B \in G$. By linearity of ordering on $\mathbf{L}$, either $F \subseteq G$ or $G \subseteq F$. In the former case $A, B \in F$, in the latter case $A, B \in G$ so either $A \cap B \in F$ or $A \cap B \in G$ and in either case $A \cap B \in S$ proving the condition (i). Similarly (ii, (iii) may be proved so we conclude that S is a filter. Clearly, $S = sup\mathbf{L}$. Thus, any chain in $\mathbf{F}$ has a supremum and the maximum principle implies the existence of a maximal element in $\mathbf{F}$ which is clearly an ultrafilter extending $\mathcal{F}$. $\qquad\square$

The dual result on the existence of a maximal ideal extending a given ideal follows by duality.

As one more important example of a filter, we may consider the set N of natural numbers and the family $cofN = \{A \subseteq N : N \setminus A \text{ is finite}\}$ i.e. the family of all co–finite subsets in N. That $cofN$ is a filter follows easily as the union of any two finite sets is finite and any subset of a finite set is finite. The filter $cofN$ is called *the Fréchet filter*. By the last proposition, there exists on N an ultrafilter $cofN^*$ extending the Fréchet filter.

We give now a characterization of ultrafilters in terms of set operations.

Proposition 6.26. *For a filter $\mathcal{F}$ the following are equivalent:*

1. *$\mathcal{F}$ is an ultrafilter*
2. *If a subset $Y \subseteq X$ is such that $Y \cap A \neq \emptyset$ for each $A \in \mathcal{F}$ then $Y \in \mathcal{F}$*
3. *For any A, B if $A \cup B \in \mathcal{F}$ then either $A \in \mathcal{F}$ or $B \in \mathcal{F}$*
4. *For any A either $A \in \mathcal{F}$ or $X \setminus A \in \mathcal{F}$.*

Proof. Property 1 implies Property 2: assume $\mathcal{F}$ an ultrafilter and $Y \cap A \neq \emptyset$ for each $A \in \mathcal{F}$. Was $Y \notin \mathcal{F}$, we could define $\mathcal{F}^*$ by adding to $\mathcal{F}$ all sets containing as a subset an intersection of the form $Y \cap A$ where $A \in \mathcal{F}$. Then clearly, $\mathcal{F}^*$ does satisfy (F1), (F2) and (F3) so $\mathcal{F}^*$ is a filter. Also $\mathcal{F} \subset \mathcal{F}^*$ as $Y \in \mathcal{F}^* \setminus \mathcal{F}$ but this is a contradiction as $\mathcal{F}$ is an ultrafilter.

Property 2 implies Property 3: assume $A \cup B \in \mathcal{F}$ but neither $A \in \mathcal{F}$ nor $B \in \mathcal{F}$. By Property 2 there exist $C, D \in \mathcal{F}$ with $C \cap A = \emptyset = D \cap B$. But then $(A \cup B) \cap (C \cap D) = \emptyset$ and $C \cap D \in \mathcal{F}$, a contradiction with (i).

Property 3 implies Property 4: as $A \cup X \setminus A = X \in \mathcal{F}$ either $A \in \mathcal{F}$ or $X \setminus A \in \mathcal{F}$ by Property 3.

Property 4 implies Property 1: assume to the contrary that $\mathcal{F}$ is not an ultrafilter so $\mathcal{F} \subset \mathcal{G}$ for some filter $\mathcal{G}$ hence there is $A \in \mathcal{G} \setminus \mathcal{F}$ i.e. $A \notin \mathcal{F}$. But also $X \setminus A \notin \mathcal{F}$: otherwise, we would have $X \setminus A \in \mathcal{G}$ and $A \cap X \setminus A = \emptyset \in \mathcal{G}$, a contradiction to (iii). $\qquad\square$

6.13 Equivalence, tolerance

Tasks of classification require that elements in a universe be clustered into aggregates within which the elements are identical or similar with respect to some properties while elements in distinct aggregates are clearly discernible with respect to some of those properties.

A simplest class of classifying relations are *equivalence relations*; we have already met equivalence functor in the sentential calculus: the functor E of equivalence is defined as follows: Epq has the same truth value as $Cpq \land Cqp$ and it follows that Epq has truth value 1 if and only if p, q take the same truth value.

Converting the logical equivalence into a relation comes easily: on the set M of meaningful expressions of the sentential calculus, we define the relation R_E by letting $\alpha R_E \beta$ if and only if $E\alpha\beta$. Then the relation R_E satisfies the following

1. R_E is reflexive: $\alpha R_E \alpha$ for every $\alpha \in M$

2. R_E is symmetric: $\alpha R_E \beta$ implies $\beta R_E \alpha$ for each pair $\alpha, \beta \in M$

3. R_E is transitive: $\alpha R_E \beta, \beta R_E \gamma$ imply $\alpha R_E \gamma$ for each triple $\alpha, \beta, \gamma \in M$

Generalizing this example, we call a relation R on a set X an *equivalence relation* if and only if R is reflexive, symmetric and transitive.

We recall the basic property of equivalence relations concerned with clustering of elements of X into classes of indiscernible objects. For an element $x \in X$, we let

$$[x]_R = \{y \in X : xRy\} \tag{6.47}$$

and we call $[x]_R$ the *equivalence class of x with respect to R*. Then we have

Proposition 6.27. *For each pair $x, y \in X$, classes $[x]_R, [y]_R$ are either disjoint or equal.*

We include a proof of this well–known fact. If $z \in [x]_R \cap [y]_R$ then by symmetry xRz, yRz hence xRz, zRy and by transitivity xRy; for $w \in [x]_R$ we have wRx hence wRy and it follows that $[x]_R \subseteq [y]_R$. By symmetry $[y]_R \subseteq [x]_R$ and finally $[x]_R = [y]_R$.

We will denote the set $\{[x]_R : x \in X\}$ by the symbol X/R and we will call it the *quotient set* of the set X by the relation R. There is a function $q_R : X \to X/R$ defined by letting $q_R(x) = [x]_R$, and called the *quotient function* which maps X onto the quotient set X/R.

A dual way of introducing equivalence relations is by means of *partitions*.

A *partition* on a set X is a family $\mathcal{F}$ of its subsets such that

(P1) $\mathcal{F}$ is non–empty

(P2) $\bigcup \mathcal{F} = X$

(P3) *If* $A, B \in \mathcal{F}$ *then* $A \cap B = \emptyset$

By the last proposition and the fact implied by reflexivity of R that $x \in [x]_R$ it follows that equivalence classes of the relation R form a partition $\mathbf{P}_R$ of the set X.

On the other hand, any partition $\mathbf{P}$ of X induces an equivalence relation R_P viz. we let $x R_P y$ if and only if there exists a set $A \in \mathbf{P}$ such that $x, y \in A$. The correspondence between equivalence relations and partitions is canonical i.e.

$$\mathbf{P}_{R_P} = \mathbf{P}, R_{\mathbf{P}_R} = R \qquad\qquad (6.48)$$

We may also compare equivalence relations on the set X: for equivalence relations R, S on X, we say that R is *finer* than S in symbols $R \trianglelefteq S$ if and only if $x R y$ implies $x S y$ for any pair x, y of elements of X. We have obvious

Proposition 6.28. $R \trianglelefteq S$ *if and only if* $[x]_R \subseteq [x]_S$ *for each* $x \in X$.

We say in case $R \trianglelefteq S$ that the partition $\mathbf{P}_R = \{[x]_R : x \in X\}$ is a *refinement* of the partition $\mathbf{P}_S = \{[x]_S : x \in X\}$.

The relation $\trianglelefteq$ is a complete ordering of the set $Eq(X)$ of all equivalence relations on X.

Proposition 6.29. *For any family* $\mathcal{R}$ *of equivalence relations on a set* X, *there exist equivalence relations* $inf\mathcal{R}, sup\mathcal{R}$ *with respect to* $\trianglelefteq$ *i.e. the family* $Eq(X)$ *of all equivalence relations on the set* X *is completely ordered by* $\trianglelefteq$.

Clearly, the intersection $\bigcap \mathcal{R}$ of all relations in $\mathcal{R}$ is an equivalence relation hence it is $inf\mathcal{R}$. For supremum, let us consider the relation $S = \bigcup \{R \in Eq(X) : \forall A : (A \in \mathcal{R} \Rightarrow A \subseteq R)\}$. By the part already proven, S is an equivalence relation and for any relation $R \in Eq(X)$ with $A \subseteq R$ for each $A \in \mathcal{R}$ we have $S \subseteq R$ so $S = sup\mathcal{R}$.

Assuming that $R \trianglelefteq S$, we may define on the quotient set X/R a new relation S/R by letting $[x]_R S/R[x']_R$ if and only if $x S x'$. It is evident that this definition does not depend on the choice of elements in $[x]_R, [x']_R$. It is also easy to see that $[[x_R]]_{S/R} = [x]_S$. We may write down this fact in a concise form as follows:

$$(X/R)/(S/R) = X/S \qquad\qquad (6.49)$$

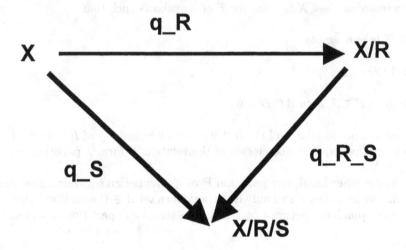

Figure 6.1: The commutative diagram: $q_S = q_{R/S} \circ q_R$

The following Figure 1 shows the commutative diagram of this situation.

Given equivalence relations $R \subseteq X \times X$ and $S \subseteq Y \times Y$, a function $f : X \to Y$ is an $R, S-morphism$ if the condition $xRy \Rightarrow f(x)Sf(y)$ holds. Then we have a function $\overline{f} : X/R \to Y/S$ with $\overline{f}([x]_R) = [f(x)]_S$; indeed, if zRx then $f(z)Sf(x)$ hence $f([x]_R) \subseteq [f(x)]_S$ and $\overline{f}$ is defined uniquely. Let us observe that $q_S \circ f = \overline{f} \circ q_R$. Again, in Figure 2 below, we have the commutative diagram for this situation

6.13.1   Tolerance relations

The idea of similarity is a weakening of the idea of equivalence; a similarity may be rendered in many ways leading to various formal notions. One of these ways is related to problems of visual perception where a similarity of objects may be interpreted as inability to distinguish between them e.g. the distance between the objects may be smaller then a given discernibility threshold: if points x, y in the real line are regarded as *similar* when their distance is less than δ then we may have x, y, and y, z similar while x, z need not be similar (cf. [Poincaré05]). Thus such a relation of similarity lacks the transitivity property.

The idea of a *tolerance relation* was introduced formally by C.E. Zeeman [Zeeman65] to capture this intuitive understanding of similarity. Following Zeeman, we call a relation R on a set X a *tolerance relation* when R is (i) reflexive (ii) symmetric. Thus a tolerance relation lacks the transitivity property of equivalence relations. This makes the analysis of tolerance relations

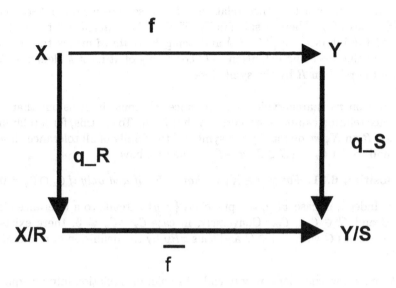

Figure 6.2: The commutative diagram: $q_S \circ f = \overline{f} \circ q_R$

more difficult and their structure more intricate.

If we have tried to mimic the case of equivalence relations then we would look at the sets $R(x)=\{y \in X : xRy\}$. However, due to the lack of transitivity, we cannot say for instance whether two elements $y, z \in R(x)$ are in the relation R and a fortiori there is nothing essential we may say about the structure of $R(x)$. So, we should with equivalence classes in mind, consider sets with the property that any two of its elements are in the relation R. Such sets are called *tolerance pre–classes*. Formally , a set $A \subseteq X$ is a tolerance pre–class when xRy for any pair $x, y \in A$. In particular, any singleton $\{x\}$ is a tolerance pre–class, and if xRy then $\{x, y\}$ is a tolerance preclass.

We may easily see that any union of a linearly ordered by inclusion family of tolerance pre—classes is a tolerance pre–class. From the maximum principle it follows that

Proposition 6.30. *Any tolerance pre–class A is a subset of a maximal with respect to inclusion tolerance pre–class.*

Maximal tolerance pre–classes are called *tolerance classes*. Within a tolerance class C, any two elements are in the relation R and for each element $y \notin C$ there is $x \in C$ such that $\neg yRx$.

An example of a tolerance relation is the following: given a family of non-empty sets $\mathcal{F}$, define a relation R on $\mathcal{F}$ via ARB if and only if $A \cap B \neq \emptyset$.

Then R is clearly a tolerance relation. Its classes are maximal 2-centered sub–families of $\mathcal{F}$ where a sub–family $F' \subseteq F$ is centered if for any finite $G \subseteq F'$ the intersection $\bigcap G \neq \emptyset$ and being k-centered means the specialization of the preceding definition to G consisting of at most k elements. We denote the relation R by the symbol $\tau_\cap$.

The relation $\tau_\cap$ is archetypical for tolerance relations: it turns out that any tolerance relation can be represented in that form. To see this, for a tolerance relation R on X, we denote by the symbol $\mathcal{C}$ the family of all tolerance classes of R and we let $\mathcal{C}_x = \{C \in \mathcal{C} : x \in C\}$. Then we have

Proposition 6.31. *For $x, y \in X$, we have xRy if and only if $\mathcal{C}_x \cap \mathcal{C}_y \neq \emptyset$.*

Proof. Indeed, in case xRy, the pre–class $\{x, y\}$ extends to a tolerance class $C \in \mathcal{C}$ and $C \in \mathcal{C}_x \cap \mathcal{C}_y$. Conversely, in case $\mathcal{C}_x \cap \mathcal{C}_y \neq \emptyset$, there exists a tolerance class C with $x, y \in C$ and thus xRy by the definition of a tolerance class. □

We may now consider a way to embed a tolerance relation into an equivalence relation. To this end, given a relation R on a set X, we define the n-th power R^n of R as the composition $R \circ R \circ R \circ .... \circ R$ of n copies of R. Clearly, xR^ny holds if and only if there exist $x = x_0, x_1, ..., x_n = y$ with x_iRx_{i+1} for $i = 0, .., n - 1$.

Then we define the *transitive closure* R^+ of the relation R as the union $\bigcup\{R^n : n \in N\}$. Then clearly $R \subseteq R^+$ and R^+ is transitive: given xR^+y, yR^+z, we have xR^ny, yR^mz for some $n, m \in N$ and then $xR^{n+m}z$. It follows from the construction of R^+ that it is the least transitive relation extending R.

We may notice that

Proposition 6.32. *For a tolerance relation R, the transitive closure R^+ is an equivalence relation. For any $x \in X$, the union of tolerance classes of R containing x is a subset of the equivalence class of x by R^+: $\bigcup \mathcal{C}_x \subseteq [x]_{R^+}$.*

Proof. As R^+ is reflexive since $R \subseteq R^+$ and symmetric by the symmetry of R, R^+ is an equivalence relation because we have checked its transitivity. For any tolerance class $C \ni x$, $C \times C \subseteq R$ so $C \times C \subseteq R^+$ i.e. $C \subseteq [x]_{R^+}$. It follows that $\bigcup \mathcal{C}_x \subseteq [x]_{R^+}$. □

The relation R^+ is constructed by inflating tolerance classes until they become disjoint or identical; we may think also of a way of constructing an equivalence by restricting tolerance classes. To this end, we first introduce a new notion: given a family $\mathcal{F}$ of subsets of a set X, we consider a new family of sets constructed in the following way. For any sub–family $\mathcal{G} \subseteq \mathcal{F}$, we let $A_\mathcal{G} = \bigcap \mathcal{G} \setminus \bigcup(\mathcal{F} \setminus \mathcal{G})$. We let $\mathcal{C}(\mathcal{F}) = \{A_\mathcal{G} : \mathcal{G} \subseteq \mathcal{F}\}$. Sets of the form $A_\mathcal{G}$ are called *components* of $\mathcal{F}$. Let us observe that

Proposition 6.33. *For $\mathcal{G}, \mathcal{G}' \subseteq \mathcal{F}$, if $\mathcal{G} \neq \mathcal{G}'$ then $A_G \cap A_{G'} = \emptyset$. Thus the family $\mathcal{C}(\mathcal{F})$ of components of $\mathcal{F}$ is a partition of $Y = \bigcup \mathcal{F}$.*

Indeed, for $\mathcal{G} \neq \mathcal{G}'$, let e.g. $A \in \mathcal{G} \setminus \mathcal{G}'$. Then $A_G \subseteq A$ and $A_{G'} \cap A = \emptyset$. Hence $A_G \cap A_{G'} = \emptyset$.

We now apply the idea of components to the family $\mathcal{C}_R$ of tolerance classes of a tolerance relation R on a set X.

Proposition 6.34. *Components in the family $\mathcal{C}(\mathcal{C}_R)$ are equivalence classes of the equivalence relation E defined by the condition xEy if and only if $\forall z(zRx \Leftrightarrow zRy)$.*

Proof. The relation E is clearly an equivalence (we have xEy if and only if $R(x) = R(y)$). Assume xEy. Then for any tolerance class $C \ni x$, $y \in C$ as yRz for any $z \in C$. Obviously the converse holds by symmetry of R: for any tolerance class $C \ni y$, we have $x \in C$. Hence for $\mathcal{G} = \{C \in \mathcal{C}_R : x \in C\}$ we have $\mathcal{G} = \{C \in \mathcal{C}_R : y \in C\}$. It follows that x, y belong to the same component A_G. Conversely, if x, y belong to the same component A_G, then given xRz we have a tolerance class C with $x, z \in C$; but $y \in C$ hence yRz. By symmetry, if yRz then xRz. It follows that xEy.   □

Historic remarks

We have seen already the logic of general terms (Syllogistic) being a first theory of sets put into a logical form by Aristotle. In medieval logic, sets were identified with properties (e.g. Joscelin de Soissons) and the same view was shared by Georg Cantor, the founder of modern Set Theory (cf. [Cantor62]) who applied towards sets the naive approach. Bertrand Russell (cf. [Frege03]) found in 1902 a paradox (the Russell antinomy) which forced researchers in Set Theory to find axiomatic foundations for the notion of a set. This paradox may be briefly described as follows: assuming the naive idea of a set, we may form the set x consisting of all objects y with the property that $y \notin y$. The question arises : is $x \in x$ or not. If $x \in x$ then x has the defining property and thus $x \notin x$ and if $x \notin x$ then x cannot have the defining property hence $x \in x$. Thus a contradiction follows: $x \in x$ if and only if $x \notin x$.

Attempts at constructing a Set Theory free of antinomies resulted in a system of axioms proposed by Ernest Zermelo [Zermelo08]. Axiom of replacement was proposed by, among others, Fraenkel so the axiomatic theory based on those axioms is called the Zermelo–Fraenkel Set Theory. The precise notions of infinity and finiteness come from Richard Dedekind [Dedekind881] and the notion of finiteness in the Tarski sense from [Tarski24]. The Knaster–Tarski fixed point theorem is from [Knaster28]. The Cantor–Bernstein theorem was proved by Bernstein (cf. [Borel898]). The completion technique of an ordered set by means of cuts comes from Dedekind and in a generalized version from MacNeille [MacNeille37]. The theorem on well ordering is due to Zermelo [Zermelo04], the Zorn lemma comes from [Zorn35], the Birkhoff theorem from [Birkhoff67] and the Vaught lemma from [Vaught52].

Boolean algebras are so called to honor the memory of George Boole (cf. [Boole847]) who introduced algebraic approach to logic and established basic laws of algebra of sets.

Tolerance relations were formally introduced by Zeeman [Zeeman65] and they were informally discussed earlier by Poincaré [Poincaré05] as well as Menger [Menger42].

Exercises

In Exercises 1–25 we propose to verify some statements of the algebra of sets corresponding to axioms and theses of Syllogistic. The reader is advised to find the figure of Syllogistic corresponding to the given theorem of algebra of sets. All sets A, B, C below are assumed to be non–empty.

1. Verify: $A \subseteq A$

2. Prove that if $A \subseteq B$ and $B \subseteq C$ then $A \subseteq C$

3. Verify that if $A \subseteq B$ and $A \cap C \neq \emptyset$ then $B \cap C \neq \emptyset$

4. Prove that if $A \subseteq B$ then $A \setminus B = \emptyset$

5. Check that $A \setminus B = \emptyset$ implies $A \subseteq B$

6. Verify that $A \subseteq B$ and $C \subseteq A$ imply $C \cap B \neq \emptyset$

7. Prove that $A \subseteq B$ and $C \cap B = \emptyset$ imply $A \cap C = \emptyset$

8. Verify that $A \cap B = \emptyset$ and $C \subseteq A$ imply $C \cap B = \emptyset$

9. Check that $A \cap B = \emptyset$ and $C \subseteq A$ imply $C \setminus B \neq \emptyset$

10. Check that $A \subseteq B$ and $C \cap A \neq \emptyset$ imply $C \cap B \neq \emptyset$

11. Verify that $A \cap B = \emptyset$ and $C \cap A \neq \emptyset$ imply $C \setminus B \neq \emptyset$

12. Check that $A \cap B = \emptyset$ and $C \subseteq A$ imply $C \cap B = \emptyset$

13. Check that $A \cap B = \emptyset$ and $C \subseteq A$ imply $C \setminus A \neq \emptyset$

14. Verify that $A \subseteq B$ and $C \cap B = \emptyset$ imply $C \cap A = \emptyset$

15. Prove that $A \cap B = \emptyset$ and $C \subseteq A$ imply $C \setminus B \neq \emptyset$

16. Prove that $A \subseteq B$ and $C \cap B = \emptyset$ imply $C \setminus A \neq \emptyset$

17. Verify that $A \subseteq B$ and $C \cap B = \emptyset$ imply $C \cap A = \emptyset$

18. Prove that $A \cap B = \emptyset$ and $C \cap B \neq \emptyset$ imply $C \setminus A \neq \emptyset$

19. Verify that $A \subseteq B$ and $C \setminus B \neq \emptyset$ imply $C \setminus A \neq \emptyset$

20. Prove that $A \subseteq B$ and $A \subseteq C$ imply $B \cap C \neq \emptyset$

21. Verify that $A \subseteq B$ and $B \cap C = \emptyset$ imply $A \cap C = \emptyset$

22. Prove that $A \subseteq B$ and $B \cap C = \emptyset$ imply $A \setminus C \neq \emptyset$

23. Check that $A \cap B \neq \emptyset$ and $B \subseteq C$ imply $A \cap C \neq \emptyset$

24. Show that $A \cap B = \emptyset$ and $B \subseteq C$ imply $C \setminus A \neq \emptyset$

25. Verify that $A \cap B = \emptyset$ and $B \cap C \neq \emptyset$ imply $C \setminus A \neq \emptyset$

In Exercises 26–32 we propose to verify some properties of relations in particular ordering relations and equivalence relations.

26. [Kuratowski–Mostowski65] Verify that $X \times Y \circ Z \times T$ is identical to $\emptyset$ in case $Y \cap Z = \emptyset$ and to $X \times T$ in case $Y \cap Z \neq \emptyset$.

27. Prove that for a function $f : X \to Y$ we have (i) $ff^{-1}B = B$ for any $B \subseteq Y$ (ii) $A \subseteq f^{-1}fA$ for any $A \subseteq X$ (iii) $f(A \cap f^{-1}B = B \cap fA$ for any pair $A \subseteq X, B \subseteq Y$. Give an example showing that in (ii) one may have a strict inclusion.

28. [Schröder895] Assume that $\leq$ is a reflexive and transitive relation on a set X. Define a relation R on X by letting xRy if and only if $x \leq y$ and $y \leq x$. Prove that (i) R is an equivalence relation on X (ii) the relation $\leq_R$ induced on the quotient set X/R by letting $[x]_R \leq_R [y]_R$ if and only if $x \leq y$ is an ordering relation.

29. [Birkhoff67] Consider ordered sets $(X, \leq)$, $(Y, \preceq)$. A pair of functions $f : X \to Y$, $g : Y \to X$ does establish a *Galois connection* between X and Y if and only if the following requirements are satisfied (1) $x \leq y$ implies $f(x) \succeq f(y)$ (2) $x \preceq y$ implies $g(x) \geq g(y)$ (3) $x \leq g(f(x))$ (4) $y \preceq f(g(x))$.

Prove that (i) $g \circ f : X \to X$ is an isotone function on X (ii) $f \circ g : Y \to Y$ is an isotone function on Y.

30. [Birkhoff67] In the conditions of Exercise 29, prove that

$$g(f(g(f(x)))) = g(f(x)), f(g(f(g(y)))) = f(g(y)).$$

[Hint: on one hand, apply (1) and (2) to respectively (3), (4). On the other hand, substitute $x/g(y)$ in (3) and apply (2) to the result; similarly, substitute $y/f(x)$ in (4) and apply (1)]

31. [Birkhoff67] A function $c : X \to X$ on an ordered set $(X, \leq)$ is said to be a *closure* if and only if (1) $x \leq c(x)$ (2) $c(c(x)) = c(x)$. Prove that $f \circ g, g \circ f$, where f, g as in Exercises 29–30, are closures on respectively X, Y.

32. Consider a relation $R \subseteq X \times Y$. For $A \subseteq X$ define $A^+ = \{y \in Y : \forall x \in A(xRy)\}$ and similarly for $B \subseteq Y$ let $B^* = \{x \in X : \forall y \in B(xRy)\}$. Prove that functions $(.)^{+,*}$ establish a Galois connection.

In Exercises 33–34 we propose to prove some facts about *trees* in particular the *König lemma*.

A *tree* is an ordered set $(X, \leq)$ such that (1) there exists the least element $x_0 \in X$ (called *the root*) (2) for each $x \in X$ the set $\{y \in X : y \leq x\}$ is ordered linearly by $\leq$.

Levels of the tree $(X, \leq)$ are defined inductively: the level 0, lev_0 is $\{x_0\}$; assuming levels $lev_0, ..., lev_k$ already defined, we let $x \in lev_{k+1}$ if and only if there exists $y \in X$ with the properties (3) $y < x$ (4) $y \in lev_k$ (5) there is no $z \in X$ with $y < z < x$. In this case we let $y = f(x)$ defining thus a function $f : A \to X$ on a subset $A \subseteq X$ of non–root elements.

We assume of trees $(X, \leq)$ discussed here that the set X is the union of levels lev_k for $k = 1, 2, ..., n, ...$ A subset $B \subseteq X$ is said to be a *branch* of the tree $(X, \leq)$ if and only if (6) $x \in B$ implies $f(x) \in B$ for every x. The *length* of a branch is its cardinality.

33. Prove that (i) every level lev_k of a tree $(X, \leq)$ consists of elements pairwise incomparable with respect to $\leq$ (ii) for $A = lev_i$ we have $f^{-1}A = lev_{i+1}$, each i.

34. [König27] Consider a tree $(X, \leq)$ with the property that for every $n \in N$ there exists a branch of length n. Prove that under axiom of choice if all levels of the tree are finite then there exists a branch of infinite length. [Hint: let σ be a selector for the family of non–empty finite subsets of X and $E = \{x \in X : \forall n \exists y (x \in lev_k \Rightarrow y \in lev_{k+n})\}$. Define inductively a sequence $(a_n)_n$ of elements of X as follows: $a_0 = x_0$, $a_{n+1} = \sigma(E \cap f^{-1}(a_n))$,

verify the correctness of this definition and prove that $\{a_n : n \in N\}$ is an infinite branch]

In Exercise 35, we propose to look at *graphs*. A graph over a set V is a pair (V, E) where V is a set and E is a set of unordered pairs $\{x, y\}$ where $x \in V, y \in V$ with $x \neq y$. A graph (V, E) is *complete* in case E is the set $[V]^2$ of all unordered pairs $\{x, y\}$ where $x, y \in V, x \neq y$. A graph (V', E') such that $V' \subseteq V, E' \subseteq E$ is said to be a *sub–graph* of the graph (V, E). The graph $(V, [V]^2 \setminus E)$ is the *complement of the graph* (V, E) denoted $(V, E)'$.

35. [Ramsey30] Assume that (V, E) is a graph over an infinite set V. Prove that under axiom of choice either (V, E) contains a complete sub–graph over an infinite set or $(V, E)'$ contains a complete sub–graph over an infinite set. [Hint: it is sufficient to assume V infinite countable i.e. $V = \{x_n : n \in N\}$. For a finite subset $\Gamma \subseteq V$, let $K(\Gamma) = \{y \in V \setminus \Gamma : \forall x \in \Gamma(\{x, y\} \in E)\}$ and $L(\Gamma) = \{y \in V \setminus \Gamma : \forall x \in \Gamma(\{x, y\} \notin E)\}$. Denote by σ a selector on the family $2^V \setminus \{\emptyset\}$ of all non–empty subsets of V. Begin with $\{x_0\} = \Delta_0$ and assuming $\Delta_0, ..., \Delta_k$ already defined, let $\Delta_{k+1} = \Delta_k \cup \{\sigma(K(\Delta_k))\}$ if $K(\Delta_k) \neq \emptyset$. Observe that if this induction continues indefinitely, we get a complete sub–graph of (V, E) as required. Otherwise stop at $\Delta_k = \Gamma_0$. Choose first $x_i \in \Gamma_0$ with the set $L(x_i)$ infinite and repeat the above procedure in $L(x_i)$ either getting a complete sub–graph of (V, E) or producing a set Γ_1. Repeating this procedure infinitely many times by induction either arrive at a complete sub–graph of (V, E) or at a sequence $(\Gamma_k, x_k)_{k \geq 1}$ in which case $\{x_k : k \geq 1\}$ is a complete sub–graph of $(V, E)'$]

Works quoted

[Balcar– Štěpánek86] B. Balcar and P. Štěpánek, *Teorie Množin*, Academia, Praha, 1986.

[Birkhoff67] G. Birkhoff, *Lattice Theory*, 3rd ed., AMS, Providence, 1967.

[Boole847] G. Boole, *The Mathematical Analysis of Logic*, Cambridge, 1847.

[Borel898] E. Borel, *Leçons sur la Théorie des Fonctions*, Paris, 1898.

[Cantor62] G. Cantor, *Gesammelte Abhandlungen mathematischen und philosophischen Inhalts*, Hildesheim, 1962.

[Dedekind881] R. Dedekind, *Was sind und was sollen die Zahlen*, Braunschweig, 1881.

[Frege03] G. Frege, *Grundgesetze der Arithmetik 2*, Jena, 1903.

[Knaster28] B. Knaster, *Un théoréme sur les fonctions d'ensembles*, Ann. Soc. Polon. Math., 6(1928), pp. 133–134.

[König27] D. König, *Über eine Schlussweise aus dem Endlichen ins Unendliche*, Acta litt. ac sc. univ. Franc. Josephinae, Sec. Sc. Math., 3(1927), pp. 121–130.

[Kuratowski–Mostowski65] K. Kuratowski and A. Mostowski, *Set Theory*, Polish Scientific Publ., Warsaw, 1965.

[Mac Neille37] H. M. Mac Neille, *Partially ordered sets*, Trans. Amer. Math. Soc., 42(1937), pp. 416–460.

[Menger42] K. Menger, *Statistical metrics*, Proc. Natl. Acad. Sci. USA, 28 (1942), pp. 535 – 537.

[Poincaré05] H. Poincaré, *Science et Hypothèse*, Paris, 1905.

[Ramsey30] F. P. Ramsey, *On a problem of formal logic*, Proc. London. Math. Soc., 30(1930), pp. 264–286.

[Schröder895] E. Schröder, *Algebra der Logik*, Leipzig, 1895.

[Tarski24] A. Tarski, *Sur les ensembles finis*, Fund. Math., 6(1924), pp. 45–95.

[Vaught52] R. L. Vaught, *On the equivalence of the axiom of choice and the maximal principle*, Bull. Amer. Math. Soc., 58(1952), p. 66.

[Zeeman65] E. C. Zeeman, *The topology of the brain and the visual perception*, in: *Topology of 3-manifolds and Selected Topics*, K. M. Fort (ed.), Prentice Hall, Englewood Cliffs, NJ, 1965, pp. 240–256.

[Zermelo08] E. Zermelo, *Untersuchungen über die Grundlagen der Mengenlehre I*, Math. Annalen, 65(1908), pp. 261–281.

[Zermelo04] E. Zermelo, *Beweiss das jede Menge wohlgeordnet werden kann*, Math. Annalen, 59(1904), pp. 514–516.

[Zorn35] M. Zorn, *A remark on method in transfinite algebra*, Bull. Amer. Math. Soc., 41(1935), pp. 667–670.

Chapter 7

Topological Structures

[...] I could be bounded in a nutshell and count myself a king of infinite space [...]

Shakespeare, *Hamlet*, II

7.1   Introduction

Topology is a theory of certain set structures which have been motivated by attempts to generalize geometric reasoning and replace it by more flexible schemes. In many schemes of reasoning one resorts to the idea of a *neighbor* with the assumption that reasonably selected neighbors of a given object preserve its properties in satisfactory degree (cf. methods based on the notion of the *nearest neighbor*.) The notion of a neighbor as well as a more general notion of a *neighborhood* are studied by topology.

Basic topological notions were originally defined by means of a notion of *distance* which bridges geometry to more general topological structures. The analysis of properties of distance falls into theory of *metric spaces* with which we begin our exposition of fundamentals of topology.

7.2   Metric spaces

Consider a copy $\mathbf{R}^1$ of the set of *real numbers*. The distance between real numbers is defined usually by means of the *natural metric* $|x - y|$. Denoting $|x - y|$ by $\rho(x, y)$, we may write down the essential properties of the natural metric well-known from elementary mathematics:

1. $\rho(x, y) \geq 0$ and $\rho(x, y) = 0$ if and only if $x = y$

2. $\rho(x, y) = \rho(y, x)$

3. $\rho(x,z) \leq \rho(x,y) + \rho(y,z)$ (the *triangle inequality*)

Properties 1.–3. are taken as characterizing any distance function (called also a *metric*). Thus, a *metric* ρ on a set X is a function $\rho : X^2 \to R^1$ which satisfies the conditions 1–3. A set X endowed with a metric ρ is called a *metric space*. Elements of a metric space are usually called *points*.

The natural metric is an example of a metric. Other examples are for instance

the function $\rho = \sqrt{\sum_{i=1}^{k}(x_i^2 - y_i^2)}$ on the vector space $\mathbf{R}^k$ (the *Euclidean metric*)
the function $\rho(x,y) = max_i|x_i - y_i|$ on $\mathbf{R}^k$ (the *Manhattan metric*)
the function $\rho(x,y) = \sum_i |x_i - y_i|$ on $\mathbf{R}^k$

An example of a metric on an arbitrary set X may be the *discrete metric* $\rho(x,y)$ defined as $\rho(x,y) = 1$ when $x \neq y$ and $\rho(x,y) = 0$ when $x = y$.

In general, in order to construct a non–trivial metric we would need a more refined set–theoretic context.

Now, we will examine set structures induced by a metric. So assume ρ is a metric on a set X. The elementary notion of topology is the notion of a *neighborhood*. To derive this notion, we first introduce the notion of an *open ball*. For $x \in X$ and a positive $r \in \mathbf{R}^1$, the *open ball of radius r about x* is the set $B(x,r) = \{y \in X : \rho(x,y) < r\}$.

Let us record the basic property of open balls.

Proposition 7.1. *Assume that* $y \in B(x,r) \cap B(z,s)$. *Then there exists* $t \in R^1$ *with the property that* $B(y,t) \subseteq B(x,r) \cap B(z,s)$.

Proof. Let $t = min\{r - \rho(x,y), s - \rho(z,y)\}$ and assume that $w \in B(y,t)$. By the triangle inequality, $\rho(w,x) \leq \rho(w,y) + \rho(y,x) < r - \rho(x,y) + \rho(x,y) = r$ and similarly $\rho(w,z) \leq \rho(w,y) + \rho(y,z) < s - \rho(z,y) + \rho(z,y) = s$. It follows that $w \in B(x,r) \cap B(z,s)$ and it follows further that $B(y,t) \subseteq B(x,r) \cap B(z,s)$. $\square$

An *open* set in the metric space (X, ρ) is a union of a family of open balls; formally $U \subseteq X$ is *open* when $U = \bigcup \mathcal{B}$ where $\mathcal{B}$ is a family of open balls. We have

Proposition 7.2. *In any metric space* (X, ρ)

1. *The intersection of any two open balls is an open set*
2. *The intersection of any finite family of open sets is an open set*
3. *The union of any family of open sets is an open set.*

Proof. (1) follows from Proposition 7.1 as for any point in the intersection of two balls there exists an open ball about that point contained in the intersection so the intersection is a union of open balls. For (2), consider open sets U, W with $U = \bigcup \mathcal{B}$, $W = \bigcup \mathcal{C}$ where $\mathcal{B}, \mathcal{C}$ are families of open balls. Then $U \cap W = \bigcup \mathcal{B} \cap \bigcup \mathcal{C} = \bigcup \{B \cap C : B \in \mathcal{B}, C \in \mathcal{C}\}$ and as any intersection $B \cap C$ of open balls is a union of open balls by (1), it follows that $U \cap W$ is a union of a family of open balls hence it is an open set. By induction on n one can then prove that any intersection $U_1 \cap U_2 \cap \ldots \cap U_n$ of open sets is an open set. For (3), if $T = \bigcup \mathcal{U}$ where $\mathcal{U}$ is a family of open sets so each $U \in \mathcal{U}$ is a union of a family of open balls, $U = \bigcup \mathcal{B}_U$, then $T = \bigcup \{B : \exists U \in \mathcal{U} \ (B \in \mathcal{B}_U)\}$ i.e. T is the union of a family of open balls and so T is an open set. $\square$

We denote by the symbol $\mathcal{O}(X)$ the family of all open sets in a metric space (X, ρ). By the last proposition this family is closed with respect to finite intersections and all unions. Clearly, $\emptyset$ and X are open sets, the former as the union of the empty family of open balls the latter as e.g. the intersection of the empty family of open balls.

A dual notion of a *closed set* is also of principal interest. A set $K \subseteq X$ is *closed* in case its complement $X \setminus K$ is open. Thus the family $\mathcal{C}(X)$ of all closed sets in X is closed with respect to finite unions and arbitrary intersections.

As not any subset of X is open or closed, one may try to characterize the position of sets in X with respect to open or closed sets. To this end, we introduce the operation of *interior* of a set as follows.

For a set $A \subseteq X$, we let

$$Int A = \bigcup \{B(x, r) : B(x, r) \subseteq A\} \tag{7.1}$$

i.e. $Int A$ is the union of all open balls contained in A; clearly, $Int A$ is an open set, moreover by its definition it is the largest open subset of A. The operator Int is the *interior* operator and its value $Int A$ is the *interior of the set A*. This operation has the following properties

Proposition 7.3. *In any metric space (X, ρ):*

1. *$Int \emptyset = \emptyset$*

2. *$Int A \subseteq A$*

3. *$Int(A \cap B) = Int A \cap Int B$*

4. *$Int(Int A) = Int A$*

Proof. As $\emptyset$ is the only open subset of itself, so property 1 holds. Clearly $intA \subseteq A$ by definition i.e. Property 2 is fulfilled. For Property 3, $x \in IntA \cap IntB$ if and only if $B(x, r) \subseteq A$ and $B(x, s) \subseteq B$ for some $r, s > 0$ if and only if $B(x, t) \subseteq A \cap B$ where $t = min\{r, s\}$ if and only if $x \in Int(A \cap B)$. Finally, Property 4 follows as $IntA$ is an open set and thus $Int(IntA) = IntA$. $\square$

The dual operator Cl of *closure* is defined via the formula

$$Cl A = X \setminus Int(X \setminus A) \tag{7.2}$$

Thus, by duality, ClA is a closed set containing A and it is the smallest among such sets. From the previous proposition, the following properties of Cl follow.

Proposition 7.4. *For any metric space* X, ρ

1. $ClX = X$

2. $A \subseteq ClA$

3. $Cl(A \cup B) = ClA \cup ClB$

4. $Cl(ClA) = ClA$

Proof. For Property 1 we have $ClX = X \setminus Int(X \setminus X) = X \setminus Int\emptyset = X \setminus \emptyset = X$. In case of Property 2, as $Int(X \setminus A) \subseteq X \setminus A$, we have $A = X \setminus (X \setminus A) \subseteq X \setminus Int(X \setminus A) = ClA$. Similarly Property 3 follows: $Cl(A \cup B) = X \setminus Int(X \setminus (A \cup B)) = X \setminus Int((X \setminus A) \cap (X \setminus B)) = X \setminus (Int(X \setminus A) \cap Int(X \setminus B)) = X \setminus (Int(X \setminus A)) \cup X \setminus (Int(X \setminus B)) = ClA \cup ClB$. Finally, ClA is a closed set hence it is identical to its closure: $Cl(ClA) = ClA$. $\square$

The difference $ClA \setminus IntA = BdA$ is a closed set called the *boundary* of A. In case $BdA = \emptyset$ we say that A is *closed–open* (or, *clopen* for short).

For any $x \in X$, a set $A \subseteq X$ is called a *neighborhood* of x if and only if $x \in IntA$. It is easy to see that the intersection of two neighborhoods of x is a neighborhood of x and that any set containing a neighborhood of x is itself a neighborhood of x i.e. the family $N(x)$ of all neighborhoods of x is a *filter*. Usually one does not need to look at all neighborhoods of x. It is sufficient to look at a *co–final* sub–family of $N(x)$ viz. we call a sub–family $B(x)$ of $N(x)$ a *neighborhood basis* if and only if for any $A \in N(x)$ there exists $B \in B(x)$ with the property that $B \subseteq A$. Thus, $B(x)$ contains *arbitrarily small* neighborhoods of x.

Proposition 7.5. *In any metric space* (X, ρ), *any point* x *has a countable neighborhood basis.*

Proof. It suffices to select for each $n \in N$ the open ball $B(x, 1/n)$ and let $B(x) = \{B(x, 1/n) : n \in N\}$. Indeed, given any $A \in N(x)$, as $x \in IntA$, there exists an open ball $B(x, r)$ with $r > 0$ and $B(x, r) \subseteq A$. For $0 < 1/n < r$, we have $B(x, 1/n) \subseteq B(x, r)$. $\qquad\square$

The fact that neighborhood bases of all points in a metric space may be chosen countable, makes it possible to describe metric topology by means of *sequences*. A *sequence* in a set X is the image $f(N)$ of the set N of natural numbers under a function $f : N \to X$. Usually, we write down a sequence as the set $\{f(n) : n \in N\}$ or in a concise form as $(x_n)_n$ disregarding the function f (which after all may be fully restored from the latter notation as $f(n) = x_n$, each n).

A sequence $(x_n)_n$ is *converging to a limit* $x \in X$ if and only if each neighborhood A of x contains *almost all* (meaning: all but a finite number) members x_n of the sequence. In this case we write $x = lim x_n$. More formally, $x = lim x_n$ if and only if for each $r > 0$ there exists $n(r) \in N$ such that $x_m \in A$ for each $m \geq n(r)$. A canonical example of a sequence converging to x may be obtained via axiom of choice by selecting a point $x_n \in B(x, 1/n)$ for each $n \in N$.

We may now characterize interiors and closures in a metric space by means of neighborhoods as well as by means of sequences.

Proposition 7.6. *In any metric space* (X, ρ)

1. *$x \in IntA$ if and only if there exists a neighborhood $P \in N(x)$ such that $P \subseteq A$*

2. *$x \in IntA$ if and only if for any sequence $(x_n)_n$ with $x = lim x_n$ there exists $n_0 \in N$ such that $x_m \in A$ for each $m \geq n_0$*

3. *$x \in ClA$ if and only if for every $P \in N(x)$ we have $P \cap A \neq \emptyset$*

4. *$x \in ClA$ if and only if there exists a sequence $(x_n)_n \subseteq A$ with $x = lim x_n$*

Proof. Property 1 follows by definition of $IntA$. Property 2 follows by definition of a limit of a sequence: if $x \in IntA$ and $x = lim x_n$ then almost every $x_n \in IntA$ a fortiori almost every $x_n \in A$. Assume conversely that $x \notin IntA$; then for any $n \in N$, we have that it is not true that $B(x, 1/n) \subseteq A$ so we may choose $x_n \in B(x, 1/n) \setminus A$ for $n \in N$. Obviously, $x = lim x_n$ and $\{x_n : n \in N\} \cap A = \emptyset$ proving 2.

Property 3 follows from Property 1 by duality : $x \in ClA$ if and only if $x \notin Int(X \setminus A)$. As for Property 4, if $x \in ClA$ then $B(x, 1/n) \cap A \neq \emptyset$ so selecting $x_n \in B(x, 1/n) \cap A$ for each n, we produce a sequence $(x_n)_n \subseteq A$ with $x = lim x_n$.

Conversely if $x = lim x_n$ for a sequence $(x_n)_n \subseteq A$, then for any neighborhood $P \in N(x)$ we have $P \cap \{x_n : n \in N\} \neq \emptyset$ a fortiori $P \cap A \neq \emptyset$ so $x \in ClA$ in virtue of 3.   □

7.3   Topological Cartesian products

It may be desirable at this point to introduce some techniques for constructing more intricate metric spaces of general importance. We now begin with a discussion of topologies on Cartesian products. For a family $\mathcal{F}$ $= \{F_i : i \in I\}$ of sets, the *Cartesian product* $\Pi_i F_i$ is the set of all functions $f : I \to \bigcup \mathcal{F}$ with the property that $f(i) \in F_i$ for each $i \in I$. We refer to a function $f \in \Pi_i F_i$ as to a *thread* and we usually write down this thread in the form $(f_i)_i$ i.e. as a *generalized sequence*.

Assume now that each F_i is a metric space with a metric ρ_i; we consider a question: is there a way to introduce a metric on the Cartesian product $\Pi_i F_i$ so it be compatible in a sense with metrics ρ_i? It turns out that in order to answer this question positively we should restrict ourselves to *countable* products in which case the set I is countable.

In case we have a Cartesian product $\Pi_n F_n$ of a countable family $\mathcal{F} = \{F_n : n \in N\}$ of metric spaces (F_n, ρ_n), we first *normalize* metrics ρ_n by letting $\eta_n(x, y) = min\{1, \rho_n(x, y)\}$ for $x, y \in F_n$, each n. Then

Proposition 7.7. *For each n, metrics ρ_n, η_n induce the same topology in the sense that families of open sets are identical in both metrics. Hence, convergence of sequences is identical with respect to both metrics.*

Indeed, for any set $A \subseteq F_n$, if A is open with respect to ρ_n i.e. $B_\rho(x, r(x)) \subseteq A$ for each $x \in A$ then $B_\eta(x, min\{1, r(x)\}) \subseteq A$ for each $x \in A$ i.e. A is open with respect to η. The converse follows on same lines.

We say that metrics ρ, η are *equivalent*. We may assume that ρ_n is already bounded by 1 for each n. Then we let for $f = (f_i)_i, g = (g_i)_i \in \Pi_i F_i$:

$$\rho(f, g) = \sum_{n=0}^{\infty} \frac{1}{2^n} \cdot \rho_n(f_n, g_n) \qquad (7.3)$$

and we declare ρ to be a *product metric* on the Cartesian product $\Pi_n F_n$. It is evident that ρ is a metric.

A classical example of a metric Cartesian product is the *Cantor cube* $\mathcal{C}$ $= \Pi_n \{0, 1\}_n$ where $\{0, 1\}_n = \{0, 1\}$ for each n. Using the symbol **2** for the set $\{0, 1\}$, we may use for the Cantor cube the symbol 2^N as any thread $f \in \mathcal{C}$ may be regarded as the *characteristic function* of a set $A \subseteq N$ i.e.

$A = \{n \in N : f(n) = 1\}$ and thus C is the set of function codes for subsets of the set of natural numbers.

Regarding the set $\mathbf{2} = \{0, 1\}$ as a metric space with the discrete metric, we have the product metric

$$\rho(f, g) = \sum_{n=0}^{\infty} \frac{1}{2^n} \cdot |f_n - g_n| \qquad (7.4)$$

on $\mathbf{2}^N$.

We may be interested in basic neighborhoods in this metric space. Let us observe that any set of the form

$$B(i_1, i_2, ..., i_n) = \{f \in \mathbf{2}^N : f_k = i_k, k = 1, 2, .., n\} \qquad (7.5)$$

where $i_1, i_2, ..., i_n \in \mathbf{2}$ is open: given $f \in B(i_1, i_2, ..., i_n)$ we have $B(f, \frac{1}{2^{n+1}})$ $\subseteq B(i_1, i_2, ..., i_n)$; indeed, $g \in B(f, \frac{1}{2^{n+1}})$ implies $g_i = f_i$ for $i = 1, 2, ..., n$. Conversely, any open ball $B(f, r)$ contains a neighborhood of f of the form $B(i_1, i_2, ..., i_n)$: it suffices to take the least n with the property that $\sum_{j=n+1}^{\infty} \frac{1}{2^j}$ $< r$ and then $B(f_1, ..., f_n) \subseteq B(f, r)$. We have verified

Proposition 7.8. *Open sets of the form $B(i_1, i_2, ..., i_n)$ form a basis for the product metric topology of the Cantor cube: each open set is the union of a family of sets of the form $B(i_1, i_2, ..., i_n)$.*

In particular, for any $f \in \mathbf{2}^N$, we have $N(f) = \{B(f_1, f_2, ..., f_n) : n \in N\}$.

Using the Cantor cube as an example, we may exhibit two important properties of topological, in particular metric, spaces viz. *completeness* and *compactness*.

To begin with, let us look at an arbitrary sequence $(x_n)_n$ in the Cantor cube. As each x_n may take at each coordinate i only one of two values $0, 1$, there exists $y \in \mathbf{2}^N$ with the property that for each $j \in N$ the set $A_j = \{x_n : x_{n_i} = y_i, i = 1, 2, ..., j\}$ is infinite. Let n_j be the first natural number n such that $x_n \in A_j$. Then the *sub-sequence* $(x_{n_j})_j$ converges to y: for any basic neighborhood $B(y_1, y_2, .., y_j)$ of y we have $x_n \in B(y_1, y_2, .., y_j)$ for $n \geq n_j$. We have proved that

Proposition 7.9. *Any sequence in the Cantor cube contains a convergent subsequence.*

Of a metric space with this property we say that it is *compact*. Let us establish other characterizations of compactness, more convenient in case of general topological spaces.

7.4   Compactness in metric spaces

We say that a family $\mathcal{U}$ of open sets in a metric space X is an *open covering* of X if and only if $X = \bigcup \mathcal{U}$. A family $\mathcal{U}' \subseteq \mathcal{U}$ is a *sub–covering* in case $X = \bigcup \mathcal{U}'$.

A family of sets $\mathcal{F}$ is *centered* if and only if for any finite sub–family $\mathcal{F}' \subseteq \mathcal{F}$ the intersection $\bigcap \mathcal{F}'$ is non–empty.

Let us observe that in a compact metric space X there exists a countable base for open sets: to this end we observe that compact metric spaces have the property of having *finite nets*.

Proposition 7.10. *In any compact metric space X, ρ, for any $r > 0$ there exists a finite subset $X_r = \{x_1, ..., x_{k_r}\}$ with the property that for any $y \in X$ there exists $j \le k_r$ with $\rho(y, x_j) < r$.*

Proof. Assume to the contrary that for some $r > 0$ no finite subset $Y \subseteq X$ satisfies the condition in the proposition. Let us select $x_0 \in X$. There exists $x_1 \in X$ with $\rho(x_0, x_1) \ge r$ and in general once we find $x_0, x_1, .., x_m$ with $\rho(x_i, x_j) \ge r$ for $i \ne j, i, j \le m$ then we can find $x_{m+1} \in X$ with $\rho(x_{m+1}, x_j) \ge r$ for $j = 1, 2, ..., m$. By the principle of mathematical induction we define thus a sequence $(x_n)_n$ in X with $\rho(x_i, x_j) \ge r$ for $i \ne j$ which clearly cannot have any convergent subsequence. $\square$

A set X_r is called an *r–net*; thus, in compact metric spaces there exist r-nets for each $r > 0$. Now, let us select for each $n \in N$ a $1/n$-net X_n. Let $Y = \bigcup \{X_n : n \in N\}$. Then for each $x \in X$ and any $r > 0$, there exists $y \in Y$ with $\rho(x, y) < r$.

It follows that the family $\{B(y, 1/m) : y \in Y, m \in N\}$ is a *countable base for open sets*: given an open ball $B(x, r)$ and $z \in B(x, r)$, we may find $y \in Y$ and $m \in N$ such that $2/m < r - \rho(x, z)$ and $\rho(z, y) < 1/m$ and then $z \in B(y, 1/m) \subseteq B(x, r)$. We have proved

Proposition 7.11. *In any compact metric space X there exists a countable base for open sets.*

It follows that in case of any compact metric space X we may restrict ourselves to countable coverings: any open covering contains a countable sub–covering; indeed, given an open covering $\mathcal{U}$ of X and a countable base $\mathcal{B}$ for open sets, the sub–family $\mathcal{U}'$, defined by selecting for each $B \in \mathcal{B}$ a $U \in \mathcal{U}$ such that $B \subseteq U$, is a countable covering of X. Now we may turn to promised characterizations.

Proposition 7.12. *For a metric space (X, ρ) the following statements are equivalent*

  1. X is compact

2. *any countable centered family of closed sets in X has a non–empty intersection*

3. *any countable open covering of X has a finite sub–covering*

Proof. Assume Property 1. Let $\mathcal{F} = \{F_n : n \in N\}$ be a centered family of closed sets. Hence each intersection $F_1 \cap ... \cap F_n$ is non–empty and we may select a point $x_n \in F_1 \cap ... \cap F_n$ for each n. There exists a convergent sub–sequence $(x_{n_k})_k$ with $x = \lim x_{n_k}$. Then $x \in F_n$ for each n and thus $x \in \bigcap \mathcal{F}$. This proves 2.

Now, 3 follows from 2 by duality: given a countable open covering $\{U_n : n \in N\}$ without any finite sub–covering, the family $\mathcal{F} = \{F_n = X \setminus U_n : n \in N\}$ is a centered family of closed sets so by 2 there exists $x_0 \in \bigcap \mathcal{F}$. But then $x_0 \notin \bigcup \mathcal{U}$, a contradiction. $\square$

Finally, Property 3 implies Property 1, as if we had a sequence $(x_n)_n$ without any convergent sub–sequence then we would have for each $x \in X$ an open neighborhood U_x with the property that $x_n \in U_x$ for finitely many n only. Then the open covering $\{U_x : x \in X\}$ contains a finite sub–covering $U_{x_1} \cup ... \cup U_{x_k} = X$ which misses infinitely many elements of the sequence $(x_n)_n$, a contradiction.

The last proposition indicates that compactness may be defined in purely topological terms of open, respectively closed, sets regardless of a specific metric from the class of equivalent metrics inducing the given topology. Compactness allows for existential arguments based on the convergent sub–sequences.

We have started with an example of the Cantor cube as a compact metric space. To give more examples, let us observe that the unit interval $I = [0,1]$ is a compact metric space in natural metric topology induced by the metric $\rho(x,y) = |x - y|$; indeed, assume that $\mathcal{U}$ is an open covering of I (we may also assume that elements of $\mathcal{U}$ are open intervals of the form $(a, b), [0, a)$, or $(a, 1]$). Was I not compact, we could consider a subset $C \subseteq I$ defined as follows: $x \in C$ if and only if the interval $[0, x]$ may be covered by a finite number of elements of $\mathcal{U}$. As under our assumption, $1 \notin C$, there exists $s = \sup C < 1$. We may pick an element $s \in (a, b) \in \mathcal{U}$ along with some $a < t < s$. Then the interval $[0, t]$ may be covered by finitely many elements $U_1, .., U_k$ of $\mathcal{U}$ hence $U_1, .., U_k, (a, b)$ cover any interval of the form $[0, w]$ for $s < w < b$, contrary to the definition of s. It follows that $\sup C = 1$ and a fortiori $[0, 1]$ is compact. Same argument will show that each interval $[a, b]$ in the real line is compact.

7.5 Completeness in metric spaces

A weakening of compactness is the condition of *completeness*. This notion depends on the notion of a *fundamental sequence*. A sequence $(x_n)_n$

is *fundamental* if and only if for each $r > 0$ there exists $n_r \in N$ such that $\rho(x_{n_r+k}, x_{n_r}) < r$ for each k. Let us observe that each convergent sequence is fundamental.

A metric space (X, ρ) is *complete* if and only if each fundamental sequence converges in X.

Proposition 7.13. *If a fundamental sequence $(x_n)_n$ contains a convergent sub-sequence then the sequence $(x_n)_n$ converges. Hence any compact metric space is complete.*

Indeed, assuming a sub-sequence $(x_{n_k})_k$ to be convergent to a point x, we have $\rho(x, x_n) \leq \rho(x, x_{n_k}) + \rho(x_{n_k}, x_n)$ and it is sufficient to observe that $\rho(x, x_{n_k})$, $\rho(x_{n_k}, x_n)$ tend to 0 as n, n_k tend to ∞. Hence $\rho(x, x_n)$ approaches 0 as n tends to ∞ and this implies $x = \lim x_n$. Thus, in a compact metric space any fundamental sequence converges implying completeness.

As any fundamental sequence in the real line $\mathbf{R}^1$ is bounded i.e. it is contained in a closed interval of the form $[a, b]$ which is compact, it contains a convergent subsequence and it does converge itself by Proposition 7.13. Thus the real line in natural metric is a complete metric space. One may prove (cf. Exercise 26) that the set of irrational real numbers is also a complete metric space under the natural metric.

As with compactness, we may attempt at characterizing completeness in topological terms. Contrary to the case of compactness, we will not get rid of the metric.

For a subset $A \subseteq X$ of a metric space (X, ρ), we define the *diameter of A, diam$A$* as follows: $diamA = sup\{\rho(x, y) : x, y \in A\}$. We say that a family $\mathcal{A}$ of subsets of X contains *arbitrarily small sets* if and only if for each $r > 0$ there exists $A \in \mathcal{A}$ with $diamA \leq r$.

Proposition 7.14. *(Cantor) A metric space (X, ρ) is complete if and only if each centered countable family $\mathcal{F}$ of closed non-empty sets containing arbitrarily small sets has a non-empty intersection.*

Proof. Assume (X, ρ) is complete. We may also assume that $\mathcal{F} = \{F_n : n \in N\}$ satisfies the condition $F_{n+1} \subseteq F_n$ for each n. Indeed, otherwise we could consider the family $\mathcal{G} = \{F_1 \cap ... \cap F_n : F_i \in \mathcal{F}\}$ as $\bigcap \mathcal{G} = \bigcap \mathcal{F}$. Selecting $x_n \in F_n$ for each n, we produce a sequence $(x_n)_n$ which is fundamental: $\rho(x_{n+k}, x_n) \leq diamF_n$ and $\lim diamF_n = 0$. Hence there exists a convergent sub-sequence $(x_{n_k})_k$ with the limit $x \in X$. Clearly, $x \in ClF_n = F_n$ for each n so $x \in \bigcap \mathcal{F}$.

Now, we assume that the condition on $\mathcal{F}$ is satisfied and we prove completeness. Given a fundamental sequence $(x_n)_n$ in X, we consider $F_n =$

$Cl\{x_k : k \geq n\}$ and $\mathcal{F} = \{F_n : n \in N\}$. Clearly, $F_{n+1} \subseteq F_n$ for each n and $limdiam F_n = 0$. Thus, there exists $x \in \bigcap \mathcal{F}$, a fortiori, $x = lim x_n$: for each ball $B(x, r)$ we have $F_n \subseteq B(x, r)$ for almost every n hence $x_n \in B(x, r)$ for almost every n. $\qquad\qquad\qquad\qquad\qquad\qquad\qquad\qquad\qquad\qquad\qquad\qquad\quad$ $\square$

Complete metric spaces proved to be very important due to their properties of which we recall here the two most fundamental and very often recalled in applications viz. the *Baire category theorem* and the *Banach fixed point theorem*.

The Baire category theorem basically states that a complete metric space cannot be covered by "small" subsets. We should define the meaning of "small".

We say that a subset $A \subseteq X$ is *meager* when $Int A = \emptyset$ and A is *nowhere dense* in case $Int Cl A = \emptyset$. When $B = \bigcup \{A_n : n \in N\}$ and each A_n is nowhere dense then we call B a *set of* 1st *category*. An archetypical example of a meager set is $Cl A \setminus A$ for any A; indeed, was $Int(Cl A \setminus A) \neq \emptyset$ we would have $G \subseteq Cl A \setminus A$ for some open non–empty G implying $G \cap A = \emptyset$ on one hand and $G \cap A \neq \emptyset$ on the other.

Proposition 7.15. *(the Baire category theorem) In any complete metric space* (X, ρ), *any set of the* 1st *category is meager.*

Proof. Consider a set $Y = \bigcup \{Y_n : n \in N\}$ of the 1st category. For an arbitrary $x_0 \in X$, we define inductively closed sets F_n for $n \in N$ such that $x_0 \in Int F_n$ and

$$F_{n+1} \subseteq Int F_n$$

for each n and $\mathcal{F} = \{F_n : n \in N\}$ contains arbitrarily small sets.

To begin with, as $Int Cl Y_0 = \emptyset$, taking an arbitrary open ball $B(x_0, s_0)$ we find

$$B(y_0, r_0) \subseteq Cl B(y_0, r_0) \subseteq B(x_0, s_0) \setminus Cl Y_0 \qquad (7.6)$$

with $r_0 < 1/2$. Assuming that open balls $B(y_0, r_0), ..., B(y_k, r_k)$ with

$$\begin{aligned} &(i) \ Cl B(y_{i+1}, r_{i+1}) \subseteq B(y_i, r_i) \ for \ i = 1, 2, .., k-1 \\ &(ii) \ Cl B(y_i, r_i) \cap Cl Y_i = \emptyset \ for \ i \leq k \\ &(iii) \ r_i < \tfrac{1}{2^{i+1}} \ for \ i \leq k \end{aligned} \qquad (7.7)$$

have been defined, we repeat our argument with Y_{k+1} in place of Y_0 and $B(y_k, r_k)$ in place of $B(x_0, s_0)$ to produce

$$B(y_{k+1}, r_{k+1}) \subseteq Cl B(y_{k+1}, r_{k+1}) \subseteq B(y_k, r_k) \setminus Cl Y_{k+1}$$

with $r_{k+1} < \frac{1}{2^{k+2}}$.

By induction, we arrive at the sequence $(F_k = ClB(y_k, r_k))_k$ satisfying (7.7) and thus $F_{k+1} \subseteq F_k$, $F_k \subseteq X \setminus ClY_k$, each k, and $lim diam F_k = 0$. It follows by completeness that there exists $x \in \bigcap\{F_k : k \in N\}$ and clearly $x \notin Y$. As $x \in B(x_0, s_0)$, we have proved that $\bigcup\{Y_k : k \in N\}$ contains no non–empty open set i.e. it is meager. $\square$

It follows from the proof that the difference $X \setminus Y$ is *dense* in X: any open non–empty ball contains a point from $X \setminus Y$. In other form: $Cl(X \setminus Y) = X$.

The Baire category theorem is an important tool in proving the existence (although non–effectively) of certain objects and moreover in proving that these objects form a large (not of the 1st category) set (like e.g. a continuous nowhere differentiable real function etc.).

The Banach fixed point theorem also allows for proving existential facts. It has to do with *contracting functions*. A function $f : X \to X$ on a metric space (X, ρ) is a *contracting function* with a *contraction factor* $0 < \lambda < 1$ if and only if $\rho(f(x), f(y)) \leq \lambda \cdot \rho(x, y)$. Then we claim that f has a *fixed point* c i.e. $f(c) = c$.

Proposition 7.16. *(the Banach fixed point theorem) Any contracting function on a complete metric space (X, ρ) has a (unique) fixed point.*

Proof. Let us begin with an arbitrary $x_0 \in X$. We define inductively a sequence $(x_n)_n$ by letting $x_{n+1} = f(x_n)$ for each n. This sequence is fundamental: letting $K = \rho(x_0, x_1) = \rho(x_0, f(x_0))$, we have $\rho(x_1, x_2) = \rho(f(x_0), f(x_1)) \leq \lambda \rho(x_0, x_1) = \lambda \cdot K$. Analogously, $\rho(x_2, x_3) \leq \lambda^2 \cdot K$, and in general $\rho(x_n, x_{n+1}) \leq \lambda^n \cdot K$. Thus, we have

$$\rho(x_{n+k}, x_n) \leq K \cdot \sum_{i=n}^{i=n+k-1} \lambda^i = K \cdot \lambda^n \frac{1 - \lambda^k}{1 - \lambda} \qquad (7.8)$$

and hence $\rho(x_{n+k}, x_n)$ tends to 0 as $n, n + k$ tend to ∞. By completeness, there exists $c = lim x_n$. As $\rho(f(c), f(x_n)) \leq \lambda \cdot \rho(c, x_n)$ we have $f(c) = lim f(x_n) = lim x_{n+1} = c$ so c is a fixed point of f. Observe that f has a unique fixed point: was d a fixed point other than c, we would have $\rho(c, d) = \rho(f(c), f(d)) \leq \lambda \cdot \rho(c, d)$, a contradiction.

Passing with k to ∞ in the formula (7.8) and noticing that $c = lim x_{n+k}$, we have

$$\rho(c, x_n) \leq K \cdot \lambda^n \frac{1}{1 - \lambda} \qquad (7.9)$$

which allows for an estimate of the depth of procedure necessary to approximate c with x_n with a given accuracy. In order to have $\rho(c, x_n) \leq \delta$ it suffices by (7.9) to have

$$K \cdot \lambda^n \frac{1}{1 - \lambda} \leq \delta \qquad (7.10)$$

hence it suffices that

$$n = \lfloor \frac{logK^{-1} \cdot (\delta \cdot (1 - \lambda))}{log\lambda} \rfloor \qquad (7.11)$$

We now pass to general topological spaces.

7.6 General topological spaces

By a *topological space*, we mean a set X along with a *topology* τ defined usually as a family $\tau \subseteq 2^X$ of sets in X which satisfies the conditions for open sets familiar to us from Proposition 7.2 viz.

1. for any finite sub–family $\tau' \subseteq \tau$, we have $\bigcap \tau' \in \tau$

2. for any sub–family $\tau' \subseteq \tau$, we have $\bigcup \tau' \in \tau$

Thus, τ is closed with respect to operations of finite unions and arbitrary intersections; let us notice that $\emptyset \in \tau$ as the union of the empty family and $X \in \tau$ as the intersection of the empty family.

Elements of τ are *open sets*. As with metric spaces, we introduce the operator Int of *interior* defined as follows:

$$IntA = \bigcup \{U \in \tau : U \subseteq A\} \qquad (7.12)$$

Thus $IntA$ is the union of all open subsets of A and a fortiori it is the greatest open subset of A. By closeness of τ on finite intersections and unions, we obtain the following properties of Int

Proposition 7.17. *(O1)* $Int\emptyset = \emptyset$
(O2) $IntA \subseteq A$
(O3) $Int(A \cap B) = IntA \cap IntB$
(O4) $Int(IntA) = IntA$

Indeed, 1, 2 follow by remarks following the definition of Int, 3 follows by the finite intersection property of τ and 4 does express obvious fact that any open set is its interior. Let us observe that by (O4) we have also

(O5) $A \subseteq B$ implies $IntA \subseteq IntB$.

Conversely, open sets may be introduced via the interior operator Int satisfying properties (O1)–(O4).

Proposition 7.18. *If an operator $I : 2^X \to 2^X$ does satisfy (O1)–(O4) then $\tau = \{A \subseteq X : A = IA\}$ is a topology on X and I is the interior operator Int with respect to τ.*

Proof. We call A such that $A = IA$ open. As $I(A \cap B) = IA \cap IB$ by (O3), the intersection of two open sets is open and by induction it follows that τ is closed with respect to finite intersections. For a family $\{A_i : i \in I\}$ where $IA_i = A_i$, we have by (O5) $A_i = IA_i \subseteq I\bigcup_i A_i$ for each i hence $\bigcup_i A_i \subseteq I\bigcup_i A_i$ and finally by (O2) $\bigcup_i A_i = I\bigcup_i A_i$ i.e. $\bigcup_i A_i \in \tau$. $\square$

By duality, we may define the *closure* operator Cl via

$$ClA = X \setminus Int(X \setminus A) \tag{7.13}$$

The closure operator does satisfy dual conditions

(C1) $ClX = X$
(C2) $A \subseteq ClA$
(C3) $Cl(A \cup B) = ClA \cup ClB$
(C4) $Cl(ClA) = ClA$

and we may define the family $\mathcal{C}(X)$ of *closed sets* in X by letting $\mathcal{C}(X) = \{A \subseteq X : ClA = A\}$. By duality, this family is closed with respect to finite unions and arbitrary intersections. The family $\tau = \{X \setminus A : A \in \mathcal{C}(X)\}$ is clearly the family of open sets with respect to topology induced by the operator Cl.

Operators Int and Cl give rise to a topological calculus on sets. Let us observe that both these operators are monotone: for $A \subseteq B$, we have $IntA \subseteq IntB$ ((O5)) and

(C5) $ClA \subseteq ClB$

by the dual to (O5) property of Cl. We begin with a useful lemma of this calculus.

Proposition 7.19. *For an open set P and an arbitrary set A we have: $P \cap ClA \subseteq Cl(P \cap A)$.*

Indeed, for $x \in P \cap ClA$ and any open set $Q \ni x$, we have $(Q \cap P) \cap A \neq \emptyset$ as $Q \cap P$ is a neighborhood of x and $x \in ClA$, hence $Q \cap (P \cap A) \neq \emptyset$ i.e. $x \in Cl(P \cap A)$.

Various algebras of sets may be defined with help of these operations. We begin with two of the most important.

7.7 Regular sets

We say that a set $A \subseteq X$ is *regular open* if and only if $A = Int(ClA)$ i.e. A is the interior of its closure.

Similarly we say that a set $A \subseteq X$ is *regular closed* if and only if $A = Cl(IntA)$ i.e. A is the closure of its interior.

Both types of sets are related to each other by duality: A is regular open if and only if $X \setminus A$ is regular closed. Indeed, if $A = IntClA$ then $X \setminus A = X \setminus IntClA = X \setminus Int(X \setminus (X \setminus ClA)) = Cl(X \setminus ClA) = Cl(Int(X \setminus A))$.

We now prove a basic fact that *regular open sets* form a Boolean algebra, denoted $RO(X)$. In order to prove this fact, we introduce a new auxiliary symbol, and we let $A^{\perp} = X \setminus ClA$. Then A is regular open if and only if $A = A^{\perp\perp}$. Indeed, $A^{\perp\perp} = X \setminus Cl(X \setminus ClA) = IntClA$. We sum up properties of the operation $A^{\perp}$.

Proposition 7.20. *The following are properties of* $A^{\perp}$

1. *if $A \subseteq B$ then $B^{\perp} \subseteq A^{\perp}$*

2. *if A is an open set then $A \subseteq A^{\perp\perp}$*

3. *if A is an open set then $A^{\perp} = A^{\perp\perp\perp}$ hence $A^{\perp\perp} = A^{\perp\perp\perp\perp}$*

4. *if A, B are open sets then $(A \cap B)^{\perp\perp} = A^{\perp\perp} \cap B^{\perp\perp}$*

5. *$(A \cup B)^{\perp} = A^{\perp} \cap B^{\perp}$*

6. *if A is an open set then $(A \cup A^{\perp})^{\perp\perp} = X$*

Proof. Property 1 follows immediately: $A \subseteq B$ implies $ClA \subseteq ClB$ implies $X \setminus ClB \subseteq X \setminus ClA$. For Property 2, A open implies $A \subseteq IntClA = A^{\perp\perp}$ as $A \subseteq ClA$. In case of Property 3, we get $A^{\perp} \subseteq A^{\perp\perp\perp}$ from 2 by substituting $A^{\perp}$ for A; also from 2 we get by applying 1: $A^{\perp\perp\perp} \subseteq A^{\perp}$ hence 3 follows. For Property 4 we have by applying 1 twice to the inclusion $A \cap B \subseteq A$ that $(A \cap B)^{\perp\perp} \subseteq A^{\perp\perp}$ and similarly $(A \cap B)^{\perp\perp} \subseteq B^{\perp\perp}$ hence $(A \cap B)^{\perp\perp} \subseteq A^{\perp\perp} \cap B^{\perp\perp}$. □

For the converse, we apply Proposition 7.19 and we observe that its statement $A \cap ClB \subseteq Cl(A \cap B)$ may be paraphrased as

$$A^{\perp\perp} \cap B \subseteq (A \cap B)^{\perp\perp} \tag{7.14}$$

Hence for A, B open, we have $A \cap ClB \subseteq Cl(A \cap B)$ i.e.

$$(A \cap B)^{\perp} = X \setminus Cl(A \cap B) \subseteq X \setminus (A \cap ClB) = (X \setminus A) \cup B^{\perp} \tag{7.15}$$

From $(A \cap B)^{\perp} \subseteq (X \setminus A) \cup B^{\perp}$, we get

$$A \cap B^{\perp\perp} = ((X \setminus A) \cup B^{\perp})^{\perp} \subseteq (A \cap B)^{\perp\perp} \qquad (7.16)$$

Substituting in

$$A \cap B^{\perp\perp} \subseteq (A \cap B)^{\perp\perp} \qquad (7.17)$$

$A^{\perp\perp}$ for A, and applying the second statement in 3, we come at

$$A^{\perp\perp} \cap B^{\perp\perp} \subseteq (A^{\perp\perp} \cap B)^{\perp\perp} \subseteq (A \cap B)^{\perp\perp\perp\perp} = (A \cap B)^{\perp\perp} \qquad (7.18)$$

proving Property 4.

Property 5 follows immediately by duality. To check Property 6, let us observe that in case A open we have $A \cup A^{\perp} = Cl A \setminus A$ hence

$$(A \cup A^{\perp})^{\perp} = IntCl(Cl A \setminus A) = Int(Cl A \setminus A) = \emptyset \qquad (7.19)$$

as $Cl A \setminus A$ is meager. Thus $(A \cup A^{\perp})^{\perp\perp} = X$. $\qquad \square$

Let us observe that from the second part of 3 above: $A^{\perp\perp} = A^{\perp\perp\perp\perp}$ it follows that $IntClIntCl A = IntCl A$ and $ClIntClInt A = ClInt A$ for any A. A consequence of this is that one may obtain from a given set A by applying $Int, Cl, \setminus$ only at most 14 distinct sets. The reader will easily write them down. From this property also follows immediately that any set of the form $A^{\perp\perp}$ is regular open, i.e. any set of the form $IntCl A$ is regular open. By duality, any set of the form $ClInt B$ is regular closed.

Now, we pass to the family $RO(X)$ of regular open sets.

Proposition 7.21. $RO(X)$ *is a Boolean algebra under operations* $\wedge, \vee$, *defined as follows*

1. $A \vee B = (A \cup B)^{\perp\perp} = IntCl(A \cup B)$

2. $A \wedge B = A \cap B$

3. $A' = A^{\perp} = X \setminus Cl A$

and with constants $0 = \emptyset, 1 = X$.

Proof. All boolean operations listed above give regular open sets by properties of $(.)^{\perp}$ listed above in Proposition 7.20. It remains to check that axioms of a Boolean algebra are satisfied. Commutativity laws $A \vee B = B \vee A, A \wedge B = B \wedge A$ are satisfied evidently. The laws $A \vee 0 = A, A \wedge 1 = A$ are also manifest. We have $A \wedge A' = A \cap A^{\perp} = A \setminus Cl A = \emptyset = 0$ as well as $A \vee A' = (A \cup A^{\perp\perp})^{\perp\perp}) = X = 1$. The distributive laws $A \vee (B \wedge C) = (A \vee B) \wedge (A \vee C)$ as well as $A \vee (B \wedge C) = (A \vee C) \wedge (A \vee C)$ hold by Property 5 and dualization. $\qquad \square$

A particular sub–algebra of $RO(X)$ is the algebra $CO(X)$ of *clopen sets* in X. In case of $CO(X)$ boolean operations $\vee, \wedge, \prime$ specialize to usual set–theoretic operations $\cup, \cap, \setminus$ i.e. $CO(X)$ is a field of sets.

The basic distinction between $RO(X)$ and $CO(X)$ is the fact that $RO(X)$ is a *complete boolean algebra* for any X while $CO(X)$ is not always such. We prove this important fact.

Proposition 7.22. *The boolean algebra $RO(X)$ is complete for any topological space X.*

Proof. Let us observe that the boolean ordering relation $\leq$ is in this case the inclusion $\subseteq$. Consider $\mathcal{A} \subseteq RO(X)$. Let $s(\mathcal{A}) = (\bigcup \mathcal{A})^{\perp\perp}$; we check that $s(\mathcal{A})$ is the supremum of $\mathcal{A}$. Indeed, for $A \in \mathcal{A}$, we have $A \in \bigcup \mathcal{A}$ hence $A = A^{\perp\perp} \subseteq (\bigcup \mathcal{A})^{\perp\perp}$ i.e. $A \leq s(\mathcal{A})$. It follows that $s(\mathcal{A})$ is an upper bound for $\mathcal{A}$. Now, assume that $B \in RO(X)$ is an upper bound for $\mathcal{A}$ i.e. $A \subseteq B$ for each $A \in \mathcal{A}$. Hence $\bigcup(\mathcal{A}) \subseteq B$ and thus $(\bigcup \mathcal{A})^{\perp\perp} \subseteq B^{\perp\perp} = B$ i.e. $s(\mathcal{A}) \leq B$ proving that $s(\mathcal{A})$ is the supremum of $\mathcal{A}$. Finally, by duality it follows that $i(\mathcal{A}) = (\bigcap \mathcal{A})^{\perp\perp}$ is the infimum of $\mathcal{A}$. $\square$

By duality applied to the family $RC(X)$ of regular closed sets in X, we obtain a dual proposal

Proposition 7.23. *$RC(X)$ is a boolean algebra under operations $\wedge, \vee, \prime$ defined as follows*

1. $A \vee B = A \cup B$

2. $A \wedge B = ClInt(A \cap B)$

3. $A' = X \setminus IntA$

and with constants $0 = \emptyset, 1 = X$. The algebra $RC(X)$ is complete.

We now pass to a study of compactness in general spaces.

7.8 Compactness in general spaces

Now it is time for exploring notions we have met with in the metric case in the general topological context. We begin with the notion of *compactness* as the most important by far and yielding most easily to generalizations.

As a definition of compactness in general spaces, we adopt the open covering condition proved for metric spaces in proposition. Thus, we say that a topological space (X, τ) is *compact* if and only if any open covering $\mathcal{U}$ of X contains a finite sub–covering $\{U_1, .., U_k\}$. The following characterizations follow easily.

For a filter $\mathcal{F}$, we say that $x \in X$ is a *cluster point* of $\mathcal{F}$ if $P \cap F \neq \emptyset$ for any $F \in \mathcal{F}$ and any neighborhood P of x. Let us observe that in case $\mathcal{F}$ is an ultrafilter, and x its cluster point, we have $N(x) \subseteq \mathcal{F}$ i.e. $\mathcal{F}$ has as elements all neighborhoods of x. In this case we say that x is a *limit* of $\mathcal{F}$, in symbols $x = lim\mathcal{F}$.

Proposition 7.24. *For a topological space (X, τ) the following are equivalent*

1. *X is compact*

2. *any centered family of closed subsets of X has a non–empty intersection*

3. *any filter has a cluster point*

4. *any ultrafilter has a limit point*

Proof. Properties 1 and 2 are equivalent as they are dual statements with respect to duality open – closed. Consider a filter $\mathcal{F}$. As any finite intersection of elements of $\mathcal{F}$ is non–empty, it follows that

$$\overline{\mathcal{F}} = \{ClF : F \in \mathcal{F}\} \tag{7.20}$$

is a centered family of closed sets hence

$$\bigcap \overline{\mathcal{F}} \neq \emptyset \tag{7.21}$$

Any $x \in \bigcap \overline{\mathcal{F}}$ is a cluster point of $\mathcal{F}$ proving Property 3. Property 4 follows from Property 3 as for any ultrafilter its any cluster point is a limit point. Now assume Property 4 and consider a centered family $\mathcal{F}$ of closed sets. Letting

$$\mathcal{F}^* = \{A \subseteq X : \exists F \in \mathcal{F}(F \subseteq A)\} \tag{7.22}$$

we define a filter $\mathcal{F}^*$ which extends to an ultrafilter $\mathcal{G}$. By Property 4, there exists $x = lim\mathcal{G}$; hence, any neighborhood of x does intersect any set in $\mathcal{G}$ and moreover any set in $\mathcal{F}$ thus $x \in ClF = F$ for any $F \in \mathcal{F}$ hence $x \in \bigcap \mathcal{F}$ proving 2 which as we know is equivalent to 1. $\square$

Compact spaces have the Baire property: any countable union of nowhere–dense sets is meager. The proof of this fact mimics the proof of the corresponding result for complete metric spaces.

When discussing compactness, in general context, two important findings come fore. The first is the *Tikhonov theorem* stating that any Cartesian product of compact spaces is a compact space and the second is the *Stone duality theorem* allowing to represent any Boolean algebra B as the field

$CO(X)$ of clopen sets in a compact space X. We give the proofs of these both facts of fundamental importance.

First, we have to define topology of a Cartesian product $\Pi_i X_i$ of a family $\{X_i : i \in I\}$ of topological spaces (X_i, τ_i). To this end, we call a *box* defined by coordinates $i_1, .., i_k$ and open sets $U_i \in \tau_i, i = 1, 2, .., k$, the set

$$O(i_1, .., i_k, U_1, .., U_k) = \{f \in \Pi_i X_i : f_i \in U_i, i = 1, 2, .., k\} \qquad (7.23)$$

We define a topology τ on $\Pi_i X_i$ by accepting the family of all boxes as a base for τ i.e. $U \in \tau$ if and only if there exists a family $\{B_i : i \in I\}$ of boxes with $U = \bigcup \{B_i : i \in I\}$.

Proposition 7.25. *(the Tikhonov theorem) A Cartesian product $X = \Pi_i X_i$ of a family $\{X_i : i \in I\}$ of compact spaces with the topology τ is a compact space.*

Proof. Consider *projections* $p_i : X \to X_i$ defined as follows: $p_i(f) = f_i$ for $f \in X, i \in I$. For an ultrafilter $\mathcal{F}$ on X, consider filters $\mathcal{F}_i$ induced by $p_i(\mathcal{F})$ in each X_i.

By compactness, there exists a cluster point x_i of $\mathcal{F}_i$ in X_i and we may define $x = (x_i)_i$ in X. Now for any neighborhood G of x in X there exists a box $x \in O(i_1, .., i_k, U_1, .., U_k) \subseteq G$ hence x_i in U_i for $i = i_1, .., i_k$. It follows that $U_i \cap F \neq \emptyset$ for each $F \in \mathcal{F}_i$ and thus $G \cap F \neq \emptyset$ for any $F \in \mathcal{F}$. Thus x is a cluster point hence a limit point for $\mathcal{F}$.   □

By the Tikhonov theorem, e.g. cubes $[0, 1]^n$ in Euclidean n–spaces ($n = 2, 3, ...$) as well as the *Hilbert cube* $[0, 1]^N$ are compact metric spaces.

Now we will consider a Boolean algebra $\mathbf{B}$ and we define a set $S(\mathbf{B})$ as follows. Elements of $S(\mathbf{B})$ are ultrafilters on $\mathbf{B}$. We define a topology σ on $S(\mathbf{B})$ by letting the sets of the form

$$F_a = \{f \in S(\mathbf{B}) : a \in f\} \qquad (7.24)$$

for $a \in \mathbf{B}$ to make a basis for open sets.
Thus

$$\sigma = \{\bigcup \{F_a : a \in A\} : A \subseteq \mathbf{B} \qquad (7.25)$$

Let us observe that $F_{a \wedge b} = F_a \cap F_b$ hence indeed σ is a topology.
Let us consider an ultrafilter Φ on $S(\mathbf{B})$ and along with it we look at the set

$$A_\Phi = \{a \in S(\mathbf{B}) : F_a \in \Phi\} \qquad (7.26)$$

Then A_Φ is a filter as $F_a \subseteq F_b$ when $a \leq b$ and $F_{a \wedge b} = F_a \cap F_b$. Actually, A_Φ is an ultrafilter as, dually, for a filter $\mathcal{G}$ on $\mathbf{B}$ the family $\{F_a : a \in \mathcal{G}\}$ is a filter on $S(\mathbf{B})$. Then we have duality

$$A_\Phi \in F_a \leftrightarrow a \in A_\Phi \leftrightarrow F_a \in \Phi \qquad (7.27)$$

i.e. $A_\Phi = lim\Phi$. Thus any ultrafilter on $S(\mathbf{B})$ converges proving this space to be compact.

The compact space $S(\mathbf{B})$ is called the *Stone space* of the Boolean algebra **B**.

Let us observe that we have $F_a = S(\mathbf{B}) \setminus F_{a'}$ for each a thus proving that each base set F_a is closed so finally it is clopen. It follows that the function $h : \mathbf{B} \to S(\mathbf{B})$ which satisfies already established conditions

$$\begin{aligned} h(a \wedge b) &= h(a) \cap h(b) \\ h(a \vee b) &= h(a) \cup h(b) \\ h(a') &= S(\mathbf{B}) \setminus h(a) \end{aligned} \qquad (7.28)$$

does establish an isomorphism between **B** and $CO(S(\mathbf{B}))$. Thus any Boolean algebra may be represented as a field of clopen sets in a compact space with a clopen base for open sets (such a space is called 0–*dimensional*).

One may ask about the Stone space in case of a complete Boolean algebra e.g. $RO(X)$: does it have some additional properties? It turns out that completeness of the Boolean algebra **B** is rendered in Stone spaces by the property of *extremal disconnectedness*: this means that closure of any open set is open, in symbols, $IntClIntA = ClIntA$ for each $A \subseteq S(\mathbf{B})$.

Proposition 7.26. *For any complete Boolean algebra* **B**, *the Stone space* $S(\mathbf{B})$ *is extremally disconnected.*

Proof. Consider $U = \bigcup \{U_a : a \in A\}$. For $A \subseteq \mathbf{B}$, there exists $c = supA$. Then $U_a \subseteq U_c$ for each $a \in A$ so $U \subseteq U_c$ and thus $ClU \subseteq U_c$. Assume $f \in U_c \setminus ClU$ for some ultrafilter f. Then there exists U_b with $f \in U_b$, $U_b \cap U_a = \emptyset$ for $a \in A$. It follows that $a \wedge b = 0$ for $a \in A$ hence $a \leq b'$ for $a \in A$ implying that $supA = c \leq b'$. But then $f \in U_{b'}$, contradicting $f \in U_b$. It follows that $ClU = U_c$ i.e. ClU is clopen. $\square$

This fact allows for a *completion* of a Boolean algebra via topological arguments: given a Boolean algebra **B**, we have an isomorphism $h : \mathbf{B} \to CO(S(\mathbf{B}))$. Now, we embed $CO(S(\mathbf{B}))$ into complete Boolean algebra $RO(S(\mathbf{B}))$ of regular open sets in the Stone space via identity embedding $i : A \to A$. The composition $i \circ h$ embeds the algebra **B** into $RO(S(\mathbf{B}))$. Thus any Boolean algebra **B** has a completion $\mathbf{B}^*$ i.e. a complete Boolean algebra into which it can be embedded as a dense set (it is so as any open set contains a basic clopen set).

As the Stone space of a Boolean algebra is compact it does satisfy the Baire theorem. A counterpart of this theorem was formulated in the language of Boolean algebras.

Proposition 7.27. *(Rasiowa, Sikorski) In any Boolean algebra* **B**, *given a countable sequence* $(A_n)_n$ *of dense sets and* $0 \neq a \in$ **B**, *there exists an ultrafilter* f *with the properties that* $a \in f$ *and* $f \cap A_n \neq \emptyset$ *for each* n.

Proof. We translate the assumptions into the language of topological spaces. For each A_n, the set $W_n = \bigcup \{ U_a : a \in A_n \}$ is open dense in the Stone space $S(\mathbf{B})$. Hence the intersection $W = \bigcap \{ W_n : n \in N \}$ is a dense set by the Baire theorem (Proposition 7.15) hence there exists $f \in W \cap U_a$. This f is as desired. $\qquad\square$

7.9 Continuity

To this point, we have been interested in the inner properties of topological spaces, without any attempt to relate spaces one to another. In order to compare or relate topological structures, we have to define an appropriate type of functions or *morphisms* among topological spaces. Of such a morphism we would require that it represents the topological structure of one space in the other. It has turned out that proper morphisms in topological case are *continuous functions*. A function $f : X \to Y$ of a topological space (X, τ) into a topological space (Y, σ) is said to be *continuous* if and only if the condition $f(ClA) \subseteq Cl(f(A))$ holds for every $A \subseteq X$.

It means that f maps closures of subsets in X into closures of their images in Y. It may be convenient to have some characterizations of continuity in terms of open or closed sets; to this end we have to look at inverse images in X of sets in Y.

Proposition 7.28. *For* $f : X \to Y$, *the following are equivalent*

1. f *is continuous*

2. $Cl f^{-1}(B) \subseteq f^{-1}(ClB)$ *for every* $B \subseteq Y$

3. $f^{-1}(IntA) \subseteq Int f^{-1}(A)$ *for every* $A \subseteq Y$

4. *the set* $f^{-1}(A)$ *is open for each open set* $A \subseteq Y$

5. *the set* $f^{-1}(A)$ *is closed for each closed set* $A \subseteq Y$

Proof. Property 1 implies 2. In the condition $f(ClA) \subseteq Cl(f(A))$, we substitute $f^{-1}(B)$ for A so we get $f(Cl f^{-1}(B)) \subseteq Cl(B))$ hence $Cl f^{-1}(B) \subseteq f^{-1}(ClB)$ i.e. 2 holds. Assuming 2, we have $f^{-1}(IntA) = f^{-1}(Y \setminus Cl(Y \setminus A))$

$= X \setminus f^{-1}(Cl(Y \setminus A)) \subseteq X \setminus Clf^{-1}(Y \setminus A) = X \setminus Cl(X \setminus f^{-1}(A) = Intf^{-1}(A)$
i.e. Property 3 follows from 2.

Now, Property 4 follows from 3 as in case A is an open set in Y we have
$f^{-1}(A) = f^{-1}(IntA) \subseteq Intf^{-1}(A)$ hence $f^{-1}(A) \subseteq Intf^{-1}(A)$ so $f^{-1}(A) = Intf^{-1}(A)$ i.e. $f^{-1}(A)$ is open. Property 5 follows from 4 by duality.

Finally, assume 5. We have for $A \subseteq X$ that $A \subseteq f^{-1}(f(A))$ hence $A \subseteq f^{-1}(Clf(A))$. As the set $f^{-1}(Clf(A))$ is closed by 5, we have $ClA \subseteq f^{-1}(Clf(A))$ hence $f(ClA) \subseteq Clf(A)$ i.e. f is continuous. $\qquad\square$

In case $f : X \to Y$ is a bijection and f as well as f^{-1} are continuous, we say that f is a *homeomorphism* of X onto Y.

A homeomorphism establishes thus a $1-1$ correspondence between open sets in X and open sets in Y via $A - f(A)$ as well as between closed sets in either space via the same assignment. Thus homeomorphic spaces are identical from general topological point of view.

Let us finally observe that the characterization of continuity in case of metric spaces X, Y may be given in terms of converging sequences. As $x \in ClA$ if and only if $x = limx_n$ for a sequence $(x_n)_n \subseteq A$, we may conjecture that the condition $f(limx_n) = limf(x_n)$ for any convergent sequence $(x_n)_n \subseteq X$ is equivalent to continuity as it does guarantee that $f(ClA) \subseteq Cl(f(A))$ for any $A \subseteq X$.

Proposition 7.29. *For metric spaces X, Y, a function $f : X \to Y$ is continuous if and only if the condition*

$$f(limx_n) = limf(x_n) \qquad\qquad (7.29)$$

holds for any convergent sequence $(x_n)_n \subseteq X$.

Proof. As already observed, the condition (7.29) implies continuity. If f is continuous then for any convergent sequence $(x_n)_n$ with $x = limx_n$ and any open $G \subseteq Y$ with $f(x) \in G$ we have $x \in f^{-1}(G)$ and as $f^{-1}(G)$ is open, we have $x_n \in f^{-1}(G)$ hence $f(x_n) \in G$ for almost every n i.e. $f(x) = limf(x_n)$. $\qquad\square$

In particular, any contracting map on a metric space X is continuous. An equivalent characterization of continuity which will be in use in Chapter 13 (cf. Exercise 10) may be given in the $\varepsilon - \delta$ language viz. $f : (X, \rho) \to (Y, \eta)$ is continuous at $x \in X$ if and only if

(Cont) *for each $\varepsilon > 0$ there exists $\delta > 0$ with the property that $\rho(x, y) < \delta$ implies $\eta(f(x), f(y)) < \varepsilon$*

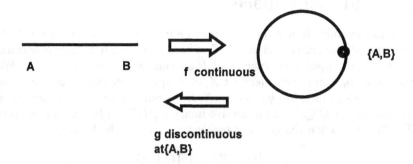

Figure 7.1: f is continuous while g is not continuous: breaking the circle is discontinuous, gluing into a common unity is continuous

and f is continuous if and only if (Cont) holds with every $x \in X$. In Figure 1 below the reader will find an archetypical example of a continuous and discontinuous functions: the function f which glues together points A, B in the interval to make a circle (e.g. $f(t) = (cos2\pi \cdot \frac{t-A}{B-A}, sin2\pi \cdot \frac{t-A}{B-A}))$ is continuous while its pseudo– inverse g (i.e. $g(f(t)) = t$ for $t \in (A, B])$is not continuous as we may find sequences $(f(t_n) = (x_n, y_n))_n$ with $y_n > 0$ as well as with $y_n < 0$ both converging to $\{A, B\}$ while in the former case $(t_n)_n$ does converge to A and in the latter case $(t_n)_n$ converges to B.

7.10   Topologies on subsets

Assume that (X, τ) is a topological space and $A \subseteq X$. One may introduce a topology on A in a natural way by declaring the natural injection $i : A \to X$ a continuous function. This implies that for any open set $G \subseteq X$, the intersection $A \cap G$ is open in A. Thus the family $\tau_A = \{A \cap G : G \in \tau\}$ is a topology on A. Operators Cl_A of closure and Int_A of interior with respect to the topology τ_A are related to operators Cl, Int with respect to the topology τ as described below

Proposition 7.30. *For any subset $Y \subseteq A$*
1. $Cl_A Y = A \cap Cl Y$
2. $A \cap Int Y \subseteq Int_A Y$

Proof. For 1., clearly $Cl_A Y \subseteq A \cap Cl Y$ as the latter set is closed in A. Conversely, for $x \in A \cap Cl Y$, we have that $A \cap (G \cap Y) = (A \cap G) \cap Y \neq \emptyset$ hence $x \in Cl_A Y$ as $A \cap G$ runs over open neighborhoods of x in A. In case of 2., as $A \cap Int Y$ is open in A we have $A \cap Int Y \subseteq Int_A Y$. The converse does not need to hold. $\square$

7.11   Quotient spaces

We now examine how a topology may be induced on a quotient set X/R for an equivalence relation R on X cf. Chapter 6, p. 236ff. Let us assume that (X, τ) is a topological space and R an equivalence relation on X. We define a topology on the quotient set X/R of equivalence classes by requiring that the quotient function $q_R : X \to X/R$ be continuous. This means that for each open set $G \subseteq X/R$ the inverse image $q_R^{-1}(G) = \bigcup G$ should be open in X. Thus we define the *quotient topology* τ_R on X/R by letting

$$\tau_R = \{G \subseteq X/R : \bigcup G \in \tau\}.$$

Then we have

Proposition 7.31. *1. If $f : X \to Y$ is a continuous function defined on a space X with an equivalence relation R into a space Y with an equivalence relation S and xRy implies $f(x)Sf(y)$ then $\overline{f} : X/R \to Y/S$ defined as $\overline{f}[x]_R = [f(x)]_S$ is continuous*
2. If R, S are equivalence relations on X with $S \subseteq R$ then the quotient spaces $(X/S)/(R/S)$ and X/R are homeomorphic.

Proof. In case of 1., for an open in Y/S set P of equivalence classes of S we have $q_R^{-1}(\overline{f}^{-1}(P)) = f^{-1}(q_S^{-1}(P)$ and as the latter set is open it follows that $q_R^{-1}(\overline{f}^{-1}(P))$ is an open set in X hence $\overline{f}^{-1}(P)$ is open in X/R witnessing that $\overline{f}$ is continuous.

For 2., we consider the function $q : (X/S)/(R/S) \to X/R$ defined by letting $q([[x]_S]_{R/S}) = [x]_R$. Then q is a bijection and q is continuous: for any open $P \subseteq X/R$ we have

$$q_S^{-1}(q_{R/S}^{-1}(q^{-1}(P))) = q_R^{-1}(P) \tag{7.30}$$

and the latter set is open in X hence $q_S^{-1}(q_{R/S}^{-1}(q^{-1}(P)))$ is open and finally $q^{-1}(P)$ is open witnessing continuity of q. We have to prove that q^{-1} is continuous which amounts to showing that q is *an open function* i.e. the image $q(Q)$ is open in X/R for each open in $(X/S)/(R/S)$ set Q. As we have

$$q_R^{-1}(q(Q)) = q_S^{-1}(q_{R/S}^{-1}(Q)) \tag{7.31}$$

and the latter set is open, it follows that $q_R^{-1}(q(Q))$ is open i.e. $q(Q)$ is open in X/R. Thus q is a homeomorphism. $\square$

7.12   Hyperspaces

In many applications, e.g. in *morphology, control theory, mathematical economy and game theory* we have to consider spaces whose points are subsets

in some underlying space. Then we speak of *hyperspaces*. The problem of
inducing topologies on such new spaces will be our subject in this section.
Consider a topological space (X, τ) and a family $\mathcal{F}$ of subsets of X. It is our
aim to define a topology on $\mathcal{F}$ related to the topology τ.

In order to do this, we consider subsets of $\mathcal{F}$ of the form $C(P, \mathcal{V})$ where

$$C(P, \mathcal{V}) = \{A \in \mathcal{F} : A \subseteq P \wedge \forall V \in \mathcal{V}(A \cap V \neq \emptyset)\} \qquad (7.32)$$

Let us observe that

$$C(P, \mathcal{V}) \cap C(Q, \mathcal{W}) = C(P \cap Q, \mathcal{V} \cup \mathcal{W}) \qquad (7.33)$$

hence if P, Q come from a family of subsets of X having the finite intersection property and $\mathcal{V}, \mathcal{W}$ are sub–families of a family of sets in X closed
on arbitrary unions then sets of the form $C(P, \mathcal{V})$ have the finite intersection
property and may be taken as a basis for open sets for a topology on $\mathcal{F}$.
Clearly, the family $\mathcal{O}\,(X)$ of open sets in X poses itself as a candidate from
whom sets P as well as elements of $\mathcal{V}$ could be taken.

Accepting that candidate we arrive at the *hit–or–miss topology* on $\mathcal{F}$ induced
by an open base consisting of sets of the form $C(P, \mathcal{V})$ where P is an open set
in X and $\mathcal{V}$ is a *finite* collection of open sets in X. Although the collection
$\mathcal{F}$ may in principle consist of arbitrary sets in X yet the most important
cases are when $\mathcal{F}$ is the family of all closed sets in X or it is the family of all
compact sets in X.

We will look at topologies on closed subsets of X.

7.12.1   Topologies on closed sets

The *hit–or–miss topology* on the family $\kappa(X)$ of closed sets in X is generated by the collection of sets of the form $C(P, \mathcal{V})$, where P is open and $\mathcal{V}$ is a
finite collection of open sets, taken as an open basis hence each open set is of
the form $\bigcup_{i \in I} C(P_i, \mathcal{V}_i)$. It is most important for applications to determine
the case when this topology is metrizable i.e. when it is identical to a metric
topology induced by some metric.

We thus consider a metric space (X, ρ). We assume that ρ is *bounded by
1* i.e. $\rho(x, y) = min\{1, \rho(x, y)\}$; as we already know this has no impact on
the induced metric topology. Now we define a metric δ_ρ on closed sets, called
the *Hausdorff–Pompéju metric*. For two closed sets M, N in X we calculate
subsequently the following

1. for each $x \in M$, we calculate $dist_\rho(x, N) = inf\{\rho(x, y) : y \in N\}$ called
   the *distance from x to N*

2. we calculate $sup\{\rho(x, N) : x \in M\}$

3. for each $y \in N$, we calculate $dist_\rho(y, M) = inf\{\rho(x, y) : x \in M\}$ (the distance from y to M)

4. we calculate $sup\{\rho(y, M) : y \in N\}$

5. we let $\delta_\rho(M, N) = max\{sup\{\rho(x, N) : x \in M\}, sup\{\rho(y, M) : y \in N\}\}$

The function δ_ρ is defined for all pairs M, N of closed sets due to our assumption that the metric ρ is bounded. We should check that δ_ρ is a metric. It is manifest that

$$\delta_\rho(M, N) = 0 \; iff \; M = N$$
$$\delta_\rho(M, N) = \delta_\rho(N, M) \tag{7.34}$$

so it remains to prove only that the triangle inequality holds for δ_ρ. Consider closed sets M, R, N; we are going to prove that

$$\delta_\rho(M, N) \leq \delta_\rho(M, R) + \delta_\rho(R, N) \tag{7.35}$$

For $x \in M, y \in N$, we have by the triangle inequality that

$$\rho(x, y) \leq \rho(x, r) + \rho(r, y) \tag{7.36}$$

for any $r \in R$. Taking infima with respect to r gives us

$$\rho(x, y) \leq dist(x, R) + dist(y, R) \leq \delta_\rho(M, R) + \delta_\rho(R, N) \tag{7.37}$$

whence

$$sup\{\rho(x, N) : x \in M\} \leq \delta_\rho(M, R) + \delta_\rho(R, N) \tag{7.38}$$

follows. By symmetry, we get

$$sup\{\rho(y, M) : y \in N\} \leq \delta_\rho(M, R) + \delta_\rho(R, N) \tag{7.39}$$

and from the last two inequalities (7.35) follows. Thus δ_ρ is a metric, indeed.

It turns out that in case of a compact metric space the two topologies: the hit-or-miss topology and the metric topology induced by the Hausdorff–Pompéju metric, coincide. We prove this important fact.

Proposition 7.32. For any compact metric space (X, ρ), the hit- or-miss topology coincides with the metric topology induced by the Hausdorff–Pompéju metric.

Proof. Consider a basic open set $C(P, \mathcal{V})$ in the hit–or–miss topology where $\mathcal{V} = \{V_1, .., V_k\}$ along with $M \in C(P, \mathcal{V})$, M a closed set. As $M \subseteq P$, the closed sets $M, X \setminus P$ are disjoint. Hence, by compactness of X, there exists $\varepsilon > 0$ such that

$$\rho(x, y) > \varepsilon \qquad (7.40)$$

for any $x \in X, y \in Y$. Letting

$$B(M, \varepsilon) = \{x \in X : dist(x, M) < \varepsilon\} \qquad (7.41)$$

we have that $B(M, \varepsilon) \subseteq P$. Similarly, as $M \cap V_i \neq \emptyset$ for $i = 1, 2.., k$ there exist $x_1, .., x_k$ with $x_i \in M \cap V_i$ each $i \leq k$ and for each x_i we may find $\varepsilon_i > 0$ with

$$B(x_i, \varepsilon_i) \subseteq V_i \qquad (7.42)$$

for each $i \leq k$.
Let $\varepsilon_0 = min\{\varepsilon, \varepsilon_1, ..., \varepsilon_k\}$ and consider a closed set N with

$$\delta_\rho(M, N) < \varepsilon_0 \qquad (7.43)$$

Then $N \subseteq B(M, \varepsilon)$ hence $N \subseteq P$; also, for each $i \leq k$, we find $y_i \in N$ with $\rho(x_i, y_i) < \varepsilon_i$ implying $N \cap V_i \neq \emptyset$. It follows that $N \in C(P, \mathcal{V})$ hence

$$B_{\delta_\rho}(M, \varepsilon_0) \subseteq C(P, \mathcal{V}) \qquad (7.44)$$

implying that $C(P, \mathcal{V})$ is open in the metric topology of δ_ρ.

For the converse, consider an open ball $B_{\delta_\rho}(M, \varepsilon)$ along with the set $P = B(M, \varepsilon)$ which is open in the topology of X. For any closed $N \subseteq P$ we have clearly $dist(y, M) < \varepsilon$ for each $y \in N$ which settles one half of the condition for $\delta_\rho < \varepsilon$. To settle the other half, i.e. to assure that $dist(x, N) < \varepsilon$ for $x \in X$, let us take in M an $\varepsilon/2$–net i.e. a set $x_1, ..., x_m$ with the property that for any $x \in M$ we find x_i such that $\rho(x, x_i) < \varepsilon/2$.

Letting $V_i = B(x_i, \varepsilon/2)$ for $i \leq k$ we observe that if $N \cap V_i \neq \emptyset$ for each $i \leq k$ then $dist(x, N) < \varepsilon$ for each $x \in M$. Thus for every $N \in C(P, \{V_1, V_2, .., V_k\})$ we have

$$\delta_\rho(M, N) < \varepsilon \qquad (7.45)$$

i.e.

$$C(P, \{V_1, V_2, .., V_k\}) \subseteq B_{\delta_\rho}(M, \varepsilon) \qquad (7.46)$$

proving that basic sets in the metric topology are open in the hit–or–miss topology thus showing both topologies to be identical and concluding the proof.   □

This result tells us that the topology induced by the Hausdorff–Pompéju metric is in fact independent of the metric on X in case X is metric compact.

As we know, a weaker but very important and useful substitute of compactness is completeness. It turns out that in case (X, ρ) is a complete metric space, the space of closed sets in X with the Hausdorff–Pompéju metric is complete as well. To discuss this fact, we need some new notions related to sequences of sets.

For a sequence $(A_n)_n$ of sets in X, we define the *lower limit* LiA_n of this sequence by letting

$$LiA_n = \{x \in X : limdist(x, A_n) = 0\}$$

i.e. $x \in LiA_n$ if and only if for each ball $B(x, \varepsilon)$ we have

$$B(x, \varepsilon) \cap A_n \neq \emptyset \tag{7.47}$$

for all but finitely many A_n i.e. if and only if every neighborhood of x does intersect almost all sets A_n. Let us observe that LiA_n is a closed set: indeed, if $y \notin LiA_n$ then some neighborhood $P \ni y$ misses infinitely many A_n's and then clearly $P \cap LiA_n = \emptyset$.

Similarly, we define the *upper limit* LsA_n of the sequence $(A_n)_n$ as

$$LsA_n = \{x \in X : liminf dist(x, A_n) = 0\}$$

i.e. $x \in LsA_n$ if and only if $limdist(x, A_{n_k}) = 0$ for a subsequence $(A_{n_k})_k$ i.e. if and only if every neighborhood of x does intersect infinitely many A_n's. As with LiA_n, the upper limit LsA_n is a closed set which may be shown by the same argument. It follows immediately that $LiA_n \subseteq LsA_n$.

In case $LiA_n = LsA_n$, we say that the sequence $(A_n)_n$ *converges* to the *limit* $LimA_n = LiA_n = LsA_n$. Now, we turn to the completeness property.

Let us first observe that in case a sequence $(A_n)_n$ converges in the metric δ_ρ i.e.

$$lim\delta_\rho(A_n, B) = 0 \tag{7.48}$$

for some closed set B, we have $B = LimA_n$.

Indeed, given $\varepsilon > 0$, there is $n_\varepsilon \in N$ with $A_n \subseteq B(B, \varepsilon)$ for $n \geq n_\varepsilon$. Hence, for each $x \in B$ we have $B(x, \varepsilon) \cap A_n \neq \emptyset$ for $n \geq n_\varepsilon$. It follows that $B \subseteq LiA_n$. On the other hand, given $x \in LsA_n$, we may define inductively a sequence $(a_{n_k})_k$ with $a_{n_k} \in A_{n_k}$ and $x = lima_{n_k}$. Passing eventually to a subsequence, in virtue of 7.48 we have $\rho(a_{n_k}, b_k) < 1/k$ with some $b_k \in B$ for each k. Thus $x = limb_k$ hence $x \in B$ and finally $LsA_n \subseteq B$. Therefore $B = LimA_n$.

Proposition 7.33. *For a complete metric space (X, ρ), the space $\mathcal{C}(X)$ of closed sets in X with the Hausdorff–Pompéju metric δ_ρ is complete.*

Proof. Assume $(A_n)_n$ is a fundamental sequence with respect to δ_ρ. Once we exhibit B with (7.48) we have $B = Lim A_n$. It is the easiest to consider $B = Ls A_n$. Given $\varepsilon > 0$, we have $n(\varepsilon)$ such that

$$\delta_\rho(A_{n(\varepsilon)}, A_n) < \varepsilon \tag{7.49}$$

for $n \geq n(\varepsilon)$. We consider $x \in Ls A_n$. Then $B(x, \varepsilon)$ intersects infinitely many A_n's hence for some $n > n(\varepsilon)$ and some $y \in A_n$ we have $\rho(x, y) < \varepsilon$ and thus $dist(x, A_n) < \varepsilon$. By (7.48), we have

$$dist(x, A_{n(\varepsilon)}) \leq 2 \cdot \varepsilon \tag{7.50}$$

hence

$$sup_x dist(x, A_{n(\varepsilon)}) \leq 2 \cdot \varepsilon \tag{7.51}$$

On the other hand, for $z \in A_{n(\varepsilon)}$, we may find a sequence $(z_{n_k})_k$ with

$$\begin{array}{l} (i) \ z_{n_k} \in A_{n_k} \\ (ii) \ n_k > n(\varepsilon) \\ (iii) \ \rho(z_{n_k}, z_{n_{k+1}}) < 1/2^k \cdot \varepsilon \end{array} \tag{7.52}$$

This sequence may be defined by induction as $(A_n)_n$ is a fundamental sequence. As the sequence $(z_{n_k})_k$ is fundamental with respect to ρ, there exists $w = lim z_{n_k}$. Clearly, $w \in Ls A_n$ and $\rho(z, w) \leq 2 \cdot \varepsilon$ hence

$$\begin{array}{l} (iv) \ dist(z, Ls A_n) \leq 2 \cdot \varepsilon \\ (v) \ sup_z \{dist(z, Ls A_n)\} \leq 2 \cdot \varepsilon \end{array} \tag{7.53}$$

Thus

$$\delta_\rho(A_{n(\varepsilon)}, Ls A_n) \leq 2 \cdot \varepsilon \tag{7.54}$$

and by the triangle inequality we have

$$\delta_\rho(A_n, Ls A_n) \leq 3 \cdot \varepsilon \ for \ n > n(\varepsilon) \tag{7.55}$$

This proves that $lim \delta_\rho(A_n, Ls A_n) = 0$. Hence $(A_n)_n$ converges to $Ls A_n = Lim A_n$. Completeness of the space $(\mathcal{C}(X), \delta_\rho)$ has been established. $\qquad \square$

7.13 Čech topologies

In many cases that arise in real problems, we have to deal with a weaker form of topology. Such situation for instance takes place when we try to define closure operator with respect not to the full family of open sets but with respect to a certain sub–family e.g. a covering of a space. Assume that we are given a covering $\mathcal{U}$ of a set X and for each $A \subseteq X$, we let

$$Cl_c A = \{x \in X : \forall G \in \mathcal{U}(x \in G \Rightarrow G \cap A \neq \emptyset)\}.$$

Then we have the following properties of the operator Cl_c.

Proposition 7.34. *The operator Cl_c does satisfy*

1. $Cl_c \emptyset = \emptyset$

2. $A \subseteq Cl_c A$ for each $A \subseteq X$

3. $A \subseteq B \Rightarrow Cl_c A \subseteq Cl_c B$ for each pair A, B of subsets of X

All these properties follow immediately from the definition of Cl_c. Let us observe that the operator Cl_c does not need to observe the properties of topological closure operator, notably, in general the properties that $Cl_c(A \cup B) = Cl_c A \cup Cl_c B$ and $Cl_c Cl_c A = Cl_c A$ do not hold and are replaced by weaker consequences of 1-3 viz.

(4) $Cl_c(A \cap B) \subseteq Cl_c A \cap Cl_c B$

(5) $Cl_c A \subseteq Cl_c Cl_c A$.

We will call the operator Cl_c satisfying 1-3, a fortiori also (4),(5) the *Čech closure operator*. A most important case when a Čech closure operator does arise is that of a *tolerance relation* when we consider e.g. a covering formed by tolerance sets of the form $\tau(x) = \{y \in X : x\tau y\}$. We denote by $\mathcal{T}(\tau)$ the set of all tolerance sets of τ. Then for any $A \subseteq X$, we have

$$Cl_c A = \{x \in X : \forall T \in \mathcal{T}(\tau)(x \in T \Rightarrow T \cap A \neq \emptyset \qquad (7.56)$$

The *Čech topology* generated by the Čech closure operator is the family of all *open sets* with respect to Cl_c viz. we declare a set $A \subseteq X$ to be *closed* in case $Cl_c A = A$ and we declare a set $A \subseteq X$ to be *open* in case the complement $X \setminus A$ is a closed set. Then we have

Proposition 7.35. *In any Čech topological space:*

1. *any intersection of a family of closed sets is a closed set*
2. *any union of a family of open sets is an open set.*

Proof. Assume that $\mathcal{F}$ is a family of closed sets in a Čech topological space X. Let $K = \bigcap \mathcal{F}$. As $K \subseteq F$ for any $F \in \mathcal{F}$, we have $Cl_c K \subseteq Cl_c F = F$ for any $F \in \mathcal{F}$ hence $Cl_c K \subseteq K$ and finally $Cl_c K = K$ i.e. K is closed. Thesis 2. follows immediately by duality. $\qquad\square$

We now confine ourselves to Čech topologies generated from coverings by tolerance sets in the way described above. We may observe that

$$(i) \ Cl_c A \ \textit{need not be a closed set}$$
$$(ii) \ Cl_c(A \cap B) \subseteq Cl_c A \cap Cl_c B \qquad (7.57)$$
$$(iii) \ Cl_c A \cup Cl_c B \subseteq Cl_c(A \cup B)$$

hold but converse statements to (ii), (iii) need not to hold.

We may characterize $Cl_c A$ via open sets as in case of topological spaces. For a point $x \in X$, we say that a set $Q \subseteq X$ is a *neighborhood* of x if and only if $x \notin Cl_c(X \setminus Q)$. Clearly, then $x \in Q$. Then we have

Proposition 7.36. *For* $x \in X$, $x \in Cl_c A$ *if and only if* $Q \cap A \neq \emptyset$ *for any neighborhood* Q *of* x.

Proof. If for some neighborhood Q of x we have $Q \cap A = \emptyset$, then $A \subseteq X \setminus Q$ and we have $Cl_c A \subseteq Cl_c(X \setminus Q)$ hence $x \notin Cl_c A$. Thus, $x \in Cl_c A$ implies $Q \cap A \neq \emptyset$ for any neighborhood Q of x. On the other hand, $x \notin Cl_c A$ implies $x \notin Cl_c(X \setminus (X \setminus A))$ hence $X \setminus A$ is a neighborhood of x disjoint to A. $\qquad\square$

Let us consider a sequence of closure operators $(Cl_c^n)_n$ defined recurrently for any $A \subseteq X$ by

$$Cl_c^0 A = A$$
$$Cl_c^{n+1} A = Cl_c(Cl_c^n A) \ \textit{for } n \geq 1 \qquad (7.58)$$

Then for any $A \subseteq X$ we have

$$Cl_c^0 A = A \subseteq Cl_c A \subseteq ... \subseteq Cl_c^n A \subseteq Cl_c^{n+1} A \subseteq .... \qquad (7.59)$$

In our setting of a tolerance relation, subsequent values $Cl_c^n A$ correspond to saturation of A with tolerance sets: $S^0 A = A$, $S^{n+1} A = S(S^n A)$ where $SA = \bigcup \{\tau(x) : \tau(x) \cap A \neq \emptyset\}$ i.e. we have $Cl_c^n A = S^n A$, any n.

Let us define the closure operator Cl^* by letting

$$Cl^* A = \bigcup \{Cl_c^n A : n \in N\}.$$

Then we have

Proposition 7.37. *For any tolerance relation* τ *on a set* X, $Cl^* A$ *is a topological closure operator. Moreover,* Cl^* *is identical to the Čech closure operator induced by the equivalence relation* $\bar{\tau}$, *the transitive closure of the relation* τ.

Proof. Clearly, $Cl^*\emptyset = \emptyset$, $A \subseteq Cl^*A$ hold by definition of Cl^*. To proceed further we observe that

(i) $x \in Cl^*A$ if and only if there is a finite sequence $x_0 \in A, x_1, ..., x_k = x$ with

$$x_i\tau x_{i+1} \ for \ i = 0, 1, ..., k-1 \qquad (7.60)$$

and we assume that $x \in Cl^*Cl^*A$. Then there is a finite sequence $y_0 \in Cl^*A$, $y_1, ..., y_m = x$ for some $m \in N$ with $y_i\tau y_{i+1}$ for $i = 0, .., m-1$ and thus we have a sequence $z_0 \in A, z_1, ..., z_p = y_0$ for some $p \in N$ with $z_i\tau z_{i+1}$ for $i = 0, ..., p-1$. It follows that $x \in Cl_c^{m+p}A$ hence $x \in Cl^*A$. Thus, $Cl^*Cl^*A \subseteq Cl^*A$ and finally $Cl^*Cl^*A = Cl^*A$.

It remains to prove that $Cl^*A \cup Cl^*B = Cl^*(A \cup B)$. Clearly, $Cl^*A \cup Cl^*B \subseteq Cl^*(A \cup B)$. Assume that $x \in Cl^*(A \cup B)$ hence there is a sequence $x_0 \in A \cup B, x_1, ..., x_k = x$ with $x_i\tau x_{i+1}$ for $i = 0, 1, ..., k-1$. Then either $x_0 \in A$ or $x_0 \in B$; in the former case, $x \in Cl^*A$ and in the latter case $x \in Cl^*B$. Thus $Cl^*(A \cup B) \subseteq Cl^*A \cup Cl^*B$ and finally the equality of both sides follows.

We have proved that Cl^* is a topological closure operator. For the second part of the proposition, indeed, we have $x \in Cl^*A$ if and only if $A \cap \overline{\tau}(x) \neq \emptyset$ i.e. the closure operator Cl^* and the Čech closure operator $Cl_{\overline{\tau}}$ coincide. $\square$

Remaining for a while with the case of an equivalence relation, we may observe that in that case the tolerance sets are equivalence classes, and the resulting Čech topology is a topology induced by taking the set of equivalence classes as an open basis for this topology. Thus, any open set is a union of a family of equivalence classes a fortiori any closed set is a union of a family of equivalence classes. Hence, in this topology, any open set is closed as well (we call such topology a *clopen* topology). Moreover, any intersection of a family of open sets is an open set.

Let us observe that, conversely, given any clopen topology with the property that the intersection of any family of open sets is an open set, we may introduce an equivalence relation which induces this topology: it is sufficient to let xRy if and only if x and y belong in the same open sets i.e.

$$xRy \leftrightarrow \forall Q = IntQ(x \in Q \leftrightarrow y \in Q).$$

Then any equivalence class $[x]_R$ is the intersection of all open sets containing x hence it is clopen. Clearly, any open set Q is the union of the equivalence classes $[x]_R$ for $x \in Q$. So we have

Proposition 7.38. *The Čech topology is generated by the set of equivalence classes of an equivalence relation if and only if (i) any open set is closed (ii) the intersection of each family of open sets is an open set.*

Historic remarks

Closure operation as a primitive topological notion was proposed by Riesz [Riesz09] (cf. also [Kuratowski22]). The notion of a closed set was defined by Cantor [Cantor883]. The notion of an open neighborhood as a basic primitive of topology was proposed by Hausdorff [Hausdorff14]. Open sets as a basic notion were used by Alexandrov [Alexandrov25] and Fréchet [Fréchet28].

The notion of a metric space comes from Hausdorff [Hausdorff14], properties defining a metric were specified by Fréchet [Frechet06]. The Hausdorff–Pompéju metric was defined by Hausdorff [Hausdorff14] and in a slightly different form by Pompéju [Pompéju05].

Compactness defined in terms of open coverings in general spaces was introduced by Alexandrov and Urysohn [Alexandrov–Urysohn29]. The property of open coverings defining compactness is known as the Borel–Lebesgue property [Lebesgue05]. The Tikhonov theorem was proved in [Tikhonov35].

The notion of a meager set (called a set of 1st category) was introduced and the Baire theorem proved in [Baire899]. The characterization of completeness in metric spaces by decreasing sequences of closed sets was found by Cantor [Cantor880].

The topology on the family of closed sets was defined by Vietoris [Vietoris21]. The completeness of the exponential space 2^X in case of a complete metric X was proved by Hahn [Hahn33].

The Banach fixed point theorem is from [Banach22]. The representation theorem for Boolean algebras was proved by Stone [Stone36]. The Rasiowa–Sikorski Lemma is from [Rasiowa–Sikorski50].

Čech topologies were introduced and studied by Čech [Čech66]. Čech topologies induced by tolerance relations were introduced into rough set theory in [Polkowski–Skowron–Zytkow95] and [Marcus94].

Exercises

In Exercises 1–8, we are concerned with topological calculus on sets and basic properties like to be boundary, nowhere–dense, of 1st category.

1. Prove that $ClA \setminus ClB \subseteq Cl(A \setminus B)$

2. Prove that the boundary operator $BdA = ClA \cap Cl(X \setminus A)$ has the following properties (i) $BdClA \subseteq BdA$ (ii) $BdIntA \subseteq BdA$ (iii) $BdA = A \cap Cl(X \setminus A) \cup (ClA \setminus A)$.

3. A set $A \subseteq X$ is *boundary* if and only if $IntA = \emptyset$. Give an example showing that BdA need not be any boundary set.

4. Prove that for each set $A \subseteq X$, sets $A \cap Cl(X \setminus A)$, $ClA \setminus A$ are boundary and infer from Exercise 3 that the union of two boundary sets need not be

any boundary set.

5. A set $A \subseteq X$ is *nowhere–dense* if and only if $IntClA = \emptyset$. Prove that $IntBdA = \emptyset$ i.e. BdA is nowhere–dense in case $A \subseteq X$ is either open or closed. [Hint: observe that in each case one of sets in decomposition described in Exercise 4. is empty]

6. Employ properties of the operation $\perp$ (cf. Proposition 7.20) to list all 14 distinct sets which may be obtained by means of $Int, Cl, \backslash$.

7. A point $x \in X$ is a *cluster point* of a set $A \subseteq X$ if and only if $x \in Cl(X \backslash A)$. The set of all cluster points of a set A is denoted by the symbol A^d. Prove that (i) $ClA = A \cup A^d$ (ii) $ClA^d = A^d$ (iii) $A^{dd} \subseteq A^d$ (iv) $(A \cup B)^d = A^d \cup B^d$ (v) $A \subseteq B$ implies $A^d \subseteq B^d$.

8. Prove that the set of irrational numbers P in the real line $\mathbf{R}^1$ with the natural topology is not any set of 1st category. [Hint: assume to the contrary that $P = \bigcup_n P_n$, each P_n nowhere–dense and observe that then $\mathbf{R}^1$ is a union of countably many nowhere–dense sets contradicting the Baire theorem]

In Exercises 9–15, we introduce spaces T_0, T_1, T_2.

9. A topological space (X, τ) is T_0 if and only if the following condition holds: $x \neq y$ implies $Cl\{x\} \neq Cl\{y\}$. Prove that if (X, τ) is T_0 then for each pair $x \neq y$ there exists an open set A such that $x \in A \leftrightarrow y \notin A$.

10. In an ordered set $(X, \leq)$ introduce a topology by declaring an open basis the family of sets $\{x^+ = \{y \in X : x \leq y\} : x \in X\}$. Prove that X is T_0.

11. A topological space (X, τ) is T_1 if and only if $Cl\{x\} = \{x\}$ for every $x \in X$. Prove that X of Exercise 10 is T_1 if and only if any two distinct elements are incomparable with respect to $\leq$.

12. Prove that any T_1 topology on a finite set X is discrete.

13. A topological space (X, τ) is T_2 (or, *Hausdorff*) if and only if for each pair $x \neq y$ there exist disjoint open sets A, B with $x \in A, y \in B$. Prove that any space T_2 is T_1.

14. Prove that any metric space is T_2.

15. Prove that (i) any compact subset of a T_2 space is closed (ii) any closed subset of a compact space is a compact space in the topology of a sub–space (iii) any continuous image of a compact space is compact. [Hint: in case (iii)

apply definition of compactness via open coverings]

In Exercises 16–17, we look at some properties of Cartesian products and exponential spaces.

16. Consider a Cartesian product $X \times Y$ of topological spaces X, Y. For $A \subseteq X, B \subseteq Y$, we consider the product $A \times B$. Prove that (i) $Int(A \times B) = IntA \times IntB$ (ii) $Cl(A \times B) = ClA \times ClB$ (iii) $Bd(A \times B) = BdA \times ClB \cup ClA \times BdB$.

17. Consider the space 2^X of closed subsets of a T_1 space X with the hit–or–miss (exponential) topology. For $A \subseteq X$ we denote by the symbol 2^A the set of all closed in X sets contained in A. Prove that

(i) $Int2^A = 2^{IntA}$

(ii) $Cl2^A = 2^{ClA}$

[Hint: show first that $A \subseteq B$ implies $2^A \subseteq 2^B$ and infer that $2^{IntA} \subseteq 2^A$ hence $2^{IntA} \subseteq Int2^A$ and $2^A \subseteq 2^{ClA} = Cl2^{ClA}$ hence $Cl2^A \subseteq 2^{ClA}$. To prove the converse, resort to the form of the open basis in 2^X]

In Exercises 18–22 , we are interested in equivalence relations in topological spaces in particular in compact spaces.

18. An equivalence relation R on a topological space X is *upper semicontinuous* if and only if the set $\bigcup\{[x]_R : [x]_R \subseteq A\}$ is open whenever the set A is open. Prove that R is upper semicontinuous if and only if the quotient function $q_R : X \to X/R$ is *closed* i.e. the image $q_R(A)$ is closed whenever $A \subseteq X$ is closed.

19. For a function $f : X \to Y$ between topological spaces X, Y consider the equivalence relation R_f defined as follows: $xR_f y$ if and only if $f(x) = f(y)$. Prove that the induced function $f_R : X/R_f \to Y$ is continuous if and only if f is continuous.

20. Prove that in a compact T_2 space X, for any two disjoint closed sets K, L there exist open disjoint sets A, B with $K \subseteq A, L \subseteq B$. [Hint: for $x \in K$, look at all $y \in Y$ and for each of them choose open disjoint $A(x,y), B(x,y)$ with $x \in A(x,y), y \in B(x,y)$; then find a finite family $\{B(x,y_i)\}$ covering L and observe that $x \in \bigcap\{B(x,y_i)\} = A(x)$, $L \subseteq \bigcup\{B(x,y_i)\} = B(x)$. By compactness of K select a finite covering $\{A(x_j)\}$ of K and verify that $A = \bigcup\{A(x_j)\}$, $B = \bigcap\{A(x_j)\}$ are as desired]

21. Prove that if X is a compact T_2 space and the equivalence relation R on X is upper semicontinuous then the quotient space X/R is a compact T_2. [Hint: apply Exercise 15(iii) to prove compactness of X/R. Given $a \neq b$

in X/R consider closed disjoint sets $K = q_R^{-1}(a), L = q_R^{-1}(b)$ and find open disjoint A, B as in Exercise 20. Let $U = Y \setminus q_R(X \setminus A), W = Y \setminus q_R(X \setminus B)$ and verify that U, W are open disjoint sets containing a, b respectively. In verifying that U, W are open apply Exercise 18]

22. Prove that if $f : X \to Y$ maps a compact T_2 space X onto a compact T_2 space Y then the quotient space X/R_f and Y are homeomorphic. [Hint. Define $g : X/R_f \to Y$ via $g([x]_{R_f}) = f(x)$ and apply Exercise 19. To show that the inverse g^{-1} is continuous prove equivalently that g maps closed sets onto closed sets applying results of Exercise 15 to show that f has this property and then verifying that g has this property as $gA = f(q_{R_f}^{-1}(A))$ for $A \subseteq X/R_f$.]

In Exercises 23–26, we look at properties of ultrafilters and Stone spaces.

23. Consider the topological space N of natural numbers with the topology induced from the real line i.e. with the discrete topology. Prove that (i) if an ultrafilter $\mathcal{F}$ on N contains a finite set then $\mathcal{F}$ is principal (ii) if an ultrafilter $\mathcal{F}$ on N contains no finite set then $\mathcal{F}$ is free i.e. $\bigcap \mathcal{F} = \emptyset$ (iii) Deduce from (i) and (ii) that an ultrafilter on N is free if and only if it contains no finite set.

24. In the setting of Exercise 23, prove that (i) if an ultrafilter $\mathcal{F}$ on N is free, $A \in \mathcal{F}$ and a set B is such that $A \triangle B$ is finite then $B \in \mathcal{F}$ (ii) if ultrafilters $\mathcal{F}_1, \mathcal{F}_2$ on N are free, $A \in \mathcal{F}_1, B \in \mathcal{F}_2$ then $A \triangle B$ is infinite. [Hint: apply Exercise 23 (iii)]

25. Prove that each ultrafilter $\mathcal{F}$ on N which extends the Fréchet filter is free.

26. Consider the set $Q = (q_n)_n$ of rational numbers in the real line and to each irrational number p assign a sequence $(q_{n_k})_k$ in Q such that $p = \lim q_{n_k}$ and let $A_p = \{n_k : k \in N\}$. Verify that the family $\{A_p : p \in P\}$ is an uncountable family of infinite sets any two distinct of them having a finite intersection. Deduce from this fact that there exists an uncountable family of distinct free ultrafilters on N and in consequence the Stone space of the complete Boolean algebra 2^N is uncountable (having actually a basis of clopen sets of cardinality equal to the cardinality of the real line). [Hint: each set A_p induces an ultrafilter as well as an element of the clopen basis of the Stone space]

Works quoted

[Alexandrov25] P. S. Alexandrov, *Zur Begründung der n–dimensionalen mengentheoretischen Topologie*, Math. Ann., 94(1925), pp. 296–308.

[Alexandrov–Urysohn29] P. S. Alexandrov and P. S. Urysohn, *Mémoire sur les espaces topologiques compacts*, Verh. Konink. Akad. Amsterdam, 14(1929).

[Baire899] R. Baire, Ann. di Math., 3(1899).

[Banach22] S. Banach, *Sur les opérations dans les ensembles abstraits et leur application aux équations intégrales*, Fund. Math., 3(1922), pp. 133–181.

[Cantor883] G. Cantor, Math. Ann., 21(1883).

[Cantor880] G. Cantor, Math. Ann., 17(1880).

[Čech66] E. Čech, *Topologicke' prostory*, in: E. Čech, *Topological Spaces*, Academia, Praha, 1966.

[Fréchet28] M. Fréchet, *Les espaces abstraits*, Paris, 1928.

[Fréchet06] M. Fréchet, *Sur quelques points du Calcul fonctionnel*, Rend. Circ. Matem. di Palermo, 22(1906).

[Hahn32] H. Hahn, *Reelle Funktionen I*, Leipzig, 1932.

[Hausdorff14] F. Hausdorff, *Grundzüge der Mengenlehre*, Leipzig, 1914.

[Kuratowski22] C. Kuratowski, *Sur l'operation $\overline{A}$ de l'Analysis Situs*, Fund. Math., 3(1922), pp.182–199.

[Lebesgue05] H. Lebesgue, J. de Math., 6(1905).

[Marcus94] S. Marcus, *Tolerance rough sets, Čech topologies, learning processes*, Bull. Polish Acad. Sci. Tech., 42 (1994), pp. 471–487.

[Polkowski–Skowron–Zytkow94] L. Polkowski, A. Skowron, and J. Zytkow, *Tolerance based rough sets*, in: T. Y. Lin and M. Wildberger (eds.), *Soft Computing: Rough Sets, Fuzzy Logic, Neural Networks, Uncertainty Management, Knowledge Discovery*, Simulation Councils, Inc., San Diego, 1995, pp. 55–58.

[Pompéju05] D. Pompéju, Ann. de Toulouse, 7(1905).

[Rasiowa–Sikorski50] H. Rasiowa and R. Sikorski, *A proof of the completeness theorem of Gödel*, Fund. Math., 37(1950), pp. 193–200.

[Riesz09] F. Riesz, *Stetigskeitbegriff und abstrakte Mengenlehre*, Atti IV Congr.

Int. Mat., Rome, 1909.

[Stone36] M. H. Stone, *The theory of representations for Boolean algebras*, Trans. Amer. Math. Soc., 40(1936), pp. 37–111.

[Tikhonov35] A. N. Tikhonov, *Über einen Funktionenraum*, Math. Ann., 111(1935).

[Vietoris21] L. Vietoris, Monat. Math. Ph., 31(1921), pp. 173–204.

Chapter 8

Algebraic Structures

Ineffective and empty–handed is the hunter who goes backward rather than forward

Marsilio Ficino, *Letter to Niccolo degli Albizzi*

8.1   Introduction

We will be interested here in algebraic structures defined within the realm of ordered sets. These structures play an eminent role in semantics of various logical calculi. In developing a theory of these structures, two ways are possible, either to begin with the most perfect structure i.e. Boolean algebras and relax gradually its requirements descending to less organized structures or to begin with the least perfect structures and gradually add requirements to ascend to more organized structures. Either approach has its merits, and here we settle with the latter so we begin with the minimal sound structure and subsequently add more features.

8.2   Lattices

We begin with a set X in which a relation $R \subseteq X \times X$ is defined such that (i) xRx for every $x \in X$ (ii) xRy, yRz imply xRz for every triple $x, y, z \in X$ i.e. R is reflexive and transitive. Clearly, such R need not be an ordering on X i.e. R may lack the weak anti–symmetry property. For instance, the relation $\rightarrow$ on formulae of propositional calculus defined via $\alpha \rightarrow \beta$ if and only if $C\alpha\beta$ is a thesis is clearly reflexive and transitive (as $C\alpha\alpha$, $CC\alpha\beta CC\beta\gamma C\alpha\gamma$ are theses) but for instance $C\alpha NN\alpha$, $CNN\alpha\alpha$ are theses which does not imply that α and $NN\alpha$ are identical. In order to produce from R an ordering, one should turn R into a weakly anti–symmetric relation. To this end, we introduce a relation $\simeq_R$ by means of the following recipe

$$x \simeq_R y \Leftrightarrow xRy \wedge yRx \tag{8.1}$$

Then we have

Proposition 8.1. *(Schröder)*

1. $\simeq_R$ is an equivalence relation
2. the relation R^* defined on the quotient set $X/\simeq_R$ via

$$[x]_{\simeq_R} R^* [y]_{\simeq_R} \ iff \ xRy \tag{8.2}$$

is an ordering on the quotient set $X/\simeq_R$.

Proof. See Exercise 28 in Chapter 6 on Set Theory.

A reflexive and transitive relation R is called a relation of *pre–order* as it does induce an ordering on the lines of the last proposition. We will assume from now on that a relation R of ordering on a non–empty set X is given; we denote R with a more suggestive symbol $\leq$ and we recall that an element $u \in X$ is said to be the *least upper bound* (l.u.b. for short) of elements x, y in case $x \leq c, y \leq c$ and $x \leq d, y \leq d$ imply $c \leq d$. Similarly, $l \in X$ is the *greatest lower bound* (g.l.b. for short) of x, y when $l \leq x, l \leq y$ and $d \leq x, d \leq y$ imply $d \leq l$.

It is not always that l.u.b. and g.l.b. exist for every pair $x, y \in X$. For instance, in the set $X = \{\{1, 2\}, \{1, 3\}, \{2, 3\}\}$ ordered by inclusion, neither l.u.b. nor g.l.b. exists for any pair x, y of distinct elements. Thus we single out this property of ordered sets.

We call a *lattice* any ordered set $(X, \leq)$ with the property that l.u.b. and g.l.b. exist for every pair x, y of elements of X. For instance, any ordered by inclusion $\subseteq$ family of subsets of a fixed set Y is a lattice when it is closed on finite unions and intersections; such are families of open sets as well as closed sets in any topological space. The g.l.b. of sets A, B is their intersection $A \cap B$ and the l.u.b. of sets A, B is the union $A \cup B$.

Motivated by this example, we denote g.l.b. with the symbol $\cap$ and the l.u.b. with the symbol $\cup$ calling them respectively the *meet* and the *join*. The properties of meets and joins which follow from their definitions are summed up in

Proposition 8.2. *For any pair x, y of elements of a lattice X we have*

1. $x \cup x = x, x \cap x = x$

2. $x \cup y = y \cup x$ and $x \cap y = y \cap x$

3. $x \cup (y \cup z) = (x \cup y) \cup z$ and $x \cap (y \cap z) = (x \cap y) \cap z$

4. $y \cup (x \cap y) = y$ and $x \cap (x \cup y) = x$

These properties follow immediately from definitions (cf. Exercise 1, below).

Let us observe a duality property: any statement in the above list is converted into its dual by substituting $\cup/\cap$ and $\cap/\cup$ and valid statements are converted into valid statements with the ordering $\leq$ changed into $\geq$.

The reverse situation when we introduce an ordering compatible with given $\cup, \cap$ may be described as follows

Proposition 8.3. *Assume that a set X is endowed with two operations $\cup, \cap$ which satisfy 1.–4. Then the relation $x \leq y$ defined via one of equivalent conditions: $x \cap y = x$, $x \cup y = y$ is an ordering on X and the operations g.l.b. and l.u.b. induced by $\leq$ coincide with respectively $\cap$, $\cup$.*

Proof. By properties 4 and 2 above if $x \cap y = x$ then $x \cup y = y$; similarly, $x \cup y = y$ implies $x \cap y = x$. Thus, either of $x \cup y = y, x \cap y = x$ may be taken as the definition of $x \leq y$.

We should check that $\leq$ is an ordering. To this end, given x we have $x \cap x = x$ by Property 1. Thus, $x \leq x$. Assume now that $x \leq y$ and $y \leq x$. Thus $x \cap y = x, y \cap x = y$ hence $x = y$ by Property 2. Finally, if $x \leq y, y \leq z$ then by Property 3 we have $x \cap y = x, y \cap z = y$ hence $x = x \cap y = x \cap (y \cap z) = (x \cap y) \cap z = x \cap z$ i.e. $x \leq z$.

Finally, we should show that operations g.l.b. and l.u.b. coincide with $\cap, \cup$ respectively. By duality, we show this for g.l.b. As $x \cap (x \cap y) = (x \cap x) \cap y = x \cap y$, we have $x \cap y \leq x$; similarly, $x \cap y \leq y$. Assume now that $c \leq x, c \leq y$ for some $c \in X$. Then $c \cap (x \cap y) = (c \cap x) \cap y = c \cap y = c$ i.e. $c \leq x \cap y$. Thus $x \cap y = g.l.b.(x, y)$. $\qquad \square$

Observe that we have proved the following identities for any lattice $X, \cup, \cap$:

5. $x \leq x \cup y$

6. $x \cap y \leq x$

A further step in adding requirements may be concerned with neutral elements with respect to operations of join and meet viz. we may ask whether an element $\sqrt{} \in X$ exists such that $x \cap \sqrt{} = x$ for every $x \in X$, respectively, whether an element $\sqcap \in X$ exists such that $x \cup \sqcap = x$ for every $x \in X$. Such elements need not to exist: consider for instance, the set $(0,1)$ of real numbers with the natural ordering $\leq$; we have in this case: $x \cup y = max\{x, y\}$, $x \cap y = min\{x, y\}$ so $((0,1), \leq)$ is a lattice, however without neutral elements. However, adding real numbers $0, 1$ to this lattice gives the lattice $([0,1], \leq)$

which has both elements: $\sqrt{} = 1$ and $\sqcap = 0$.

The existence of neutral elements should therefore be assured by additional requirements. Assuming that they exist we have the following properties
7. $x \cap \sqrt{} = x$ for every x
8. $x \cup \sqcap = x$ for every x
9. $x \leq \sqrt{}$ for every x
10. $\sqcap \leq x$ for every x
11. $x \cup \sqrt{} = \sqrt{}$ for every x
12. $x \cap \sqcap = \sqcap$ for every x

Indeed, 7 and 8 are definitions, 9, and 10 follow by definition of the ordering $\leq$ and 11, 12 again express 9, 10 in a dual way. It follows that $\sqcap$ is the *least element* in the lattice and $\sqrt{}$ is the *greatest element* in the lattice.

The element $\sqrt{}$ when exists is called the *unit* element and similarly $\sqcap$ is called the *zero* or *null* element. These elements are often denoted **0** (*zero*), **1** (*unit*). A lattice in which zero and unit elements exist is called *bounded*.

8.3   Distributive lattices

In the lattice $((0,1), \leq)$ defined above, the *distributive laws* hold viz.

13. $x \cap (y \cup z) = x \cap y \cup x \cap z$
14. $x \cup (y \cap z) = x \cup y \cap x \cup z$.

Indeed, to prove 13. observe that on one hand

$$min\{x, z\}, min\{x, y\} \leq min\{x, max\{y, z\}\} \qquad (8.3)$$

hence

$$max\{min\{x, z\}, min\{x, y\}\} \leq min\{x, max\{y, z\}\} \qquad (8.4)$$

(cf. in this respect Exercise 3, below).
On the other hand, by symmetry with respect to y, z of 13, we need to check each of three cases (i) $x \leq y, z$ (ii) $y \leq x \leq z$ (iii) $y, z \leq x$ and it is a simple matter to verify that in either case we have

$$max\{min\{x, z\}, min\{x, y\}\} \geq min\{x, max\{y, z\}\} \qquad (8.5)$$

It is equally simple to verify 14.
Simple examples show that these laws need not hold in any lattice e.g. the lattice P (see Figure 1 below)

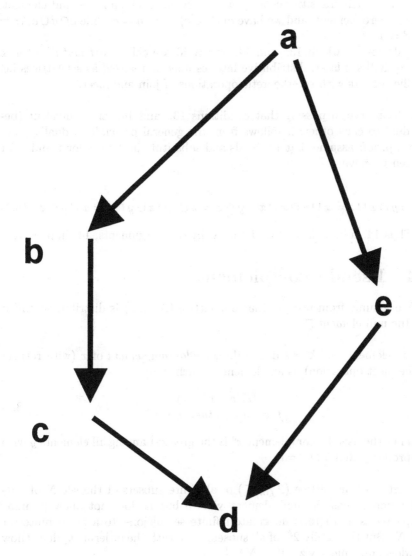

Figure 8.1: The lattice on 5 elements (a pentagon) which is not distributive

consisting of elements a, b, c, d, e subject to the ordering given by

$$a \geq b, a \geq e, b \geq c, c \geq d, e \geq d \qquad (8.6)$$

and all their transitive consequences is such. There, a is the unit element, d is the zero element, and we have $c \cap (d \cup e) = c \cap a = c$ while $c \cap d \cup c \cap e = d \cup d = d$.

Lattices for which 13. and 14. are valid are called *distributive lattices*. As we shall see later, distributive lattices may be realized as set lattices i.e. families of sets with set–theoretic operations of join and meet.

Let us observe in passing that conditions 13. and 14. are equivalent (being dual to each other, it follows from the general principle of duality); we offer a proof: assume that 13. holds and substitute in it $x \cup y$ for x and x for y; then we have

$$x \cup y \cap x \cup z = x \cup y \cap x \cup x \cup y \cap z = x \cup x \cap z \cup y \cap z = x \cup y \cap z \quad (8.7)$$

Thus 14. follows from 13. Similarly, by dual arguments, 14. implies 13.

8.4 Pseudo–complement

We assume from now on that our lattice $(X, \cap, \cup)$ is distributive and it has the null element $\sqcap$.

For an element $x \in X$, we define the *pseudo–complement* of x (with respect to the meet operation) as an element x^c such that

$$\begin{array}{c} (i) \; x \cap x^c = \sqcap \\ if \; x \cap y = \sqcap \; then \; y \leq x^c \end{array} \qquad (8.8)$$

Thus the pseudo–complement x^c is the greatest among all elements y with the property that $x \cap y = \sqcap$.

For instance, the lattice $(\mathcal{F}_{fin}, \subseteq)$ of all finite subsets of the set N of natural numbers has the null element viz. $\emptyset$ but it does not allow pseudo–complements as there is no greatest finite set disjoint to a given finite set $A \subseteq N$. But the family 2^N of all subsets of N with the ordering $\subseteq$ does allow pseudo–complement viz. $A^c = N \setminus A$.

The notion of a pseudo–complement may be taken also relative to an element viz. given elements $x, y \in X$, we call an element z a *pseudo–complement of x relative to y* if and only if z is the greatest among elements u which satisfy the requirement that $x \cap u \leq y$. The pseudo–complement of x relative to y is denoted by the symbol $x \Rightarrow y$.

As an example we consider the family $\mathcal{O}\ (X)$ of open subsets of a topological space X. Given a subset $A \in \mathcal{O}\ (X)$, we find the pseudo–complement A^c to A. The condition $A \cap B = \emptyset$ for an open set B implies $B \subseteq X \setminus A$ and as B is open we have $B \subseteq Int(X \setminus A)$. It follows that $Int(X \setminus A)$ is the greatest among open sets disjoint to A and thus $A^c = Int(X \setminus A)$.

In a similar way we may verify the existence of a relative pseudo–complement $A \Rightarrow B$ for each pair A, B of open sets viz. it follows from $A \cap Z \subseteq B$ that $Z = Z \cap A \cup Z \cap (X \setminus A)$ hence $Z \subseteq B \cap A \cup X \setminus A$ and thus $Z \subseteq Int(A \cap B \cup X \setminus A)$. It follows that $A \Rightarrow B = Int(A \cap B \cup X \setminus A)$. In the particular case when $A = B$, we get $A \Rightarrow A = Int(A \cup X \setminus A) = IntX = X$ i.e. $A \Rightarrow A$ is the unit element.

Let us put here for the record some basic observations on $\Rightarrow$.

15. $z \leq x \Rightarrow y$ if and only if $z \cap x \leq y$ for every triple x, y, z of elements of X

16. $y \leq x \Rightarrow y$; $x \leq x \Rightarrow y$ if and only if $x \leq y$

A distributive lattice X in which the relative pseudo–complement $x \Rightarrow y$ exists for every pair x, y of elements is said to be *relatively pseudo–complemented*: let us observe that this assumption forces the existence of the unit element $\sqrt{}$: indeed, $\sqrt{} = x \Rightarrow x$ for every $x \in X$ as $x \cap z \leq x$ for every z hence the unique element $x \Rightarrow x$ is the greatest element in X i.e. the element $\sqrt{}$. If in addition X has the zero element $\sqcap$ then $x \Rightarrow \sqcap$ is the pseudo–complement x^c.

Relatively–pseudo–complemented lattices with the unit and the null elements are called *pseudo–Boolean algebras* (or, *Heyting algebras*). In any pseudo–Boolean algebra, the pseudo–complement x^c exists for every x; indeed, $x^c = x \Rightarrow \sqcap$. From the above discussion it follows that the lattice $\mathcal{O}$ (X) of open sets in a topological space X is a pseudo–Boolean algebra.

Let us put for the record here basic properties of pseudo–complement $(\cdot)^c$.

Proposition 8.4. *The operation* $(\cdot)^c$ *has the following properties*

1. $x \leq x^{cc}$

2. $x \leq y$ *does imply* $y^c \leq x^c$

3. $x^c = x^{ccc}$

Proof. Property 1 follows from the identity $x \cap x^c = \sqcap$ which implies $x \leq x^{cc}$. For Property 2, $x \leq y$ means $x \cap y = x$ hence $x \cap y^c = x \cap y \cap y^c = x \cap \sqcap = \sqcap$ which implies that $y^c \leq x^c$. From properties 1, 2, we infer that $x^{ccc} \leq x^c$ and substituting in Property 1 x^c for x, we get $x^c \leq x^{ccc}$ so finally Property 3 follows.

8.5   Stone lattices

A distributive pseudo–complemented lattice with the unit element is said to be a *Stone lattice* in case the condition $x^c \cup x^{cc} = \sqrt{}$ for every x holds.

We have then

Proposition 8.5. *In any Stone lattice, we have* $(x \cap y)^c = x^c \cup y^c$.

Proof. First we have $x \cap y \cap (x^c \cup y^c) = \sqcap$. Assume $x \cap y \cap z = \sqcap$ so that $y \cap z \leq x^c$ and hence $y \cap z \cap x^{cc} \leq x^c \cap x^{cc} = \sqcap$ which implies that $z \cap x^{cc} \leq y^c$. By the condition $x^c \cup x^{cc} = \sqrt{}$, we have $z = z \cap x^c \cup z \cap x^{cc} \leq x^c \cup y^c$. It follows that $x^c \cup y^c$ is the pseudo–complement of $x \cap y$. □

Let us observe that the lattice $\mathcal{O}(X)$ of open sets in a topological space X need not be any Stone lattice. Given an open A in X, we have

$$A^c \cup A^{cc} = Int(X \setminus A) \cup Int(X \setminus Int(X \setminus A)) \qquad (8.9)$$

and the latter set is identical to the unit X if and only if A is clopen which is not always the case.

8.6   Complement

For an element x of a distributive bounded lattice $(X, \cap, \cup)$, we say that an element y is the *complement to* x if and only if
17. $x \cap y = \sqcap$
18. $x \cup y = \sqrt{}$.

We denote the complement to x with the symbol $--x$.

Then we may check that

Proposition 8.6. $--x$ *is the greatest element among* y *satisfying* $x \cap y = \sqcap$ *and the least element among* y *that satisfy* $x \cup y = \sqrt{}$.

Proof. Assume that $x \cap y = \sqcap$; then $y = y \cap \sqrt{} = y \cap (x \cup --x) = y \cap x \cup y \cap --x = y \cap --x$ hence $y \leq --x$. Similarly, if $x \cup y = \sqrt{}$ then $y = y \cup \sqcap = y \cup (x \cap --x) = y \cup x \cap y \cup --x = y \cup --x$ hence $--x \leq y$. □

It follows that the complement $--x$ is indeed defined uniquely for every x. In particular we have

19. $--\sqcap = \sqrt{}; \quad --\sqrt{} = \sqcap$

Let us observe that

Proposition 8.7. *If the complement* $--x$ *exists then the relative pseudo–complement* $x \Rightarrow y$ *exists and it is equal to* $--x \cup y$ *for each pair* x, y.

Proof. We have $x \cap --x \cup y = x \cap --x \cup x \cap y = x \cap y \leq y$ i.e. $--x \cup y$ does satisfy the definition of the relative pseudo–complement. Assume that $a \cap x \leq y$; then $a = a \cap \sqrt{} = a \cap (x \cup --x) = a \cap x \cup a \cap --x \leq y \cup --x = --x \cup y$ so indeed $--x \cup y$ is the greatest element among z such that $z \cap x \leq y$. $\square$

We have then some new properties of a relative pseudo–complement.

20. $x \Rightarrow x = \sqrt{}$; $x \Rightarrow y = \sqrt{}$ if and only if $x \leq y$
21. $x \Rightarrow \sqcap = --x$

Among properties of complement we may mention the following

Proposition 8.8. *1.* $--(x \cup y) = --x \cap --y$ *provided* $--x, --y$ *exist*

2. $--(x \cap y) = --x \cup --y$ *provided* $--x, --y$ *exist*

3. $--(--x) = x$ *provided* $--x$ *exists*

4. $x \leq y$ *if and only if* $--y \leq --x$ *provided* $--x, --y$ *exist*

Proof. For Property 1, if $z \cap (x \cup y) = \sqcap$ then $z \cap x \cup z \cap y = \sqcap$ hence $z \cap x = \sqcap$ and $z \cap y = \sqcap$ i.e. $z \leq --x$ and $z \leq --y$ and thus $z \leq --x \cap --y$. As $x \cup y \cap --x \cap --y = \sqcap$ it follows that $--x \cap --y$ is the pseudo–complement of $x \cup y$. It remains to observe that $x \cup y \cup --x \cap --y = \sqrt{}$ to conclude 1. Property 2 follows similarly by duality. Property 3 follows from 17 and 18 by symmetry of conditions. For Property 4, assume $x \leq y$ hence $x \cup y = y$ so $--y = --x \cap --y$ by 1, meaning that $--y \leq --x$. The last inequality implies $x \leq y$ by 3. $\square$

Let us mention without proof that every lattice with the property that $x \Rightarrow y$ exists for every pair x, y of its elements is distributive (cf. Exercise 2, below).

In a pseudo–Boolean algebra, not any element needs to have the complement. As we know the lattice $\mathcal{O}(X)$ of open sets in a topological space X is a pseudo–Boolean algebra and the pseudo–complement A^c is equal to $Int(X \setminus A)$. However, the join $A \cup Int(X \setminus A)$ is equal to the unit X if and only if A is clopen. Thus the pseudo–complement need not be any complement.

Let us state some properties that hold true in pseudo–Boolean algebras.

Proposition 8.9. *The below properties are observed in any pseudo–Boolean algebra* $(X, \cap, \cup)$:
22. *If* $x \leq y$ *then* $y^c \leq x^c$
23. $x \leq x^{cc}$
24. $x^c = x^{ccc}$
25. $(x \cup y)^c = x^c \cap y^c$

26. $x^c \cup y^c \leq (x \cap y)^c$

Proof. Properties 22–25 are already proved above. Property 26, follows from $(x \cap y) \cap (x^c \cup y^c) = \sqcap$ by which we have $x^c \cup y^c \leq (x \cap y)^c$. $\qquad\square$

8.7 Boolean algebras

The ultimate stage in developing theory of lattices is thus to consider a distributive bounded lattice with the property that the complement $--x$ exists for every of its elements. Any such lattice is called a *Boolean algebra*. In the case of a Boolean algebra properties 22, 23, 26 may be strengthened viz.

22*. $x \leq y$ if and only if $y^c \leq x^c$
23*. $x = x^{cc}$
26*. $x^c \cup y^c = (x \cap y)^c$.

A natural example of a Boolean algebra is any *field of sets* i.e. a set lattice $(\mathcal{F}, \subseteq)$ where $\mathcal{F} \subseteq 2^X$ for some set X is such that $X \setminus A \in \mathcal{F}$ for every $A \in \mathcal{F}$, and $A \cup B, A \cap B \in \mathcal{F}$ for every pair $A, B \in \mathcal{F}$. For instance, the set lattice $(\mathcal{F}_{finf}, \subseteq)$ consisting of those $A \subseteq N$ for which either A is finite or $N \setminus A$ is finite is a field of sets.

Boolean algebras may be realized in weaker structures. For instance, consider a Stone algebra S along with its subset $C(S) = \{x^c : x \in S\}$. Then

27. For $x, y \in C(S)$ we have:
   (i) $x \cap y \in C(S)$
   (ii) $x = x^{cc}$

and the last identity characterizes elements in $C(S)$.

Indeed, we know (cf. Property 24) that $y^c = y^{ccc}$ for every $y \in S$ and thus in case $x \in C(S)$, we have $x = y^c$ for some $y \in S$ hence $x = y^c = y^{ccc} = x^{cc}$. Conversely, if $x = x^{cc}$ then clearly $x \in C(S)$. Now, for $x, y \in C(S)$, we have $x = x^{cc}, y = y^{cc}$ hence $x \geq (x \cap y)^{cc}$ and $y \geq (x \cap y)^{cc}$ thus $x \cap y \geq (x \cap y)^{cc}$. As always $x \cap y \leq (x \cap y)^{cc}$, we have $x \cap y = (x \cap y)^{cc}$ hence $x \cap y \in C(S)$.

We have also:

28. For $x, y \in C(S)$, the element $(x^c \cap y^c)^c$ is the join of $\{x, y\}$ in $C(S)$.

Indeed, $x \leq (x^c \cap y^c)^c$ and $y \leq (x^c \cap y^c)^c$ by Property 26 and given z with $x, y \leq z$, we have $x^c, y^c \geq z^c$ which by Property 22 imply $z \geq (x^c \cap y^c)^c$.

We let $x \uplus y = (x^c \cap y^c)^c$; thus $C(S)$ with $\cap, \uplus$ is a bounded lattice. It remains to check that

29. For $x \in C(S)$, x^c is the complement of x with respect to $\cap, \uplus$.

The claim follows from: $x \uplus x^c = (x^c \cap x^{cc})^c = \cap^c = \sqrt{}$ and $x \cap x^c = \cap$.

Finally, we check that $C(S)$ with $\cap, \uplus$ is distributive.

30. $C(S)$ with $\cap, \uplus$ is distributive.

To prove distributivity, observe that given $x, y, z \in C(S)$, we have $x \cap z, y \cap z \leq x \uplus (y \cap z)$ hence $z \cap (x \uplus (y \cap z))^c \leq x^c, y^c$ i.e. $z \cap (x \uplus (y \cap z))^c \leq x^c \cap y^c$. Thus, $z \cap (x^c \cap y^c)^c \leq (x \uplus (y \cap z))^{cc}$ which means that $z \cap (x \uplus y) \leq x \uplus (y \cap z)$. To complete the proof, we need only to consider

30'. $z \cap (x \cup y) \leq x \cup (y \cap z)$

and verify that

31. In any lattice $(A, \cup, \cap)$, properties 13 (as well as 14) and 30' are equivalent meaning distributivity.

We hint at the proof. In any lattice, we have (i) $(x \cup y) \cap (x \cap z) \geq x \cap (y \cup z)$ which may be checked directly. Assuming 13 (and 14 as well), we have $x \cup (y \cap z) = (x \cup y) \cap (x \cup z) \geq (x \cup y) \cap z$ i.e. 30' follows. The converse goes on similar lines.

Combining the results of 27.–30'., we arrive at

Proposition 8.10. *For each Stone algebra S, the lattice $(C(S), \uplus, \cap)$ is a Boolean algebra.*

8.8 Filters on lattices

The notion of a filter plays an important role in the study of lattices. In set lattices, a filter is a non–empty family $\mathcal{F} \subseteq 2^X$ of subsets of a set X such that

(Fl1) $\emptyset \notin \mathcal{F}$
(Fl2) $A \cap B \in \mathcal{F}$ whenever $A, B \in \mathcal{F}$
(Fl3) $B \in \mathcal{F}$ whenever $A \in \mathcal{F}$ and $A \subseteq B$.

Thus, a set filter is any non–empty family of subsets of a set X which contains the intersection of any pair of its elements, contains any super–set of any of its elements, and does not contain the empty set. For instance, the set *CofN* of sets in the lattice $(\mathcal{F}_{finf}, \subseteq)$ consisting of those $A \subseteq N$ for which $N \setminus A$ is finite is a filter (called the *Frechét filter*). Another classical example is the family $\mathcal{N}(x)$ of all subsets A of a topological space (X, τ) such that $x \in Int A$ called the *filter of neighborhoods of $x \in X$*.

By analogy with set – theoretic filters, for any lattice $(X, \cap, \cup)$ we call a *filter* any family $\mathcal{F}$ of elements of X such that

(Filt1) $\mathcal{F} \neq \emptyset$
(Filt2) $x, y \in \mathcal{F}$ imply $x \cap y \in \mathcal{F}$
(Filt3) $x \in \mathcal{F}$ and $x \cup y = y$ imply $y \in \mathcal{F}$
(Filt4) $\sqcap \notin \mathcal{F}$ whenever X has the null element $\sqcap$.

Let us observe that (Filt3) is equivalent to

(Fil3t*) $x \in \mathcal{F}$ and $x \leq y$ imply $y \in \mathcal{F}$

Thus, for any element $a \neq \sqcap$ in the lattice X, the family $\mathcal{F}_a = \{x \in X : x \leq a\}$ is a filter in X, called the *principal filter generated by a*.

A filter $\mathcal{F}$ is said to be *maximal* whenever from $\mathcal{F} \subseteq \mathcal{G}$ it follows that $\mathcal{F} = \mathcal{G}$ for any filter $\mathcal{G}$. As the union of any linearly ordered by inclusion family of filters is a filter, it follows by the maximum principle (Proposition 14 in Chapter 6) that any filter is contained in a maximal filter.

A useful property of maximal filters can be given in the realm of distributive lattices.

We say that a filter $\mathcal{F}$ in a lattice X is *prime* whenever the condition holds

(Filt5) $x \in \mathcal{F}$ or $y \in \mathcal{F}$ whenever $x \cup y \in \mathcal{F}$

which is a partial converse to (Filt3).

Then we have the announced result

Proposition 8.11. *In any distributive lattice $(X, \cap, \cup)$, any maximal filter is a prime filter.*

Proof. Assume that a filter $\mathcal{F}$ is maximal but not prime i.e. there exist $a, b \in X$ such that $a \cup b \in \mathcal{F}$, $a \notin \mathcal{F}$ and $b \notin \mathcal{F}$. Thus neither $a \leq b$ nor $b \leq a$. Hence we may look at the set $\mathcal{G}$ of those x for which there exists an element $u \in \mathcal{F}$ such that $a \cap u \leq x$. Then it follows easily that $a \in \mathcal{G}$ and $b \notin \mathcal{G}$ hence $\mathcal{F}$ is a proper subset of $\mathcal{G}$, contradicting maximality of $\mathcal{F}$.   □

It turns out that one may exploit prime filters to show that any distributive lattice is actually a set lattice in a sense. To specify this sense, we define the notion of a *homomorphism*. Given two lattices A, B, a function $h : A \to B$ is said to be a *homomorphism* from A into B if and only if (hom) $h(x \cap y) = h(x) \cap h(y)$, $h(x \cup y) = h(x) \cup h(y)$

A homomorphism h is an *isomorphism* if and only if h is a bijection. If $h : A \to B$ is a homomorphism then the image $h(A) \subseteq B$ is closed on lattice operations $\cap, \cup$ i.e. it is a lattice in its own right. It is said to be a *sublattice* of B.

In order to show that any distributive lattice is in a sense a set lattice, it will suffice to demonstrate an isomorphism of this lattice onto a set lattice. To this end, we consider a distributive lattice X along with the set **M** of all prime filters on X.

To each element $x \in X$, we assign the subset **M**$_x$ of all prime filters which contain x i.e. for any prime filter $\mathcal{F}$ we have $\mathcal{F} \in$ **M**$_x$ if and only if $x \in \mathcal{F}$. We let **M**$_X$ to denote the family of sets **M**$_x$ i.e. **M**$_X = \{$**M**$_x : x \in X\}$.

We have clearly **M**$_{x \cap y} =$ **M**$_x \cap$ **M**$_y$ as any filter containing $x \cap y$ has to contain x and y and **M**$_{x \cup y} =$ **M**$_x \cup$ **M**$_y$ as our filters are prime. It follows that **M**$_X$ is a set lattice.

We may define a function $h : X \to$ **M**$_X$ by letting $h(x) =$ **M**$_x$. Then h is a homomorphism onto and it remains only to show that it is an injective mapping i.e. $x \neq y$ implies that $h(x) \neq h(y)$.

Assume then that $x \neq y$ hence e.g. it is not true that $x \leq y$. We have to find a prime filter $\mathcal{F}$ such that

(*) $y \notin \mathcal{F}$ and $x \in \mathcal{F}$.

The principal filter $\mathcal{F}_x$ does satisfy Condition (*) and as (*) is preserved by the union of any chain of filters, it follows by the maximum principle that there exists a maximal filter $\mathcal{F}^*$ in the collection of all filters satisfying (*). Clearly, $\mathcal{F}^*$ need not be a maximal filter so we have to check whether $\mathcal{F}^*$ is prime. Assume not i.e. there are c, d with $c \cup d \in \mathcal{F}^*$ and $c, d \notin \mathcal{F}^*$. Consider filters $\mathcal{M}, \mathcal{N}$ defined by

$$\begin{aligned} \mathcal{M} &= \{u : u \geq c \cap v \text{ for some } v \in \mathcal{F}^*\} \\ \mathcal{N} &= \{u : u \geq d \cap v \text{ for some } v \in \mathcal{F}^*\} \end{aligned} \qquad (8.10)$$

Then clearly $u \geq c \cap u, d \cap u$ for every $u \in \mathcal{F}^*$ hence $\mathcal{F}^* \subseteq \mathcal{M}, \mathcal{N}$ in particular $x \in \mathcal{M}, \mathcal{N}$.

Let us observe that: *either* $y \notin \mathcal{M}$ *or* $y \notin \mathcal{N}$;
indeed, was $y \in \mathcal{M} \cap \mathcal{N}$, we would have

$$y \geq c \cap u, d \cap w \text{ for some } u, w \in \mathcal{F}^* \qquad (8.11)$$

hence

$$y \geq c \cap u \cup d \cap w \geq (c \cup d) \cap (u \cap w) \qquad (8.12)$$

and as the latter element is in $\mathcal{F}^*$ we would have $y \in \mathcal{F}^*$ contrary to the definition of $\mathcal{F}^*$.

So we may assume that e.g. $y \notin \mathcal{M}$ and thus $\mathcal{M}$ satisfies Condition (*) contradicting the maximality of $\mathcal{F}^*$ in the set of all prime filters satisfying (*), as

$$c \in \mathcal{M} \setminus \mathcal{F}^* \qquad (8.13)$$

It follows that $\mathcal{F}^*$ is prime.

We have proved that for any pair $x \neq y$ there exists a prime filter which contains exactly one of x, y. Thus

$$x \neq y \text{ implies } h(x) \neq h(y) \qquad (8.14)$$

i.e. the homomorphism h is an isomorphism. It is called the *Stone isomorphism*.

We may sum up this discussion as follows

Proposition 8.12. *(Stone) Any distributive lattice* $(X, \cup, \cap)$ *is isomorphic to the set lattice* $St(X) = \{h(x) : x \in X\}$ *via the isomorphism* $h : X \to St(X)$ *defined as* $h(x) = \mathbf{M}_x$ *which is the set of all prime filters on* X *containing* x.

The lattice $St(X)$ is called the *Stone lattice* and the set $\mathbf{M}$ with the topology in which the open basis is the set of all finite intersections of sets of the form $h(x)$ is called the *Stone space* of the distributive lattice X.

We have proved in Chapter 7 a specialization of this result to the case of Boolean algebras in which case the Stone space is a compact 0–dimensional space and the Stone lattice is the lattice of clopen sets in the Stone space. In case of a complete Boolean algebra, the Stone space becomes extremally disconnected (cf. Chapter 7).

8.9   Filters on Boolean algebras

Passing on to more organized lattices, we stop at the class of Boolean algebras. Given a Boolean algebra $\mathbf{B}$ and a filter $\mathcal{F}$ on $\mathbf{B}$, we consider a relation $\approx_{\mathcal{F}}$ on $\mathbf{B}$ defined as follows

$$x \approx_{\mathcal{F}} y \text{ iff } --x \cup y \in \mathcal{F} \text{ and } --y \cup x \in \mathcal{F} \qquad (8.15)$$

Then

Proposition 8.13. $\approx_{\mathcal{F}}$ *is an equivalence relation on* **B**

Indeed, $x \approx_{\mathcal{F}} x$ as $- - x \cup x = \sqrt{} \in \mathcal{F}$; clearly, the relation $\approx_{\mathcal{F}}$ is symmetric. Transitivity follows easily as $- - x \cup y, - - y \cup x, - - y \cup z, - - z \cup y \in \mathcal{F}$ imply $- - x \cup z \geq (- - x \cup y) \cap (- - y \cup z) \in \mathcal{F}$ and similarly it follows that $- - z \cup x \in \mathcal{F}$.

Equivalence class of x by this relation will be denoted by the symbol $||x||$.

The quotient set **B** $/ \approx_{\mathcal{F}}$ will be denoted **B** $/\mathcal{F}$. Let us observe that **B** $/\mathcal{F}$ can be made into a Boolean algebra via the ordering relation $||x|| \preceq ||y||$ if and only if $- - x \cup y \in \mathcal{F}$. Then we have

Proposition 8.14. *1.* $||x \cup y|| = ||x|| \cup ||y||$

2. $||x \cap y|| = ||x|| \cap ||y||$

3. $|| - -x|| = - - ||x||$

4. $||x|| = \sqrt{}$ *if and only if* $x \in \mathcal{F}$

5. $|| \cap || = \cap_{\mathbf{B}} {}_{/\mathcal{F}}$

Proof. Indeed, to prove Property 1, we observe that $||x|| \preceq ||x \cup y||$ as $- - x \cup x \cup y = \sqrt{}$; similarly, $||y|| \preceq ||x \cup y||$ hence $||x|| \cup ||y|| \preceq ||x \cup y||$. On the other hand, if $||x||, ||y|| \preceq ||z||$ then $- - -x \cup z, - - y \cup z \in \mathcal{F}$ hence $-(x \cup y) \cup z = (- - x \cup z) \cap (- - y \cup z) \in \mathcal{F}$ i.e. $||x \cup y|| \preceq ||z||$. It follows that $||x \cup y|| \preceq ||x|| \cup ||y||$ and 1 follows. Dual proof of Property 2 is similar. For Property 4, $- - y \cup x \in \mathcal{F}$ for every y if and only if $x \in \mathcal{F}$ and then $||y|| \preceq ||x||$ for every $y \in \mathcal{F}$. Similarly for Property 5: $|| \cap || \preceq ||x||$ for every x as $- - \cap \cup x = \sqrt{} \cup x = \sqrt{} \in \mathcal{F}$. From 1, 2, 4, 5, we infer Property 3 as $\sqrt{} = ||x \cup - -x|| = ||x|| \cup || - -x||$ and $\cap_{\mathbf{B}/\mathcal{F}} = ||x \cap - -x|| = ||x|| \cap || - -x||$. $\square$

Let us observe that in any Boolean algebra **B**

(M) *a filter* $\mathcal{F}$ *is maximal if and only if for every* $x \in \mathbf{B}$*, exactly one of elements* $x, - - x$ *is in* $\mathcal{F}$

Indeed, let us observe that maximality of $\mathcal{F}$ demands that any element $z \in \mathbf{B}$ such that $z \cap x \neq \cap$ for every $x \in \mathcal{F}$ be in $\mathcal{F}$. Otherwise, we could define a greater filter by considering the set

$$\{u \in \mathbf{B} : \exists x \in \mathcal{F} \ u \geq z \cap x\}.$$

Assuming then that for some $x \in \mathbf{B}$ neither $x \in \mathcal{F}$ nor $- - x \in \mathcal{F}$, we could find $a, b \in \mathcal{F}$ such that

$$x \cap a = \cap = - - x \cap b$$

Then $\sqcap = x \cap a \cup -- x \cap b \geq (x \cup -- x) \cap (a \cap b) = a \cap b$ a contradiction as $a \cap b \in \mathcal{F}$.

Conversely, assuming that $\mathcal{F}$ has the property (M), assume also that $\mathcal{F}$ is not maximal i.e. $\mathcal{F} \subseteq \mathcal{G}$ for a filter $\mathcal{G}$ distinct from $\mathcal{F}$. Then there is some $a \in \mathbf{B}$ with $a \in \mathcal{G} \setminus \mathcal{F}$. As $a \notin \mathcal{F}$, we have $-- a \in \mathcal{F}$ hence $-- a \in \mathcal{G}$ and thus $\sqcap = a \cap -- a \in \mathcal{G}$, a contradiction. It follows that $\mathcal{F}$ is maximal.

As a corollary we obtain that

Proposition 8.15. *In any Boolean algebra* $\mathbf{B}$, *any filter is maximal if and only if it is prime.*

Proof. Indeed, we have already seen that maximal filters are prime filters (cf. Proposition 8.11) in any distributive lattice. It remains now to prove the converse. But if $\mathcal{F}$ is a prime filter then $x \cup -- x = \sqrt{} \in \mathcal{F}$ hence either $x \in \mathcal{F}$ or $-- x \in \mathcal{F}$ so $\mathcal{F}$ is maximal. $\square$

In consequence, in case of a Boolean algebra, we may use maximal filters when constructing the Stone isomorphism.

Let us observe also that in case of a maximal filter $\mathcal{F}$ in a Boolean algebra $\mathbf{B}$, we have

Proposition 8.16. *The Boolean algebra* $\mathbf{B} /_{\mathcal{F}}$ *consists of two elements viz.* $\sqcap, \sqrt{}$.

Indeed, for every $x \in \mathbf{B}$ we have either $||x|| = \sqrt{}$ in case $x \in \mathcal{F}$ or $||--x|| = \sqrt{}$ in case $-- x \in \mathcal{F}$ hence $||x||$ is either $\sqrt{}$ or $\sqcap$ for every x. Let us also observe that this property does characterize maximal filters : if for every $x \in \mathbf{B}$ we have that $||x||$ is either $\sqcap$ or $\sqrt{}$ then this means that $x \in \mathcal{F}$ or $-- x \in \mathcal{F}$ for every $x \in \mathbf{B}$ i.e. the filter $\mathcal{F}$ is prime hence maximal.

8.10   Pseudo–Boolean algebras

We have already observed that the lattice of open sets in a topological space is a pseudo–Boolean algebra; the converse also holds: any pseudo–Boolean algebra may be represented as a pseudo–Boolean algebra of open sets in a topological space [McKinsey–Tarski44]. We now insert a proof of this fact (cf. [Rasiowa–Sikorski63]). To this end, we should extend our notion of a Boolean algebra by means of topology: a Boolean algebra $\mathbf{B}$ endowed with an operator I satisfying the conditions

1. $I(x) \leq x$

2. $I(x \cap y) = I(x) \cap I(y)$

3. $I(I(x)) = I(x)$

4. $I(\sqrt{}) = \sqrt{}$

is called a *topological Boolean algebra*; a standard example will be the field of sets 2^X endowed with a topology τ on X inducing the interior operator *Int*.

The operator I is called the *interior* operator and we have without doubt noticed that 1–4 above reflect the properties of the topological interior operator in the lattice–theoretical setting.

Continuing the analogy, we call an element $a \in \mathbf{B}$ *open* if and only if $I(a) = a$. From Condition 2 above it follows immediately that $a \cap b$ is open whenever a and b are open. Similarly, for open a, b, the element $a \cup b$ is open; indeed, by Condition 1, $I(a \cup b) \leq a \cup b$ and from $a \leq a \cup b, b \leq a \cup b$ it follows by 1 that $a \cup b = I(a) \cup I(b) \leq I(a \cup b)$ so finally $I(a \cup b) = a \cup b$. We also have $I(\sqcap) = \sqcap$ which follows from 1.

Moreover, letting $C(a) = -I(-a)$, we define the *closure* of the element a for each $a \in \mathbf{B}$ with properties dual to 1–4 i.e.

$1'.\ x \leq C(x)$
$2'.\ C(x \cup y) = C(x) \cup C(y)$
$3'.\ C(C(x)) = C(x)$
$4'.\ C(\sqcap) = \sqcap$
and a is *closed* in case $C(a) = a$.

We consider the subset $\mathcal{O}(\mathbf{B}) = \{a \in \mathbf{B} : I(a) = a\}$ of $\mathbf{B}$ consisting of *open elements* in $\mathbf{B}$. Clearly, $\mathcal{O}(\mathbf{B})$ is a bounded lattice and given $a, b, x \in \mathcal{O}(\mathbf{B})$ we have $a \cap x \leq b$ if and only if $x \leq a \Rightarrow b$ if and only if $x = I(x) \leq I(a \Rightarrow b)$ hence $I(a \Rightarrow b) = I(a \cap)b\cup--a$, (where $\Rightarrow$ is the relative pseudo–complement in $\mathbf{B}$), is the relative pseudo–complement in $\mathcal{O}(\mathbf{B})$ which we denote by the symbol $\Rightarrow_O$. The pseudo–complement $(\cdot)^{co}$ in $\mathcal{O}(\mathbf{B})$ is then defined as $a^{co} = a \Rightarrow_O \sqcap = I(a \Rightarrow \sqcap) = I(--a)$.

Following [McKinsey–Tarski44] (cf. [Rasiowa–Sikorski63]), for a given pseudo–Boolean algebra $\mathcal{L}$, we define a topological Boolean algebra $\mathbf{B}$ with the property that $\mathcal{L} = \mathcal{O}(\mathbf{B})$. This will provide the required representation of $\mathcal{L}$ as a pseudo–Boolean algebra of open sets in a topological space.

We know already that $\mathcal{L}$ may be represented as a set lattice i.e. there exists a set X and a bounded lattice $\mathcal{L}^*$ of its subsets isomorphic to $\mathcal{L}$.

Then we may consider the least field of sets $\mathbf{B}$ containing $\mathcal{L}^*$; it is a matter of direct checking that $\mathbf{B}$ is the set of all subsets of X of the form

$$(-a_1 \cup b_1) \cap (-a_2 \cup b_2) \cap \ldots \cap (-a_n \cup b_n)$$

where $a_1, b_1, a_2, b_2, \ldots, a_n, b_n$ are elements of $\mathcal{L}^*$ and $-a$ is the complement in the field 2^X of all subsets of X (cf. Chapter 10, p. 352).

Denoting by the symbol $\Rightarrow$ the relative pseudo–complementation in $\mathcal{L}^*$ and by the symbol $\Rightarrow_B$ the pseudo–complementation in $\mathbf{B}$, we have clearly that $x \Rightarrow y \le x \Rightarrow_B y$ as $\mathcal{L}^* \subseteq \mathbf{B}$.

Let us observe that $-a \cup b = a \Rightarrow_B b$. It follows that

$$(-a_1 \cup b_1) \cap (-a_2 \cup b_2) \cap \ldots \cap (-a_n \cup b_n) \le -a \cup b \qquad (8.16)$$

i.e.

$$(a_1 \Rightarrow_B b_1) \cap (a_2 \Rightarrow_B b_2) \cap \ldots \cap (a_n \Rightarrow_B b_n) \le a \Rightarrow_B b \qquad (8.17)$$

implies

$$(a_1 \Rightarrow b_1) \cap (a_2 \Rightarrow b_2) \cap \ldots \cap (a_n \Rightarrow b_n) \le a \Rightarrow b \qquad (8.18)$$

Indeed,

$$(a_1 \Rightarrow_B b_1) \cap (a_2 \Rightarrow_B b_2) \cap \ldots \cap (a_n \Rightarrow_B b_n) \le a \Rightarrow_B b \qquad (8.19)$$

is equivalent to

$$(a_1 \Rightarrow_B b_1) \cap (a_2 \Rightarrow_B b_2) \cap \ldots \cap (a_n \Rightarrow_B b_n) \cap a \le b \qquad (8.20)$$

which implies

$$(a_1 \Rightarrow b_1) \cap (a_2 \Rightarrow b_2) \cap \ldots \cap (a_n \Rightarrow b_n) \cap a \le b \qquad (8.21)$$

equivalent in turn to

$$(a_1 \Rightarrow b_1) \cap (a_2 \Rightarrow b_2) \cap \ldots \cap (a_n \Rightarrow b_n) \le a \Rightarrow b \qquad (8.22)$$

It follows that given an element

$$b = (-a_1 \cup b_1) \cap (-a_2 \cup b_2) \cap \ldots \cap (-a_n \cup b_n) \in \mathbf{B} \qquad (8.23)$$

the element

$$(a_1 \Rightarrow b_1) \cap (a_2 \Rightarrow b_2) \cap \ldots \cap (a_n \Rightarrow b_n) \in A^* \qquad (8.24)$$

is defined uniquely so we may let

$$I(b) = (a_1 \Rightarrow b_1) \cap (a_2 \Rightarrow b_2) \cap \ldots \cap (a_n \Rightarrow b_n) \qquad (8.25)$$

It is again a matter of direct checking that I is an interior operator in $\mathbf{B}$ making it into a topological Boolean algebra. It is also clear that $I(a) = a$ for every $a \in A^*$ and that $I : \mathbf{B} \to A^*$ so finally we conclude that $A^* = O(\mathbf{B})$.

Historic remarks

Lattice theory goes back to Boole's [Boole847] where foundations for algebraic analysis of logic were laid. The Schröder book [Schröder895] summed up the development of the Algebra of Logic in the second half of 19 century. Lattice theoretical ideas proved to unify and simplify many particular results in diverse areas of Mathematics as witnessed e.g. in [Stone36], [Birkhoff67] and [McKinsey–Tarski44]. The algebraic tendency in logical investigations was highlighted in e.g. [Rasiowa–Sikorski63] and [Rasiowa74] where a throughout discussion of lattice theory will be found.

Exercises

In Exercises 1–6 we propose to consider basic properties of lattices.

1. Prove that in any lattice we have $y \cup (x \cap y) = y, y \cap (x \cup y) = y$.

2. Prove that in any lattice $x \cap y \cup x \cap z \leq x \cap (y \cup z)$.

3. [Birkhoff67] For a lattice X, we call a *polynomial* any meaningful expression built from variables $x, y, z, \ldots$ and $\cup, \cap$ (formally, any variable x is a polynomial and if α, β are polynomials then $\alpha \cup \beta, \alpha \cap \beta$ are polynomials, nothing else is any polynomial). Prove that any polynomial α regarded as a function on the lattice X is isotone i.e. $x \leq y$ implies $\alpha(x) \leq \alpha(y)$.

4. Prove that the lattice $X = (\{a, b, c, d, e\}, \leq)$ where $a \geq b, c, d, e$; $b, d, e \geq c$ is not distributive.

5. Prove that any relatively pseudo–complemented lattice is distributive. [Hint: consider $x \cap y \cup x \cap z$; as $x \cap y, x \cap z \leq x \cap y \cup x \cap z$ it follows that $y, z \leq x \Rightarrow x \cap y \cup x \cap z$ hence $y \cup z \leq x \Rightarrow x \cap y \cup x \cap z$ and $x \cap (y \cup z) \leq x \cap y \cup x \cap z$. For the converse, cf. Exercise 2]

6. Prove that in any pseudo–complemented lattice we have $x^{cc} \cup y^{cc} \leq (x \cup y)^{cc}$.

In Exercises 7–23, we look at relative pseudo–complementation, pseudo – complementation, and some special elements in pseudo–Boolean algebras.

7. [Tarski38] In a pseudo–Boolean algebra $\mathcal{A}$, an element x is *dense* if and only if $x^c = \sqcap$. Prove that (i) in any pseudo–Boolean algebra of open sets in a topological space (X, τ), an element A is dense if and only if $ClA = X$ (ii) an element $x \in \mathcal{A}$ is dense if and only if $x \cap y \neq \sqcap$ for each $y \neq \sqcap$. [Hint: for

(i) recall that $A^c = Int(X \setminus A)$. For (ii) observe that $y \cap x = \sqcap$ is equivalent to $y \leq x^c$]

8. [Tarski38] Prove that the set of all dense elements in a pseudo–Boolean algebra $\mathcal{A}$ is a filter. [Hint: use Exercise 7(ii) to prove that the meet of two dense elements is dense]

9. Consider a relatively – pseudocomplemented lattice $\mathcal{A}$ along with a filter $\mathcal{F}$ on $\mathcal{A}$. Define a relation R on $\mathcal{A}$ by letting xRy if and only if $x \Rightarrow y \in \mathcal{F}$ and $y \Rightarrow x \in \mathcal{F}$. Prove that R is an equivalence relation on $\mathcal{A}$. [Hint: when proving transitivity, observe that $x \Rightarrow y \cap y \Rightarrow z \leq x \Rightarrow z$]

10. Consider the quotient space $\mathcal{A} / R$ denoted usually $\mathcal{A}/\mathcal{F}$. The equivalence class $[x]_R$ will be denoted by the symbol $|x|$. Prove that $\mathcal{A}/\mathcal{F}$ is partially ordered by the relation $\leq_R$ defined as follows: $|x| \leq_R |y|$ if and only if $x \Rightarrow y \in \mathcal{F}$. [Hint: when proving transitivity, use the hint to Exercise 9]

11. In the notation of Exercise 10, prove that $(\mathcal{A}/\mathcal{F}, \leq_R)$ is a lattice with respect to the join $|x| \cup |y| = |x \cup y|$ and the meet $|x| \cap |y| = |x \cap y|$. [Hint: to prove that $|x \cup y| = |x| \cup |y|$, observe that $|x|, |y| \leq_R |x \cup y|$ and if $|x|, |y| \leq_R |z|$ then $x \Rightarrow z, y \Rightarrow z \in \mathcal{F}$ and apply the fact that $x \cup y \Rightarrow z = x \Rightarrow z \cap y \Rightarrow z$. The statement about the join follows by duality]

12. In the notation of Exercises 10, 11 , prove that the lattice $(\mathcal{A}/\mathcal{F}, \leq_R)$ is relatively pseudo–complemented with the pseudo–complement $\Rightarrow_R$ in $(\mathcal{A}/\mathcal{F}, \leq_R)$ related to the pseudo–complement $\Rightarrow$ in $\mathcal{A}$ via $|x| \Rightarrow_R |y| = |x \Rightarrow y|$. [Hint: check that $x \cap y \Rightarrow z = y \Rightarrow x \Rightarrow z$ and deduce from it that $|x| \cap |y| \leq_R |z|$ if and only if $|y| \leq_R |x \Rightarrow z|$]

13. Verify that the unit $\sqrt{}_R$ of the relatively pseudo–complemented lattice $(\mathcal{A}/\mathcal{F}, \leq_R)$ is equal to any $|x|$ where $x \in \mathcal{F}$. [Hint: verify that $x \in \mathcal{F}$ implies $y \Rightarrow x \in \mathcal{F}$ for each $y \in \mathcal{F}$]

14. Prove that if $\mathcal{A}$ is a pseudo–Boolean algebra then $\mathcal{A}/\mathcal{F}$ is a pseudo–Boolean algebra with the null element $\sqcap_R = |\sqcap|$. [Hint: $\sqcap \Rightarrow x = \sqrt{}$ for each $x \in \mathcal{F}$]

15. [Tarski38] We denote by the symbol $\mathcal{D}$ the filter of all dense elements in a pseudo–Boolean algebra $\mathcal{A}$. Prove that $\mathcal{A}/\mathcal{D}$ is a Boolean algebra. [Hint: check that by Proposition 8.9, $x \cup x^c$ is dense hence $|x| \cup |x^c| = \sqrt{}_R$ i.e. $-|x| = |x^c|$ is the complement of $|x|$]

16. [McKinsey–Tarski44] An element x of a pseudo–Boolean algebra $\mathcal{A}$ is called *regular* if and only if $x = x^{cc}$. Prove that x^c is regular for each $x \in \mathcal{A}$.

[Hint: apply Proposition 8.9]

17. [McKinsey–Tarski44] Prove that the meet $x \cap y$ of two regular elements x and y is regular. [Hint: apply Proposition 8.9]

18. [McKinsey–Tarski44] Prove that x^{cc} is the least element among all regular elements of $\mathcal{A}$ greater than x. [Hint: apply Proposition 8.9]

19. [McKinsey–Tarski44] Denote by the symbol $\mathcal{R}$ the set of all regular elements in $\mathcal{A}$. Prove that $\mathcal{R}$ is a lattice under the meet $x \cap y$ and the join $x \cup_R y = (x \cup y)^{cc}$. [Hint: apply Exercises 17 and 18]

20. [McKinsey–Tarski44] Prove that for each $|x| \in \mathcal{A}/\mathcal{F}$ there exists regular y with $|x| = |y|$. [Hint: check $y = x^{cc}$ applying the fact that $x^{cc} \Rightarrow x$ is dense]

21. [McKinsey–Tarski44] Infer from Exercise 20 that the quotient function $q : \mathcal{A} \to \mathcal{A}/\mathcal{D}$ maps the lattice $\mathcal{R}$ of regular elements onto $\mathcal{A}/\mathcal{D}$.

22. [McKinsey–Tarski44] Prove that $\mathcal{R}$ is a Boolean algebra. [Hint: prove that $x \cup_R x^c = \sqrt{}$ (cf. Exercise 19) for any regular x. To this end, verify that $x \cup x^c$ is dense applying Proposition 8.9]

23. [McKinsey–Tarski44] In conditions of Exercise 21, prove that q does establish an isomorphism of $\mathcal{R}$ onto $\mathcal{A}/\mathcal{D}$. [Hint: verify that for regular x, y from $|x| \leq_R |y|$ it follows that $x \leq y$. To this end, observe that $x \leq y$ implies $x \Rightarrow y$ dense. Observe also that $x \Rightarrow y$ is regular (to this end, apply Exercise 16 to represent y as z^c and observe that $x \Rightarrow y = (x \cap z)^c$. Finally observe that the only element dense and regular is the unit element]

24. Prove results of Exercises 7–23 for open as well as regular sets in a topological space (X, τ).

Works quoted

[Birkhoff67] G. Birkhoff, *Lattice Theory*, AMS, Providence, 1940 (3rd ed., 1967).

[Boole847] G. Boole, *The Mathematical Analysis of Logic*, Cambridge, 1847.

[McKinsey–Tarski44] J. C. C. McKinsey and A. Tarski, *The algebra of topology*, Annals of Mathematics, 45(1944), pp. 141–191.

[Rasiowa74] H. Rasiowa, *An Algebraic Approach to Non–Classical Logics*, North Holland, 1974.

[Rasiowa–Sikorski63] H. Rasiowa and R. Sikorski, *The Mathematics of Meta-mathematics*, PWN-Polish Scientific Publishers, Warszawa, 1963.

[Schröder895] E. Schröder, *Algebra der Logic*, Leipzig, 1890–1895.

[Stone36] M. H. Stone, *The theory of representations for Boolean algebras*, Trans. Amer. Math. Soc., 40(1936), pp. 37–111.

[Tarski38] A. Tarski, *Der Aussagenkalkül und die Topologie*, Fund. Math., 31(1938), pp. 103–134.

Chapter 9

Predicate Calculus

All things are related – related in different and complicated ways. But all things are not one. The word "everything" should mean simply the total (a total to be reached, if we knew enough, by enumeration) of all the things that exist at a given moment.

C. S. Lewis, *Miracles*, Epilogue

9.1 Introduction

We have already met with the fact that relations i.e. objects of set–theoretic nature are expressed in symbolic form as *predicates* especially when we refer to them irrespective of the underlying set. The aim of calculus of predicates is to give a formal rigorous description of logical aspects of predicates and of theories based on them. Usage of predicates i.e. expressions of the form $P(x_1, .., x_m)$ which formally render statements we use in our reasoning like $x < y$, $x^2 + y^2 > 1$, $IND_A(x, y)$ etc. is here accompanied with the use of *quantifiers* $\forall$, $\exists$. In the Chapter on Set Theory, we witnessed the use of quantifiers in an informal way so here we only complete the description by giving an outline of their formal theory.

As with propositional calculus, we discern in the structure of predicate calculus various sets of symbols used to denote various types of variables, and various types of objects we reason about and then we have some rules to form meaningful expressions of the language of this calculus. Next, we have some axioms i.e. expressions we accept as true on the basis of their syntax only and derivation rules used to produce in a systematic, as it were mechanical way, theses of the system.

We are also taking into account the semantic aspect by defining meanings of expressions in order to select true expressions of the system. Finally, we may compare semantic and syntactic aspects via e.g. completeness properties.

9.2   A formal predicate calculus

We introduce the components described above. We begin with symbols used in this calculus. In its general form it does aim to encompass all objects one may come across i.e. relations as well as functions. We call here generally this calculus the *predicate calculus*; it is also called the *first order theory, first order functional calculus* etc. and then the term *(pure) predicate calculus* is reserved for a theory in which no functional symbols intervene.

We will make use of the following sets of symbols

1. a countable set V of individual variables

2. a countable set F of function symbols

3. a countable set P of predicate symbols

4. the set of propositional connectives $C, N, \vee, \wedge$

5. the set of *quantifier symbols* $Q = \{\forall, \exists\}$

6. a set of auxiliary symbols $\{(,), [,], \{,\}, \}$

From symbols, more complex expressions are constructed. We represent the set F of function symbols as the union $\bigcup\{F_n : \in N\}$ and we call elements of the set F_m $m-ary$ function symbols for each $m \in N$. From function symbols, parentheses and individual variable symbols expressions called *terms* are constructed; formally, a term is an expression of the form $\phi(\tau_1, ..., \tau_m)$ where $\phi \in F_m$ and $\tau_1, ..., \tau_m$ have already been declared terms. In order to make this idea into a formal definition, we resort to induction on complexity of formulae involved and we define the set $TERM$ of terms as the intersection of all sets X such that

$$
\begin{aligned}
&(i) \ all \ individual \ variable \ symbols \ are \ in \ X \\
&(ii) \ all \ 0 - ary \ function \ symbols \ are \ in \ X \\
&(iii) \ for \ each \ m \geq 1 \ if \ \tau_1, ..., \tau_m \ are \ in \ X \ and \ \phi \ is \ an \\
&\qquad\qquad m - ary \ function \\
&\qquad symbol \ then \ \phi(\tau_1, ..., \tau_m) \ is \ in \ X
\end{aligned}
\qquad (9.1)
$$

Thus the set $TERM$ is the least set satisfying (9.1). Having terms defined, we proceed on to form more complex expressions viz. *formulae* and again we will perform this task in a few steps. So we define the set $FORM$ of formulae as the intersection of all sets Y such that

(FORM1) for each $m \geq 1$, if $\tau_1, ..., \tau_m$ are terms and ρ is an m-ary predicate symbol then $\rho(\tau_1, ..., \tau_m)$ is in Y

(FORM2) if $\alpha, \beta \in Y$ then $N\alpha, C\alpha\beta \in Y$

(FORM3) if $\alpha \in Y$ then $\forall x\alpha, \exists x\alpha \in Y$

Now, as with propositional calculus, we have to assign meanings to formulae (meaningful expressions) of predicate calculus. Unlike in the former case, we cannot assume the existence of two states for variables i.e. truth and falsity. To account for individual variables, we have to interpret them in a set. Thus, we introduce the notion of an *interpretation frame* as a non–empty set I. Given I, we define an *interpretation* in I as a function M_I with properties

(INT1) for each $m \in N$, M_I maps each m-ary function symbol ϕ onto a function $f_\phi : I^m \to I$ (in case $m = 0$, f_ϕ is a constant i.e. a fixed element in I)

(INT2) for each $m \in N$, M_I maps each m-ary predicate symbol ρ onto a relation $R_\rho \subseteq I^m$

Let us observe that under the function M_I, formulae of predicate calculus become *propositional functions* when in addition individual variable symbols are interpreted as variables ranging over the set I. This means that to every term of the form $\phi(x_1, ..., x_m)$, where $x_1, ..., x_m$ are individual variable symbols, the function M_I assigns the function

$$f_\phi(x_1, ..., x_m) : I^m \to I$$

where $x_1, ..., x_m$ are variables ranging over I. Clearly, in case of a term of the form $\phi(\tau_1, ..., \tau_m)$, where $\tau_1, ..., \tau_m$ have been assigned functions $g_1, ..., g_m$, with $g_j : I^{m_j} \to I$, the term $\phi(\tau_1, ..., \tau_m)$ is assigned the composition

$$f_\phi(g_1(x_1^1, ..., x_{m_1}^1), ..., g_m(x_1^m, ..., x_{m_m}^m)).$$

Then, to each formula of the form $\rho(\tau_1, ..., \tau_m)$, the function M_I assigns the propositional function

$$R_\rho(g_1(x_1^1, ..., x_{m_1}^1), ..., g_m(x_1^m, ..., x_{m_m}^m))$$

which becomes a proposition after substituting the specific elements of I for variables x_i^j. We will call such assignments *elementary*.

Now, in order to check the truth of formulae, we have to interpret in the set I variables. To this end, we introduce the notion of a *valuation* being a function $v : V \rightarrow I$; thus, v assigns to each individual variable symbol x_i an element $v(x_i) \in I$. Under v, each elementary assignment becomes a proposition

$$R_\rho(g_1(v(x_1^1), ..., v(x_{m_1}^1)), ..., g_m(v(x_1^m), ..., v(x_{m_m}^m)));$$

for an elementary assignment α, we denote by the symbol $[[\alpha]]_M^v$ the logical value (true/false) of α under the valuation v. In the sequel, when no confusion is possible, we will denote this value with the symbol $[[\alpha]]^v$.

Having done with elementary assignments, we will define the truth value of more complex formulae under any valuation v. To this end, we let

1. $[[N\alpha]]^v = 1$ if and only if $[[\alpha]]^v = 0$

2. $[[C\alpha\beta]]^v = 1$ if and only if either $[[\alpha]]^v = 0$ or $[[\beta]]^v = 1$

3. $[[\forall x\alpha]]^v = 1$ if and only if $[[\alpha]]^{v(x/c)} = 1$ for every choice of $c \in I$ where $v(x/c)$ is the valuation v in which the value $v(x)$ is changed to c

4. $[[\exists x\alpha]]^v = 1$ if and only if $[[\alpha]]^{v(x/c)} = 1$ for some choice of $c \in I$

Let us observe that valuation of a formula obtained by means of propositional connectives corresponds to the case of propositions and interpretation of quantifiers in symbolic formulae is done by a common–sense interpretation of phrases "every choice", "some choice".

Now, we will call a formula α *true under* M_I *in* I when $[[\alpha]]^v = 1$ for every valuation v. The formula α is said to be *true in* I when it is true under every function M_I. Finally, α is *true* if and only if it is true in every interpretation frame I.

We now apply this notion of truth towards the exhibition of basic derivation rules in formal syntax of predicate calculus.

We formulate the detachment rule as usual i.e. if $\alpha, C\alpha\beta$ are theses then β is accepted as a thesis. The substitution rule will be modified adequately to the predicate calculus case as follows. For a formula $\forall x\alpha$ as well as for a formula $\exists x\alpha$, if x occurs in α then we say that x is *bound* in α; any variable y not bound by a quantifier is *free* in α. To denote free variables in α we will write $\alpha(x_1, .., x_k)$ in place of α and all other variables not listed in $\alpha(x_1, .., x_k)$ are either bound or do not occur in α.

(S) (*Substitution for free individual variables*) If $\alpha(x_1, ..., x_k)$ is a thesis and $\tau_1, ..., \tau_k$ are terms then the formula $\alpha(\tau_1, ..., \tau_k)$ obtained from $\alpha(x_1, ..., x_k)$ by the substitution $x_1/\tau_1, ..., x_k/\tau_k$ is accepted as a thesis.

We observe that

Proposition 9.1. *The detachment rule and the substitution rule give true formulae when applied to true formulae.*

Proof. For the case of detachment, if $\alpha, C\alpha\beta$ are true then for every I, M_I and every valuation v we have $[[\alpha]]^v = 1$ hence as $[[C\alpha\beta]]^v = 1$ we have $[[\beta]]^v = 1$ i.e. β is true.

For substitution, if $[[\alpha(x_1, ..., x_k)]]^v = 1$ for every v then clearly $[[\alpha(\tau_1, ..., \tau_k)]]^v = 1$ for every v as in the latter formula values taken by $\tau_1, ..., \tau_k$ form a subset of I^k which is the set of values taken by $(x_1, ..., x_k)$ in the former formula.   □

Now, we turn to rules governing use of quantifiers; these rules describe, respectively, admissible cases in which we introduce, respectively, omit quantifier symbols.

(*Introduction of existential quantifier (IEQ)*) For formulae $\alpha(x), \beta$ with y not bound in $\alpha(x)$ if $C\alpha(x)\beta$ is true then $C\exists y\alpha(y)\beta$ is true.

We justify this rule. If $C\alpha(x)\beta$ is true i.e. either $[[\alpha(x)]]^v = 0$ or $[[\beta]]^v = 1$ for every valuation v then either there is a valuation v with $[[\alpha(x)]]^v = 1$ hence $[[\exists y\alpha(y)]]^v = 1$ as witnessed by the valuation $v(y/v(x))$ and then $[[\beta]]^v = 1$ so $[[C\exists y\alpha(y)\beta]]^v = 1$ or $[[\alpha(x)]]^v = 0$ for every valuation v hence $[[\beta]]^v = 1$ and $[[C\exists y\alpha(y)\beta]]^v = 1$ for every valuation v.

In a similar manner we may justify the remaining rules which we list now.

(*Introduction of universal quantifier (IUQ)*) For formulae $\alpha, \beta(x)$ with no occurrence of x in α and y not bound in α if $C\alpha\beta(x)$ is true then $C\alpha\forall x\beta(x)$ is true.

(*Elimination of existential quantifier (EEQ)*) For formulae $\alpha(x), \beta$ with no occurrence of y in α if $C\alpha(x)\beta$ is true then $C\exists y\alpha(y)\beta$ is true.

(*Elimination of universal quantifier (EUQ)*) For formulae $\alpha, \beta(x)$ with no occurrence of x in $\beta(y)$ if $C\alpha\forall x\beta(x)$ is true then $C\alpha\beta(y)$ is true.

We will define formally syntax of predicate calculus. We will take as axioms the formulae

(T1) $CC\alpha\beta CC\alpha\gamma C\beta\gamma$

(T2) $CCN\alpha\alpha$

(T3) $C\alpha CN\alpha\beta$

and we use detachment (MP), substitution for free variables (S), (IEQ), (IUQ), (EEQ), and (EUQ) as derivation rules.

Let us observe that – in opposition to the propositional calculus case – we will regard (T1)–(T3) as *axiom schemata* i.e. we will admit as an axiom any substitution of a meaningful expression for equiform symbols in anyone of (T1)–(T3); any such substitution will be called an *axiom instance*. It should be clear – as any axiom instance is true – that any thesis of propositional calculus derived from (T1)–(T3) will be true after any substitution of meaningful expressions for equiform symbols. In the sequel we will invoke this fact without explicit comments.

Let us denote by the symbol **2** the Boolean algebra consisting of truth values $0, 1$ with Boolean operations $x \cup y = max\{x, y\}$, $x \cap y = min\{x, y\}$, $- - x = 1 - x$, $x \Rightarrow y = - - x \cup y$ and distinguished elements $0 < 1$.

We now introduce a Boolean structure, in fact, a Boolean algebra called the *Lindenbaum–Tarski* algebra which owes its inception to an idea of Lindenbaum to treat formulae of predicate calculus as elements of an algebra.

9.3   The Lindenbaum-Tarski algebra

As a first step in algebraization of predicate calculi, one may consider an interpretation of propositional connectives $V, \wedge, C, N$ as operations, respectively of $\cup, \cap, \Rightarrow, --$ in an algebra of formulae of predicate calculus. However, these operations e.g. $\cup, \cap$ are neither commutative nor associative for instance formulae $\alpha \vee \beta$, $\beta \vee \alpha$ are distinct; a remedy for this is to consider a relation $\approx$ on formulae defined as follows

$$\alpha \approx \beta \iff C\alpha\beta \wedge C\beta\alpha \tag{9.2}$$

Then we have

$$\alpha \vee \beta \approx \beta \vee \alpha \tag{9.3}$$

by the thesis $C(\alpha \vee \beta)(\beta \vee \alpha)$ and its converse $C(\beta \vee \alpha)(\alpha \vee \beta)$.

The relation $\approx$ is an equivalence (cf. the Schröder theorem, Chapter 8); indeed $\alpha \approx \alpha$ by the thesis $C\alpha\alpha$ for each α.
   If

$$\alpha \approx \beta \text{ and } \beta \approx \gamma \tag{9.4}$$

then

$$\alpha \approx \gamma \tag{9.5}$$

by a thesis

$$CC\alpha\beta C\beta\gamma C\alpha\gamma$$

and its converse along with detachment. Finally, $\approx$ is symmetrical by its very definition. We denote by $FORM/\approx$ the resulting quotient set and by the symbol q the quotient function from $FORM$ onto $FORM/\approx$.

We also denote by the symbol $||\alpha||$ the equivalence class of α with respect to $\approx$. The Lindenbaum–Tarski algebra of predicate calculus is a Boolean algebra (cf. a historic note in [Rasiowa–Sikorski63, pp. 209, 245–6]).

Proposition 9.2. *The algebra $FORM/\approx$ is a Boolean algebra with operations*

(i) $||\alpha|| \cup ||\beta|| = ||\alpha \vee \beta||$
(ii) $||\alpha|| \cap ||\beta|| = ||\alpha \wedge \beta||$
(iii) $||\alpha|| \Rightarrow ||\beta|| = ||C\alpha\beta||$
(iv) $--||\alpha|| = ||N\alpha||$

Moreover, for each formula of the form $\alpha(x)$, we have

(v) $||\exists y\alpha(y)|| = sup\{||\alpha(\tau)|| : \tau \in TERM\}$ *where $\alpha(\tau)$ comes from $\alpha(x)$ as a result of substitution of a term τ for the free variable x*
(vi) $||\forall y\alpha(y)|| = inf\{||\alpha(\tau)|| : \tau \in TERM\}$

Proof. We denote by the symbol $\preceq$ the ordering relation in $FORM/\approx$ defined via

$$||\alpha|| \preceq ||\beta||$$

if and only if $C\alpha\beta$ is a thesis. We first show that $||\alpha \vee \beta||$ is the join of classes $||\alpha||, ||\beta||$; the thesis

$$C\alpha(\alpha \vee \beta)$$

implies that

$$||\alpha|| \preceq ||\alpha \vee \beta|| \tag{9.6}$$

and similarly

$$||\beta|| \preceq ||\alpha \vee \beta|| \tag{9.7}$$

hence

$$||\alpha|| \cup ||\beta|| \preceq ||\alpha \vee \beta|| \tag{9.8}$$

On the other hand, if $||\alpha|| \preceq ||\gamma||$ and $||\beta|| \preceq ||\gamma||$ i.e. $C\alpha\gamma$ and $C\beta\gamma$ are theses then applying detachment twice to the thesis

$$CC\alpha\gamma CC\beta\gamma C(\alpha \vee \beta)\gamma$$

we get $C(\alpha \vee \beta)\gamma$ i.e.

$$||\alpha \vee \beta|| \preceq ||\gamma|| \qquad (9.9)$$

It follows that

$$||\alpha|| \cup ||\beta|| = ||\alpha \vee \beta|| \qquad (9.10)$$

A similar proof shows that

$$||\alpha|| \cap ||\beta|| = ||\alpha \wedge \beta|| \qquad (9.11)$$

To show distributivity it suffices to prove the existence of a relative pseudo–complementation (cf. Exercise 5, Chapter 8).

For classes $||\alpha||, ||\beta||$, we prove that

$$||\alpha|| \Rightarrow ||\beta|| = ||C\alpha\beta|| \qquad (9.12)$$

To this end, consider any $||\gamma||$ with the property that

$$||\alpha|| \cap ||\gamma|| \preceq ||\beta||$$

i.e. such that $C(\alpha \wedge \gamma)\beta$ is a thesis. Then it follows from a thesis

$$CC(\alpha \wedge \gamma)\beta C\gamma C\alpha\beta$$

via detachment that $C\gamma C\alpha\beta$ is a thesis i.e. $||\gamma|| \preceq ||C\alpha\beta||$. To conclude this part of the proof it suffices to consider the thesis $CC\alpha(\beta \wedge \alpha)\beta$ which implies that

$$||C\alpha(\beta \wedge \alpha)|| = ||C\alpha\beta|| \cap ||\alpha|| \preceq ||\beta|| \qquad (9.13)$$

It follows that $||C\alpha\beta||$ is the relative pseudo–complement $||\alpha|| \Rightarrow ||\beta||$.

In consequence of the last proved fact, there exists the unit element in this lattice, $\sqrt{}$ and

$$\sqrt{} = ||C\alpha\alpha|| = ||\alpha|| \Rightarrow ||\alpha|| \qquad (9.14)$$

for each α.

As $C(\alpha \wedge N\alpha)\beta$ is a thesis, we have that

$$||\alpha \wedge N\alpha|| = ||\alpha|| \cap ||N\alpha|| \preceq ||\beta||$$

for any formula β i.e. $||\alpha \wedge N\alpha||$ is the null element $\sqcap$. Thus

$$||\alpha|| \cap ||N\alpha|| = \sqcap \qquad (9.15)$$

i.e. $||N\alpha|| \preceq --||\alpha||$. The thesis $CC\alpha(\alpha \wedge N\alpha)N\alpha$ implies that

$$||C\alpha(\alpha \wedge N\alpha)|| = ||\alpha|| \Rightarrow \sqcap = --||\alpha|| \preceq ||N\alpha|| \qquad (9.16)$$

hence $||N\alpha|| = --||\alpha||$. Finally, the thesis $\alpha \vee N\alpha$ renders the equality

$$||\alpha|| \cup --||\alpha|| = \sqrt{} \qquad (9.17)$$

for each formula α i.e. $--||\alpha||$ is the complement of $||\alpha||$ and $FORM/ \preceq$ is a Boolean algebra.

There remain statements (v) and (vi) to verify. Let us look at (v) and consider a formula $\alpha(x)$. As the set of individual variable symbols is infinite, there exists a variable symbol y which does not occur in $\alpha(x)$ and no quantifier in $\alpha(x)$ binds y; as $C\gamma\gamma$ is a thesis of predicate calculus for every formula γ, also the formula $C\exists y\alpha(y)\exists y\alpha(y)$ is a thesis and the derivation rule (EEQ) (of elimination of existential quantifier) yields the thesis $C\alpha(x)\exists y\alpha(y)$. Substitution of a term τ for the free variable x in the last thesis via the substitution rule (S) gives us the thesis $C\alpha(\tau)\exists y\alpha(y)$. Thus $||\alpha(\tau)|| \preceq ||\exists y\alpha(y)||$ for every term $\tau \in TERM$. It follows that

$$sup\{||\alpha(\tau)|| : \tau \in TERM\} \preceq ||\exists y\alpha(y)|| \qquad (9.18)$$

For the converse, let us assume that for some formula γ we have

$$||\alpha(\tau)|| \preceq ||\gamma|| \qquad (9.19)$$

for every term $\tau \in TERM$. We may find a variable z which does not occur in γ; as we have $||\alpha(z)|| \preceq ||\gamma||$ i.e. $C\alpha(y)\gamma$, it follows by the rule (IEQ) (of introduction of existential quantifier) that $C\exists y\alpha(y)\gamma$ is a thesis i.e.

$$||\exists y\alpha(y)|| \preceq ||\gamma|| \qquad (9.20)$$

Thus

$$||\exists y\alpha(y)|| = sup\{||\alpha(\tau)|| : \tau \in TERM\} \qquad (9.21)$$

The statement (vi) may be proved on similar lines, with rules (EUQ), (IUQ) in place of rules (EEQ), (EIQ) (these results are due to [Rasiowa51, 53], [Henkin49] among others (cf. [Rasiowa–Sikorski63, pp. 251])). $\square$

The equalities

1. $||\exists y\alpha(y)|| = sup\{||\alpha(\tau)|| : \tau \in TERM\}$

2. $||\forall y\alpha(y)|| = inf\{||\alpha(\tau)|| : \tau \in TERM\}$

are called (Q)–*joins* and *meets*, respectively cf. [Rasiowa–Sikorski63].

9.4   Completeness

Now, we may produce a proof of the main result bridging syntax and semantics of predicate calculus.

Proposition 9.3. *A formula α of predicate calculus is a thesis if and only if $||\alpha|| = \sqrt{}$.*

Proof. Assume that α is a thesis. By the thesis $CaCCa\alpha\alpha$ of predicate calculus, it follows via detachment that $CCa\alpha\alpha$ is a thesis hence $||Ca\alpha|| \preceq ||\alpha||$ i.e. $\sqrt{} \preceq ||\alpha||$ and thus $\sqrt{} = ||\alpha||$.

Conversely, in case $||\alpha|| = \sqrt{}$ we have $||\beta|| \preceq ||\alpha||$ for each β hence in particular $||Ca\alpha|| \preceq ||\beta||$ i.e. $CCa\alpha\alpha$ is a thesis whence it follows by detachment that α is a thesis. $\square$

We should now extend the last result to any interpretation. Consider an interpretation frame I along with an interpretation M_I and a valuation v. Then we have the following result.

Proposition 9.4. *For any formulae α, β of predicate calculus, if $||\alpha|| = ||\beta||$ then $[[\alpha]]_M^v = [[\beta]]_M^v$.*

Proof. Assume that $||\alpha|| = ||\beta||$ i.e. the formulae $C\alpha\beta$ and $C\beta\alpha$ are theses of predicate calculus. Then the conditions hold

1. either $[[\alpha]]_M^v = 0$ or $[[\beta]]_M^v = 1$

2. either $[[\beta]]_M^v = 0$ or $[[\alpha]]_M^v = 1$

The conjunction of conditions 1 and 2 results in the condition

$$[[\alpha]]_M^v = 0 = [[\beta]]_M^v \vee [[\alpha]]_M^v = 1 = [[\beta]]_M^v$$

i.e. $[[\alpha]]_M^v = [[\beta]]_M^v$. $\square$

This result shows that the function $[[\cdot]]_M^v$ defined by the valuation v from the set $FORM$ of formulae of predicate calculus into the Boolean algebra **2** can be factored through the Boolean algebra $FORM/\preceq$:

$$[[\cdot]]_M^v = h \circ q(\cdot)$$

where $q : FORM \rightarrow FORM/\preceq$ is the quotient function and $h : FORM/\preceq \rightarrow$ **2** is defined by the formula

$$h(||\alpha||) = [[\alpha]]_M^v.$$

Actually we can prove more viz.

Proposition 9.5. *The function $h : FORM/\preceq \rightarrow$ **2** is a homomorphism preserving (Q) –joins and –meets.*

Proof. Given formulae α, β of predicate calculus, we have

$$h(||\alpha|| \cup ||\beta||) = h(||\alpha \vee \beta||) = [[\alpha \vee \beta]]_M^v = [[\alpha]]_M^v \cup [[\beta]]_M^v =$$

$$h(||\alpha|| \cup h(||\beta||);$$

similarly we prove that

$$h(||\alpha|| \cap ||\beta||) = h(||\alpha||) \cap h(||\beta||)$$

and

$$h(--||\alpha||) = --h(||\alpha||).$$

Moreover

$$h(\sqrt{}) = h(||\alpha \vee N\alpha||) = [[\alpha]]_M^v \cup [[N\alpha]]_M^v = 1.$$

For (Q)–joins, we have

$$h(||\exists y\alpha(y)||) = [[\exists y\alpha(y)]]_M^v = sup_x[[\alpha(y)]]_M^{v(y/x)} = sup_\tau[[\alpha(\tau)]]_M^v =$$

$$sup_\tau h(||\alpha(\tau)||).$$

Similarly, the result for meets follows. $\square$

Motivated by the above fact, we will enlarge the domain of our interpretations viz. we will say that a *generalized* interpretation frame is a non–empty set I along with a homomorphism h preserving (Q)-joins and meets of $FORM/\preceq$ into a Boolean algebra **B** and then an interpretation of a formula α is $h(||\alpha||)$. In particular interpretation frames discussed above are of this type.

By the *canonical interpretation frame* we will understand the set $TERM$ of terms along with the interpretation M_{TERM} which sends each term τ to itself and which assigns each formula α the value $||\alpha||$ in the algebra $FORM/\preceq$.

We may now sum up latest developments in a version of completeness theorem.

Proposition 9.6. *[Rasiowa–Sikorski63] (A completeness theorem for predicate calculus) The following statements are equivalent for each formula α of predicate calculus*

1. *α is a thesis of predicate calculus*

2. *$h(||\alpha||) = 1_{\mathbf{B}}$ where $1_{\mathbf{B}}$ is the unit element in the algebra **B** for every generalized interpretation frame $I, h, \mathbf{B}$*

3. *$||\alpha|| = \sqrt{}$*

Indeed, 1 implies 2 by Propositions 9.3, 9.4 and 2 implies 3 obviously. That 3 implies 1 is one part of Proposition 9.3.

As with propositional calculus, completeness allows us for verifying whether a formula is a thesis on semantic grounds. For instance,

Proposition 9.7. *The following are theses of predicate calculus (cf. Exercise section, below).*

1. $C\forall x\alpha(x) \lor \beta\forall x(\alpha(x) \lor \beta)$

2. $C\forall x(\alpha(x) \lor \beta)\forall x\alpha(x) \lor \beta$

3. $C\exists x\alpha(x) \lor \beta\exists x(\alpha(x) \lor \beta))$

4. $C\exists x(\alpha(x) \lor \beta)\exists x\alpha(x) \lor \beta$

5. $C\forall x\alpha(x) \land \beta\forall x(\alpha(x) \land \beta)$

6. $C\forall x(\alpha(x) \land \beta)\forall x\alpha(x) \land \beta$

7. $C\exists x\alpha(x) \land \beta\exists x(\alpha(x) \land \beta)$

8. $C\exists x(\alpha(x) \land \beta)\exists x\alpha(x) \land \beta$

9. $C\forall x\alpha(x)\exists x\alpha(x)$

10. $C\forall x C\alpha(x)\beta C\exists\alpha(x)\beta$

11. $CC\exists\alpha(x)\beta\forall x C\alpha(x)\beta$

12. $C\forall x C\alpha(x)\beta C\exists x\alpha(x)\beta$

13. $CC\exists x\alpha(x)\beta\forall x C\alpha(x)\beta$

14. $C\forall x C\alpha\beta(x)C\alpha\forall x\beta(x)$

15. $C\forall x C\alpha(x)\beta(x)C\forall x\alpha(x)\forall x\beta(x)$

16. $C\forall x C\alpha(x)\beta(x)C\exists x\alpha(x)\exists x\beta(x)$

17. $C\forall x\alpha(x) \land \beta(x)\forall x\alpha(x) \land \forall x\beta(x)$

18. $C\forall x\alpha(x) \land \forall x\beta(x)\forall x\alpha(x) \land \beta(x)$

19. $C\exists x\alpha(x) \lor \beta(x)\exists x\alpha(x) \lor \exists x\beta(x)$

20. $C\exists x\alpha(x) \lor \exists x\beta(x)\exists x\alpha(x) \lor \beta(x)$

21. $C\forall x\alpha(x) \lor \forall x\beta(x)\forall x\alpha(x) \lor \beta(x)$

22. $C\exists x\alpha(x) \land \beta(x)\exists x\alpha(x) \land \exists x\beta(x)$

From the completeness theorem some consequences follow of which we would like to remark on following.

Remarks.

1. Predicate calculus is *consistent* (meaning that there is no formula α such that both α and $N\alpha$ are theses); indeed as $\|\alpha\| = \sqrt{}$ it implies that $\|N\alpha\| = \sqcap$

2. For a formula $\alpha(x_1, .., x_k)$ with free variable symbols $x_1, ..., x_k$, we call *syntactic closure* of $\alpha(x_1, .., x_k)$ the formula

$$\forall y_1 \forall y_2 ... \forall y_k \alpha(x_1/y_1, .., x_k/x_k).$$

It easily follows from the semantic interpretation that $\alpha(x_1, .., x_k)$ is a thesis of predicate calculus if and only if the syntactic closure

$$\forall y_1 \forall y_2 ... \forall y_k \alpha(x_1/y_1, .., x_k/y_k)$$

is a thesis of predicate calculus

3. A formula α is *prenex* if all quantifier operations in α are preceded by predicate operations i.e. α is of the form

$$Qx_1 ... Qx_m \beta(x_1, .., x_m)$$

where $\beta(x_1, ..., x_m)$ contains no quantifier symbol occurrence (and other variables may appear in β as well) and the symbol Q stands for $\forall, \exists$. A prenex formula α which is equivalent to a formula γ i.e. both $C\gamma\alpha$ and $C\alpha\gamma$ are theses is said to be a *prenex form* of the formula γ.

We have

Proposition 9.8. *For every formula γ there exists a prenex form α of γ.*

Proof. Indeed, we can construct a prenex form of α by structural induction. Elementary formulae of the form $\rho(\tau_1, ..., \tau_m)$ are already prenex. If formulae α, β are in prenex forms, respectively,

$$Qx_1 ... Qx_m \gamma(x_1, ..., x_m), Qy_1 ... Qy_k \delta(y_1, ..., y_k),$$

where variables x_i do not occur in δ and variables $y_1, .., y_k$ do not occur in γ, then $\alpha \lor \beta$ i.e.

$$Qx_1 ... Qx_m \gamma(x_1, ..., x_m) \lor Qy_1 ... Qy_k \delta(y_1, ..., y_k) \qquad (9.22)$$

may be represented equivalently by virtue of Proposition 9.7 1–4 as

$$Qx_1 [Qx_2 ... Qx_m \gamma(x_1, ..., x_m) \lor Qy_1 ... Qy_k \delta(y_1, ..., y_k)] \qquad (9.23)$$

and by successive repeating of 1–4 we arrive finally at

$$Qx_1...Qx_mQy_1...Qy_k[\gamma(x_1,...,x_m) \vee \delta(y_1,...,y_k)] \qquad (9.24)$$

which is a prenex form of $\alpha \vee \beta$. A similar argument using theses 5–8 of Proposition 9.7 in place of theses 1–4 will give a prenex form of $\alpha \wedge \beta$. Letting $Q* = \forall$ in case $Q = \exists$ and $Q* = \exists$ in case $Q* = \forall$, we may write down the formula $N\alpha$ in the prenex form $Q * x_1...Q * x_m N\gamma(x_1,..,x_m)$. Finally if $\alpha*$ is a prenex form of α then $Qx\alpha*$ is a prenex form of $Qx\alpha$. □

9.4.1 Calculus of open expressions

A meaningful expression is *open* when there are no occurrences of quantifier symbols in it. Denoting by the symbol $FORM^0$ the set of all open meaningful expressions of predicate calculus, we may define a thesis of this calculus as an expression which may be derived from axioms by means of detachment and substitution rules only. Thus, a thesis of the calculus of open expressions is a thesis of predicate calculus as well. Denoting by $||\alpha||^0$ the element of the Lindenbaum–Tarski algebra of the open expressions calculus, we have the following

Proposition 9.9. *The function h assigning to the class $||\alpha||^0$ the class $||\alpha||$ is an isomorhism of the Lindenbaum–Tarski algebra of calculus of open expressions and the Lindenbaum–Tarski algebra of predicate calculus.*

9.5 Calculus of unary predicates

We now consider a particular case when all predicates are *unary* i.e. they require a single argument, so they are of the form $P(x)$. It turns out that in this case when verifying truth of formulae, we may restrict ourselves to interpretation frames of small cardinality viz. we have

Proposition 9.10. *For a formula $\alpha(P_1,...,P_k;x_1,...,x_n)$ where $P_1,...,P_k$ are all (unary) predicate symbols occurring in α and $x_1,...,x_n$ are all variable symbols in α, if $\alpha(P_1,...,P_k;x_1,...,x_n)$ is true in an interpretation frame M then it is also true in an interpretation frame $M^* \subseteq M$ such that cardinality $|M^*|$ is at most 2^k.*

Proof. We may assume that α is in prenex form

$$Qx_1...Qx_m\beta(P_1,...,P_k;x_1,...,x_n)$$

and for each sequence $I = i_1...i_k$ of $0's$ and $1's$ we may form the set $M(I) = \{x \in M : [[P_m(x)]] = i_m; m = 1,...,k\}$. Then the family $\mathcal{P} = \{M(I) : I \in \{0,1\}^k\}$ is a partition of the set M. We may select then for each $x \in M$ the partition element $\mathcal{P}(x)$ containing x and for each non–empty set $\mathcal{P}(x)$ we may select an element $s(x) \in \mathcal{P}(x)$. Then the set $M^* = \{s(x) : x \in M\}$ has cardinality at most that of the set $\{0,1\}^k$ i.e. 2^k.

Let us observe that formulae

$$
\begin{aligned}
&(i) \ \beta(P_1, ..., P_k; x_1, ..., x_n) \\
&(ii) \ \beta(P_1, ..., P_k; s(x_1), ..., s(x_n))
\end{aligned}
\tag{9.25}
$$

are equivalent hence

$$[[(Qx_1...Qx_m\beta(P_1, ..., P_k; x_1, ..., x_n]]_M^v =$$

$$[[Qx_1...Qx_m\beta(P_1, ..., P_k; s(x_1), ..., s(x_n)]]_M^{s(v)} =$$

$$[[Qx_1...Qx_m\beta(P_1, ..., P_k; x_1, ..., x_n)]]_{M^*}^{s(v)}.$$

It follows that α is true in M if and only if α is true in M^*. □

In consequence of the above result, in order to verify the truth of α it is sufficient to check its truth on a set M with $|M| \leq 2^k$. In this case α reduces to the proposition e.g. in case α is $\forall x P(x)$ the corresponding proposition is $P(x_1) \land ... \land P(x_{2^k})$. Thus to verify the truth of this proposition we may employ the method of truth tables known in propositional calculus. This method requires a finite number of steps, in our exemplary case at most 2^{2^k}. It follows that calculus of unary predicates is *decidable*: for a given formula of this calculus we may decide in a finite number of steps whether this formula is a thesis. Let us observe that in general predicate calculus is undecidable.

9.6 Fractional truth values

We close our discussion of unary predicate calculus with an outline of the idea of assigning fractional truth values to meaningful expressions of calculus of unary predicates due to Jan Łukasiewicz [Lukasiewicz13]. Let us consider unary predicates $\alpha(x), \beta(x), ...$ along with a fixed finite interpretation frame I and a fixed interpretation M_I. Any valuation v may be regarded then as a constant $v(x) \in I$. We assign to $\alpha(x)$ the number $w(\alpha)$ defined as follows

$$w(\alpha) = \frac{s_\alpha}{|I|} \tag{9.26}$$

where s_α is the number of valuations v that satisfy $\alpha(x)$. Thus $w(\alpha)$ is a fraction of valuations which satisfy the formula $\alpha(x)$. Clearly, $\alpha(x)$ is true in I, M_I if and only if $w(\alpha) = 1$; similarly, under the definition that $\alpha(x)$ is *false* when no valuation does satisfy it, we have that $\alpha(x)$ is false if and only if $w(\alpha) = 0$.

Thus $0 < w(\alpha) < 1$ does indicate that $\alpha(x)$ is neither true nor false; we may say that $\alpha(x)$ is *indefinite* or, in Birkhoff's phrase *undecidable* (cf. [Birkhoff67]). It follows immediately from our semantic considerations above that

Proposition 9.11. *(Lukasiewicz) For meaningful expressions $\alpha(x)$, $\beta(x)$ we have with respect to fixed I, M_I that*

1. $w(\alpha) = 0$ *if and only if $\alpha(x)$ is false*

2. $w(\alpha) = 1$ *if and only if $\alpha(x)$ is true*

3. $w(\alpha) + w(N\alpha \wedge \beta) = w(\beta)$ *whenever $C\alpha(x)\beta(x)$ is true (is a thesis)*

Let us state further consequences of 1–3.

Proposition 9.12. *(Lukasiewicz) We have moreover*
4. *If $C\alpha\beta$ and $C\beta\alpha$ are true then $w(\alpha) = w(\beta)$*
5. $w(\alpha) + w(N\alpha) = 1$
6. $w(\alpha \vee \beta) = w(\alpha) + w(\beta) - w(\alpha \wedge \beta)$
7. $\alpha \wedge \beta = 0$ *implies $w(\alpha \vee \beta) = w(\alpha) + w(\beta)$*
8. $w(C\alpha\beta) = 1 - w(\alpha) + w(\alpha \wedge \beta)$

Proof. For 4, if $C\beta\alpha$ is true then $\beta \wedge N\alpha = 0$ and 3 implies that $w(\beta) = w(\alpha)$. Claim 5 follows from 3 when we substitute 1 for β and apply 2. For 6, as $C\alpha \wedge \beta\beta$ is true, we have by 3 that $w(\beta) = w(\alpha \wedge \beta) + w(N(\alpha \wedge \beta) \wedge \beta = w(\alpha \wedge \beta) + w(N\alpha \wedge \beta)$. Similarly, as $C\alpha\alpha \vee \beta$ is true we get by 3 that $w(\alpha \vee \beta) = w(\alpha) + w(N\alpha \wedge \alpha \vee \beta) = w(\alpha) + w(N\alpha \wedge \beta)$. From (i) $w(\alpha \vee \beta) = w(\alpha) + w(N\alpha \wedge \beta)$ and (ii) $w(\beta) = w(\alpha \wedge \beta) + w(N\alpha \wedge \beta)$ it follows by elimination of $w(N\alpha \wedge \beta)$ that $w(\alpha \vee \beta) = w(\alpha) + w(\beta) - w(\alpha \wedge \beta)$.

Case 7 follows directly from 6. Finally, 8 follows since by 4 we have $w(C\alpha\beta) = w(N\alpha \vee \beta)$ thus by 6 we get $w(C\alpha\beta) = w(N\alpha) + w(\beta) - w(N\alpha \wedge \beta)$ and as $C\beta\alpha \wedge \beta \vee N\alpha \wedge \beta$ and $C\alpha \wedge \beta \vee N\alpha \wedge \beta\beta$ are true and $\alpha \wedge \beta \wedge N\alpha \wedge \beta = 0$ it follows by 7 that $w(C\alpha\beta) = w(N\alpha) + w(\alpha \wedge \beta)$ and finally 5 implies that $w(C\alpha\beta) = 1 - w(\alpha) + w(\alpha \wedge \beta)$. $\square$

Lukasiewicz goes on to introduce *relative truth values* viz. he defines $w_\alpha(\beta)$ as the fraction $\frac{w(\alpha \wedge \beta)}{w(\alpha)}$ provided $w(\alpha)$ is not 0. We introduce a connective $|$ and a statement $\beta|\alpha$ will be satisfied by a valuation v if and only if $[[\beta]]^v = 1$ given that $[[\alpha]]^v = 1$. Then $w_\alpha(\beta) = w(\beta|\alpha)$.

The notion of *independence* of meaningful expressions may be introduced now viz. α and β are *independent* if and only if $w_\alpha(\beta) = w_{N\alpha}(\beta)$. Thus α and β are independent if and only if $\frac{w(\alpha \wedge \beta)}{w(\alpha)} = \frac{w(N\alpha \wedge \beta)}{w(N\alpha)}$. Then we have

Proposition 9.13. *(Lukasiewicz) For meaningful expressions α, β*
1. *If α, β are independent then (i) $w_\alpha(\beta) = w_{N\alpha}(\beta) = w(\beta)$*
(ii) $w(\alpha \wedge \beta) = w(\alpha) \cdot w(\beta)$
2. *If α, β are independent then $w_\beta(\alpha) = w_{N\beta}(\alpha) = w(\alpha)$*

Proof. The first equality in (i) is valid by definition. As $\frac{w(\alpha \wedge \beta)}{w(\alpha)} = \frac{w(N\alpha \wedge \beta)}{w(N\alpha)}$ we have $\frac{w(\alpha \wedge \beta)}{w(\alpha)} = \frac{w(N\alpha \wedge \beta)}{w(N\alpha)} = \frac{w(\alpha \wedge \beta) + w(N\alpha \wedge \beta)}{w(\alpha) + w(N\alpha)} = \frac{w(\beta)}{1} = w(\beta)$ where in the fore–last equality property 7 was applied to the numerator. From (i) thus proved, (ii) follows immediately.

Assuming 1., we have $\frac{w(\alpha \wedge N\beta)}{w(N\beta)} = \frac{w(\alpha) - w(\alpha \wedge \beta)}{1 - w(\beta)}$; applying 1(ii) to the numerator in the last fraction, we get $\frac{w(\alpha \wedge N\beta)}{w(N\beta)} = w(\alpha)$ i.e. $w_{N\beta}(\alpha) = w(\alpha)$. That $w_\beta(\alpha) = w(\alpha)$ follows by 1(ii). $\qquad \square$

The import of 1(ii) is an important characterization of independence and 2. shows that independence is a symmetric relation.

The reader has undoubtedly noticed the correspondence between the theory of truth values and elementary probability theory based on classical approach to counting of probabilities. We may conclude development of this parallelism by stating and proving the counterpart to the well – known Bayes theorem.

Proposition 9.14. *(Lukasiewicz) Assume that meaningful expressions* $\beta_1, ...,$ β_n *are such that* $\beta_1 \vee ... \vee \beta_n$ *is true and* $\beta_i \wedge \beta_j = 0$ *whenever* $i \neq j$. *Then*

$$w_\alpha(\beta_j) = \frac{w(\beta_j) \cdot w_{\beta_j}(\alpha)}{\sum_{i=1}^{n} w(\beta_i) \cdot w_{\beta_i}(\alpha)}.$$

Proof. As $w_\alpha(\beta_j) = \frac{w(\alpha \wedge \beta_j)}{w(\alpha)}$ the theorem follows from two following facts. Fact 1: $w(\alpha \wedge \beta_j) = w(\beta_j) \cdot w_{\beta_j}(\alpha)$ Fact 2. As $\beta_1 \vee ... \vee \beta_n = 1$ and theses α, $\alpha \wedge 1$ are equivalent we have by property 4 that $w(\alpha) = w(\alpha \wedge (\beta_1 \vee ... \vee \beta_n))$ and as $\alpha \wedge \beta_i \wedge \alpha \wedge \beta_j = 0$ in case $i \neq j$, property 7 implies that $w(\alpha) = \sum_{i=1}^{n} w(\alpha \wedge \beta_i)$ whence the final formula follows by 2. $\qquad \square$

9.7 Intuitionistic propositional logic

When discussing predicate calculus, we exploited in the proof that the Lindenbaum–Tarski algebra of predicate calculus is a Boolean algebra the following theses of the propositional calculus:

1. $C\alpha\alpha$

2. $CC\alpha\beta CC\beta\gamma C\alpha\gamma$

3. $C\alpha\alpha \vee \beta$

4. $C\beta\alpha \vee \beta$

5. $CC\alpha\gamma CC\beta\gamma C\alpha \vee \beta\gamma$

6. $C\alpha \wedge \beta\alpha$

7. $C\alpha \wedge \beta\beta$

8. $CC\gamma\alpha CC\gamma\beta C\gamma\alpha \wedge \beta$

9. $CC\alpha C\beta\gamma C\alpha \wedge \beta\gamma$

10. $CC\alpha \wedge \beta\gamma C\alpha C\beta\gamma$

11. $C\alpha \wedge N\alpha\beta$

12. $CC\alpha\alpha \wedge N\alpha N\alpha$

13. $\alpha \vee N\alpha$

Of these theses, expression 13 was responsible for complementation in the resulting Boolean algebra. Without 13, the Lindenbaum–Tarski algebra is a pseudo–Boolean algebra. Expressions like 13 have been a subject to careful inspection from philosophical as well as methodological points of view. Its truth was put in doubt by a direction in foundational studies called *intuitionism* originated by Luitzen E. J. Brouwer in the 1900's whose basic objection to truth of principles of logic was founded on principles of effectiveness: for instance, the implication $C\alpha\beta$ was hold to be true if a proof of the formula β could be obtained from a proof of α. Similarly the truth of the formula $\exists x\alpha(x)$ required that an object c be found such that $\alpha(c)$ was true. These views led to *intuitionistic logic* which rejected some theses of classical propositional calculi. In its formalization, intuitionistic propositional calculus may be based on theses 1–12 above: the validity of 13 was questioned by proponents of intuitionistic logic on the ground of effectiveness again; given the formula $\alpha \vee N\alpha$, its truth would require that a method be proposed to establish which of $\alpha, N\alpha$ was actually true. Thus intuitionistic propositional calculus is based on axiom schemes 1–12 above and on detachment as the derivation rule.

It follows from our discussion that the Lindenbaum–Tarski algebra of intuitionistic propositional calculus is a pseudo–Boolean algebra. It is therefore isomorphic to a pseudo–Boolean algebra of open sets in a topological Boolean algebra. Thus we may specialize the general completeness result to the intuitionistic case as follows (the proof is exactly as in the case of Proposition 13.31)

Proposition 9.15. *The following are equivalent for every meaningful expression α of intuitionistic calculus*

1. *α is a thesis of intutionistic calculus*

2. *for any valuation $v(\alpha)$ in the pseudo–Boolean algebra $O(\mathbf{B})$ of open sets in a topological Boolean algebra $\mathbf{B}$ we have $v(\alpha) = \sqrt{}$*

3. *for the valuation $||\alpha||$ of α in the Lindenbaum–Tarski algebra of intuitionistic calculus we have $||\alpha|| = \sqrt{}$*

9.7.1 Gentzen–type formalization of predicate and intuitionistic calculi

As with propositional calculus (Chapter 2), we have an alternative method for a formalization of the predicate as well as intuitionistic calculi by means of sequents (cf. [Gentzen34], [Kanger57]). We refer to Chapter 2 for a discussion of sequents and a method for proving completeness which may be easily adapted to these new cases (op.cit., op.cit.). We recall here the axiom as well as inference rules for predicate calculus in the version of [Gentzen34], better suited to our future purpose (cf. Chapter 12).

The axiom

$$\Gamma, A, \Delta \longrightarrow \Gamma, A, \Delta$$

Inference rules

Inference rules in the Gentzen original formalization are divided into two groups. The first group contains some schemata for forming sequents from sequents by adding, deleting, or permuting some sequences and a schema called *cut* essentially equivalent to the *Hauptsatz* (cf. Chapter 2, Exercise 43). As in Chapter 2, we denote with capital letters $A, B, C, \ldots$ meaningful expressions of predicate calculus.

(Ver A) $\dfrac{\Gamma \longrightarrow \Delta}{A, \Gamma \longrightarrow \Delta}$

(Ver S) $\dfrac{\Gamma \longrightarrow \Delta}{\Gamma \longrightarrow \Delta, A}$

(Zus A) $\dfrac{A, A, \Gamma \longrightarrow \Delta}{A, \Gamma \longrightarrow \Delta}$

(Zus S) $\dfrac{\Gamma \longrightarrow \Delta, A, A}{\Gamma \longrightarrow \Delta, A}$

(Vert A) $\dfrac{\Delta, A, B, \Gamma \longrightarrow \Theta}{\Delta, B, A, \Gamma \longrightarrow \Theta}$

(Vert S) $\dfrac{\Gamma \longrightarrow \Delta, A, B, \Theta}{\Gamma \longrightarrow \Delta, B, A, \Theta}$

(Cut) $\dfrac{\Gamma \longrightarrow \Delta, A; A, \Theta \longrightarrow \Lambda}{\Gamma, \Theta \longrightarrow \Delta, \Lambda}$

The second group contains – like in Chapter 2 – schemes for transforming sequents by means of logical functors and quantifiers.

$(\wedge \longrightarrow)$ $\dfrac{\Gamma \longrightarrow \Theta, A; \Gamma \longrightarrow \Theta, B}{\Gamma \longrightarrow \Theta, A \wedge B}$

$(\wedge \longleftarrow)$ $\dfrac{A, \Gamma \longrightarrow \Theta}{A \wedge B, \Gamma \longrightarrow \Theta}$

$(\vee \longrightarrow)$ $\dfrac{\Gamma \longrightarrow \Theta, A}{\Gamma \longrightarrow \Theta, A \vee B}$

$(\vee \longleftarrow)$ $\dfrac{A,\Gamma\longrightarrow\Theta;B,\Gamma\longrightarrow\Theta}{A\vee B,\Gamma\longrightarrow\Theta}$

$(N \longrightarrow)$ $\dfrac{A,\Gamma\longrightarrow\Theta}{\Gamma\longrightarrow\Theta,NA}$

$(N \longleftarrow)$ $\dfrac{\Gamma\longrightarrow\Theta,A}{NA,\Gamma\longrightarrow\Theta}$

$(C \longrightarrow)$ $\dfrac{A,\Gamma\longrightarrow\Theta,B}{\Gamma\longrightarrow\Theta,CAB}$

$(C \longleftarrow)$ $\dfrac{\Gamma\longrightarrow\Theta,A;B,\Delta\longrightarrow\Lambda}{CAB,\Gamma,\Delta\longrightarrow\Theta,\Lambda}$

$(\forall S)$ $\dfrac{\Gamma\longrightarrow\Theta,\alpha(c)}{\Gamma\longrightarrow\Theta,\forall x\alpha(x)}$

$(\exists A)$ $\dfrac{\alpha(c),\Gamma\longrightarrow\Theta}{\exists x\alpha(x),\Gamma\longrightarrow\Theta}$

where in the rules $(\forall S)$, $(\exists A)$ the variable c does not occur in $\Gamma,\Theta,\alpha(x)$

$(\forall A)$ $\dfrac{\alpha(c),\Gamma\longrightarrow\Theta}{\forall x\alpha(x),\Gamma\longrightarrow\Theta}$

$(\exists S)$ $\dfrac{\Gamma\longrightarrow\Theta,\alpha(c)}{\Gamma\longrightarrow\Theta,\exists x\alpha(x)}$

The notions of a proof, a thesis are defined as in Chapter 2, and the proof of completeness goes on similar lines (cf. [Kanger57]).

We now look at theses 1–13 above of the propositional calculus of which 1–12 may be adopted as axioms for intuitionistic propositional calculus (cf. [Rasiowa–Sikorski63], p.379). The question of the modification of the notion of a proof as to single out theses of the intuitionistic propositional calculus from among all theses of the sentential calculus given by the sequent calculus just outlined was solved by Gentzen (cf. [Gentzen34], p. 192).

This solution is based on the analysis of proofs of 1–13. The proof of 13 is as follows (cf. [Gentzen34], p. 193):

$\dfrac{p\longrightarrow p}{\longrightarrow p,Np}$; $\dfrac{\longrightarrow p,Np}{\longrightarrow p,p\vee Np}$; $\dfrac{\longrightarrow p,p\vee Np}{\longrightarrow p\vee Np,p}$; $\dfrac{\longrightarrow p\vee Np,p}{\longrightarrow p\vee Np,p\vee Np}$; $\dfrac{\longrightarrow p\vee Np,p\vee np}{\longrightarrow p\vee Np}$ where we listed consecutive applications of rules of inference $(N \longrightarrow)$, $(\vee \longrightarrow)$, (Vert S), $(\vee \longrightarrow)$, (Zus S).

In this proof the expression $p\vee Np$, one of expressions listed in the succedent of the sequent $\longrightarrow p \vee Np$ to be proved occurs twice in the succedent of one of the proof sequents. This is peculiar to the proof of only this from among theses 1–13 of the sentential calculus. We include here a proof of the thesis 2 in the form $CCpqCCqrCpr$; for simplicity, we do not mark the application of the permutation schema (Vert A, S):

$\dfrac{q\longrightarrow q;r\longrightarrow r}{q,Cqr\longrightarrow r}$ (by $(C \longleftarrow)$);

$\dfrac{p\longrightarrow p;q,Cqr\longrightarrow r}{Cpq,p,Cqr\longrightarrow r}$ (by $(C \longleftarrow)$);

$$\frac{p,Cpq,Cqr \longrightarrow r}{Cqr,Cpq \longrightarrow Cpr} \ \text{(by } (C \longrightarrow));$$

$$\frac{Cqr,Cpq \longrightarrow Cpr}{Cpq \longrightarrow CCqrCpr} \ \text{(by } (C \longrightarrow));$$

$$\frac{Cpq \longrightarrow CCqrCpr}{\longrightarrow CCpqCCqrCpr} \ \text{(by } (C \longrightarrow)).$$

Other theses are proposed to be proved as an exercise.

We have the Gentzen theorem. In a proof Π of a sequent S, we refer to an expression being a member of a sequence in S as to an *S–expression* and by a Π*–sequent* we mean a sequent appearing in Π.

Proposition 9.16. *(Gentzen) Theses A of the intuitionistic propositional calculus correspond to theses $S : \longrightarrow A$ of the sequent calculus which have the following property : in the proof Π of S there is no Π–sequent in whose succedent there are at least two occurrences of an S–expression.*

Historic remarks

Quantifiers were introduced to logic by Charles S. Peirce. Completeness of the predicate calculus was proved by Kurt Gödel [Gödel30]. The algebraic approach presented above is due to [Rasiowa–Sikorski50], [Rasiowa–Sikorski63]. A proof of the completeness theorem based on the idea of consistent sets of meaningful expressions was proposed by Henkin [Henkin49] and simplified by Hasenjaeger [Hasenjaeger53] and in [Rasiowa–Sikorski50] a proof based on the Rasiowa – Sikorski lemma (cf. Chapter 7) was proposed. The calculus of fractional truth values is due to Łukasiewicz [Łukasiewicz13]. Principles of intuitionistic logic were formulated by Brouwer [Brouwer08] cf. [Heyting56]. The sequent calculus was introduced in [Gentzen34] cf. also [Kanger57].

Exercises

In Exercises 1–22, use the Completeness Theorem to verify that respective meaningful expressions are theses of predicate calculus.

1. $C\forall x\alpha(x) \vee \beta\forall x(\alpha(x) \vee \beta)$

2. $C\forall x(\alpha(x) \vee \beta)\forall x\alpha(x) \vee \beta$

3. $C\exists x\alpha(x) \vee \beta\exists x(\alpha(x) \vee \beta))$

4. $C\exists x(\alpha(x) \vee \beta)\exists x\alpha(x) \vee \beta$

5. $C\forall x\alpha(x) \wedge \beta\forall x(\alpha(x) \wedge \beta)$

6. $C\forall x(\alpha(x) \wedge \beta)\forall x\alpha(x) \wedge \beta$

7. $C\exists x\alpha(x) \wedge \beta\exists x(\alpha(x) \wedge \beta)$

8. $C\exists x(\alpha(x) \wedge \beta)\exists x\alpha(x) \wedge \beta$

9. $C\forall x\alpha(x)\exists x\alpha(x)$

10. $C\forall xC\alpha(x)\beta C\exists \alpha(x)\beta$

11. $CC\exists \alpha(x)\beta \forall xC\alpha(x)\beta$

12. $C\forall xC\alpha(x)\beta C\exists x\alpha(x)\beta$

13. $CC\exists x\alpha(x)\beta \forall xC\alpha(x)\beta$

14. $C\forall xC\alpha\beta(x)C\alpha\forall x\beta(x)$

15. $C\forall xC\alpha(x)\beta(x)C\forall x\alpha(x)\forall x\beta(x)$

16. $C\forall xC\alpha(x)\beta(x)C\exists x\alpha(x)\exists x\beta(x)$

17. $C\forall x\alpha(x) \wedge \beta(x)\forall x\alpha(x) \wedge \forall x\beta(x)$

18. $C\forall x\alpha(x) \wedge \forall x\beta(x)\forall x\alpha(x) \wedge \beta(x)$

19. $C\exists x\alpha(x) \vee \beta(x)\exists x\alpha(x) \vee \exists x\beta(x)$

20. $C\exists x\alpha(x) \vee \exists x\beta(x)\exists x\alpha(x) \vee \beta(x)$

21. $C\forall x\alpha(x) \vee \forall x\beta(x)\forall x\alpha(x) \vee \beta(x)$

22. $C\exists x\alpha(x) \wedge \beta(x)\exists x\alpha(x) \wedge \exists x\beta(x)$

In Exercises 23–24 we address the intuitionistic propositional calculus versus the classical propositional calculus.

23. Prove that for a meaningful expression α of classical propositional calculus, $||\alpha \cup N\alpha||$ is a dense element in the Lindenbaum–Tarski algebra $\mathcal{L}$ of intuitionistic propositional calculus. [Hint: apply Exercise 15 (Chapter 8) and observe that for $q : \mathcal{L} \to (\mathcal{L}/\mathcal{D}$ we have $q(||\alpha \cup N\alpha||) = \surd$ hence $||\alpha \cup N\alpha|| \in \mathcal{D}]$

24. [Tarski38] Prove that a meaningful expression α of intuitionistic propositional calculus is a thesis of the classical propositional calculus if and only if $||\alpha||$ is a dense element in the Lindenbaum–Tarski algebra of intuitionistic propositional calculus. [Hint: argue as in Exercise 23]

In Exercises 25–27, we go back to the calculus of fractional truth values of Łukasiewicz.

25. Prove that $w(C\alpha\beta) = w(N\alpha) + w(\alpha \wedge \beta)$

26. Prove that if $CN\alpha\beta$ is a thesis, then $w(\alpha \vee \beta) = 1$ and $w(\alpha) + w(\beta) = 1 + w(\alpha \wedge \beta)$.

27. Prove that $w(\alpha) = 0$ implies that $w(C\alpha\beta) = 1$.

The last exercise is devoted to the Gentzen–type intuitionistic calculus.

28. Find proofs in the sequent calculus of theses corresponding to axioms 3–12 of the intuitionistic propositional calculus.

Works quoted

[Birkhoff67] G. Birkhoff, *Lattice Theory*, AMS, Providence, 1940 (3rd ed., 1967).

[Brouwer08] L. E. J. Brouwer, *De onbetrouwbaarheid der logische principes*, Tijdschrift voor wijsbegeerte 2(1908), pp. 152–158.

[Gentzen34] G. Gentzen, *Untersuchungen über das logische Schliessen. I,II.*, Mathematische Zeitschrift, 39 (1934–5), pp. 176–210, 405–431.

[Gödel30] K. Gödel, *Die Vollstandigkeit der Axiome des Logischen Funktionen-kalküls*, Monats. Math. Phys., 37(1930), pp. 349–360.

[Hasenjaeger53] G. Hasenjaeger, *Eine Bemerkung zu Henkin's Beweis fuer die Vollstandigkeit des Prädikatenkalküls des Ersten Stufe*, J. Symb. Logic, 18(1953), pp. 42–48.

[Henkin49] L. Henkin, *The completeness of the first–order functional calculus*, J. Symb. Logic, 14(1949), pp. 159–166.

[Heyting56] A. Heyting, *Intuitionism, an Introduction*, North Holland, Amsterdam, 1956.

[Kanger57] S. Kanger, *Provability in Logic*, Acta Universitatis Stockholmiensis. Stockholm Studies in Philosophy I, Stockholm, 1957.

[Łukasiewicz13] J. Łukasiewicz, *Die Logische Grundlagen der Wahrschein-*

lichkeitsrechnung, Cracow, 1913 [English translation in: [Borkowski70], pp. 16–63].

[Rasiowa53] H. Rasiowa, *On satisfiability and deducibility in non–classical functional calculi*, Bull. Polish Acad.Sci. Math., (Cl.III), 1(1953), pp. 229–231.

[Rasiowa51] H. Rasiowa, *A proof of the Skolem–Löwenheim theorem*, Fund. Math., 38 (1951), pp. 230–232.

[Rasiowa–Sikorski63] H. Rasiowa and R. Sikorski, *The Mathematics of Meta-mathematics*, PWN-Polish Scientific Publishers, Warszawa, 1963.

[Rasiowa–Sikorski50] H. Rasiowa and R. Sikorski, *A proof of the completeness theorem of Gödel*, Fund. Math., 37(1950), pp. 193–200.

[Tarski38] A. Tarski, *Der Aussagenkalkül und die Topologie*, Fund. Math., 31(1938), pp. 103–134.

PART 3:
MATHEMATICS of
ROUGH SETS

Chapter 10

Independence, Approximation

And so from the old soil – as men say – a new harvest sprouts

Chaucer, a paraphrase (cf. C.S. Lewis, *The Discarded Image*)

10.1 Introduction

In this Chapter, we address rough sets from the set–theoretic point of view. Given an information system $\mathcal{A} = (U, A)$, a family $\{IND_B : B \subseteq A\}$ of indiscernibility relations can be generated, the relation IND_B defined as follows

$$IND_B(x, y) \Leftrightarrow \forall a \in B(a(x) = a(y)).$$

Indiscernibility relations of the form IND_B serve a manifold purpose. First, they allow for defining *reducts* of the system $\mathcal{A}$ as minimal subsets B with the property that $IND_B = IND_A$; this definition realizes a form of the notion of *(functional) independence* of sets of attributes. Next, classes of relations of the form IND_B are regarded as elementary pieces of knowledge from which more complex knowledge descriptions may be built; one form of such descriptions is *dependencies* among attributes expressed in the form of *rules* $B \mapsto C$ meaning that each class of IND_B is a subset of a class of IND_C. Finally, classes of indiscernibility relations allow for constructing more complex *concepts* e.g. by taking the unions of families of those classes; in this way, *exact* concepts are defined. Other concepts, which are *inexact*, may be *approximated* by the exact ones from below and from above leading to the *lower–* respectively the *upper–* approximations. The universe U of the system $\mathcal{A}$ along with lower and upper approximation operators is then considered as an abstract algebra, *the approximation space*.

In what follows, we give an analysis of those three aspects of an information system, set in an abstract mathematical context. First, we discuss the notion of independence.

10.2   Independence

The notion of independence permeates many areas of mathematics, its particular expression depending on the specific context and language of a given area. Probably best known is the notion of algebraic independence applied in the theory of vector spaces. Let us recall it as a starting point.

Given a vector space V, e.g. over the set of real numbers $\mathbf{R}^1$, we say that a set $W \subseteq V$ of vectors is *(linearly) independent* in case the condition $r_1 \cdot v_1 + ... + r_k \cdot v_k = 0$ implies $r_1 = 0, ..., r_k = 0$ for any finite subset $\{v_1, ..., v_k\} \subseteq W$ and any set $\{r_1, ..., r_k\}$ of real numbers. The independence of W means then that no vector $v \in W$ can be expressed as a linear combination of some vectors in $W \setminus \{v\}$. On the other hand, was W not independent, we would have $v = r_1 \cdot v_1 + ... + r_m \cdot v_m$ for some $v, v_1, ..., v_m$ in W and some real numbers $r_1, ..., r_m$. One may say that dependence of a set of vectors means expressibility within that set, while independence means inexpressibility.

Let us also mention that in case V is a finite–dimensional vector space, the independence of W means that

$$dim W^* > dim(W \setminus \{v\})^* \qquad\qquad (10.1)$$

for each $v \in W$ where the symbol W^* denotes the vector space generated by W. On the contrary, in case W is not independent we have

$$dim W^* = dim(W \setminus \{v\})^* \qquad\qquad (10.2)$$

for some $v \in W$. Thus dependence means preserving value of some function (in this case, the algebraic dimension dim) over a proper subset; we may say that in this case dependence means redundancy of some vector with respect to the function dim.

These simple examples convey to us some basic meanings associated with the notion of independence as well as some basic methods for introducing a notion of independence.

Passing to another area, we may ask how to define independence in set theory. Given a set M, let us consider a finite collection $\mathcal{F} \subseteq 2^M$ of its subsets. For a set $A \subseteq M$, we let

$$\begin{aligned} A^0 &= A \\ A^1 &= M \setminus A \end{aligned} \qquad\qquad (10.3)$$

Then we may select for each $A \in \mathcal{F}$ an index $i_A \in \{0, 1\}$. For each selection i thus defined, we define the set

$$C_i(\mathcal{F}) = \bigcap \{A^{i_A} : A \in \mathcal{F}\} \tag{10.4}$$

called the *component* of $\mathcal{F}$ (with respect to i).

Then we have

Proposition 10.1. *For any set M and any (finite) family $\mathcal{F}$ of its subsets, the set of components of $\mathcal{F}$ is a partition of the set M; more generally, for any sub–family $\mathcal{G}$ of $\mathcal{F}$, the set of components of $\mathcal{F} \setminus \mathcal{G}$ is a partition of any of components of $\mathcal{G}$.*

Proof. Any two distinct components of $\mathcal{F}$ are disjoint: given selections $i \neq j$, we have $A^{i_A} \neq A^{j_A}$ for some $A \in \mathcal{F}$ thus

$$C_i(\mathcal{F}) \cap C_j(\mathcal{F}) \subseteq A^{i_A} \cap A^{j_A} = A \cap (M \setminus A) = \emptyset \tag{10.5}$$

For $x \in M$, we let $i_A(x) = 0$ in case $x \in A$ and $i_A(x) = 1$, otherwise. Then, $x \in C_{i_A(x)}(\mathcal{F})$. Thus components of $\mathcal{F}$ form a partition of M. The proof in general case follows the same pattern. $\square$

We will say that the family $\mathcal{F}$ is *independent* when every component of $\mathcal{F}$ is non–empty. More generally, we would say that a (not necessarily finite) family $\mathcal{K}$ of subsets of M is *independent* when any finite sub–family $\mathcal{F} \subseteq \mathcal{K}$ is independent.

Let us observe that components are minimal with respect to inclusion sets among all sets which may be obtained from sets in $\mathcal{F}$ by means of operations of intersection and complement. Moreover

Proposition 10.2. *The family $\mathcal{F}^*$ of unions of components of $\mathcal{F}$ is closed under operations of union, intersection and complement and it does contain the empty–set as well as M hence it is a field of sets.*

Proof. It suffices to mention that the empty set is the union of the empty family of components, the set M is the union of the family of all components of $\mathcal{F}$ and other claims follow immediately from the fact that components are pair–wise disjoint.

A proposition follows.

Proposition 10.3. *Given subsets $S_1, S_2, ..., S_k$ of the set M, the field of sets generated by $S_1, S_2, ..., S_k$ (i.e. the smallest field of sets containing $S_1, S_2, ..., S_k$) coincides with the field of sets $\{S_1, S_2, ..., S_k\}^*$.*

Indeed, the result follows from the last two propositions.

We have also as a particular case (applied in Sect. 10, Chapter 8)

Proposition 10.4. *For a family $\mathcal{K}$ of subsets of a set M closed with respect to the union and intersection, and containing the empty set and the set M, the field of sets $\mathcal{K}^*$ generated by $\mathcal{K}$ is the collection of all sets of the form*

$$(-D_1 \cup E_1) \cap (-D_2 \cup E_2) \cap ... \cap (-D_k \cup E_k)$$

where $D_1, E_1, ..., D_k, E_k \in \mathcal{K}$ and $-$ denotes the complement operation in M.

Proof. $\mathcal{K}^*$ is the collection of unions of components of $\mathcal{K}$ or, equivalently, passing to complements, it is the collection of intersections of unions of sets each of which is either A or $M \setminus A$ for $A \in \mathcal{K}$. Thus, it remains to show that any union of sets each of which is either A or $M \setminus A$ is of the form $-D \cup E$ i.e.

$$-A_1 \cup -A_2 \cup ... \cup -A_r \cup A_{r+1} \cup ... \cup A_k = -D \cup E \qquad (10.6)$$

But to satisfy this condition it suffices to let $D = A_1 \cap ... \cap A_r$ and $E = A_{r+1} \cup ... \cup A_k$.   □

10.2.1 Functional dependence

For families of functions, the notion of independence may be formulated along the lines of set independence. To this end, we may observe that any function $f : X \to Y$ is determined by its *fibres* $f^{-1}(f(x))$ for $x \in X$. Thus, we may define independence of a finite set $\mathcal{F}$ of functions from a set X into a set Y as the independence for every $x \in X$ of the family $\mathcal{F}^{-1}(x) = \{f^{-1}(x) : f \in \mathcal{F}\}$ of sets.

This condition when negated, however, does not set apart any function in $\mathcal{F}$ as dependent on other functions and for this reason we may need a stronger version of independence/dependence viz. we will say that a function f is *dependent functionally* on functions $f_1, ..., f_k$ if and only if for every $x \in X$ the following inclusion

$$\bigcap_{i=1}^{k} f_i^{-1}(f_i(x)) \subseteq f^{-1}(f(x)) \qquad (10.7)$$

holds. We may express this dependence in a functional form

Proposition 10.5. *A function f depends functionally on functions $f_1, ..., f_k$ if and only if a function $\phi : Y^k \to Y$ exists with the property that*

(i) $f(x) = \phi(f_1(x), .., f_k(x))$ for every $x \in X$.

Proof. In case ϕ as above exists, we have

$$\bigcap_{i=1}^{k} f_i^{-1}(f_i(x)) \subseteq f^{-1}(f(x)) \tag{10.8}$$

for every x as $f_i(y) = f_i(x)$ for $i = 1, 2, ..., k$ imply

$$f(y) = \phi(f_1(y), ..., f_k(y)) = \phi(f_1(x), ..., f_k(x)) = f(x),$$

for each pair x, y.

Conversely, when

$$\bigcap_{i=1}^{k} f_i^{-1}(f_i(x)) \subseteq f^{-1}(f(x)) \tag{10.9}$$

holds for every x, we may define ϕ by letting

$$\phi(f_1(x), ..., f_k(x)) = f(x)$$

for $x \in X$ and $\phi(y_1, ..., y_k) = y_0$ in case $(y_1, ..., y_k)$ is distinct from $(f_1(x), ..., f_k(x))$ for all x where y_0 is an arbitrarily chosen element in Y. $\square$

We extend the notion of functional dependence to infinite families of functions: we will say that a function f depends on a family $\mathcal{F}$ of functions if and only if f depends on a finite sub–family $\mathcal{K} \subseteq \mathcal{F}$. Let us sum up basic properties of functional dependence

Proposition 10.6.   *1. If f depends on $\mathcal{F}$ and $\mathcal{F} \subseteq \mathcal{K}$ then f depends on $\mathcal{K}$*

2. *If f depends on $\mathcal{F}$ then there exists a minimal with respect to inclusion sub–family $\mathcal{K} \subseteq \mathcal{F}$ such that f depends on $\mathcal{K}$*

3. *If $\mathbf{L}$ is a linearly ordered with respect to inclusion family of sets of functions and a function f depends on $\bigcup \mathbf{L}$ then there exists $\mathcal{L} \in \mathbf{L}$ such that f depends on $\mathcal{L}$*

Proof. Property (1) is obviously true, in case (2), there exists a finite $\mathcal{G} \subseteq \mathcal{F}$ on which f does depend and then examining finitely many subsets of $\mathcal{G}$, we find a subset $\mathcal{K} \subseteq \mathcal{G}$ such that f depends on $\mathcal{K}$ and f does not depend on $\mathcal{H}$ for any proper subset $\mathcal{H} \subseteq \mathcal{K}$.

For (3), there exists a finite $\mathcal{H}$, a subset of $\bigcup \mathbf{L}$, f depends on. $\mathcal{H}$ being finite, there exists $\mathcal{L} \in \mathbf{L}$ with $\mathcal{H} \subseteq \mathcal{L}$ hence by (1) f depends on $\mathcal{L}$. $\square$

The dual notion of independence of a set of functions comes naturally now: for a family $\mathcal{K}$ of functions, we will say that $\mathcal{K}$ is *independent* if and only if no $f \in \mathcal{K}$ is dependent on $\mathcal{K} \setminus \{f\}$. Then we have the following consequences of Proposition 10.6.

Proposition 10.7.   *1. If $\mathcal{K}$ is independent and $\mathcal{H} \subseteq \mathcal{K}$ then $\mathcal{H}$ is independent*

    2. Any independent $\mathcal{H} \subseteq \mathcal{K}$ can be extended to a maximal independent $\mathcal{G}$ $\subseteq \mathcal{K}$

Proof. Assuming that in case (1) $\mathcal{H}$ is not independent, we get a function $f \in \mathcal{H}$ such that f depends on $\mathcal{H} \setminus \{f\}$ hence by Proposition 10.6, f depends on $\mathcal{K} \setminus \{f\}$, a contradiction proving (1). The claim (2) follows from the maximum principle (Proposition 14, Chapter 6) and Proposition 10.6, which in dual form states that for any chain $\mathbf{L}$ of independent families of functions, the union $\bigcup \mathbf{L}$ is independent. $\square$

We now would like to extend this notion of dependence in order to express dependence relations between families of functions. We first introduce a new notion concerning families of functions.

For sets X, Y and a family of functions $\mathcal{F}$ from X into Y, we call the *fiber product* of $\mathcal{F}$ the function $\Delta_{\mathcal{F}} : X \to Y^{|\mathcal{F}|}$ (where $|\mathcal{F}|$ denotes the cardinality of $\mathcal{F}$) defined via the formula

$$\Delta_{\mathcal{F}}(x) = (f(x) : f \in \mathcal{F})$$

i.e. $\Delta_{\mathcal{F}}(x)$ is an element (a thread) in the Cartesian product (cube) of $|\mathcal{F}|$–many copies of Y.

For families $\mathcal{F}, \mathcal{G}$ of functions from X into Y, we will say that $\mathcal{G}$ *depends functionally on* $\mathcal{F}$ if and only if there exists a function $\phi : Y^{|\mathcal{F}|} \to Y^{|\mathcal{G}|}$ such that $\Delta_{\mathcal{G}}(x) = \phi(\Delta_{\mathcal{F}}(x))$ for each $x \in X$.

We will use the symbol $\mathcal{F} \mapsto \mathcal{G}$ to denote that $\mathcal{G}$ depends on $\mathcal{F}$. We collect below the basic properties of the dependence relation $\mapsto$.

Proposition 10.8. *The relation $\mapsto$ has the following properties*

    1. If $\mathcal{G} \subseteq \mathcal{F}$ then $\mathcal{F} \mapsto \mathcal{G}$

    2. If $\mathcal{F} \mapsto \mathcal{G}$ and $\mathcal{K}$ is a family of functions from X into Y then $\mathcal{F} \cup \mathcal{K} \mapsto \mathcal{G}$

    3. If $\mathcal{F} \mapsto \mathcal{G}$ and $\mathcal{G} \mapsto \mathcal{K}$ then $\mathcal{F} \mapsto \mathcal{K}$

    4. If $\mathcal{F} \mapsto \mathcal{G}$ and $\mathcal{F} \mapsto \mathcal{K}$ then $\mathcal{F} \mapsto \mathcal{G} \cup \mathcal{K}$

Proof. The projection function $pr_{F,G}$ defined for $x \in X$ via the formula

$$pr_{F,G}(\Delta_{\mathcal{F}}(x)) = \Delta_{\mathcal{G}}(x) \qquad\qquad (10.10)$$

does establish the relation $\mathcal{F} \mapsto \mathcal{G}$ proving 1.

We will assume that in all cases when $\mathcal{F} \mapsto \mathcal{G}$ holds, this fact is established by means of a function ϕ as above. In case of 2., the formula

$$(f(x) : f \in \mathcal{G}) = \phi(pr_{F,F \cup K}((f(x) : f \in \mathcal{F} \cup \mathcal{K}))) \qquad (10.11)$$

does witness the fact that $\mathcal{F} \cup \mathcal{K} \mapsto \mathcal{G}$.

In case of 3., assuming that the function ψ does witness that $\mathcal{G} \mapsto \mathcal{K}$, we find that the composition $\psi \circ \phi$ witnesses $\mathcal{F} \mapsto \mathcal{K}$ i.e.

$$(f(x) : f \in \mathcal{K}) = \psi(\phi((f(x) : f \in \mathcal{F}))) \qquad (10.12)$$

for $x \in X$.

In case of 4., assuming that ψ witnesses $\mathcal{F} \mapsto \mathcal{K}$ we have (we may assume that $\mathcal{G}$, $\mathcal{K}$ are disjoint, otherwise we would replace $\mathcal{K}$ with $\mathcal{K} \setminus \mathcal{G}$) that

$$(f(x) : f \in \mathcal{G} \cup \mathcal{K}) = \phi((f(x) : f \in \mathcal{F})) \cup \psi((f(x) : f \in \mathcal{F})) \qquad (10.13)$$

for $x \in X$. □

From Proposition 10.8, other properties follow

Proposition 10.9. *The following properties of $\mapsto$ are implied by properties 1-4 in Proposition 10.8*

5. $\mathcal{F} \mapsto \mathcal{F}$
6. If $\mathcal{F} \mapsto \mathcal{G}$ then $\mathcal{F} \cup \mathcal{K} \mapsto \mathcal{G} \cup \mathcal{K}$
7. If $\mathcal{F} \mapsto \mathcal{G}$ and $\mathcal{K} \subseteq \mathcal{G}$ then $\mathcal{F} \mapsto \mathcal{K}$
8. If $\mathcal{F} \mapsto \mathcal{G}$ and $\mathcal{G} \cup \mathcal{L} \mapsto \mathcal{K}$ then $\mathcal{F} \cup \mathcal{L} \mapsto \mathcal{K}$
9. If $\mathcal{F} \mapsto \mathcal{G} \cup \mathcal{K}$ and $\mathcal{K} \mapsto \mathcal{L}$ then $\mathcal{F} \mapsto \mathcal{G} \cup \mathcal{K} \cup \mathcal{L}$
10. If $\mathcal{F} \mapsto \mathcal{G}$ and $\mathcal{K} \mapsto \mathcal{L}$ then $\mathcal{F} \cup \mathcal{K} \mapsto \mathcal{G} \cup \mathcal{L}$

Proof. 5 follows from 1 as a particular case. For 6, apply 2 to get $\mathcal{F} \cup \mathcal{K} \mapsto \mathcal{G}$ and 1 to get $\mathcal{F} \cup \mathcal{K} \mapsto \mathcal{K}$ and then $\mathcal{F} \cup \mathcal{K} \mapsto \mathcal{G} \cup \mathcal{K}$ follows by 4.

In case of 7, apply 1 to get $\mathcal{G} \mapsto \mathcal{K}$ and then 3 to get $\mathcal{F} \mapsto \mathcal{K}$. In case of 8, apply 2 to get $\mathcal{F} \cup \mathcal{L} \mapsto \mathcal{K}$ and 1 to get $\mathcal{F} \cup \mathcal{L} \mapsto \mathcal{L}$ and then 4 to get $\mathcal{F} \cup \mathcal{L} \mapsto \mathcal{G} \cup \mathcal{L}$ from which 3 yields $\mathcal{F} \cup \mathcal{L} \mapsto \mathcal{K}$.

For 9, it suffices by 4 to prove that $\mathcal{F} \mapsto \mathcal{L}$ and the last instance follows by 3 from $\mathcal{G} \cup \mathcal{K} \mapsto \mathcal{L}$ which in turn follows by 2 from $\mathcal{K} \mapsto \mathcal{L}$.

In case of 10, it follows by 3 that $\mathcal{F} \cup \mathcal{K} \mapsto \mathcal{G}$ and $\mathcal{F} \cup \mathcal{K} \mapsto \mathcal{L}$ hence by 4 it follows that $\mathcal{F} \cup \mathcal{K} \mapsto \mathcal{G} \cup \mathcal{L}$. □

10.2.2   An abstract view on independence

We will now present an abstract view on the above introduced notion of independence, formulated in the language of semi–lattices and congruences on them. The results presented below were proved and presented in [Novotný – Pawlak opera cit.] and [Novotný opera cit.].

Semi–lattices and congruences

We will begin with a weaker notion than that of a lattice (cf. Chapter 9) viz. with a semi–lattice. By a semi–lattice we will understand a set A endowed with an operation $\star : A \times A \to A$ of which we will require that it should satisfy the following

(SM1) $x \star x = x$
(SM2) $x \star y = y \star x$
(SM3) $x \star (y \star z) = (x \star y) \star z$

Clearly, any lattice $(A, \cup, \cap)$ defines two semi–lattices $(A, \cup)$ and $(A, \cap)$. The semi–lattice operation $\star$ does induce an ordering $\preceq$ on A viz. we let $x \preceq y$ if and only if $x \star y = y$. Indeed, $x \preceq x$ follows by (SM1), transitivity $x \preceq y \wedge y \preceq z \Rightarrow x \preceq z$ follows from $x \star z = x \star (y \star z) = (x \star y) \star z = y \star z = z$, where (SM3) was applied, and in case $x \preceq y$ and $y \preceq x$ we have $x \star y = y$ and $y \star x = x$ hence $x = y$ follows by (SM2).

In the sequel we will sometimes discuss the semi–lattice A with the ordering $\preceq$ in place of the operation $\star$. We may also introduce the operation of join $\cup$ letting $x \cup y = x \star y$; more generally, inducting on k, we may prove that $\cup\{x_1, ..., x_k\} = x_1 \star ... \star x_k$.

As with lattices, the notion of a *homomorphism* of semi–lattices is introduced viz. for semi–lattices $(A, \star)$ and $(B, *)$ we say that a function $h : A \to B$ is a homomorphism of A into B if and only if the condition $h(x \star y) = h(x) * h(y)$ holds for each pair x, y of elements of A. It is easy to see that this condition is equivalent in case h is an onto function to the condition $h(\cup\{x_1, ..., x_k\}) = \cup\{h(x_1), ..., h(x_k)\}$ for each finite subset $\{x_1, ..., x_k\} \subseteq A$.

Congruences

We consider a semi–lattice $(A, \star)$ or, equivalently, $(A, \cup)$, along with an equivalence relation R on A; we will be interested in equivalence relations related to semi–lattice operation $\cup$ viz. we will say that the relation R is a *congruence* on A if and only if the condition

$$xRy \wedge x'Ry' \Rightarrow x \cup x' R y \cup y' \qquad (10.14)$$

holds for each pair x, y and each pair x', y' of elements of A.

For a congruence R on A, we may consider the quotient set A/R of equivalence classes with the induced operation $\star_R$:

$$[x]_R \star_R [y]_R = [x \star y]_R.$$

Clearly, the definition of $\star_R$ is independent of the choice of representative in each class: if $[x]_R = [x']_R, [y]_R = [y']_R$ then $[x \star y]_R = [x' \star y']_R$ by the definition of a congruence. Moreover, the quotient set A/R endowed with the operation $\star_R$ is a semi–lattice: $[x]_R \star_R [x]_R = [x \star x]_R = [x]_R$, $[x]_R \star_R [y]_R = [x \star y]_R = [y \star x]_R = [y]_R \star_R [x]_R$, and similarly the associativity condition (SM3) follows. In terms of the operation $\cup$ we have for the induced operation $\cup_R$ that

$$[x]_R \cup_R [y]_R = [x]_R \star_R [y]_R = [x \star y]_R = [x \cup y]_R \qquad (10.15)$$

An example of a semi–lattice is clearly the power set 2^A of a set A together with the union of sets operation $\cup$; assuming that $h : 2^A \to S$ is a homomorphism of 2^A into a semi–lattice $(S, \star)$ we may define a congruence R_h by letting $X R_h Y$ if and only if $h(X) = h(Y)$; that R_h is indeed a congruence may be checked in a straightforward manner:
$X R_h Y, X' R_h Y'$ imply $h(X) = h(Y), h(X') = h(Y')$ hence

$$h(X \cup X') = h(X) \star h(X') = h(Y) \star h(Y') = h(Y \cup Y') \qquad (10.16)$$

An instance of this situation arises naturally when A is a set of functions on a set U and we consider the power set $2^{2^{2^U}}$ whose elements are relations on U as a semi–lattice with the intersection operation $\cap$ as the semi–lattice operation.

The homomorphism $h : 2^A \to 2^{2^{2^U}}$ is constructed as follows. Given $X \subseteq U$, we consider the relation IND_X on U defined as follows:

$$(IND) \quad u\,IND_X v \Leftrightarrow \forall a \in X (a(u) = a(v))$$

for each pair u, v of elements of U. Then we let $h(X) = IND_X$. As

$$h(X \cup Y) = IND_{X \cup Y} = IND_X \cap IND_Y = h(X) \cap h(Y) \qquad (10.17)$$

h is indeed a homomorphism of semi–lattices 2^A and $2^{2^{2^U}}$.

The congruence R_h then satisfies

$$X R_h Y \Leftrightarrow IND_X = IND_Y \qquad (10.18)$$

and thus from $IND_X = IND_{X'}, IND_Y = IND_{Y'}$ it follows that

$$IND_{X \cup X'} = IND_X \cap IND_{X'} = IND_Y \cap IND_{Y'} = IND_{Y \cup Y'} \quad (10.19)$$

providing a direct proof that the relation $IND_X = IND_Y$ is a congruence on the semi–lattice $(2^A, \cup)$.

This example justifies our interest in congruences on semi–lattices as a natural framework for a study of independence in functional spaces.

10.2.3   Dependence spaces

Following Novotný and Pawlak, we call a pair (A, R), where A is a semi–lattice and R a congruence on A, a *dependence space*. We begin with an account of some basic congruences one may introduce here and we will focus on the most important particular case related to information systems of a semi–lattice 2^A (where A may be regarded as a set of attributes of objects in a universe U) with the union of sets operation $\cup$.

As with lattices, we invoke here the notion of a *closure* operator; a function $C : S \to S$ from a semi–lattice $(S, \preceq)$ with the semi–lattice operation $\star$ into itself is said to be a *closure* on S if and only if the following hold

(C1) $x \preceq C(x)$
(C2) $x \preceq y$ implies $C(x) \preceq C(y)$
(C3) $C(C(x)) = C(x)$

Any closure operator C on a semi–lattice S induces a congruence R_C on S defined as follows

$$xR_C y \Leftrightarrow C(x) = C(y) \quad (10.20)$$

That R_C is indeed a congruence, may be shown as follows.
For x, y, u, v with $xR_C u, yR_C v$ i.e. with $C(x) = C(u), C(y) = C(v)$ we infer from $x \preceq C(x) = C(u)$ and $u \preceq u \star v$ by (C1) and (C2) that $x \preceq C(x) = C(u) \preceq C(u \star v)$. Similarly, $y \preceq C(y) = C(v) \preceq C(u \star v)$. Thus, $x \star y \preceq C(u \star v)$ hence by (C2) and (C3)

$$C(x \star y) \preceq C(C(u \star v)) = C(u \star v) \quad (10.21)$$

By symmetry, we have also $C(u \star v) \preceq C(x \star y)$ and finally $C(x \star y) = C(u \star v)$ i.e. $x \star y R_C u \star v$ which means that R_C is a congruence.

We would like to settle the converse: whether a given congruence R does induce a closure? The answer turns out to be in affirmative: the closure operator, say, C_R, may be induced by means of R as follows.

For $x \in S$, consider the equivalence class $[x]_R$ of x and select in this class a maximal with respect to $\preceq$ element, say $C_R(x)$. Let us observe that in the finite case (i.e. when S is a finite set) the element $C_R(x)$ is uniquely determined as $x_1 \star x_2 \star ... \star x_k$ where $[x]_R = \{x_1, ..., x_k\}$.

Indeed, as xRx_i for $i = 1, 2, ..., k$, we have $x \star x \star ... \star xRx_1 \star \star... \star x_k$ i.e. $xRx_1 \star \star... \star x_k$ so $x_1 \star \star... \star x_k \in [x]_R$ and clearly $x_1 \star \star... \star x_k = \cup[x]_R$. It follows that $x \preceq C_R(x)$ so (C1) holds.

For x, y with $x \preceq y$, we have $xRC_R(x), yRC_R(y)$ hence $x \star yRC_R(x) \star C_R(y)$ i.e. $yRC_R(x) \star C_R(y)$. Thus, $C_R(x) \star C_R(y) \preceq C_R(y)$ i.e. $C_R(x) \preceq C_R(y)$ proving that (C2) holds.

Finally, as $xRC_R(x)$ we have

$$C_R(x)RC_R(C_R(x)) \tag{10.22}$$

and thus

$$xRC_R(C_R(x)) \tag{10.23}$$

which implies that

$$C_R(C_R(x)) \preceq C_R(x) \tag{10.24}$$

By already proven (C1), the equality

$$C_R(C_R(x)) = C_R(x) \tag{10.25}$$

follows so (C3) holds.

The congruence R_C and the closure operator C_R introduced above are canonical in the following sense

Proposition 10.10. *For a congruence R and a closure C on a semi–lattice* $(A, \star)$, *we have*

$$R_{C_R} = R, C_{R_C} = C$$

Proof. For a congruence R, we have $xR_{C_R}y$ if and only if $C_R(x) = C_R(y)$ which implies that xRy. Thus $R_{C_R} \subseteq R$.

Conversely, xRy implies $C_R(x) = C_R(y)$ as $[x]_R = [y]_R$ and thus $xR_{C_R}y$ hence $R \subseteq R_{C_R}$ and finally $R = R_{C_R}$.

Now assume that $y = C(x)$. Then $C(y) = C(C(x)) = C(x)$ hence xR_Cy i.e. $xR_CC(x)$ which implies that $C(x) \preceq C_{R_C}(x)$. On the other hand, if xR_Cy then $C(y) = C(x)$ i.e. $C(x) = C_{R_C}(x)$.   □

10.2.4   Independence

We apply the results presented above to a study of independence in an abstract setting (after Novotn'y–Pawlak). We will consider a semi–lattice S along with a congruence R on it. Given $x \in S$, we call x an R-*independent* element in S if and only if x is a minimal with respect to $\preceq$ element in its equivalence class $[x]_R$. Then clearly

Proposition 10.11. *An element x is R-independent if and only if from $y \prec x$ it follows that it is not true that xRy*

A deeper result is presented in the following proposition. We assume that the semi–lattice $(S, \cup)$ is endowed with the relative additive pseudo–complement $x \to y$ i.e. we have $y \cup x \to y = x$ for each pair x, y in S. This is obviously the case for semi–lattices of the form 2^A.

Proposition 10.12. *If x is R–independent and $y \preceq x$, then y is R–independent as well.*

Proof. Clearly, by Proposition 10.11, either $y = x$ and there is nothing to prove or $y \prec x$ and then not yRx. Thus, $[y]_R \neq [x]_R$ and it remains to prove that y is minimal in its class. Assume, to the contrary, that $[z]_R = [y]_R$ for some $z \prec y$. Then, as R is a congruence, zRy implies that $z \cup x \to yRy \cup x \to y$ i.e. $z \cup x \to yRx$. As clearly $z \cup x \to y \prec x$ we arrive at contradiction with minimality of x. It follows that y is minimal in its class, a fortiori it is R–independent. $\qquad\qquad\qquad\qquad\qquad\qquad\qquad\qquad\qquad\qquad\qquad\quad\square$

We denote by the symbol $I(R)$ the set of all R–independent elements in S. The above proposition showing that R-independence is hereditary may be strengthened as to give an insight into the structure of minimal sets.

To this end, we introduce a relative notion of R-independence viz. for $x \in S$ we say that $y \in S$ is R-independent relative to x in case y is R-independent and yRx i.e. minimality of y is witnessed by the class of x; clearly, if y is R–independent relative to x and zRx then y is R–independent relative to z.

In particular, any R–independent element is R–independent relative to itself. In accordance with rough set terminology, we will also call any element R–independent relative to x an R-reduct of x. The set of all R–reducts of x will be denoted with the symbol $RED(x, R)$.

To have an insight into the structure of this set, we will introduce some auxiliary notions. We will assume that the semi–lattice S is of the form 2^A for some set A i.e. we have in S the 0 element, the intersection and the complement of sets are also defined (although our semi–lattice operation still remains the union of sets) and each element of S is a union of one–element sets. Clearly, the relation $\preceq$ is identical to the inclusion relation $\subseteq$. We now will say that y is R-redundant in x in case $x \to yRx$, otherwise y is R-non-redundant. Then we have

Proposition 10.13. *The intersection of all reducts of x is identical to the set of all non–redundant one-element subsets of x.*

Proof. We denote with the symbol $CR(x, R)$ the set of all non–redundant singleton subsets of x (called the *core* of x). Assume that a singleton z is such that $z \in \bigcap RED(x, R)$ and $x \to zRx$. Thus there exists $y \in RED(x, R)$

with $y \preceq x \to z$, in consequence $z \notin y$, a contradiction proving that $\bigcap RED(x, R) \subseteq CR(x, R)$.

Conversely, assume that $z \in CR(x, R)$ and there exists $y \in RED(x, R)$ with $z \notin y$. Hence

$$y \preceq x \to z$$

and thus $x \to zRx$, contrary to non–redundancy of z which shows that

$$CR(x, R) \subseteq RED(x, R) \tag{10.26}$$

$\square$

It is useful to look also at the set $\{y : y \preceq x \land y \in I(R)\}$ whose elements are called *sub–reducts* of x. Clearly, every reduct of x is a sub–reduct of x and any sub–reduct y of x with the property that yRx is a reduct of x. We denote the set of all sub–reducts of x with the symbol $SUBRED(x, R)$. We will also say that $y \in S$ is *hereditarily (R, x)–inaccessible* if and only if zRx holds for no $z \preceq y$.

We have (cf. [Novotný98[a,b]])

Proposition 10.14. *For each $x \in S$, the difference*

$$y = x - \bigcup RED(x, R)$$

is hereditarily (R, x)–inaccessible.

Hence every $x \in S$ can be represented as the union

$$y \cup \bigcup RED(x, R)$$

of a hereditarily (R, x)–inaccessible element and all of its reducts.

Proof. Obvious by the definition of a reduct.

We will say that an element $y \in S$ is *R–almost zero* if and only if $yR0$. Then we have the following proposition in case $S = 2^A$ with a *finite* set A.

Proposition 10.15. *1. For each each $x \in S$ there exists a maximal y^* among elements y with properties $y \preceq x$ and y is R–almost zero*

2. If y^* and x are as in (1) then y^*Ry for every $y \preceq y^*$

3. If y^* and x are as in (1) then $x - y^*$ contains no R–almost zero element save 0

4. If y^* and x are as in (1) then $x - y^*$ is the union of all sub–reducts contained in x

Proof. For (1), observe that if y, z are R–almost zero then $y \cup z$ is R–almost zero as well; indeed, from $yR0, zR0$ it follows that $y \cup zR0 \cup 0$ i.e. $y \cup zR0$. Thus in the finite case, having a finite family $\mathcal{Z}$ of R–almost zero subsets of x, we get that the union $y^* = \bigcup \mathcal{Z}$ is an R–almost zero set which is clearly maximal among R–almost zero subsets of x. By construction of y^* it follows that $x - y^*$ contains no non–zero R–almost zero element which proves (3).

For (2), as y^*R0 it follows that $(y^* \cup z)R(0 \cup z)$ for every z in particular y^*Rz for every $z \preceq y^*$. Let $z = x - y^*$ and consider the family $SUBRED(x, R)$. By (2), no sub–reduct of x save 0 can meet y^* hence $z_1 = z - \bigcup SUBRED(x, R)$ must be 0, otherwise from $z_1 \neq 0$ and $non(z_1 R0)$ it would follow that there would exist some $w \in SUBRED(x, R)$ with $w \preceq z$, $w \neq 0$, a contradiction.

Hence it must be $z_1 = 0$ and $x - y^* = \bigcup SUBRED(x, R)$ proving (4). □

The upshot of this last proposition is a representation result

Proposition 10.16. *For each $x \in S$, we have*

$$x = y \cup \bigcup SUBRED(x, R)$$

where y is an R–almost zero subset of x.

Indeed, it does follow from (1) and (4) in the preceding Proposition 10.16.

10.2.5   Dependence

In a dependence space (S, R) where S a semi–lattice and R a congruence, we introduce the notion of a dependence which will be an abstract counterpart of the notion of functional dependence discussed above.

We will say that y *depends on* x if and only if $C_R(y) \preceq C_R(x)$ where – we recall – $C_R(x)$ is the closure of x (via the closure operator C_R) defined as the maximal element in the equivalence class $[x]_R$ of x. We denote by the symbol $x \mapsto_R y$ the fact that y depends on x. We will check that the dependence $\mapsto_R$ does satisfy the basic postulates 1–4 established in case of functional dependence cf. Proposition 10.8.

Proposition 10.17. *The following properties are satisfied with the relation $\mapsto_R$*

1. *$y \preceq x$ implies $x \mapsto_R y$*

2. *$x \mapsto_R y$ implies $x \cup z \mapsto_R y$*

3. *$x \mapsto_R y$ and $y \mapsto_R z$ imply $x \mapsto_R z$*

4. *$x \mapsto_R y$ and $x \mapsto_R z$ imply $x \mapsto_R y \cup z$*

Proof. From $y \preceq x$ it follows that $C_R(y) \preceq C_R(x)$ i.e. $x \mapsto_R y$ proving 1.

From $x \mapsto_R y$ i.e. $C_R(y) \preceq C_R(x)$ it follows that $C_R(y) \preceq C_R(x \cup z)$ i.e. $x \cup z \mapsto_R y$ proving 2.

The property 3 follows immediately by transitivity of the closure operator C_R.

Finally, $x \mapsto_R y$ and $x \mapsto_R z$ imply $C_R(y) \preceq C_R(x), C_r(z) \preceq C_R(x)$ hence $y \preceq C_R(y), z \preceq C_R(z)$ imply $y \cup z \preceq C_R(x)$ whence $C_R(y \cup z) \preceq C_R(C_R(x)) = C_R(x)$ i.e. $x \mapsto_R y \cup z$ which proves 4. $\qquad\square$

As with functional dependence, the following properties of $\mapsto_R$ are implied by properties 1–4.

Proposition 10.18. *The following properties of $\mapsto_R$ are valid*

5. $x \mapsto_R x$
6. *If $x \mapsto_R y$ then $x \cup z \mapsto_R y \cup z$*
7. *If $x \mapsto_R y$ and $z \preceq y$ then $x \mapsto_R z$*
8. *If $x \mapsto_R y$ and $y \cup z \mapsto_R w$ then $x \cup z \mapsto_R w$*
9. *If $x \mapsto_R y \cup z$ and $z \mapsto_R w$ then $x \mapsto_R y \cup z \cup w$*
10. *If $x \mapsto_R y$ and $z \mapsto_R w$ then $x \cup z \mapsto_R y \cup w$*

We observe some other properties of the dependence relation $\mapsto_R$. The property (13) characterizes reducts of any x as minimal elements on which x depends.

Proposition 10.19. *11. xRy if and only if $x \mapsto_R y$ and $y \mapsto_R x$*
12. The relation $\mapsto_R$ is reflexive and transitive i.e. a pre–ordering on S
13. $y \in RED(x, R)$ if and only if y is minimal in the set $\{z : z \preceq x \wedge z \mapsto_R x\}$

Proof. xRy if and only if $C_R(x) = C_R(y)$ if and only if $x \mapsto_R y$ and $y \mapsto_R x$ so (11) is verified.

(12) follows by properties 1 and 3 of $\mapsto_R$ and if $y \in RED(x, R)$ then yRx hence $y \mapsto_R x$. Was $z \preceq x$, $z \prec y$ with $z \mapsto_R x$ we would have $C_R(x) \preceq C_R(z)$, $C_R(z) \preceq C_R(x)$ hence $C_R(z) = C_R(y) = C_R(z)$ and thus zRx, contradicting the minimality of y. Thus the first half of (13) is proved.

If y is minimal in the set $\{z : z \preceq x \wedge z \mapsto_R x\}$ then clearly $C_R(x) = C_R(y)$ hence yRx and thus $y \in RED(x, R)$ proving (13). $\qquad\square$

We close our discussion of dependence relations with a characterization of those relations.

Proposition 10.20. *If ρ is a relation in a semi–lattice $(S, \cup)$ which satisfies properties 1–4 of Proposition 10.8 then there exists a congruence R on S with the property that $\rho = \mapsto_R$.*

Proof. We let $R = \rho \cap \rho^{-1}$; then clearly R is an equivalence. For xRy, uRv, we have by property (10) which results from 1–4 that $x \cup uRy \cup v$ i.e. R is a congruence on S. Assume $x \mapsto_R y$ i.e. $C_R(y) \preceq C_R(x)$ whence $C_R(x)\rho C_R(y)$ by property (2) and $xRC_R(x), yRC_R(y)$ imply

$$x\rho C_R(x), C_R(y)\rho y$$

so finally it follows via property (3) that $x\rho y$.

Assume now that $x\rho y$; as

$$xRC_R(x), yRC_R(y)$$

and $R = \rho \cap \rho^{-1}$ we have by property (3) that $C_R(x)\rho C_R(y)$ hence by property (4) we arrive at

$$C_R(x)\rho C_R(x) \cup C_R(y) \tag{10.27}$$

Similarly, by property (2), we get

$$C_R(x) \cup C_R(y)\rho C_R(x) \tag{10.28}$$

i.e.

$$C_R(x)\rho^{-1} C_R(x) \cup C_R(y)$$

and thus finally

$$C_R(x)RC_R(x) \cup C_R(y) \tag{10.29}$$

which implies that

$$C_R(x) \cup C_R(y) \preceq C_R(x) \tag{10.30}$$

i.e. $C_R(y) \preceq C_R(x)$ which means that $x \mapsto_R y$ concluding the proof.   □

10.2.6   Interpretation in information systems

For an information system (U, A), we consider the semi–lattice $(2^A, \cup)$ with the congruence $R = R_h$ where $h : 2^A \to 2^{2^U}$ is the homomorphism $h(X) = IND_X$ where IND_X is the X–indiscernibility relation defined as $uIND_Xv$ if and only if $a(u) = a(v)$ for every $a \in X$. Thus XRY if and only if $IND_X = IND_Y$ for $X, Y \subseteq A$.

We examine the results presented above in this particular context. Let us observe that a reduct Y of X is a minimal element in the set $\{Z \subseteq A : IND_Z = IND_X\}$. For the notion of an R – almost 0 element, we have the following: a set $Y \subseteq A$ is R–almost 0 if and only if $IND_Y = IND_0 = U \times U$. In case no attribute $a \in A$ is constant, i.e. $|V_a| \geq 2$ for every $a \in A$, we get the following: if $Y \subseteq A$ is R–almost zero then $Y = 0 = \emptyset$.

For the notion of a sub–reduct, if $Y \subseteq X$ is in $SUBRED(X, R)$ then Y is minimal in its class $[Y]_R$ hence Y is in $RED(Z, R)$ for any $Z \supseteq Y$, $IND_Z = IND_Y$; hence the set $SUBRED(X, R)$ coincides with the set $\{Y : Y \subseteq X \wedge \exists Z.Y \in RED(Z, R)$. Proposition 10.16 yields in this particular case the following

Proposition 10.21. *Assume no attribute is constant. Then each $0 \neq X \subseteq A$ is the union of all reducts it does contain.*

Concerning dependency, we have $X \mapsto_R Y$ in case $C_R(Y) \subseteq C_R(X)$ i.e. in case $IND_X = IND_{C_R(X)} \subseteq IND_{C_R(Y)} = IND_Y$. Clearly, the condition $IND_X \subseteq IND_Y$ is equivalent to the condition $\{(u, v) : uIND_X v\} \subseteq \{(u, v) : uIND_Y v\}$ which in turn is equivalent to the condition $\{(u, v) : \forall a \in X.a(u) = a(v)\} \subseteq \{(u, v) : \forall a \in Y.a(u) = a(v)\}$ and this is equivalent finally to the condition $\forall u \in U. \bigcap_{a \in X} a^{-1}(a(u)) \subseteq \bigcap_{a \in Y} a^{-1}(a(u))$. It follows that

Proposition 10.22. *For $X, Y \subseteq A$, the dependence $X \mapsto_R Y$ holds if and only if the set of attributes Y depends functionally on the set of attributes X.*

Combining this last result with the characterization of a reduct cf. Proposition 10.19, we obtain the following

Proposition 10.23. *Each reduct X of A determines functionally every subset $Y \subseteq A$.*

Indeed, it follows from (13) in Proposition 10.19 that $X \mapsto_R A$ and then property (1) of the relation $\mapsto_R$ implies that $X \mapsto_R Y$ for every $Y \subseteq A$.

10.3 Classification/approximation spaces

Consider a semi–lattice $(2^A, \cup)$ with a relation R of equivalence. The triple $(2^A, \cup, R)$ is called a *classification space*. The semi–lattice 2^A is also endowed with operations of the intersection $\cap$ and of complement – which allow for a Boolean algebra (field of sets) structure. Given a classification space, we may introduce a *closure operator* on 2^A which we denote with the symbol C^R by letting

$$(CR) \; C^R(X) = \bigcup \{[Y]_R : [Y]_R \cap X \neq \emptyset\}.$$

It is easy to check that C^R is a closure operator i.e. it does satisfy properties (C1)–(C3). Indeed, $X \subseteq C^R(X)$ follows by the fact that classes $[Y]_R$ form a partition of the set A; that $C^R(Y) \subseteq C^R(X)$ whenever $Y \subseteq X$ is also evident. Finally, $C^R(C^R(X)) = C^R(X)$, again by the fact that classes $[Y]_R$ form a partition.

The closure operator C^R induces on 2^A a congruence R_{C^R} defined via the formula

$$XR_{C^R}Y \Leftrightarrow C^R(X) = C^R(Y) \tag{10.31}$$

The congruence R_{C^R} is called also the rough top equality and we will denote it with the symbol $\simeq$. It follows from the definition of the closure C^R and of congruence R_{C^R} that

Proposition 10.24.   *1.* $C^R(X \cup Y) = C^R(X) \cup C^R(Y)$

  2. $C^R(X \cap Y) \subseteq C^R(X) \cap C^R(Y)$

  3. $C^R(X) \setminus C^R(Y) \subseteq C^R(X \setminus Y)$

  4. $X \simeq C^R(X)$

  5. $X \simeq X$

  6. $X \simeq Y$ and $Y \simeq Z$ imply $X \simeq Z$

  7. $X \simeq Y$ and $Z \simeq W$ imply $X \cup Z \simeq Y \cup W$

  8. $C^R(A \setminus C^R(X)) = A \setminus C^R(X)$

Let us observe that

Proposition 10.25. *The set* $\mathcal{F}_C = \{X \subseteq A : X = C^R(X)\}$ *of* C^R*-closed sets is a field of sets containing* $\emptyset, A$.

Proof. That $C^R(\emptyset) = \emptyset, C^R(A) = A$ is clear by definition. From property (1) it follows that $X, Y \in \mathcal{F}_C$ imply $X \cup Y \in \mathcal{F}_C$. Assume that $C^R(X) = X$ and $C^R(Y) = Y$. Then by property (2) above and property (C1) we have

$$C^R(X \cap Y) \subseteq C^R(X) \cap C^R(Y) \subseteq X \cap Y \subseteq C^R(X \cap Y) \tag{10.32}$$

and thus $C^R(X \cap Y) = X \cap Y$ i.e. $X \cap Y \in \mathcal{F}_C$. $\square$

We now consider X, Y with $C^R(X) = X$ and $C^R(Y) = Y$ along with the difference $X \setminus Y$. We have

$$(i)\ (X \setminus Y) \cap C^R(Y) = \emptyset$$

$$(ii)\ (X \setminus Y) \cup C^R(Y) = X$$

and (ii) implies that

$$(iii)\ C^R(X \setminus Y) \cup C^R(Y) = C^R(X)$$

while it follows from (i) that

$$(iv)\ C^R(X \setminus Y) \cap C^R(Y) = \emptyset.$$

Combining (iii) and (iv), we get

$$C^R(X \setminus Y) = C^R(X) \setminus C^R(Y) = X \setminus Y$$

i.e. $X \setminus Y \in \mathcal{F}_C$.

This fact that $\mathcal{F}_C$ is a field of sets is related canonically to rough top equality. Given a dependence space (S, R) with $S = 2^A$ we may ask when R is a rough top equality congruence i.e. when there exists an equivalence ρ on A with the property that $R = R_{C^\rho}$. We have the following answer to this question

Proposition 10.26. $R = R_{C^\rho}$ *for an equivalence relation ρ on A if and only if the set $\mathcal{F}_R$ of C_R-closed sets is a field of sets containing $\emptyset, A$.*

Proof. The *only if* half of the proposed equivalence has been proved already so we may assume that the set $\mathcal{F}_R$ is a field of sets containing $\emptyset, A$.

As A is a finite set, the Boolean algebra $\mathcal{F}_R$ is atomic and thus each of its elements is a union of atoms of this field of sets. In particular, the set A is a union of all those atoms, i.e. it is partitioned by those atoms. Denoting by the symbol ρ the equivalence relation whose classes coincide with the atoms of the field of sets $\mathcal{F}_R$ we have that each C_R-closed set is a union of some atoms i.e. classes of ρ hence it is C^ρ-closed and it follows that XRY if and only if $C_R(X) = C_R(Y)$ if and only if $C^\rho(X) = C^\rho(Y)$ if and only if $X \simeq Y$. Thus R coincides with the rough top equality induced by ρ. $\qquad\square$

In a classification space $(2^A, R)$, the closure operator C^R does induce the dual operator D^R via the formula

$$(DR) \quad D^R(X) = A \setminus C^R(A \setminus X) \tag{10.33}$$

Then the operator D^R is an *interior operator* on 2^A and we have the following properties which follow from the definition (DR) and corresponding properties of C^R.

Proposition 10.27. *1.* $D_R(X) = \bigcup \{[Y]_R : [Y]_R \subseteq X\}$

2. $X = D^R(X)$ *if and only if* $X = C^R(X)$

3. $D^R(D^R(X)) = D^R(X); D^R(C^R(X)) = C^R(X); C^R(D^R(X)) = D^R(X)$

4. $X \subseteq Y$ *implies* $D^R(X) \subseteq D^R(Y)$

5. $D^R(X \cup Y) \subseteq D^R(X) \cup D^R(Y)$

6. $D^R(X \cap Y) = D^R(X) \cap D^R(Y)$

7. $D^R(X \setminus Y) \subseteq D^R(X) \setminus D^R(Y)$

8. $D^R(X) \subseteq C^R(X)$

By means of D^R we may introduce the dual to rough top equality relation of *rough bottom equality* R_{D^R} which we denote with the symbol $\sim$ viz. we let $X \sim Y$ if and only if $D^R(X) = D^R(Y)$. The following properties of the relation $\sim$ follow from Proposition 10.24 by duality and above properties.

Proposition 10.28.   *1.* $X \sim D^R(X)$

   2. $X \sim X$

   3. $X \sim Y$ and $Y \sim Z$ imply $X \sim Z$

   4. $X \sim Y$ and $Z \sim W$ imply $X \cap Z \sim Y \cap W$

   5. $D^R(A \setminus D^R(X)) = A \setminus D^R(X)$

The property 4. does express the fact that D^R is a congruence on the dual semi–lattice $(2^A, \cap)$.

We say that a set X is D^R–*open* if and only if $X = D^R(X)$. As each set X is C^R–closed if and only if it is D^R–open it follows that D^R–open sets form a field of sets.

Conversely, if for a relation R, the family of D^R–open sets is a field of sets then by Proposition 10.26, $R^c(X) = R_{C^\rho(U \setminus X)}$ is a rough top equality hence $R(X) = R^c(U \setminus X) = R_{D^\rho}$ is a rough bottom equality induced by ρ.

Finally, we may combine rough top and rough bottom equalities into a single notion of *rough equality*. To this end, we let

$$E^R(X) = R_{C^R}(X) \cap R_{D^R}(X)$$

for each $X \subseteq A$. It thus follows that

$$X E^R Y \Leftrightarrow C^R(X) = C^R(Y) \wedge D^R(X) = D^R(Y) \qquad (10.34)$$

We use the symbol $\equiv$ for rough equality. We may easily check that

Proposition 10.29.   *1.* $X \equiv X, X \equiv C_R(X), X \equiv D^R(X)$

   2. $X \subseteq Y \subseteq Z$ and $X \equiv Z$ imply $X \equiv Y \equiv Z$

   3. $X \equiv Y, Y \equiv Z$ imply $X \equiv Z$

In all above cases, it is instrumental that the set $\mathcal{F}_C$ of C^R–closed sets is a field of sets. For this reason, a special name is reserved for a pair $(S, \mathcal{F})$ of a semi–lattice S along with a field of sets of its subsets, containing $\emptyset, S$ viz. such a pair is called an *approximation space*. An archetypical example of approximation space is the universe U of an information system (U, A) along with the field of IND_X–closed sets induced by a set X of attributes.

10.3.1   Approximation spaces of an information system

We consider an information system $\mathcal{A} = (U, A)$ with the two induced semi–lattices: $(2^A, \cup)$ and $(2^U, \cap)$. The homomorphism $h_A : 2^A \to 2^{2^{2^U}}$ defined by means of the formula $h_A(B) = IND_B$ where IND_B is a relation on U defined by already well–familiar formula

$$u \, IND_B \, v \Leftrightarrow \forall a \in B(a(u) = a(v))$$

transforms sets in 2^A into indiscernibility relations. In this way, for each $B \subseteq A$, we assign to B the pair (U, IND_B) as a corresponding approximation space. The resulting closure operator C^{IND_B} assigns to each subset $X \subseteq U$, its closure

$$C^{IND_B}(X) = \bigcup \{ [Y]_{IND_B} : [Y]_{IND_B} \cap X \neq \emptyset \} \qquad (10.35)$$

In the context of rough set theory of information systems this closure is denoted with the symbol $\overline{B}(X)$ and it is called the B–upper approximation to X. In the light of our earlier discussion, the rough top equality becomes the equality of upper approximations i.e. $X \simeq Y \Leftrightarrow \overline{B}(X) = \overline{B}(Y)$. Clearly, the upper approximation operator has all properties of the closure operator C^R introduced earlier.

The dual operator of interior may be introduced in the similar way to that in which the interior operator D^R has been defined by means of C^R viz. we let

$$(LA) \ \underline{B}(X) = U \setminus \overline{B}(U \setminus X).$$

Then the explicit characterization of $\underline{B}$ may be extracted in an analogous way i.e.

$$\underline{B}(X) = \bigcup \{ [Y]_{IND_B} : [Y]_{IND_B} \subseteq X \qquad (10.36)$$

We collect here the basic properties of approximation operators $\underline{B}, \overline{B}$ which follow from definitions.

Proposition 10.30. 1. $\underline{B}\overline{B}(X)) = \overline{B}(X), \overline{B}\underline{B}(X)) = \underline{B}(X)$
2. $\underline{B}\underline{B}(X)) = \underline{B}(X), \overline{B}\overline{B}(X)) = \overline{B}(X)$
3. $\underline{B}(X) \subseteq X \subseteq \overline{B}(X)$
4. $\underline{B}(X) \subseteq Z \subseteq X$ imply $\underline{B}Z = \underline{B}(X)$
5. $X \subseteq Z \subseteq \overline{B}(X)$ imply $\overline{B}Z = \overline{B}(X)$
6. $X \subseteq Y$ implies $\underline{B}(X) \subseteq \underline{B}(Y)$ and $\overline{B}(X) \subseteq \overline{B}(Y)$
7. $\underline{B}(X \cap Y) = \underline{B}(X) \cap \underline{B}(Y)$
8. $\underline{B}(X) \cup \underline{B}(Y) \subseteq \underline{B}(X \cup Y)$
9. $\overline{B}(X \cup Y) = \overline{B}(X) \cup \overline{B}(Y)$
10. $\overline{B}(X \cap Y) \subseteq \overline{B}(X) \cap \overline{B}(Y)$

11. $\underline{B}(X \setminus Y) \subseteq \underline{B}(X) \setminus \underline{B}(Y)$
12. $\overline{B}(X) \setminus \overline{B}(Y) \subseteq \overline{B}(X \setminus Y)$
13. $B \mapsto_{IND} C$ implies $\underline{C}(X) \subseteq \underline{B}(X), \overline{B}(X) \subseteq \overline{C}(X)$
14. $\underline{B}(X) \subseteq \overline{B}(X)$

Proof. Indeed, properties (1,2,3,6,7,9,10,11,12) paraphrase the corresponding general properties of closure, respectively, interior operators; after all, they may be proved easily from definitions of lower- and upper- approximations.

Properties (4),(5) follow from monotonicity (6) and idempotency (1,2) of approximation operators. Property (13) follows from the fact that in case a set C of attributes depends functionally with respect to IND on a set B of attributes we have $IND_B \subseteq IND_C$ which results in inclusions (13).

Definable sets, non–definable sets (rough sets)

Following our discussion of approximations, as well as our discussion of rough equality relation

$$E^R(X) = R_{C^R}(X) \cap R_{D^R}(X)$$

for each $X \subseteq U$, denoted with the symbol $X \equiv Y$, we assign to each subset $X \subseteq U$ the pair $(C^R(X), D^R(X))$. Let us observe that in case $C^R(X) = D^R(X)$ we have $X = C^R(X)$ by property 14. and thus from $X \equiv Y$ it follows that $X = C^R(X) = C^R(Y) = D^R(Y) = Y$ hence $X = Y$ i.e. we have

$$[X]_{E^R} = \{X\}.$$

We will say that a set $X \subseteq U$ is *definable* if and only if $[X]_{E^R} = \{X\}$, otherwise we will say that X is *non–definable* or *rough*. We know that definable sets form a Boolean algebra under set – theoretical operators of $\cup, \cap, \setminus$ as well as the field of clopen sets in the topology generated by indiscernibility classes. On the other hand, rough sets may not be given a structure in so simple a way and we postpone the discussion of relevant structures for them to Chapter 11.

Let us observe here that a rough set X does invoke a pair (C, D) of definable sets with the properties that $C = C^R(X), D = D^R(X)$. The problem arises of how to characterize those pairs (C, D) which are induced by rough sets.

Clearly, the requirements that $C \neq D$ and $C \subseteq D$ pose itself. After a deeper reflection, we find that as D is the closure $D^R(X)$, no equivalence class contained in $D^R(X) \setminus C^R(X)$ may be a singleton: indeed, was a singleton $\{x\} = [x]_R \subseteq D^R(X) \setminus C^R(X)$ we would have $\{x\} \cap X \neq \emptyset$ hence $x \in X$ and thus $\{x\} \subseteq C^R(X)$, a contradiction. We denote with $B^R(X)$ the difference $D^R(X) \setminus C^R(X)$ and we let

$$\mathcal{B}(X) = \{[x]_R : [x]_R \subseteq D^R(X) \setminus C^R(X)\} \qquad (10.37)$$

Thus, we arrive at

Proposition 10.31. *A pair* (C, D) *of definable sets does characterize a rough set* X *i.e.* $C = C^R(X), D = D^R(X)$ *if and only if the following hold*
1. $C \neq D$
2. $C \subseteq D$
3. *if* $[x]_R \in \mathcal{B}(X)$ *then cardinality of* $[x]_R$ *is at least 2.*

Proof. Only the sufficiency requires a proof, so we assume that definable sets C, D satisfy the requirements (1)–(3). By (3), we may choose $y(C) \in C$ for each equivalence class $C \in \mathcal{B}(X)$ and we let

$$X = C \cup \{y(C) : C \in \mathcal{B}(X).$$

Then clearly, $C = C^R(X)$ and $D = D^R(X)$. □

The set $B^R(X)$ is called the *boundary set* of X. The sets $C^R(X)$, $D^R(X), B^R(X)$ have an informal interpretation in rough set theory viz. any $x \in C^R(X)$ is said to be *with certainty (certainly)* in X (in the sense that as $[x]_R \subseteq X$ we cannot find any object identical to x not in X); similarly, any $x \in D^R(X)$ is said to be *possibly* in X as we can find in X an object identical to X. The boundary set $B^R(X)$ collects all objects for which it is neither certain that $x \in X$ nor it is certain that $x \in U \setminus X$. A further discussion of this notion with fuzzy sets in the play is given in Chapter 14.

10.4 Partial dependence

The idea of Lukasiewicz to assign probabilities to formulae of logical calculus (cf. Chapter 9) may be invoked when we would like to discuss pairs (C, D) of attributes for which no functional dependence is present. In this case we may still want to bound these two sets by means of a weaker kind of dependence, "close to functional" or "partial" (after [Novotn'y–Pawlak88^a]). In order to introduce this kind of dependence notion, we, as with functional dependence, first associate to each subset $B \subseteq A$ the indiscernibility relation IND_B.

In the subset $\mathbf{EQ} \subseteq 2^{2^{U \times U}}$ of equivalence relations on the set U, we introduce for each $P \in \mathbf{EQ}$ a homomorphism $POS_P : \mathbf{EQ} \to 2^U$ defined by means of the formula

$$(POS) \quad POS_P(Q) = \bigcup \{[u]_P : [u]_P \subseteq [u]_Q\}$$

where $[u]_P$ denotes the equivalence class of $u \in U$ with respect to P. It follows that $POS_P(Q)$ is the maximal subset of U on which the restriction to it of Q depends functionally on the restriction to it of P i.e.

$$P|POS_P(Q) \times POS_P(Q) \mapsto_R Q|POS_P(Q) \times POS_P(Q) \qquad (10.38)$$

That POS_P is a homomorphism follows from the following:

$$POS_P(Q \cap T) = POS_P(Q) \cap POS_P(T) \qquad (10.39)$$

Indeed, for $u \in U$ we have $u \in POS_P(Q \cap T)$ if and only if $[u]_P \subseteq [u]_{Q \cap T}$ if and only if $[u]_P \subseteq [u]_Q \cap [u]_T$ if and only if $[u]_P \subseteq [u]_Q \wedge [u]_P \subseteq [u]_Y T$ if and only if $u \in POS_P(Q) \cap POS_P(T)$.

We state two more properties of the homomorphism POS

Proposition 10.32. *1. $POS_P(Q) = U$ if and only if $P \mapsto_R Q$*

2. $POS_P(Q) \cap POS_T(Q) \subseteq POS_{P \cap T}(Q)$

Both 1., 2., follow directly from definitions.

We now invoke the idea of Łukasiewicz; to this end, we will regard formally sets $A, B, \ldots$ of attributes as elementary formulae and implications of the form $B \mapsto C$ as formulae of a logical calculus. The semantic interpretation of the formulae will be realized in the set U and we accept the following convention that $u \models B$ for each B and every $u \in U$. Then we let that $u \models B \mapsto C$ if and only if $[u]_B \subseteq [u]_C$ and assign to each formula $\alpha : B \mapsto C$ its weight $w(\alpha) = |U|^{-1} \cdot |POS_B(C)|$; thus $w(\alpha)$ is the measure of the fraction of objects that satisfy both B, C relative to B.

In case $w(B \mapsto C) \geq r$ we will say that C depends on B to *degree r* and we will use the symbol $\mapsto_{R,r}$ to denote that fact. General properties of w established in Chapter 9 hold in this case so now we will establish general properties related to transitivity properties of the relation $\mapsto_{R,r}$ of partial dependence.

The reader may consult [Pawlak–Skowron94] for an extension of the basic idea to a more general one involving *rough membership functions* (cf. Chapter 1).

First, we will find a formula for expressing t in the inference scheme

$$\frac{B \mapsto_{R,r} C, C \mapsto_{R,s} D}{B \mapsto_{R,t} D} \qquad (10.40)$$

Proposition 10.33. *The following inference scheme holds*

$$(L) \quad \frac{B \mapsto_{R,r} C, C \mapsto_{R,s} D}{B \mapsto_{R,\max\{0,r+s-1\}} D}.$$

Proof. We have

$$[u]_B \subseteq [u]_C \wedge [u]_C \subseteq [u]_D \Rightarrow [u]_B \subseteq [u]_D \qquad (10.41)$$

hence if it was not

$$[u]_B \subseteq [u]_D$$

then we would have either not

$$[u]_B \subseteq [u]_C$$

or not

$$[u]_C \subseteq [u]_D.$$

In the worst case, we then have

$$1 - t \le 1 - r + 1 - s \qquad (10.42)$$

which implies $t \ge r + s - 1$ and as t is non-negative we have finally

$$t \ge max\{0, r + s - 1\} \qquad (10.43)$$

$\square$

Similar arguments validate the following

Proposition 10.34. *1.* $\dfrac{B \mapsto_{R,r} C, D \mapsto_{R,s} C}{B \cup D \mapsto_{R,max\{r,s\}} C}$

2. $\dfrac{B \mapsto_{R,r} C, B \mapsto_{R,s} D}{B \mapsto_{R,min\{r,s\}} C \cup D}$

3. $\dfrac{B \mapsto_{R,r} C, D \mapsto_{R,s} E}{B \cup D \mapsto_{R,max\{0,r+s-1\}} C \cup E}$

We would like to observe that

Proposition 10.35. $B \mapsto_{R,1} C$ *if and only if* $B \mapsto_R C$.

The reader will without doubt recognize, after reading Chapter 13, in the above formulae for partial dependence the *Lukasiewicz multiplication* $\otimes(x, y) = max\{0, x + y - 1\}$. We will see its important role yet in Chapters 13 and 14.

Historic remarks

Notions of independence as well as dependence abound in many areas of mathematics; for a general survey and a discussion of those notions see [Marczewski58] or [Głazek79]. Dependence, reducts and their applications toward induction of decision rules are discussed in [Pawlak82], [Pawlak85^a], [Pawlak91] cf. also [Buszkowski–Orłowska86], [Orłowska83], and [Grzymala–Busse86]. Dependence in information systems as well as rough top as well as bottom equalities were discussed in [Novotný–Pawlak opera cit.]. A survey of dependence spaces and rough equalities is given in [Novotný98^{a,b}]; [Novotný83] as well as [Novotný–Pawlak83] address the problem of finding information systems whose dependence spaces are isomorphic to a given one. For algebraic as well as logical analysis of dependencies in particular functional dependencies see also [Rauszer opera cit.]. For functional dependencies see also [Luxenburger98]. Algoritmic aspects of reduct generation are addressed in [Skowron–Rauszer92] and various ramifications of this notion are analyzed in [Bazan et al.00], [Stepaniuk00], [Ślęzak00].

Exercises

In Exercises 1–5, we are concerned with the notion of a reduct and notions related to it. The exercises are based on results of [Novotný–Pawlak opera cit.] and [Novotný98^{a,b}].

1. For a dependence space (A, K), and $X \subseteq A$, one says that $Y \subseteq A$ is a K-*superreduct* of X if and only if Y is a minimal set in the class $\mathcal{Z}$ of sets such that $Z \in \mathcal{Z}$ if and only if $Z \subseteq X$ and $\forall T \subseteq X \exists W \subseteq Z(W, T) \in K$. Prove that if Y is a superreduct of X then $(Y, X) \in K$.

2. For a dependence space (A, K), a relation K^* is defined as follows : $(X, Y) \in K$ if and only if $\forall T \subseteq X \exists W \subseteq Y(W, T) \in K$ and $\forall U \subseteq Y \exists V \subseteq Z(U, V) \in K$. Prove that K^* is a congruence i.e. (A, K^*) is a dependence space.

3. In the notation of Exercise 2, prove that $K^* \subseteq K$.

4. Prove that, given a dependence space (A, K), Y is a superreduct of X if and only if Y is a K^*-reduct of X.

5. Prove for a given dependence space (A, K) that any superreduct of X contains a K-reduct of X.

In Exercises 6–8, we refer to a particular dependence space (A, K_I) where (U, A) is an information system and the congruence K is defined via $(X, Y) \in K_I$ if and only if $Ind_X = Ind_Y$.

6. Give an example of an information system (U, A) and a subset $Y \subseteq A$ such that Y is a reduct of A and Y is not any superreduct of A.

7. [Novotný98^b] A subset $Q \subseteq U$ is said to be *distinguishing* for the dependence space (A, K_I) if and only if given any X, Y with $(X, Y) \notin K_I$ there exist $y_1, y_2 \in Y$ with the property that $(y_1, y_2) \in IND_X \setminus IND_Y \cup IND_Y \setminus IND_X$. Assume that cardinality of the quotient set $2^A / K_I$ is $q > 1$ and prove that there exists a distinguishing subset of U of cardinality less or equal to $2q^2$. [Hint: choose a selector S on the family $2^A / K_I$ and for any $X, Y \in S$ with $X \neq Y$ select a pair $y_1(X, Y), y_2(X, Y)$ witnessing that $IND_X \neq IND_Y$]

8. [Novotný83], [Novotný–Pawlak83] Given a dependence space (A, K) prove that there exists an information system (U, A) such that the dependence space (A, K_I) is isomorphic to (A, K). [Hint: by the result of Exercise 7,

select a set U of cardinality at least $2q^2$ where q is the cardinality of the quotient set $2^A/K$ and choose in each class $[X]_K$ of K the greatest element $C(K)(X)$. Let $C(K)$ denote the set of chosen elements. Then select a surjection $f : U \times U \to C(K)$ of the product $U \times U$ onto $C(K)$ and for each $X \subseteq A$ let $(x, y) \in IND_X$ if and only if $X \subseteq f(x, y)$]

9. Apply the result of Exercise 8 in order to construct an information system (W, U) whose dependence space is isomorphic to the dependence space $(U, \approx)$ of the rough top equality.

10. Dualize results of Exercise 9 in order to construct an information system (W, U) whose dependence space is isomorphic to the dependence space $(U, \simeq)$ of the rough bottom equality.

In Exercises 11–12, we discuss partial dependencies.

11. Paraphrase the coefficient $w(\alpha) = |U|^{-1} \cdot |POS_B(C)|$ defined for $\alpha : B \mapsto C$ as $w(B, C)$ and prove the following (1) $X \subseteq Y$ implies $w(X, Z) \leq w(Y, Z)$ (2) $w(X \cup Y, Z) \geq max\{w(X, Z), w(Y, Z)$ (3) $w(X, X) = 1$ (4) $w(X, Y) = 1 \Leftrightarrow X \mapsto_R Y$.

12. [Novotný98^b] Prove that the function $\rho(X, Y) = 1 - \frac{1}{2} \cdot (w(X, Y) + w(Y, X))$ is a metric on the quotient space $2^A/K_I$ i.e. the following hold: (1) $\rho(X, Y) = 0 \Leftrightarrow X = Y$ (2) $\rho(X, Y) = \rho(Y, X)$ (3) $\rho(X, Z) \leq \rho(X, Y) + \rho(Y, Z)$ where X, Y, Z are classes of K_I.

In Exercises 13–15, we address the subject of generalized notions of a reduct. For a detailed discussion of the current state of this topic see [Ślęzak00].

13. Generalize the notion of the homomorphism POS by considering for given disjoint subsets $B, C \subseteq A$ the function $\mu_{B,C} : U \to [0, 1]$ defined as follows: $\mu_{B,C}(u) = |POS_B([u]_{IND_C})| \cdot (|[u]_{IND_C}|)^{-1}$ for $u \in U$. In particular prove that (1) $\mu_{B_1 \cup B_2}(C) \geq max\{\mu_{B_1}(C), \mu_{B_2}(C)\} \geq \frac{1}{2} \cdot (\mu_{B_1}(C) + \mu_{B_2}(C))$.

14. In the notation of Exercise 13, define a μ–reduct of $C \subseteq A$ as a minimal with respect to inclusion subset $B \subseteq A \setminus C$ with the property that $\mu_B(C) = \mu_{A \setminus C}(C)$. Prove that if B is a μ–reduct of C and $D \subseteq A \setminus C$ then $\mu_{B \cup D}(C) = \mu_B(C)$.

15. In the notation as above, consider the *entropy function*

$$h_{B,C} = - \sum_{u \in U} \mu_{B,C} log \mu_{B,C}.$$

Prove that $h_{B,C}$ does satisfy (1) in Exercise 13 in place of μ. Define an *entropy* reduct of C as a minimal subset $B \subseteq A \setminus C$ with the property that

$h_{B,C} = h_{A\setminus C,C}$ and prove that if B is an entropy reduct of C and $D \subseteq A \setminus C$ then $hB \cup D(C) = hB(C)$.

16. Generalize results of Exercises 13–15 to the case of a function $\phi_{B,C} : U \to [0,1]$, where $B, C \subseteq A$ are disjoint, with the properties (i) $\phi_{B,C}$ is constant on each class $[u]_{IND_C}$ (ii) $\phi_{B,C}$ does satisfy condition (1) of Exercise 13 in place of μ. In particular, define a ϕ–reduct of $C \subseteq A$ as a minimal subset $B \subseteq A \setminus C$ with the property that $\phi_{B,C} = \phi_{A\setminus C,C}$ and prove that if B is a ϕ–reduct of C and $D \subseteq A \setminus C$ then $\phi_{B\cup D,C} = \phi_{B,C}$.

In Exercises 17–26, we return to functional dependencies in information systems and we propose to verify some basic facts about functional dependencies among sets of attributes by means of algebraic structures; we use these facts in Chapter 12. In particular theses in Exercise 25, 26 give an algebraic characterization of independence. The results presented here are due to [Pawlak81^a, 83] and [Pawlak–Rauszer85] (cf. also [Rauszer 84, 85^b]). We assume an information system $\mathcal{A} = (U, A)$ given; for a set $B \subseteq A$ of attributes, we consider the indiscernibility relation IND_B on the universe U. In what follows, $B, C, \dots$ stand for subsets of the attribute set A.

17. Verify that (i) $B \subseteq C$ implies $IND_C \subseteq IND_B$ (ii) $IND_{B\cup C} = IND_B \cap IND_C$.

18. A set $B \subseteq A$ is said to be *independent* if and only if there is no proper subset $C \subset B$ with $C \mapsto B$. Prove that a set B is independent if and only if there is no proper subset $C \subset B$ with $IND_C = IND_B$.

19. Verify the statement: if $B \subseteq A$ is independent then for each pair $C, D \subseteq B$ the following is true: if $IND_C \subseteq IND_D$ then $IND_D \subseteq IND_C$. [Hint: by Exercise 17(ii), $IND_{C\cup D} = IND_C$ and by Exercise 17(i), $IND_{C\cup D} \subseteq IND_D$]

20. Prove that if $B \subseteq A$ is independent then each $C \subseteq B$ is independent. [Hint: observe that by Exercise 19 we have $IND_C = IND_B$ and apply Exercise 18]

21. We say that $B \subseteq A$ is *dependent* in case B is not independent. A non–empty subset $C \subseteq B$ is said to be *redundant relative to B* if and only if $IND_B = IND_{B\setminus C}$. Prove that if B is dependent then there exists a redundant subset relative to B. [Hint: if B is dependent then there exists $D \subset B$ with $IND_B = IND_D$; consider $C = B \setminus D$]

22. For $B \subseteq A$, consider the set $\mathcal{I}_B = \{IND_C : C \subseteq B\}$ ordered via the set–theoretic inclusion. Clearly, $IND_\emptyset$ is the greatest and IND_B is the

least element in this set. Prove that the meet $\wedge$ exists in this ordered set and $IND_C \wedge IND_D = IND_{C \cup D}$. [Hint: apply Exercise 17(ii)]

23. Prove that if B is independent then the join $\vee$ exists in $\mathcal{I}_B$ and $IND_C \vee IND_D = IND_{C \cap D}$. [Hint: by Exercise 17(i), $IND_C, IND_D \leq IND_{C \cap D}$. If $IND_C, IND_D \leq IND_E$ then by Exercise 19, $IND_C = IND_E$, $IND_D = IND_E$ and by the same Exercise it follows that $IND_E = IND_{C \cap D}$]

24. The relative pseudocomplement $IND_C \Rightarrow IND_D$ in $\mathcal{I}_B$ is defined as usual, as the greatest IND_E with the property that $IND_C \cap IND_E \subseteq IND_D$. Prove that if B is independent then $IND_C \Rightarrow IND_D = IND_{(B \setminus C) \cap D}$ for each pair $C, D \subseteq B$. [Hint: if $IND_C \cap IND_E \subseteq IND_D$ then $D \subseteq C \cup E$ by independence of D (cf. Exercise 20) hence $(B \setminus C) \cap D \subseteq E$ i.e. $IND_E \subseteq IND_{(B \setminus C) \cap D}$. Verify directly that $IND_C \cap IND_{(B \setminus C) \cap D} \subseteq IND_D$ to conclude the proof]

25. Prove that if B is independent then $IND_{B \setminus C}$ is the complement $--$ IND_C to IND_C and IND_B is the null element in $\mathcal{I}_B$ hence $\mathcal{I}_B$ is a Boolean algebra of cardinality $2^{|B|}$. [Hint: apply the formula for the pseudo-complement of Exercise 24]

26. Prove the converse to the thesis of Exercise 25 i.e. if $\mathcal{I}_B$ is a Boolean algebra of cardinality $2^{|B|}$ then B is independent. [Hint: the assumption means that $IND_C \neq IND_B$ for each $C \subseteq B$]

Works quoted

[Bazan00] J. Bazan, H.S. Nguyen, S. H. Nguyen, P. Synak, and J. Wróblewski, *Rough set algorithms in classification problems*, in: [Polkowski–Tsumoto–Lin], pp. 49–88.

[Buszkowski–Orlowska86] W. Buszkowski and E. Orlowska, *On the logic of database dependencies*, Bull. Polish Acad. Sci. Math., 34(1986), pp. 345–354.

[Głazek79] K. Głazek, *Some old and new problems of independence in mathematics*, Coll. Math., 17(1979), pp. 127–189.

[Grzymala–Busse86] J. Grzymala–Busse, *On the reduction of knowledge representation systems*, in: Proceedings of the 6th Intern. Workshop on Expert Systems and Appl., Avignon, France, 1986, vol. 1, pp. 463–478.

[Luxenburger98] M. Luxenburger, *Dependencies between many–valued attributes*, in: [Orlowska 98], pp. 316–346.

[Marczewski58] E. Marczewski, *A general scheme of independence in mathematics*, Bull. Polish Acad. Math. Sci., 6(1958), pp. 731–736.

[Novotný98[a]] M. Novotný, *Dependence spaces of information systems*, in [Orłow-ska98], pp. 193–246.

[Novotný98[b]] M. Novotný, *Applications of dependence spaces*, in [Orłowska98], pp. 247–289.

[Novotný83] M. Novotný, *Remarks on sequents defined by means of information systems*, Fund. Inform., 6(1983), pp. 71–79.

[Novotný–Pawlak92] M. Novotný and Z. Pawlak, *On a problem concerning dependence spaces*, Fund. Inform., 16(1992), pp. 275–287.

[Novotný–Pawlak91] M. Novotný and Z. Pawlak, *Algebraic theory of independence in information systems*, Fund. Inform., 14(1991), pp. 454–476.

[Novotný–Pawlak90] M. Novotný and Z. Pawlak, *On superreducts*, Bull. Polish Acad. Sci. Tech., 38(1990), pp. 101–112.

[Novotný–Pawlak89] M. Novotný and Z. Pawlak, *Algebraic theory of independence in information systems*, Report 51, Institute of Mathematics of the Czechoslovak Academy of Sciences, 1989.

[Novotný–Pawlak88[a]] M. Novotný and Z. Pawlak, *Partial dependency of attributes*, Bull. Polish Acad. Sci. Math., 36(1988), pp. 453–458.

[Novotný–Pawlak88[b]] M. Novotný and Z. Pawlak, *Independence of attributes*, Bull. Polish Acad. Sci. Math., 36(1988), pp. 459–465.

[Novotný–Pawlak87] M. Novotný and Z. Pawlak, *Concept forming and black boxes*, Bull. Polish Acad. Sci. Math., 35(1987), pp. 133–141.

[Novotný–Pawlak85[a]] M. Novotný and Z. Pawlak, *Characterization of rough top equalities and rough bottom equalities*, Bull. Polish Acad. Sci. Math., 33(1985), pp. 91–97.

[Novotný–Pawlak85[b]] M. Novotný and Z. Pawlak, *On rough equalities*, Bull. Polish Acad. Sci.Math., 33(1985), pp. 99–104.

[Novotný–Pawlak85[c]] M. Novotný and Z. Pawlak, *Black box analysis and rough top equality*, Bull. Polish Acad. Sci. Math., 33(1985), pp. 105–113.

[Novotný–Pawlak85^d] M. Novotný and Z. Pawlak, *Independence of attributes*, Bull. Polish Acad. Sci. Tech., 33 (1985), pp. 459–465.

[Novotný–Pawlak83] M. Novotný and Z. Pawlak, *On a representation of rough sets by means of information systems*, Fund. Inform., 6(1983), pp. 289–296.

[Orłowska98] E. Orłowska, ed., *Incomplete Information: Rough Set Analysis*, Studies in Fuzzines and Soft Computing vol. 13, Physica Verlag, Heidelberg, 1998.

[Orłowska83] E. Orłowska, *Dependencies of attributes in Pawlak's information systems*, Fund. Inform., 6(1983), pp. 247–256.

[Pawlak91] Z. Pawlak, *Rough Sets. Theoretical Aspects of Reasoning about Data*, Kluwer, Dordrecht, 1991.

[Pawlak85^a] Z. Pawlak, *On rough dependency of attributes in information systems*, Bull. Polish Acad. Sci., Tech., 33(1985), pp. 551–559.

[Pawlak83] Z. Pawlak, *Rough classification*, Reports of the Computing Centre of the Polish Academy of Sciences, 506, Warsaw, 1983.

[Pawlak82^a] Z. Pawlak, *Rough sets*, Intern. J. Information and Computer Science, 11(1982), pp. 341–356.

[Pawlak81^a] Z. Pawlak, *Information Systems–Theoretical Foundations* (in Polish), PWN–Polish Scientific Publishers, Warsaw, 1981.

[Pawlak81^b] Z. Pawlak, *Information system theoretical foundations*, Inform. Systems, 6(1981), pp. 205–218.

[Pawlak–Rauszer85] Z. Pawlak and C. Rauszer, *Dependency of attributes in information systems*, Bull. Polish Acad. Sci. Math., 33(1985), pp.551–559.

[Pawlak–Skowron94] Z. Pawlak and A. Skowron, *Rough membership functions*, in: R.R. Yaeger, M. Fedrizzi, and J. Kacprzyk, eds., *Advances in the Dempster–Schafer Theory of Evidence*, Wiley, New York, 1994, pp. 251–271.

[Polkowski–Tsumoto–Lin00] L. Polkowski, S. Tsumoto, and T. Y. Lin, eds., *Rough Set Methods and Applications. New Developments in Knowledge Discovery in Information Systems*, Studies in Fuzzines and Soft Computing vol. 56, Physica Verlag, Heidelberg, 2000.

[Rauszer91] C. M. Rauszer, *Reducts in information systems*, Fund. Inform., 15(1991), pp. 1–12.

[Rauszer88] C. M. Rauszer, *Algebraic properties of functional dependencies*, Bull. Polish Acad. Sci. Math., 36(1988), pp. 561–569.

[Rauszer87] C. M. Rauszer, *Algebraic and logical description of functional and multi-valued dependencies*, Proceedings ISMIS'87, Charlotte, NC, North Holland, Amsterdam, 1987, pp. 145–155.

[Rauszer85^a] C. M. Rauszer, *Dependency of attributes in information systems*, Bull. Polish Acad. Sci. Math., 33(1985), pp. 551–559.

[Rauszer85^b] C. M. Rauszer, *An equivalence between theory of functional dependencies and a fragment of intuitionistic logic*, Bull. Polish Acad. Sci. Math., 33(1985), pp. 571–579.

[Rauszer84] C. M. Rauszer, *An equivalence between indiscernibility relations in information systems and a fragment of intuitionistic logic*, in: *Lecture Notes in Computer Science*, vol. 208, Springer Verlag, Berlin, 1984, pp. 298–317.

[Skowron–Rauszer92] A. Skowron and C. M. Rauszer, *The discernibility matrices and functions in information systems*, in: R. Słowiński, ed., *Intelligent Decision Support. Handbook of Applications and Advances of the Rough Set Theory*, Kluwer, Dordrecht, 1992, pp. 311–362.

[Stepaniuk00] J. Stepaniuk, *Knowledge discovery by application of rough set model*, in: [Polkowski–Tsumoto–Lin00], pp. 137–236.

[Ślęzak00] D. Ślęzak, *Various approaches to reasoning with frequency based decision reducts*, in: [Polkowski–Tsumoto–Lin00], pp. 235–288.

Chapter 11

Topology of Rough Sets

Those who wish to succeed must ask the right preliminary questions

Aristotle, *Metaphysics*, II

11.1 Introduction

We consider here rough sets that arise in an information system from the point of view of topology. It will be our aim to introduce topologies into the realm of rough sets and investigate their properties. We will be guided by facts from general topological theory exposed in Chapter 7.

We will restrict ourselves here to the most important in practice case of *countable* knowledge bases i.e. we will consider an information system $\mathcal{A} = (U, A)$ where A is a countable set of attributes, $A = \{a_n : n = 1, 2, ...\}$. We may assume that for the relation set $\mathcal{R} = \{R_n = Ind_{a_n} : n = 1, 2, ...\}$ we have $R_{n+1} \subseteq R_n$ for each n, replacing if necessary a_n with $(a_1, a_2, ..., a_n)$ whose value on an object x is $(a_1, ..., a_n)(x) = (a_1(x), ..., a_n(x))$.

We assume further that the family $\mathcal{R}$ does *separate points* of the universe U i.e. given $x \neq y$ in U, there exists an equivalence class $Q \in U/R_n$ for some n such that $x \in Q, y \notin Q$. This property may be assumed to hold, because in case it does not hold i.e. there are points $x \neq y$ for which $[x]_R = [y]_R$ where $R = \bigcap_n R_n$, we may consider instead of the universe U the quotient universe U/R which already has the desired separation property.

The family $\mathcal{R}$ does induce various topologies on the universe U. Simplest among them are topologies induced by each R_n separately. Given R_n, we denote by Π_n the topology obtained by taking the family $P_n = \{[x]_{R_n} : x \in U\}$ as an open base i.e. each open set W in the topology Π_n is a union of a sub–family $P \subseteq P_n$.

Following this idea, we define a topology Π_0 on the universe U by taking the family $\mathcal{P} = \{[x]_{R_n} : x \in X, n = 1, 2, ...\}$ of all equivalence classes of all relations R_n as an open base. Then, we define Π_0-*exact sets* and Π_0–*rough sets* as follows where Cl_{Π_0}, Int_{Π_0} is closure resp. interior with respect to Π_0

1. a set $Z \subseteq U$ is Π_0-exact if and only if $Int_{\Pi_0} Z = Cl_{\Pi_0} Z$

2. otherwise, Z is Π_0–rough i.e. in this case, $Int_{\Pi_0} Z \neq Cl_{\Pi_0} Z$

We will be interested now in Π_0–rough sets and we denote by the symbol R_{Π_0} the family of all Π_0–rough sets.

It follows from our assumption about separation of points that the singleton $\{x\}$ is Π_0–closed for any $x \in U$ (in terminology used in general topology, (U, Π_0) is a T_1–*space* (cf. Chapter 7, Exercise 11)).

Let us observe that we may introduce an equivalence relation **r** on the collection $R\Pi_0$ by letting $X\mathbf{r}Y$ if and only if $Int_{\Pi_0} X = Int_{\Pi_0} Y$ and $Cl_{\Pi_0} X = Cl_{\Pi_0} Y$. Then we may replace any Π_0–rough set X with its equivalence class $[X]_\mathbf{r}$; in the sequel, we will rather call a Π_0–rough set the equivalence class $[X]_\mathbf{r}$ instead of X itself.

11.2 Π_0–rough sets

Given a Π_0–rough set $[X]_\mathbf{r}$, we may observe that it may be represented as a pair (Q, T) of Π_0–closed sets Q, T where $Q = Cl_{\Pi_0} X$, $T = U \setminus Int_{\Pi_0} X$. The first task before us is to characterize Π_0–rough sets in terms of pairs (Q, T).

We have the following

Proposition 11.1. *A pair* (Q, T) *of* Π_0–*closed subsets in U satisfies conditions* $Q = Cl_{\Pi_0} X, T = U \setminus Int_{\Pi_0} X$ *with a rough subset* $X \subseteq U$ *if and only if* Q, T *satisfy the following conditions*

1. $U = Q \cup T$

2. $Q \cap T \neq \emptyset$

3. $Q \cap T$ *does not contain any point* x *such that the singleton* $\{x\}$ *is* Π_0–*open*

Proof. First, we observe necessity of 1.–3. Indeed, if $Q = Cl_{\Pi_0} X, T = U \setminus Int_{\Pi_0} X$ for a set $X \subseteq U$ then clearly, 1. and 2. are satisfied. Was an open $\{x\} \subseteq Q \cap T$ then as $X \subseteq Cl_{\Pi_0} X$, we would have $x \in X$ hence

$x \in Int_{\Pi_0} X$ thus $x \notin T$, a contradiction. It follows that Q, T satisfy 3.

Now, for the sufficiency. We assume that 1.–3. hold for a pair (Q, T) and we construct a subset $X \subseteq U$ which would satisfy $Q = Cl_{\Pi_0} X, T = U \setminus Int_{\Pi_0} X$.

We enumerate equivalence classes of the relation R_n as $\{x_n^t : t \in T_n\}$ for $n = 1, 2, ...$ We assume that $T_n \cap T_m = \emptyset$ whenever $n \neq m$. For each natural number n, we let

$$W_n = \{t \in T_n : x_n^t \cap T \neq \emptyset, x_n^t \cap Q \neq \emptyset\} \tag{11.1}$$

We now define inductively sets

$$A_n = \{x_{t,n} : t \in W_n, n = 1, 2, ...\} \tag{11.2}$$

and

$$B = \{y_{t,n} : t \in W_n, n = 1, 2, ...\} \tag{11.3}$$

which would satisfy the following

(a_n) if $x_n^t \cap Q$ contains $x_{t',j}$ with $j < n$ then $x_{t,n} = x_{t',j}$

(b_n) if x_n^t contains $y_{t',j}$ with $j < n$ then $y_{t,n} = y_{t',j}$

(c_n) $x_{t,n} \in x_n^t \cap Q$, $y_{t,n} \in x_n^t$ and $y_{t,n} \in x_n^t \setminus Q$ whenever $x_n^t \setminus Q \neq \emptyset$

(d_n) $y_{t,n}$ is distinct from each $x_{t'}, j$ with $j = 1, 2, ..., n$ for each n

(e_n) $x_{t,n}$ is distinct from each $y_{t'}, j$ with $j = 1, 2, ..., n - 1$ for each n

Let us comment now on the role of A, B. Once they are constructed, we let $X = (U \setminus T) \cup A$ and we check that X is as desired. To this end, we verify the following two statements.

Statement 1. For each n and each $t \in T_n$ we have $x_n^t \subseteq X$ if and only if $x_n^t \subseteq U \setminus T$

Indeed, assume that $x_n^t \subseteq X$ and $x_n^t \cap T \neq \emptyset$. Thus $t \in W_n$ and $y_{t,n} \in x_n^t$ witnesses that it is not true that $x_n^t \subseteq X$, a contradiction.
    It follows that $U \setminus T = Int_{\Pi_0} X$.

Statement 2. For each n and each $t \in T_n$ we have $x_n^t \cap X \neq \emptyset$ if and only if $x_n^t \cap Q \neq \emptyset$

Indeed, assume that $x_n^t \cap Q \neq \emptyset$; then either $x_n^t \subseteq U \setminus T$ hence $x_n^t \subseteq X$

or $x_n^t \cap T \neq \emptyset$ hence $t \in W_n$ and $x_{t,n} \in x_n^t$ witnesses that $x_n^t \cap X \neq \emptyset$.

It follows that $Q = Cl_{\Pi_0} X$ and thus the pair (Q, T) defines a rough set X and a fortiori the class $[X]_{\mathbf{r}}$.

It remains to define $x_{t,n}, y_{t,n}$ which would satisfy $(a_n) - (e_n)$. Actually, the only problem we may have is to satisfy the requirement that $x_{t,n}$ be distinct from $y_{t,n}$. Let us observe that by the condition 3., for any $t \in W_n$, the class x_n^t is infinite: otherwise, as by our assumption of point separation every singleton $\{x\}$ is Π_0–closed, each point of x_n^t would make an open singleton, contradicting 3. Thus we have freedom in choosing $y_{t,n}$ distinct from $x_{t,n}$. This concludes the proof.   □

11.3   Metrics on rough sets

We recall that the symbol R_{Π_0} stands for the family of Π_0–rough sets and we now define a metric topology on R_{Π_0}.

We begin with the relation R_n for a fixed $n = 1, 2, ...$, and we define a function $d_n : U \times U \to \mathbf{R}^+$ where $\mathbf{R}^+$ is the set of non–negative real numbers by the following recipe:

$$(RSM) \quad d_n(x, y) = \begin{array}{l} 1 \ in \ case \ [x]_n \neq [y]_n \\ 0, \quad otherwise \end{array} \tag{11.4}$$

Then the function d_n does satisfy the following

$$d_n(x, x) = 0, d_n(x, y) = d_n(y, x) \tag{11.5}$$

This done, we define a function $d : U \times U \to \mathbf{R}^+$ as follows

$$d(x, y) = \sum_n 10^{-n} \cdot d_n(x, y) \tag{11.6}$$

Then we may check that the function d has the properties

$$d(x, x) = 0 \tag{11.7}$$

$$x \neq y \Rightarrow d(x, y) > 0 \tag{11.8}$$

$$d(x, y) = d(y, x) \tag{11.9}$$

$$d(x, z) \leq d(x, y) + d(y, z) \qquad (11.10)$$

Indeed, the first and the third properties (11.7), (11.9) follow from (11.5), the second, (11.8), is a result of our assumption about point separation: if $x \neq y$ then there exists n with the property that $[x]_n \neq [y_n]$ and thus $d(x, y) \geq \sum_{j=n}^{\infty} 10^{-n} = \frac{1}{9} \cdot 10^{-n+1}$.

The triangle inequality follows easily from the definition of d: given x, z, if n is the first among j such that $[x]_j \neq [y]_j$ then for any y either $[y_n]$ is distinct from both $[x]_n, [z]_n$ or $[y_n]$ is identical to, say $[x]_n$ hence distinct from $[z]_n$. It follows that actually we may have the following better estimate

$$d(x, z) \leq max\{d(x, y), d(y, z)\} \qquad (11.11)$$

It follows that d is a *non-archimedean metric* on U (the term non-archimedean metric refers to the property (11.11)).

We have the following

Proposition 11.2. *The metric topology induced by d coincides with the topology Π_0.*

Proof. Let us consider a basic open set $B = \mathbf{x}_n^t$; for $y \in B$, let $r < 10^{-n}$ and consider the open ball $B(y, r)$. For any $z \in B(y, r)$ we have $[z]_n = [y_n]$ as otherwise $d(y, z) \geq \frac{1}{9} \cdot 10^{-n+1}$ hence $[z]_n = \mathbf{x}_n^t$ i.e. $B(y, r) \subseteq \mathbf{x}_n^t$ implying that $\mathbf{x}_n^t$ is open in the metric topology.

Conversely, for an open ball $B(x, s)$ in the metric topology, and $y \in B(x, s)$, let $r = s - d(x, y)$ and choose n such that $10^{-n} < r$. For $z \in [y]_n$, we have $d(y, z) \leq \frac{1}{9} \cdot 10^{-n} < r$ hence by the triangle inequality $d(z, x) < s$ which implies that $[y]_n \subseteq B(x, s)$. Thus $B(x, s)$ is open in the topology Π_0. $\square$

Now we introduce a new metric d_H, modeled on the Hausdorff–Pompéju metric (cf. Chapter 7), into the family $\mathcal{C}(U)$ of closed subsets of U in the topology Π_0. For closed subsets K, H of U, we let

$$(HPM) \; d_H(K, H) = max\{max_{x \in K} dist(x, H), max_{y \in H} dist(y, K)\}$$

(recall that $dist(x, H) = min\{d(x, z) : z \in H\}$ is the *distance* of x to the set H). The standard proof mimicking that of Sect. 12.1, Chapter 7 shows that d_H is a metric on $\mathcal{C}$.

We may propose an algorithm for computing of $d_H(K, H)$ based on the comparison of closures $Cl_n K, Cl_n H$ for $n = 1, 2, \ldots$ where Cl_n is the closure operator in the topology Π_n.

When $Cl_n K \neq Cl_n H$, there exists e.g. $x \in Cl_n K \setminus Cl_n H$ and thus $[x]_n \subseteq Cl_n K, [x]_n \cap Cl_n H = \emptyset$ from which it follows that $d(x, z) \geq \frac{1}{9} \cdot 10^{-n+1}$ for every $z \in Cl_n H$ implying that $d_H(K, H) \geq \frac{1}{9} \cdot 10^{-n+1}$. We have thus the following

Proposition 11.3. *For any pair K, H of closed sets, we have*

1. *if $Cl_n K = Cl_n H$ for every n then $d_H(K, H) = 0$*

2. *if n is the first among indices j such that $Cl_j K \neq Cl_j H$ then $d_H(K, H) = \frac{1}{9} \cdot 10^{-n+1}$*

We may pass to the problem of metrics on R_{Π_0}. For any pair $(Q_1, T_1), (Q_2, T_2)$ of rough sets, we let

$$D((Q_1, T_1), (Q_2, T_2)) = max\{d_H(Q_1, Q_2), d_H(T_1, T_2)\} \qquad (11.12)$$

and

$$D^*((Q_1, T_1), (Q_2, T_2)) = \\ max\{d_H(Q_1, Q_2), d_H(T_1, T_2), d_H(Q_1 \cap Q_2, T_1 \cap T_2) \qquad (11.13)$$

As d_H is a metric on closed sets, D, D^* are metrics on rough sets. From 1, 2, above, we infer that

Proposition 11.4. *For any pair $(Q_1, T_1), (Q_2, T_2)$ of rough sets*

1. *if n is the first among indices j with the property that either $Cl_j Q_1 \neq Cl_j Q_2$ or $Cl_j T_1 \neq Cl_j T_2$ then $D((Q_1, T_1), (Q_2, T_2)) = \frac{1}{9} \cdot 10^{-n+1}$; otherwise, $D((Q_1, T_1), (Q_2, T_2)) = 0$*

2. *if n is the first among indices j with the property that either $Cl_j Q_1 \neq Cl_j Q_2$ or $Cl_j T_1 \neq Cl_j T_2$ or $Cl_j(Q_1 \cap T_1) \neq Cl_j(Q_2 \cap T_2)$ then $D^*((Q_1, T_1), (Q_2, T_2)) = \frac{1}{9} \cdot 10^{-n+1}$; otherwise, $D^*((Q_1, T_1), (Q_2, T_2)) = 0$*

We now would like to investigate basic properties of metric spaces (R_{Π_0}, D) respectively (R_{Π_0}, D^*).

We know from our discussion of topology that the fundamental properties of topological spaces both from theoretical as well as application – oriented points of view are expressed in terms of completeness and compactness.

We should fill in some information about the limit conditions in our information system $\mathcal{A} = (U, A)$ viz. we should make some assumptions about equivalence classes $[x]_R$ where $\mathbf{R} = \bigcap_n R_n$ is the intersection of all relations R_n. We assume that

(C) For each descending sequence $([x_n]_n)_n$ of equivalence classes we have

$$\bigcap_n [x_n]_n \neq \emptyset \qquad (11.14)$$

(C) does express our positive assumption that the universe of our information system has no "gaps"; on the other hand (C) does imply that in the case where infinitely many of relations R_n are non–trivial the universe U would be uncountable so information systems satisfying (C) should be searched for among those constructed on sets of real numbers or real vectors.

The condition (C) allows us to demonstrate the basic completeness property of metric spaces of rough sets. The completeness property here is more complex then in the general case of a metric space as it requires that both metrics D, D^* cooperate in the following sense

Proposition 11.5. *Under (C): Each D^*–fundamental sequence $((Q_n, T_n))_n$ of rough sets converges in the metric D to a rough set*

Proof. We recall (cf. Sect. 5, Chapter 7) that a sequence $((Q_n, T_n))_n$ is D^*–fundamental if and only if for each positive real number ε there is a natural number m_ε such that $D^*((Q_i, T_i), (Q_j, T_j)) < \varepsilon$ whenever $i, j > m_\varepsilon$.

Thus, the following assertions hold with $\varepsilon = \frac{1}{n}$ and $m_n = m_{\frac{1}{n}}$

(I) there is an increasing sequence $(j_n)_n$ of natural numbers with the property that

$$Cl_{j_n} Q_m = Cl_{j_n} Q_{m_n}$$
$$Cl_{j_n} T_m = Cl_{j_n} T_{m_n} \tag{11.15}$$
$$Cl_{j_n}(Q_m \cap T_m) = Cl_{j_n}(Q_{m_n} \cap T_{m_n})$$

whenever $m \geq m_n$.

The property (I) holds by our recipe for calculating D^*. We now consider the family **DS** of all descending sequences $([x_n]_n)_n$ of equivalence classes; elements of **DS**, will be denoted for convenience with boldface: $\mathbf{x}$ etc. Then $\mathbf{x}_j$ will denote the j–th member of $\mathbf{x}$.

We define the following sets

$$Q^* = \bigcup\{\bigcap \mathbf{x} : \mathbf{x} \in \mathbf{DS}, \forall n.\ \mathbf{x}_{j_n} \cap Q_{m_n} \neq \emptyset\}$$
$$T^* = \bigcup\{\bigcap \mathbf{x} : \mathbf{x} \in \mathbf{DS}, \forall n.\ \mathbf{x}_{j_n} \cap T_{m_n} \neq \emptyset\} \tag{11.16}$$

We expect that (Q^*, T^*) is the limit of the sequence. We prove it in a sequence of statements.

Statement 1. Sets Q^*, T^* are Π_0–closed

Suppose that $x \notin Q^*$; for $\mathbf{x} = ([x]_n)_n$ we have $\mathbf{x}_{j_n} \cap Q_{m_n} = \emptyset$ for some n so that $[x]_{j_n} \cap Q^* = \emptyset$. The complement of Q^* is hence open proving that Q^* is closed. The proof in case of T^* goes on the same lines.

Statement 2. $U = Q^* \cup T^*$

Assume to the contrary that $x \notin Q^* \cup T^*$ for some x. Then for some m, n we have

$$(*)[x]_{j_n} \cap Q_{m_n} = \emptyset = [x]_{j_m} \cap T_{m_m}$$

and we may assume that $m \geq n$. Was $[x]_{j_m} \cap Q_{m_m} \neq \emptyset$ we would have $[x]_{j_n} \cap Q_{m_m} \neq \emptyset$ hence by the property (I) $[x]_{j_n} \cap Q_{m_n} \neq \emptyset$ would follow, contradicting $(*)$.

Hence $[x]_{j_m} \cap Q_{m_m} = \emptyset$ thus $x \notin Q_{m_m} \cup T_{m_m}$, a contradiction.

Statement 3. $Q^* \cap T^* \neq \emptyset$

We may choose $x_1 \in Q_{m_1} \cap T_{m_1}$; then by the property (I), there exists $x_2 \in [x_1]_{j_1}$ with the property that $[x_2]_{j_2} \cap Q_{m_2} \cap T_{m_2} \neq \emptyset$. By induction, we define a sequence $(x_k)_k$ such that

$$(**)[x_k]_{j_k} \cap Q_{m_k} \cap T_{m_k} \neq \emptyset$$

and

$$(***)[x_{k+1}]_{j_{k+1}} \subseteq [x_k]_{j_k}.$$

For $\mathbf{x} = ([x_k]_k)_k$ we have then $\emptyset \neq \bigcap \mathbf{x} \subseteq Q^* \cap T^*$.

Statement 4. $Q^* \cap T^*$ does not contain any Π_0–open singleton

Indeed, was such a singleton $\{x\}$ in $Q^* \cap T^*$, we would have $x = [x]_{j_n}$ for some n and $x \in Q_{m_n} \cap T_{m_n}$, a contradiction.

It follows that (Q^*, T^*) is a rough set.

Statement 5. For each n, the following hold

$$Cl_{j_n} Q^* = Cl_{j_n} Q_{m_n}, Cl_{j_n} T^* = Cl_{j_n} T_{m_n}$$

Assume $x \in Cl_{j_n} Q^*$; then $[x]_{j_n} \cap Q_{m_n} \neq \emptyset$ hence $x \in Cl_{j_n} Q_{m_n}$. Conversely, assume that $x_0 \in Cl_{j_n} Q_{m_n}$; by the property (I), we define inductively a sequence $(x_k)_k$ such that

$$(a)[x_k]_{j_{n+k}} \cap Q_{m_{n+k}} \neq \emptyset$$

and

$$(b)[x_{k+1}]_{j_{n+k+1}} \subseteq [x_k]_{j_{n+k}}$$

for each k. Then $\mathbf{x} = ([x_k]_k)_k$ does satisfy $\bigcap \mathbf{x} \subseteq Q^*$ implying $[x_0]_{j_n} \cap Q^* \neq \emptyset$ and thus $x_0 \in Cl_{j_n} Q^*$. The proof in case of T goes along similar lines.

Statement 5. implies – as the sequence $(j_n)_n$ is co–final in the set N of natural numbers – that the sequence $((Q_n, T_n))_n$ does converge in the metric D to the rough set (Q^*, T^*). The completeness property has been demonstrated.   □

From point of view of applications, we may often come across an information system $\mathcal{A}$ satisfying the finiteness condition

(F) *Each relation R_n induces a finite number of equivalence classes*

It turns out that

Proposition 11.6. *Under (C)+(F), the space (R_{Π_0}, D) is compact i.e. any sequence $((Q_j, T_j))_j$ of rough sets contains a sub-sequence $((Q_{j_k}, T_{j_k}))_k$ convergent in the metric D to a limit rough set (Q^*, T^*).*

Proof. By (F), we define a sequence $\mathbf{s} = (s_n)_n$ such that $s_n = (s_n(j))_j$ is an increasing sequence of natural numbers and s_{n+1} is a subsequence of s_n for each n. The sequence $\mathbf{s}$ does satisfy for any $i, j \in s_n$ the following

$$(a) \quad Cl_n Q_i = Cl_n Q_j, Cl_n T_i = Cl_n T_j, Cl_n(Q_i \cap T_i) = Cl_n(Q_j \cap T_j).$$

The existence of $\mathbf{s}$ with these properties is an immediate consequence of (F) and definition of closure.

Now we consider the diagonal sequence $\mathbf{t} = (s_n(n))_n$; clearly, $\mathbf{t}$ is increasing and by (a) we have

$$(b) \quad Cl_n Q_{t_n} = Cl_n Q_{t_m}, Cl_n T_{t_n} = Cl_n T_{t_m},$$

$$Cl_n(Q_{t_n} \cap T_{t_n}) = Cl_n(Q_{t_m} \cap T_{t_m})$$

for every n and $m \geq n$.

As (b) implies that the sequence $(Q_{t_n}, T_{t_n})_n$ is fundamental with respect to the metric D^*, it follows that it does converge to a rough set constructed in the proof of the previous proposition. $\square$

11.3.1 Some examples

Questions may arise whether the above results are best and may not be improved i.e. whether (i) each D^*–fundamental sequence converges in the metric D^* (ii) each D–fundamental sequence converges in the metric D. We disprove either conjecture .

Example 11.1. *Not every D^*–fundamental sequence needs to D^*–converge*

Our example will be constructed in the universe U being the Cantor set $\mathbf{C}$ i.e. the Cartesian product $P_{j=1}^{\infty}\{0,1\}_j$ whose elements are sequences of 0's and 1's. For sequences $\mathbf{x}, \mathbf{y} \in \mathbf{C}$, and $n \in N$, we let $\mathbf{x} \, R_n \mathbf{y}$ if and only if $x_i = y_i$ for $i \leq n$. This defines the equivalence relation $R_n = Ind_{a_n}$ for each n. We let $x_1 x_2 ... x_n$ to denote the class $[\mathbf{x}]_n$ for $\mathbf{x} = (x_j)_j$. For $i \in \{0,1\}$, we *denote by the symbol i_n the sequence of length n composed of i's solely and the symbol $\mathbf{i}$ will denote the infinite sequence of i's. Sets $i_1 i_2 ... i_n$ are referred to as n–boxes and they constitute an open basis of the topology Π_0 on $\mathbf{C}$. Each n–box $i_1 i_2 ... i_n$ splits into two $n+1$–boxes $i_1 i_2 ... i_n 0$ (the left–hand box) and $i_1 i_2 ... i_n 1$ (the right–hand box).*

We define a sequence $(Q_n, T_n)_n$ of Π_0–rough sets with $P_n = U \setminus T_n$ for each n. To begin with, we let

$$P_1 = 10 \cup 010, P_2 = 100 \cup 0100, P_n = 10_n \cup 010_n$$

for each n and then we let

$$Q_n = P_n \cup 110_{n-1}$$

for each n. Thus, P_{n+1} is obtained from P_n by going inside P_n one level up and taking the left–hand box and $Q_n \cap T_n = 110_{n_1}$ for each n. It follows that (Q_n, T_n) is a Π_0–rough set for each n.

Definitions of P_n and Q_n imply that

(a) $Cl_{n+1} Q_{n+1} = Cl_{n+1} Q_{n+k}, Cl_{n+1} T_{n+1} = Cl_{n+1} T_{n+k},$

$$Cl_{n+1}(Q_{n+1} \cap T_{n+1}) = Cl_{n+1}(Q_{n+k} \cap T_{n+k})$$

hence the sequence $(Q_n, T_n)_n$ is D^*–fundamental. Now we may construct the limit set (Q^*, T^*) by the procedure defined above letting $j_n = n + 1 = m_n$ for each n. Then $010, 10 \subseteq Q^* \cap T^*$ hence $d_H(Q^* \cap T^*, Q_{n+1} \cap T_{n+1}) = \frac{1}{9}$ for each n; it follows that the sequence $(Q_n, T_n)_n$ does not converge in the metric D^* to the set (Q^*, T^*). Our example is concluded.

Example 11.2. *D–fundamental sequences need not D–converge*

In this case our universe U will be the set $\mathbf{P}$ of irrational numbers represented as the collection f all infinite sequences of natural numbers i.e. as the Cartesian product $P_{j=1}^{\infty} N_j$ where $N_j = N$ for each j. Relations $\{R_n : n = 1, 2, ...\}$ will be defined as in Example 1: for $\mathbf{x}, \mathbf{y} \in \mathbf{P}$ we let $\mathbf{x} \, R_n \mathbf{y}$ if and only if $x_j = y_j$ for $j \leq n$. We denote by the symbol $x_1 x_2 ... x_n$ the class $[\mathbf{x}]_n$ where $\mathbf{x} = (x_j)_j$ and we will call it an n–box. Each n–box $i_1 i_2 ... i_n$ splits into infinitely many boxes of the form $i_1 i_2 ... i_n j$ for $j \in N$. The symbol i_n denotes as before the sequence of i's of length n.

We let

$$Q_n \cap T_n = \bigcup_{k \geq n} k 0_{n-1}, Q_n \setminus T_n = 0 \cup 10 \cup 200 \cup ... \cup (n-1) 0_{n-1}$$

and $T_n \setminus Q_n$ is the union of all remaining n–boxes.

It follows that (Q_n, T_n) is a Π_0–rough set for each n and

$$(a) \ Cl_n Q_n = Cl_n Q_{n+k}, Cl_n T_n = Cl_n T_{n+k}$$

for each k witnessing that the sequence $(Q_n, T_n)_n$ is D–fundamental. Constructing the limit set (Q^*, T^*) we find that $Q^* \cap T^* = \emptyset$; indeed, was $x \in Q^* \cap T^*$ we would have

$$(b) \ [x]_n \cap Q_n \neq \emptyset \neq [x]_n \cap T_n$$

for each n. As $[x]_n$ may be only one of $n0_{n-1}, (n+1)0_{n-1}, \dots$ there is no x which would satisfy (b). Thus $Q^* \cap T^* = \emptyset$ and (Q^*, T^*) is a Π_0–exact set so there is no limit for the sequence $(Q_n, T_n)_n$ in the space (R_{Π_0}, D).

11.4 Almost rough sets

In the last example we experienced a new phenomenon: a set $X \subseteq U$ may be rough with respect to each relation R_n but it may be exact with respect to the topology Π_0. It is desirable to set those cases apart as a separate case.

Following this path, we call a set $X \subseteq U$ *almost rough* if and only if it is Π_n–rough for each $n \in N$. We denote by the symbol R_{Π_ω} the collection of all almost rough sets in our information system $\mathcal{A}$.

As with rough sets, we are interested in a convenient characterization of those sets. No doubt, a representation in the form of a pair (Q, T) deserves to be preserved in this case. So let us assume that $X \subseteq U$ is almost rough. Then for each n, Π_n–closed sets $Cl_n X = Q_n$ and $T_n = U \setminus Int_n X$ satisfy the requirements set above i.e.

$$(i) \ U = Q_n \cup T_n$$
$$(ii) \ Q_n \cap T_n = \emptyset \tag{11.17}$$
$$(iii) \ Q_n \cap T_n \text{ contains no } x \text{ with } \{x\} \ \Pi_n \ \text{– –open}$$

Moreover, as $R_{n+1} \subseteq R_n$ for each n, we have

$$(iv) \ Cl_m Q_n = Q_m, Cl_m T_n = T_m \text{ for each pair } m, n \text{ with } m \leq n \tag{11.18}$$

It follows that we may assign to X the sequence $(Q_n, T_n)_n$ satisfying (i)–(iv). Let us consider an equivalence class $[y]_n$ for some y, n with $[y]_n \cap Q_n \neq \emptyset$. It follows that for some $x \in X$ we have $x \in [y_n]$ and thus $[x]_j \subseteq [y_n]$ for each $j \geq n$ with obviously $\emptyset \neq \bigcap_j [x]_j$. To make this observation formal, we consider the set **DS** of all descending sequences $([x_n]_n)_n$ of equivalence classes and we let

$$DS(Q) = \{([x_n]_n)_n \in \mathbf{DS} : \forall n [x_n]_n \cap Q_n \neq \emptyset\}$$
$$DS(T) = \{([x_n]_n)_n \in \mathbf{DS} : \forall n [x_n]_n \cap T_n \neq \emptyset\} \qquad (11.19)$$

For any equivalence class $[y_j]$, we let

$$DS(Q)(y,j) = \{([x_n]_n)_n \in \mathbf{DS} : [x_j]_j = [y]_j\}$$
$$DS(T)(y,j) = \{([x_n]_n)_n \in \mathbf{DS} : [x_j]_j = [y]_j\} \qquad (11.20)$$

Then we may state the observed last property

For each equivalence class $[y]_n$ with $DS(Q)(y,n) \neq \emptyset$, there exists a sequence $([x_m]_m)_m \in DS(Q)(y,n)$ with the property that

$$\bigcap_m [x_m]_m \neq \emptyset \qquad (11.21)$$

Similarly, for each equivalence class $[y]_n$ with $DS(T)(y,n) \neq \emptyset$, there exists a sequence $([x_m]_m)_m \in DS(T)(y,n)$ with the property that

$$\bigcap_m [x_m]_m \neq \emptyset \qquad (11.22)$$

It will be now our task to demonstrate that any sequence $(Q_m, T_m)_m$ satisfying (11.17 – 11.22) defines an almost rough set.

We let

$$Q^* = \bigcup\{\bigcap_m [x_m]_m : ([x_m]_m)_m \in DS(Q)\}$$
$$T^* = \bigcup\{\bigcap_m [x_m]_m : ([x_m]_m)_m \in DS(T)\} \qquad (11.23)$$

Then we may check the following statements.

Statement 1. $U = Q^* \cup T^*$

Indeed, assume that $x \notin Q^* \cup T^*$ for some x. Then for some m, n we have $[x]_m \cap Q_m = \emptyset = [x]_n \cap T_n$. Suppose e.g. $n \geq m$; by (11.18), $[x]_n \cap Q_n = \emptyset$ hence $x \notin Q_n \cup T_n$, contradicting (11.17,(i)).

Statement 2. $Q^* \cap T^*$ contains no Π_0–open singleton

This is obvious by Property (11.17, (iii)).

Statement 3. $Cl_m Q^* = Q_m, Cl_m T^* = T_m$ for each m

Suppose that $x \in Cl_m Q^*$; we have $[x]_m \cap Q_m \neq \emptyset$ hence $x \in Q_m$. It follows that $Cl_m Q^* \subseteq Q_m$. For the converse, assume that $x \in Q_m$. By

(11.18) , we define a sequence $([x_k]_k)_k \in \mathbf{DS}(x, m)$ inducting on k. By (11.21), there exists a sequence $([z_k]_k)_k \in \mathbf{DS}(x, m)$ with the property that $\emptyset \neq \bigcap_k [x_k]_k \subseteq Q^*$; pick a $y \in \bigcap_k [x_k]_k$. Then $y \in [x]_m$ hence $x \in Cl_m Q^*$. Thus $Q_m \subseteq Cl_m Q^*$ and finally $Cl_m Q^* = Q_m$. The proof in case of T^* goes on similar lines.

We have proved that (Q^*, T^*) defines a class $[X]_\mathbf{r}$ which is an almost rough set and either a Π_0–rough set (in case $Q^* \cap T^* \neq \emptyset$) or a Π_0–exact set (otherwise).

Following the case of $\mathbf{R}_{\Pi_0}$, we introduce into the set of almost rough sets a metric D' constructed as follows. First, we recall the function d_n defined for each n as $d_n(x, y) = 1$ in case $[x]_n \neq [y]_n$ and $d_n(x, y) = 0$, otherwise.

From this point, we take a slightly different path, and we define a Hausdorff–Pompéju type metric $d_{H,n}$ on pairs K, H of Π_n–closed sets for each n. To this end we let

$$(HPM1)\ d_{H,n}(K, H) =$$

$$max\{max_{x \in K} dist_n(x, H), max_{y \in H} dist_n(y, K)\}$$

where as usual $dist_n(x, H) = min_{y \in H} d_n(x, y)$ is the distance from x to H.

In the following step, we extend $d_{H,n}$ to the metric D'_n on pairs of Π_0–closed sets by letting

$$D'_n((Q, T), (Q', T')) = max\{d_{H,n}(Q, Q'), d_{H,n}(T, T')\} \qquad (11.24)$$

for each n. Finally, we glue together metrics D'_n into a global metric on $\mathbf{R}_{\Pi_\omega}$ denoted D' and defined via the formula

$$D'(((Q_n, T_n))_n, ((Q'_n, T'_n))_n) = \sum_n 10^{-n} \cdot D'_n((Q_n, T_n), (Q'_n, T'_n)) \qquad (11.25)$$

As on previous occasions, here also we may extract from the definition of the metric D' a procedure for computing its values.

Proposition 11.7. *Given almost rough sets* $K = ((Q_n, T_n))_n$, $H = ((Q'_n, T'_n))_n$: *if for each n, we have* $Q_n = Q'_n, T_n = T'_n$ *then* $D'(K, H) = 0$; *otherwise,* $D'(K, H) = \frac{1}{9} \cdot 10^{-n+1}$ *where n is the first among indices j such that either* $Q_j \neq Q'_j$ *or* $T_j \neq T'_j$.

We may observe that any Π_0–rough set (Q^*, T^*) is an almost rough set as well with a representation $((Q_n = Cl_n Q^*, T_n = Cl_n T^*))_n$. Thus the identity function $\mathbf{i}_{\mathcal{A}} : \mathbf{R}_{\Pi_0} \to \mathbf{R}_{\Pi_\omega}$ expressed by means of both representations as follows: $\mathbf{i}_{\mathcal{A}}((Q^*, T^*)) = ((Cl_n Q^*, Cl_n T^*))_n$, embeds $\mathbf{R}_{\Pi_0}$ into $\mathbf{R}_{\Pi_\omega}$.

A comparison of metrics D and D' results in the conclusion that

$$D'(\mathbf{i}_A(K), \mathbf{i}_A(H)) = D(K, H)$$

for each pair K, H of Π_0–rough sets. Thus the function $\mathbf{i}_A$ is the *isometric embedding* (i.e. metric preserving injection) of rough sets into almost rough sets.

We would like to explore our new metric space $(\mathbf{R}_{\Pi_\omega}, D')$ with respect to completeness and compactness properties. We observe first that

Proposition 11.8. *The metric space $(\mathbf{R}_{\Pi_\omega}, D')$ is complete i.e. each D'– fundamental sequence $(((Q_m^n, T_m^n))_m)_n$ of almost rough sets converges in the metric D' to an almost rough set.*

Proof. By definition of a fundamental sequence, there exists a strictly increasing sequence $(k_n)_n$ of natural numbers with the property that for any $m \leq n$ and $p, q \geq k_n$ we have

$$Q_m^p = Q_m^q, T_m^p = T_m^q \tag{11.26}$$

Then for $j \geq n$ we have

$$Q_n^{k_n} = Q_n^{k_j} = Cl_n Q_{k_j}^{k_j} \tag{11.27}$$

and similarly

$$T_n^{k_n} = T_n^{k_j} = Cl_n T_{k_j}^{k_j} \tag{11.28}$$

and it follows easily that the sequence $((Q_n^{k_n}, T_n^{k_n}))_n$ does represent an almost rough set. For $p \geq k_n$, we have $Q_m^p = Q_m^{k_n}, T_m^p = T_m^{k_n}$ for each $m \leq n$ hence

$$D'(((Q_m^p, T_m^p))_m, ((Q_m^{k_m}, T_m^{k_m}))_m) \leq \frac{1}{9} \cdot 10^{-n+1} \tag{11.29}$$

when $p \geq k_n$ witnessing that

$$(((Q_m^{k_m}, T_m^{k_m}))_m)_n = lim_{n \to \infty}((Q_m^p, T_m^p))_m \tag{11.30}$$

concluding the proof.   □

We know that under the condition (F), the space $(\mathbf{R}_{\Pi_0}, D)$ is compact; it turns out that under (F) spaces of rough sets and almost rough sets coincide

Proposition 11.9. *Assume (F); then every Π_0–almost rough set is a Π_0– rough set.*

Proof. For an almost rough set $((Q_n, T_n))_n$, we let

$$\mathcal{D}_n = \{[x]_n : [x]_n \cap Q_n \neq \emptyset \neq [x]_n \cap T_n$$

for each n. We order $\mathcal{D} = \bigcup_n \mathcal{D}_n$ by proper inclusion. It follows by (11.18) that any $[x]_n \in \mathcal{D}_n$ is a root of a tree under this ordering whose every level is finite and which has branches of any finite height. By the König Lemma (cf. Chapter 6, Exercise 34) and (11.21), there exists an infinite branch $([x_n]_n)_n$ in this tree hence $\emptyset \neq \bigcap_n [x_n]_n \subseteq Q^* \cap T^*$ thus $((Q_n, T_n))_n = i_{\mathcal{A}}(Q^*, T^*)$ is a Π_0–rough set.   □

11.5   Fractals, Approximate Collage

We have introduced basic topological structures on rough sets in the most important countable case. It may be useful now to point to some applications. We include here a discussion of fractals, in which we introduce a new fractal dimension related to concepts in information systems. We also address the Banach fixed point theorem, proving its rough set version.

11.5.1   Fractals

Objects called now "fractals" have been investigated since 1920's (cf. [Carathéodory14], [Hausdorff19]), yet the renewed interest in them goes back to 1970's in connection with studies of chaotic behavior, irregular non–smooth sets, dynamic systems, information compression and computer graphics [Mandelbrot75].

The basic characteristics of "fractals" are rooted in dimension theory. The topological dimension theory assigns to any subset T of a (sufficiently regular) topological space X an integer $ind\ T \geq -1$ called the *dimension* of T (cf. [Hurewicz–Wallman41]).

Assuming that (X, ρ) is a metric separable space, one assigns to X an integer $ind\ X$ in the following way

1. $ind\ X = -1$ if and only if $X = \emptyset$

2. $ind\ X \leq n$ if and only if for each $x \in X$ and every $r > 0$, there exists a neighborhood W of x with the property that $ind\ BdW \leq n - 1$

3. $ind\ X = n$ if and only if $ind\ X \leq n$ and it is not true that $ind\ X \leq n-1$

The dimension ind has among others the following properties (cf. [Hurewicz–Wallman41])

1. $X \subseteq Y$ implies that $ind\ X \leq ind\ Y$ (monotonicity)

2. If $X = \bigcup_n X_n$ where each X_n is a closed subset of X then $ind\ X = max\{ind\ X_n : n \in N\}$ (additivity)

3. $ind\ E^n = n$ for each n where E^n is the Euclidean n–space

This dimension function, however, does not reflect local structure of a fractal. For this reason, fractals are evaluated by means of other functions e.g. *Hausdorff dimension* or *Minkowski (box) dimension* better suited at capturing the peculiarities of local structure.

Many fractal objects can be generated by means of iterations of affine mappings (iterated function systems) (cf. [Hutchinson81]) hence they allow for knowledge compression algorithms (cf. [Barnsley88]).

We look at fractals from the rough set theoretical point of view. We are interested in transferring the notion of a fractal to the general framework of rough set theory and we examine here some propositions for a counterpart of fractal dimension in this general framework.

Fractal dimensions

Although there seems to not exist a general commonly accepted definition of a "fractal", yet we may adopt one of its versions according to which a "fractal" is a set in (usually) Euclidean space of n–dimensions E^n $(n = 1, 2, ...)$ whose "fractional" dimension is distinct from its topological dimension. For a set $T \subseteq E^n$, and $s \geq 0, \delta > 0$, one lets (cf. [Falconer 90^{a,b}])

$$\mathcal{H}_\delta^s(T) = inf \sum_i diam^s(Q_i) \tag{11.31}$$

the infimum taken over all families $\{Q_i : i = 1, 2, ...\}$ of sets in E^n such that (i) $T \subseteq \bigcup_i Q_i$ (ii) $diam(Q_i) \leq \delta$.

Then–as $\mathcal{H}_\delta^s$ is decreasing in δ– the limit

$$\mathcal{H}^s(T) = lim_{\delta \to 0+} \mathcal{H}_\delta^s(T) \tag{11.32}$$

exists and it follows easily that there exists a unique s^* with the property that $\mathcal{H}^s(T) = \infty$ for $s < s^*$ and $\mathcal{H}^s(T) = 0$ for $s > s^*$ (to see this, observe that

$$\mathcal{H}^{s'}(T) \leq lim_{\delta \to 0+} \delta^{s'-s} \cdot \mathcal{H}^s(T) \tag{11.33}$$

hence by passing to the limit with δ we have that either $\mathcal{H}^s(T) = \infty$ or $\mathcal{H}^{s'}(T) = 0$ for $s' > s$).

The real number s^* is the *Hausdorff dimension* of the set T, denoted $dim_\mathcal{H}(T)$.

Figure 11.1: The second approximation D_2 to the Cantor set shown in thick line

Our test fractal will be the Cantor set C i.e. the set of all infinite binary sequences formally written down as the Cartesian product $P_n\{0,1\}_n$ (cf. Chapter 7). To represent the Cantor set in a Euclidean space, let us observe that C is bijective to the set D of real numbers in the interval $[0,1]$ whose ternary expansion does not require any coefficient of 1 i.e. $x \in D$ if and only if $x = \sum_{j=1}^{\infty} \frac{a_j}{3^j}$ where $a_j \in \{0,2\}$ for each j. The function $h : C \to D$ defined by means of the formula

$$h((x_j)_j) = \sum_{j=1}^{\infty} \frac{2x_j}{3^j}$$

does establish a 1–1 correspondence between the two sets.

Moreover, we may observe that the image $h(i_1 i_2 ... i_n)$ of the basic n–box $i_1 i_2 .... i_n$ has diameter $\frac{1}{3^n}$ thus h as well as its inverse h^{-1} are continuous so h does establish a topological equivalence of C and D. Figure 1 below illustrates the process of getting the Cantor set in the interval $[0,1]$ by highlighting with the thick line the set D_2 i.e. the second approximation to the Cantor set.

We may thus rightly call D the Cantor set. There exists a simple description of the set D viz. first we divide the unit interval into the intervals of length $\frac{1}{3}$ each: $P_1 = [0,\frac{1}{3}], (\frac{1}{3},\frac{2}{3}), [\frac{2}{3},1]$ and we remove the middle one letting $D_1 = [0,\frac{1}{3}] \cup [\frac{2}{3},1]$. In the next step, we divide each of intervals $[0,\frac{1}{3}], [\frac{2}{3},1]$ into three intervals of length $\frac{1}{9}$ each obtaining a family P_2 and we remove either of middle intervals letting

$$D_2 = [0,\frac{1}{9}] \cup [\frac{2}{9},\frac{1}{3}] \cup [\frac{2}{3},\frac{7}{9}] \cup [\frac{8}{9},1].$$

This procedure of dividing each interval into three sub–intervals of same length and removing the middle one–third is repeated to yield families P_n and sets D_n for $n = 1,2,.....$. Clearly, elements of D_1 can be written in their ternary expansion without using the digit 1 in the first position, elements of D_2 can be written down in this form without using the digit 1 in the second position etc. etc. Thus we have $D = \bigcap_n D_n$. Intervals making any D_n will be called *marked*.

This in a sense visualization of $\mathcal{D}$ makes it relatively easy to calculate the Hausdorff dimension of the Cantor set. Let us observe that each D_j is the union of 2^j intervals of length 3^{-j} each. Given a covering (U_i) of $\mathcal{D}$ by sets of diameter less than δ, we may observe that first, each U_i may be assumed open, hence by the compactness of $\mathcal{D}$ the family (U_i) may be assumed finite. Next we may observe that each U_i may be assumed a union of marked intervals and finally we may observe that the function $y = x^s$ is concave for $s < 1$ i.e. for any x, y we have $\frac{x^s + y^s}{2} \leq (\frac{x+y}{2})^s$ (to see this observe that the second derivative of this function is negative for $x > 0$) which implies that we may restrict ourselves to "canonical" coverings defined for each j as the collection of marked intervals of the level j (i.e. making together the set D_j).

Under this simplification we have

$$\mathcal{H}^s(\mathcal{D}) = lim_{j \to \infty} 2^j \cdot (3^{-j})^s = lim_{j \to \infty} (\frac{2}{3^s})^j \qquad (11.34)$$

which results in 1 when $s = \frac{log2}{log3}$, it does result in 0 when $s > \frac{log2}{log3}$ and it results in ∞ when $s < \frac{log2}{log3}$. Thus

$$s^* = \frac{log2}{log3}$$

is the Hausdorff dimension of the Cantor set.

The Hausdorff dimension has many regular properties, but it is too closely related to the metric structure of the underlying space to admit any substantial abstraction. For our purposes, the other function, the *Minkowski dimension (box dimension)* seems to be better suited. This dimension has an information theoretic content and it may be transferred–with changes relaxing its geometric content–into a universe of a general information system.

For a bounded set $T \subseteq E^n$ (i.e. $diam(T) < \infty$), and $\delta > 0$, we denote by $n_\delta(T)$ the least number of n–cubes of diameter less than δ that cover T. Then we may consider the fraction

$$\frac{-logn_\delta(T)}{log\delta} \qquad (11.35)$$

and evaluate its limit. When the limit

$$lim_{\delta \to 0+} \frac{-logn_\delta(T)}{log\delta} \qquad (11.36)$$

exists, it is called the *Minkowski dimension* of the set T and it is denoted $dim_\mathcal{M}(T)$. One may interpret this dimension as an *information content* of T: the shortest description of T over an alphabet of δ–cubes has length of order of $dim_\mathcal{M}(T)$.

We measure our archetypical fractal $\mathcal{D}$ with the Minkowski dimension; clearly

$$dim_{\mathcal{M}}(\mathcal{D}) \le lim_{j\to\infty} \frac{log 2^j}{-log 3^{-j}} = \frac{log 2}{log 3} \qquad (11.37)$$

On the other hand, given $\delta > 0$ with $j = \lfloor -log_3\delta \rfloor$, any set of length less than δ does intersect at most one marked interval of the level j hence $n_\delta(\mathcal{D}) \ge 2^j$ implying that

$$dim_{\mathcal{M}}(\mathcal{D}) \ge lim j \to \infty \frac{log 2^j}{log 3^j} = \frac{log 2}{log 3} \qquad (11.38)$$

It follows that the Minkowski dimension of the Cantor set is $\frac{log 2}{log 3}$ and it is equal to the Hausdorff dimension of this set.

Although both dimensions agree on "standard" fractal objects like the Cantor set yet in general they disagree. The general relation is (cf. [Falconer90^{a,b}])

$$dim_{\mathcal{H}}(T) \le liminf_{\delta\to 0+} \frac{log - n_\delta(T)}{log\delta} \le limsup_{\delta\to 0+} \frac{log - n_\delta(T)}{log\delta} \qquad (11.39)$$

An advantage of the Minkowski dimension is that families of δ–cubes in its dimension may be selected in many ways, one among them is to consider a δ–grid of cubes of side length δ on E^n and to count the number $N_\delta(T)$ of those among them which intersect T; then we have (cf. [Falconer 90^{a,b}])

$$lim_{\delta\to 0+} \frac{-log N_\delta(T)}{log\delta} \qquad (11.40)$$

if exists is equal to $dim_{\mathcal{M}}(T)$.

We now consider an information system $\mathcal{A}_C$ on the Euclidean space E^n; this system consists of the universe $U = E^n$ and of attributes a_k for $k = 1, 2, ...$ defined via partitions $\mathcal{P}_k$ induced by relations Ind_{a_k}. The partition $\mathcal{P}_k$ consists of n–cubes of the form

$$(c) \quad \prod_{i=1}^{n} [m_i + \frac{j_i}{2^k}, m_i + \frac{j_i + 1}{2^k}) \qquad (11.41)$$

where m_i is an integer for each $i = 1, 2, ..., n$ and $0 \le j_i \le 2^k - 1$ is an integer. Thus cubes constituting $\mathcal{P}_k$ are products of left–closed and right–open intervals of length 2^{-k} and $\mathcal{P}_{k+1}$ is a sub-division of $\mathcal{P}_k$ for each k. From the definition of the Minkowski dimension it follows easily

Proposition 11.10. *If the Minkowski dimension* $dim_{\mathcal{M}}(T)$ *exists then*

$$dim_{\mathcal{M}}(T) = lim_{k \to \infty} \frac{log N'_k}{k log 2}$$

where N'_k *is the number of cubes in* $\mathcal{P}_k$ *which intersect* T.

Proof. Indeed, letting $\delta_k = 2^{-k}$ we have

$$liminf_{\delta \to 0+} \frac{-log N_\delta(T)}{log \delta} \leq liminf_{\delta_k \to 0+} \frac{-log N_{\delta_k}(T)}{log \delta_k}$$

$$\leq limsup_{\delta_k \to 0+} \frac{-log N_{\delta_k}(T)}{log \delta_k} \leq limsup_{\delta \to 0+} \frac{-log N_\delta(T)}{log \delta}$$

and in case $dim_{\mathcal{M}}(T)$ exists, the left–most and the right–most terms in the above chain of inequalities coincide hence the two middle terms coincide and the result follows from the fact that $N_{\delta_k}(T) = N'_k$. □

It now follows from the properties of fractal dimension $dim_{\mathcal{M}}$ that

Proposition 11.11. *For any* Π_A*–exact set* Z, *we have* $dim_{\mathcal{M}}(Z) = n$.

Proof. Indeed, if a set Z is $\Pi_{\mathcal{M}}$–exact then Z is a union of a family $\{Q_j : j = 1, 2, ...\}$ of n–cubes of the form (c) and thus $n \geq dim_{\mathcal{M}}(T) \geq dim_{\mathcal{M}}(Q_1) = n$ by the monotonicity and stability of $dim_{\mathcal{M}}$(cf. [Falconer90^{a,b}]). □

Thus a proposition follows

Proposition 11.12. *Any set* Z *of fractional dimension* $dim_{\mathcal{M}}$ *is a* Π_A*–rough set.*

By this proposition, fractal objects are among rough sets with respect to the topology Π_{A_c}. This fact directs us towards general information systems and rough sets resulting in them.

Fractals in information systems

The last result shows that it is possible to characterize fractal objects with respect to information systems having a direct geometric content. We now attempt at transferring the basic notions of a fractal object and fractal dimension into a general framework of information systems where as a rule no directly accessible geometric structure is present.

For an information system $\mathcal{A} = (U, A)$ with the countable set $A = \{a_n : n = 1, 2, ...\}$ of attributes such that $Ind_{a_{n+1}} \subseteq Ind_{a_n}$ for $n = 1, 2, ...$, we will define the notion of an $\mathcal{A}$–dimension, denoted dim_A.

We restrict ourselves to *bounded* subsets $Z \subseteq U$ i.e. such Z which for each n are covered by a finite number of equivalence classes of Ind_{a_n}. We may therefore assume that

(1) the number of equivalence classes of Ind_{a_1} is k_1

(2) each class of Ind_{a_n} ramifies into k_{n+1} classes of $Ind_{a_{n+1}}$

Thus the number of equivalence classes of the relation Ind_{a_n} is $m_n = \prod_{i=1}^{n} k_i$. We will say that the information system $\mathcal{A}$ is of type $\kappa = (k_i)_i$. Although these assumptions impose some regularity conditions on the information system, yet they may be satisfied by subdividing classes if necessary.

For a bounded set $T \subseteq U$, we let

$$(FDU) \quad \overline{dim_A}(T) = lim_{n \to \infty} \frac{log \prod_{i=1}^{n} l_i}{log \prod_{i=1}^{n} m_i}$$

where l_i is the number of classes of Ind_{a_i} that intersect T and m_i has been defined as the number of classes of Ind_{a_i}.

$\overline{dim_A}(T)$ is the *upper fractal dimension* of T.

Similarly, we define the *lower fractal dimension* of T viz.

$$(FDL) \quad \underline{dim_A}(T) = lim_{n \to \infty} \frac{log \prod_{i=1}^{n} r_i}{log \prod_{i=1}^{n} m_i}$$

where r_i is the number of classes of Ind_{a_i} contained in T. We observe that $\overline{dim_A}(T), \underline{dim_A}(T)$ are based on upper, respectively, lower, approximations of T. In case both are equal, we denote their common value with $dim_A(T)$ and call it the $\mathcal{A}$-*dimension* of T. Let us collect here basic properties of dim_A which are parallel to respective properties of the Minkowski (box) dimension.

Proposition 11.13.   *1.* $dim_A(Z) \le dim_A(T)$ *whenever* $Z \subseteq T$

    2. $\overline{dim_A}(Z \cup T) = max\{\overline{dim_A}(Z), \overline{dim_A}(T)\}$ *in case* $\mathcal{A}$ *is of type* κ *with* $k_i \ge 2$ *for infinitely many* i

    3. $dim_A(Z) = dim_A(Cl_{\Pi_A} Z)$

Proof. Indeed, (1) follows by the very definition of dim_A. For (2), by (1) it follows that

$$\overline{dim_A}(Z \cup T) \ge max\{\overline{dim_A}(Z), \overline{dim_A}(T)\} \tag{11.42}$$

To prove the converse let us assume that $\overline{dim_A}(Z) \ge \overline{dim_A}(T)$ and split infinite sequences of natural numbers into two classes (p_j denotes the number of classes of Ind_{a_j} intersecting Z and q_j means the same for T):

(I) a sequence $(n_j)_j$ falls here in case $p_{n_j} < q_{n_j}$ for infinitely many j and $p_{n_j} \geq q_{n_j}$ for infinitely many j

(II) a sequence falls here in case $p_{n_j} < q_{n_j}$ for almost every j

(III) a sequence falls here in case $p_{n_j} \geq q_{n_j}$ for almost every j.

We assume that l_j is the number of classes of Ind_{a_j} intersecting $Z \cup T$; clearly

$$l_j \leq p_j + q_j \tag{11.43}$$

for each j. Now consider a sub-sequence n_j for which

$$\frac{\log \prod_{i=1}^{n_j} l_i}{\log \prod_{i=1}^{n_j} m_i} \tag{11.44}$$

converges. In case it falls into (II), we have

$$l_{n_j} \leq 2q_{n_j} \tag{11.45}$$

for almost every j and thus

$$lim_{j \to \infty} \frac{\log \prod_{i=1}^{n_j} l_i}{\log \prod_{i=1}^{n_j} m_i} \leq lim_{j \to \infty} \frac{\log \prod_{i=1}^{n_j} 2q_i}{\log \prod_{i=1}^{n_j} m_i} \leq \overline{dim}_A(T) \leq \overline{dim}_A(Z) \tag{11.46}$$

Similarly in case the sequence falls into (III)

$$l_{n_j} \leq 2p_{n_j} \tag{11.47}$$

for almost every j and thus

$$lim_{j \to \infty} \frac{\log \prod_{i=1}^{n_j} l_i}{\log \prod_{i=1}^{n_j} m_i} \leq \overline{dim}_A(Z) \tag{11.48}$$

In case the sequence is in (I), by its convergence we have

$$lim \frac{\log \prod_{i=1}^{n_j} l_i}{\log \prod_{i=1}^{n_j} m_i} \leq lim_{u \to \infty} \frac{\log \prod_{i=1}^{u} 2p_i}{\log \prod_{i=1}^{u} k_i}, lim_{v \to \infty} \frac{\log \prod_{i=1}^{v} 2q_i}{\log \prod_{i=1}^{v} m_i} \leq$$
$$max\{\overline{dim}_A(Z), \overline{dim}_A(T)\} = \overline{dim}_A(Z) \tag{11.49}$$

where u, v run respectively over indices n_j where $p_{n_j} < q_{n_j}$, $p_{n_j} \geq q_{n_j}$.

Finally, (3) follows from the fact that $Q \cap Cl_{\Pi_A} Z \neq \emptyset$ if and only if $Q \cap Z \neq \emptyset$ for every Q, a class of Ind_{a_n}, any n. $\square$

Example 11.3. *We now return to our example of the Cantor set and we calculate its $\mathcal{A}$–dimension. Assuming an information system $\mathcal{A}$ partitioning the Cantor set $\mathcal{D}$ into families $\mathcal{P}_n$ described above, we have $l_j = 2^j, k_j = 3$, and $m_j = 3^j$ for each j and thus*

$$dim_{\mathcal{A}}(\mathcal{D}) = lim_{n\to\infty} \frac{log \prod_{j=1}^{n} 2^j}{log \prod_{j=1}^{n} 3^j} = \frac{log2}{log3} \tag{11.50}$$

We also recall now the metric D (cf. (11.12, 11.13)) on topological rough sets resulting in the information system $\mathcal{A}$ and we demonstrate the continuity property of the function $dim_{\mathcal{A}}$. We give the proof for the upper dimension, that for the lower one being analogous.

Proposition 11.14. *Assume that $\mathcal{A}$ is of type κ with $k_i \geq 2$ for infinitely many i and a sequence $(T_n)_n$ of rough sets is convergent in the metric D to a rough set T and $dim_{\mathcal{A}}(T_n)$ exists for each n. Then*

$$lim_{n\to\infty} dim_{\mathcal{A}}(T_n) = dim_{\mathcal{A}}(T).$$

Proof. Assume $\varepsilon > 0$; there exists $n(\varepsilon)$ with the property that $D(T_n, T) < \varepsilon$ for $n > n(\varepsilon)$ hence for a natural number M (it suffices, by Proposition 11.4, to take $M \geq -log_{10} \frac{9\varepsilon}{10}$) we have $T_{n\;m}^+ = T_m^+$ for each $m \geq M$ and $n > n(\varepsilon)$. For $n > n(\varepsilon)$ we thus have

$$lim_{n\to\infty} dim_{\mathcal{A}}(T_n) = lim_{n\to\infty} lim_{j\to\infty} \frac{log \prod_{i=1}^{j} l_i^{(n)}}{log \prod_{i=1}^{j} m_i} =$$

$$lim_{j\to\infty} [lim_{n\to\infty} \frac{log \prod_{i=1}^{j} l_i^{(n)}}{log \prod_{i=1}^{j} m_i}] =$$

$$lim_{j\to\infty} [lim_{n\to\infty} \frac{log \prod_{i=1}^{M} l_i^{(n)} \prod_{i=M+1}^{j} l_i^{(n)}}{log \prod_{i=1}^{j} m_i}] =$$

$$lim_{j\to\infty} \frac{log \prod_{i=1}^{M} l_i^{(n)} \prod_{i=M+1}^{j} l_i^{(n)}}{log \prod_{i=1}^{j} m_i} =$$

$$lim_{j\to\infty} \frac{log \prod_{i=M+1}^{j} l_i}{log \prod_{i=1}^{j} m_i} = dim_{\mathcal{A}}(T)$$

where $l_i^{(n)}, l_i$ refer respectively to sets T_n, T.   □

11.5.2   The Approximate Collage Theorem

We now propose yet another application of rough set topologies. The Collage Theorem has been proposed to be applied in fractal compression;

from technical point of view it is a version of the Banach fixed–point theorem adopted to the case of compact sets (fractals) endowed with the Hausdorff-Pompéju metric induced by the Euclidean metric in a Euclidean space (cf. [Barnsley88]). We know that those spaces are complete.

Here we consider the following case (cf. [Polkowski99]). We assume that a sequence $(F_n)_n$ of compact sets (fractals) is given in a Euclidean space E (say 2– or 3–dimensional). We denote by the symbol D_E the Hausdorff –Pompéju metric on the space of compact sets in E induced by a standard metric ρ on E (e.g. to fix our attention, we may adopt the metric

$$\rho(\mathbf{x}, \mathbf{y}) = \sum ((x_i - y_i)^2)^{\frac{1}{2}}.$$

We already know that every fractal in our sense (i.e. having a fractional dimension) is a Π_0–rough set hence an almost rough set with respect to the information system $\mathcal{A}$ on E. We may thus consider the sequence $(F_n)_n$ as a sequence in the complete space $(\mathbf{R}_\omega, D')$ and thus we may apply the Banach fixed – point theorem (or a variant of it).

We denote by the symbol $a_m^+ F$ the upper approximation of the compactum F with respect to the attribute a_m of the information system $\mathcal{A}$; we may formulate our question related to the sequence $(F_n)_n$. We assume that the sequence $(F_n)_n$ does converge in the metric D' to the limit set F. Then a question does arise

(Q) *Estimate the least natural number n with the property that $a_m + F_n = a_m^+ F$ for a given natural number m*

Let us comment on the meaning of this problem. Given F_n, F, we replace those sets with their upper approximations $a_m^+ F_n, a_m^+ F$. Let us observe that description of $a_m^+ F$ requires only a finite number of names of m–cubes in the partition P_m into indiscernibility classes of the relation Ind_{a_m} so this description is a compression of knowledge about F similar to that which results in the case of fractals generated as iterated function systems in which case the description of the fractal is encoded in the starting compact set C and in the coefficients of affine maps whose iterates applied to C generate the fractal in question. Our case here is more general as the compression in terms of upper approximations may be applied to any fractal regardless of the method of its generation.

As with the Collage Theorem we ask for the first integer n such that sets F_n and F have the same upper approximation with respect to a_m. It follows from our discussion of the metric D' that then we have $a_j^+ F_n = a_j^+ F$ for every $j \geq n$. It is a matter of a simple calculation to check that in case $a_m^+ F_n = a_m^+ F$ we also have $D_E(F_n, F) \leq 2^{-m+\frac{1}{2}}$ hence $(F_n)_n$ does converge to F in the metric D_E as well. It follows that $a_m^+ F_n = a_m^+ F$ for a sufficiently large m does assure that sets F_n and F are sufficiently close to each other with respect to metrics D' as well as D_E. The "roughification" $a_m^+ F_n$ may be thus taken as a satisfactory approximation to F.

Let us now refer to the case of fractals generated via iterated function systems. In this case in its simplest form we are given compact fractals $C_0, C_1, ..., C_k$ and a (usually) affine function $f : E \rightarrow E$ with the contraction coefficient $c \in (0, 1)$. The resulting fractal F is obtained as the limit of the sequence $(F_n)_n$ of sets defined inductively as follows:

1. $F_0 = \bigcup_{i=1}^{k} C_i$

2. $F_{k+1} = f(F_k)$

It is easy to check that the resulting mapping on sets has also the contraction coefficient c. As we know, in the context of the Banach fixed–point theorem, the distance between the limit F and the set F_n may be evaluated as

$$D_E(F_n, F) \leq c^n \cdot (1 - c)^{-1} \cdot D_E(F_0, F_1) \tag{11.51}$$

This general result implies

Proposition 11.15. *Assume that* $K = D_E(F_0, F_1)$. *Then in order to satisfy the requirement*

$$D_E(F_n, F) \leq \varepsilon$$

it is sufficient to satisfy the requirement

$$a_m^+ F_n = a_m^+ F$$

with

$$m = \lceil \frac{1}{2} - log_2 \varepsilon \rceil$$

and

$$n \geq \lceil \frac{log[2^{-m+\frac{1}{2}} \cdot K^{-1} \cdot (1 - c)]}{log c} \rceil.$$

Proof. The formula for m follows from the already mentioned fact that $a_m^+ F_n = a_m^+ F$ implies $D_E(F_n, F) \leq 2^{-m+\frac{1}{2}}$ and the corresponding inequality $\varepsilon \leq 2^{-m+\frac{1}{2}}$. Given the formula for m, we solve with respect to n the inequality

$$2^{-m+\frac{1}{2}} \geq c^n \cdot (1 - c)^{-1} \cdot K$$

which results in

$$n \geq \lceil \frac{log[2^{-m+\frac{1}{2}} \cdot K^{-1} \cdot (1 - c)]}{log c} \rceil.$$

If therefore $a_m^+ F_n = a_m^+ F$ with m, n as above then $D_E(F_n, F) \leq \varepsilon$ i.e. F_n does approximate F with the desired accuracy. $\square$

Historic remarks

Topological aspects of rough set theory were recognized early (cf. [Pawlak82[b]], [Skowron88]) in the framework of topology of partitions. General topological rough sets were defined and studied in [Wiweger88]. In [Polkowski opera cit.] topological spaces of rough sets arising in information systems were studied and characterized; those results constitute the greater part of this Chapter. Early attempt in this direction was made in [Marek–Rasiowa86]. Fractals were studied and fractal dimension introduced in [Carathéodory] and [Hausdorff19]. The recent interest in fractals was stimulated by [Mandelbrot75].

Exercises

In Exercises 1–5, we address the problems of convergence in spaces of rough sets.

1. According to [Marek–Rasiowa86], a sequence $(< Q_j^1, ..., Q_j^k >)_j$ of k–tuples of Π_0–closed sets *approximately converges* to the k–tuple $< Q^1, ..., Q^k >$ of Π_0–closed sets if and only if $Cl_{\Pi_j} Q_j^i = Cl_{\Pi_j} Q^i$ for each $j \in N$ and each $i \leq k$. Prove that if a sequence $((Q_j, T_j))_j$ of Π_0–rough sets does converge approximately to a Π_0–rough set (Q, T) then it does converge in the metric D to the limit (Q, T).

2. In the notation of Exercise 1, prove that if $((Q_j, T_j))_j$ is a sequence of Π_0–rough sets approximately converging to a pair (Q, T) of Π_0–closed sets then (Q, T) does represent a Π_0–rough set (thus this assumption in Exercise 1 is superfluous).

3. In the notation and terminology of Exercise 1, prove that if a sequence $((Q_j, T_j, Q_j \cap T_j))_j$, where each of (Q_j, T_j) is a Π_0–rough set, does approximately converge to a triple $(Q, T, Q \cap T)$ of Π_0–closed sets then the sequence $((Q_j, T_j))_j$ of Π_0–rough sets does converge in the metric D^* to a Π_0–rough set (Q, T).

4. Consider an equivalence relation r on the set of Π_0–rough sets defined via $(Q, T) r (Q', T')$ if and only if $Q \cap T = Q' \cap T'$. Denote by the symbol $[Q, T]$ the equivalence class $[(Q, T)]_r$. The symbol $[[Q, T]]$ will denote the representative $(Q \cap T, U)$ of the class $[Q, T]$. Prove that the function $D_1([Q_1, T_1], [Q_2, T_2]) = D([[Q_1, T_1]], [[Q_2, T_2]])$ is a complete metric on the quotient set R_{Π_0}/r.

5. Prove that if a sequence $(([Q_j, T_j]))_j$ does approximately converge to the pair $(Q \cap T, U)$ then the sequence $(([Q_j, T_j]))_j$ does converge to $[[Q, T]]$ in the metric D_1.

In Exercises 6–13, we are concerned with topological properties of some natural transformations on rough sets. The results enclosed in these exercises are taken from [Polkowski op.cit., op.cit.].

6. Consider the operation of *rough complement* $(.)^c$ defined on any Π_0–rough set (Q,T) via $(Q,T)^c = (T,Q)$. Prove that this operation is an isometric function on the space (R_{Π_0}, D).

7. In the notation of Exercise 6, prove that the rough complement is an isometric function on the space (R_{Π_0}, D).

8. Prove the identity: $[[Q,T]]^c = [[(Q,T)^c]]$.

9. Consider a function B defined for any Π_0–rough set (Q,T) via $B((Q,T)) = [[Q,T]]$. Prove that B is continuous as a function from (R_{Π_0}, D^*) into itself.

10. For a given equivalence class $C = [x]_{R_n}$ for some $x \in U, n \in N$, which contains no Π_0–open singleton, consider a function e_C on Π_0–rough sets defined via $e_C(Q,T) = (Q \cup C, T \cup C)$ called the C–erosion. Prove that $e_C(Q,T)$ is a Π_0–rough set whenever (Q,T) is.

11. Prove that the function e_C is continuous as a function on the space (R_{Π_0}, D).

12. For a given equivalence class $C = [x]_{R_n}$ for some $x \in U, n \in N$, which contains no Π_0–open singleton, consider a function d_C on Π_0–rough sets defined via $d_C(Q,T) = (Q \cup C, T \setminus C)$ called the C–dilation. Prove that the family $\mathcal{R} = \{(Q,T) \in R_{\Pi_0} : d_C(Q,T) \in R_{\Pi_0}\}$ is a closed set in the space (R_{Π_0}, D^*).

In [Wiweger88], the notion of a *topological rough set* was introduced viz. for a topological space (X, τ), a pair (P,Q) where P is a τ–open and Q a τ–closed set is a topological rough set whenever there exists a subset $Y \subseteq X$ with the properties that $P = Int_\tau Y, Q = Cl_\tau Y$.

13. Prove that if the set $Q \setminus P$ contains a subset T which is simultaneously dense and boundary then (P,Q) is a topological rough set. [Hint: consider $Y = P \cup T$. For a detailed analysis consult [Wiweger88]]

14. The *Sierpiński Δ–gasket* cf. Figure 2 above is the set S in the plane constructed as follows. Let Δ_0 be the equilateral triangle with side length of 1. Divide each side into halves and use the division points and vertices of Δ_0 to form 4 triangles of side length 1/2. Let P_1 be the set of these 4 triangles and Δ_1 be the union of triangles in P_1 minus the interior of the

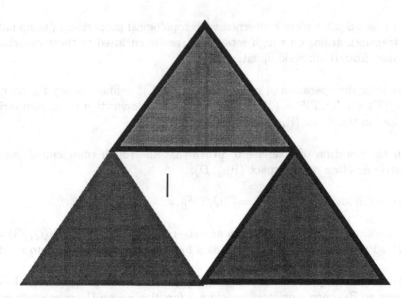

Figure 11.2: The Sierpiński 3–gasket: the second approximation shown in grey

middle one. Repeat this construction with each of remaining triangles and define inductively the sequences $(P_n)_n$ of triangles and $(\Delta_n)_n$ of sets letting finally $S = \bigcap_n \Delta_n$. Prove that $dim_{\mathcal{A}}(S) = \frac{log3}{log4}$ for an appropriate information system $\mathcal{A}$ having the property that classes of Ind_{a_n} are elements of P_n.

Works quoted

[Barnsley88] M. F. Barnsley, *Fractals Everywhere*, Academic Press, 1988.

[Carathéodory14] C.Carathéodory, *Über das lineare Mass von Punktmenge eine Verallgemeinerung des Langenbegriffs*, Nach. Gessell. Wiss. Göttingen, 1914, pp. 406-426.

[Falconer90^a] K. J. Falconer, *The Geometry of Fractal Sets*, Cambridge U. Press, 1990.

[Falconer90^b] K. J. Falconer, *Fractal Geometry. Mathematical Foundations and Applications*, Wiley and Sons, 1990.

[Hausdorff19] F.Hausdorff, *Dimension und ausseres Mass*, Math. Annalen, 79(1919), pp. 157-179.

[Hurewicz–Wallman41] W.Hurewicz and H. Wallman, *Dimension Theory*, Princeton U. Press, 1941.

[Hutchinson81] J. E. Hutchinson, *Fractals and self–similarity*, Indiana Math. Journal, 30(1981), pp. 713–747.

[Mandelbrot75] B. Mandelbrot, *Les Objects Fractals: Forme, Hasard et Dimension*, Flammarion, Paris, 1975.

[Marek–Rasiowa86] W. Marek and H. Rasiowa, *Approximating sets with equivalence relations*, Theor. Computer Sci. 48(1986), pp. 145–152.

[Pawlak91] Z. Pawlak, *Rough Sets: Theoretical Aspects of Reasoning about Data*, Kluwer, Dordrecht, 1991.

[Pawlak82^b] Z. Pawlak, *Rough sets, algebraic and topological approach*, Int. J. Inform. Comp. Sciences, 11(1982), pp. 341–366.

[Polkowski01] L. Polkowski, *On fractals defined in information systems via rough set theory*, in: Proceedings *RSTGC-2001*, Bulletin Intern. Rough Set Society 5(1/2)(2001), pp. 163–166.

[Polkowski99] L. Polkowski, *Approximation mathematical morphology*, in: S. K. Pal, A. Skowron (eds.), *Rough Fuzzy Hybridization. A New Trend in Decision Making*, Springer Verlag Singapore, 1999, pp. 151–162.

[Polkowski98] L. Polkowski, *Hit –or–miss topology*, in: *Encyclopaedia of Mathematics, Supplement 1*, Kluwer, Dordrecht, 1998, p.293.

[Polkowski94] L. Polkowski, *Concerning mathematical morphology of almost rough sets*, Bull. Polish Acad. Sci. Tech., 42(1994), pp. 141–152.

[Polkowski93^a] L. Polkowski, *Metric spaces of topological rough sets from countable knowledge bases*, Foundations of Computing and Decision Sciences, 18(1993), pp. 293–306.

[Polkowski93^b] L. Polkowski, *Mathematical morphology of rough sets*, Bull. Polish Acad. Sci. Math., 41(1993), pp. 241–273.

[Polkowski92] L. Polkowski, *On convergence of rough sets*, in: R. Słowiński (ed.), *Intelligent Decision Support. Handbook of Applications and Advances of the Rough Sets Theory*, Kluwer, Dordrecht, 1992, pp. 305–311.

[Skowron88] A. Skowron, *On topology in information systems*, Bull. Polish Acad. Sci. Math., 36(1988), pp. 477–479.

[Wiweger88] A. Wiweger, *On topological rough sets*, Bull. Polish Acad. Sci. Math., 37(1988), pp. 89–93.

Chapter 12

Algebra and Logic of Rough Sets

Certes, a shadwe hath the likeness of the thing of which it is shadwe, but shadwe is nat the same thing of which it is shadwe

Chaucer, *The Canterbury Tales*, The Parson's Tale

12.1   Introduction

In this Chapter, we discuss algebraic structures induced in collections of rough sets and we present two logical structures, rooted respectively in intuitionistic and modal logics, which reflect properties of indiscernibility and tolerance relations that arise in the attribute–value formalization of information systems. The foundations for this discussion were laid in the papers [Pawlak 81[b], 82[b], 87], [Orlowska–Pawlak 84[a,b]].

In our discussion in previous chapters, two basic structures were singled out viz. *exact sets* and *rough sets*, both relative to a given information system $\mathcal{A} = (U, A)$. We recall that, given an attribute set $B \subseteq A$ and the resulting indiscernibility relation IND_B, any subset $X \subseteq U$ is characterized by means of its approximations: the lower and the upper, which may be also topologically characterized as the interior, respectively, the closure of the set X in the clopen topology induced by the relation IND_B.

In consequence of this approach, any subset $X \subseteq U$ may be represented as a pair (I, C) where $I = Int X$, $C = Cl X$. Alternatively, we may invoke the boundary set $B = C \setminus I$ and then the set X may be represented as the pair (I, B) as the closure C is recovered via $C = I \cup B$.

Finally, letting $D = U \setminus C$, we may represent the set X as the pair (I, D).

As, clearly, $I \cap D = \emptyset$, this representation is called the *disjoint representation* of X (cf. [Pagliani98[a,b]]).

12.2 Algebraic structures via rough sets

We begin with rendering some basic algebraic structures in which we may represent rough sets. First, we recall that exact subsets of U form a Boolean algebra which we denote here with the symbol $\mathcal{E}(U)$. Making use of disjoint representation, we assign to any subset $X \subseteq U$ the pair $r(X) = (I, D)$ where $I = IntX, D = U \setminus ClX$. Then we have a simple observation that

Proposition 12.1. *A subset $X \subseteq U$ is exact if and only if $r(X) = (I, D)$ with $D = U \setminus I$.*

In the sequel we will use the complement symbol $-$ to denote $U \setminus X$ as $-X$. Thus a set $X \subseteq U$ is exact if and only if $r(X) = (I, -I)$. Contrary to that, any rough set $X \subseteq U$ is represented as the pair $r(X) = (I, D)$ with $I \cap D = \emptyset$ and $I \neq -D$. Let us observe once more that both sets I, D occurring in this representation are exact.

Formally, the function r maps the power set 2^U into the Cartesian product $\mathcal{E}(U) \times \mathcal{E}(U)$. Although we could introduce a Boolean algebra structure into this Cartesian product by performing boolean operations coordinate-wise, yet it would come to no avail, as clearly, neither interior nor closure are homomorphisms with respect to both Boolean operations $\cup, \cap$. Thus, a question arises about proper structures for representing the algebra of rough sets.

We first exploit *Nelson algebras* (called also *quasi–pseudo–Boolean algebras* or *$\mathcal{N}$–lattices* cf. [Rasiowa74]) to this end. Nelson algebras are related to *constructive logic with strong negation* (cf. [Nelson49], [Rasiowa74, Ch. XII] for a detailed discussion) but our interest in them is motivated by the fact that they provide an algebraic structure for rough sets (see [Pagliani96, 98[a]] for a deeper discussion).

12.2.1 Nelson algebras of rough sets

A Nelson algebra is a set N together with lattice operations $\vee, \wedge$, zero and unit elements $\mathbf{0}, \mathbf{1}$ and unary operations $\sim, \neg, \rightarrow$ which satisfy the following requirements, where $\leq$ is the lattice order induced by $\vee, \wedge$:

(NLS1) N with $\vee, \wedge, \mathbf{0}, \mathbf{1}$ is a bounded distributive lattice

(NLS2) $\sim\sim x = x$ for each $x \in N$

(NLS3) $\sim (x \vee y) = \sim x \wedge \sim y$ for each pair $x, y \in N$

(NLS4) $x \wedge \sim x \leq y \vee \sim y$ for each pair $x, y \in N$

(NLS5) $x \wedge y \leq \sim x \vee z$ if and only if $y \leq x \rightarrow z$ for each triple $x, y, z \in N$

(NLS6) $x \rightarrow (y \rightarrow z) = (x \wedge y) \rightarrow z$ for each triple $x, y, z \in N$

(NLS7) $\neg x = x \rightarrow \sim x = x \rightarrow 0$ for each $x \in N$

Operations $\rightarrow, \sim, \neg$ are called, respectively, *weak relative pseudo–complementation, quasi–complementation, pseudo–complementation*.

Going back to rough sets, we single out those pairs (I, D) that represent rough sets; by Proposition 31 in Chapter 10, a pair (I, D) of the form $r(X)$ does represent a rough set if and only if no atom of the algebra $\mathcal{E}(U)$ is outside the union $I \cup D$. We denote with the symbol $RS(U)$ the set of those pairs. For pairs $(I_1, D_1), (I_2, D_2) \in RS(U)$, we define the operations $\vee, \wedge, \rightarrow, \neg, \sim$ as well as constants $\mathbf{0}, \mathbf{1}$ as follows (cf. [Pagliani 98^b], Prop. 65).

(1) $\mathbf{1} = (U, \emptyset), \mathbf{0} = (\emptyset, U)$

(2) $(I_1, D_1) \wedge (I_2, D_2) = (I_1 \cap I_2, D_1 \cup D_2)$

(3) $(I_1, D_1) \vee (I_2, D_2) = (I_1 \cup I_2, D_1 \cap D_2)$

(4) $(I_1, D_1) \rightarrow (I_2, D_2) = (-I_1 \cup I_2, I_1 \cap D_2)$

(5) $\sim (I_1, D_1) = (D_1, I_1)$

(6) $\neg (I_1, D_1) = (-I_1, I_1)$

It remains to check that the set $RS(U)$ endowed with the above defined constants $\mathbf{0}, \mathbf{1}$ and operations $\vee, \wedge, \sim, \neg, \rightarrow$ is a Nelson algebra. Let us observe that in definitions of operations $\vee$ and $\wedge$ we have applied Boolean operations $\cup, \cap$ coordinate–wise hence $RS(U)$ with $\vee, \wedge, 0, 1$ is a bounded distributive lattice as the respective conditions in the definition of a lattice are fulfilled by $\cup, \cap$.

It should be also clear that $\mathbf{0}, \mathbf{1}$ defined in the point 1 above are the zero, and respectively, the unit elements of the lattice $RS(U)$. Operations $\vee, \wedge$ induce in already familiar to us way the lattice order $\leq$ viz. $(I_1, D_1) \leq (I_2, D_2)$ if and only if $I_1 \subseteq I_2$ and $D_2 \subseteq D_1$. We may thus proceed with the following

Proposition 12.2. *(Pagliani) The lattice $RS(U)$ with operations $\vee, \wedge$ and*

constants $0, 1$ *is a Nelson algebra when additionally equipped with operations* $\sim, \neg, \rightarrow$.

Proof. We may verify the requirements $(NLS1) - (NLS7)$. We have already observed that $(NLS1)$ is fulfilled.

Concerning $(NLS2)$, we have $\sim\sim (I, D) =\sim (D, I) = (I, D)$ so $(NLS2)$ holds.

For $(NLS3)$, we have $\sim [(I_1, D_1) \vee (I_2, D_2)] =\sim (I_1 \cup I_2, D_1 \cap D_2) = (D_1 \cap D_2, I_1 \cup I_2) = (D_1, I_1) \wedge (D_2, I_2) =\sim (I_1, D_1) \wedge \sim (I_2, D_2)$.

We now verify $(NLS4)$; we consider $x = (I_1, D_1), y = (I_2, D_2)$. As $x \wedge \sim x = (I_1, D_1) \wedge (D_1, I_1) = (I_1 \cap D_1, D_1 \cup I_1) = (\emptyset, D_1 \cup I_1)$ and $y \vee \sim y = (I_2, D_2) \vee \sim (I_2, D_2) = (I_2, D_2) \vee (D_2, I_2) = (I_2 \cup D_2, D_2 \cap I_2) = (I_2 \cup D_2, \emptyset)$ it follows that $x \wedge \sim x \leq y \vee \sim y$.

In case of $(NLS5)$, we consider $x = (I, D), y = (I_1, D_1), z = (I_2, D_2)$. Then $x \wedge y$ is $(I \cap I_1, D \cup D_1)$ and $\sim x \vee z$ is $(D \cup I_2, I \cap D_2)$. Thus $x \wedge y \leq\sim x \vee z$ if and only if the following requirements hold

(i) $I \cap I_1 \subseteq D \cup I_2$

(ii) $I \cap D_2 \subseteq D \cup D_1$

On the other hand, $x \rightarrow z$ is $(-I \cup I_2, I \cap D_2)$ thus $y \leq x \rightarrow z$ if and only if

(iii) $I_1 \subseteq -I \cup I_2$

(iv) $I \cap D_2 \subseteq D_1$

Let us observe that (iv) obviously implies (ii) and from (iii) it follows immediately that $I \cap I_1 \subseteq I_2$ implying clearly (i). Thus, conditions (iii), (iv) imply conditions (i), (ii). On the other hand, as $I \cap D = \emptyset$, from (ii) it follows that (iv) holds. As clearly (i) implies (iii) (again, by the fact that $I \cap D = \emptyset$), we have finally that conditions (i), (ii) are equivalent to conditions (iii), (iv) thus proving $(NLS5)$.

For $(NLS6)$, we let $x = (I, D), y = (I_1, D_1), z = (I_2, D_2)$; then we have

$$x \rightarrow (y \rightarrow z) = (-I \cup -I_1 \cup I_2, I \cap I_1 \cap D_2).$$

On the other hand, we have

$$x \wedge y \rightarrow z = (I \cap I_1, D \cup D_1) = (-I \cup -I_1 \cup I_2, I \cap I_1 \cap D_2).$$

Thus

$$x \rightarrow (y \rightarrow z) = x \wedge y \rightarrow z$$

proving $(NLS6)$.

Finally, passing to $(NLS7)$, we have with $x = (I, D)$ that $\neg x = (-I, I)$ and $x \to \sim x = (I, D) \to (D, I) = (-I \cup D, I) = (-I, I)$ hence $\neg x = x \to \sim x$ proving $(NLS7)$. $\square$

We have proved that rough sets represented in disjoint form may be given the algebraic structure of a quasi–pseudo–boolean algebra (i.e. Nelson algebra). A question may arise whether it would be possible to generate within this representation the algebraic structure of a pseudo–boolean (i.e. Heyting) algebra.

12.2.2 Heyting algebras of rough sets

As neither of weak complement operations in the Nelson algebra of rough sets presented above satisfies the requirement for a pseudo–complement, we will propose a new operation $\Rightarrow$ which will fill this gap. In this way, we turn the Nelson algebra of rough sets into a Heyting algebra of rough sets.

To define $\Rightarrow$, we let

$$(HI) \quad x \Rightarrow y = \neg x \wedge \neg \sim y \vee \sim \neg \sim x \vee y.$$

We reveal the disjoint representation of $x \Rightarrow y$.

To this end, we assume that $x = (I_1, D_1), y = (I_2, D_2)$. Then we have $\neg x = (-I_1, I_1)$ and $\neg \sim y = \neg(D_2, I_2) = (-D_2, D_2)$ hence

$$\neg x \wedge \neg \sim y = (-I_1 \cap -D_2, I_1 \cup D_2) \tag{12.1}$$

Similarly, $\sim \neg \sim x = (D_1, -D_1)$. Hence

$$x \Rightarrow y = (-I_1 \cap -D_2, I_1 \cup D_2) \vee (D_1, -D_1) \vee (I_2, D_2) \tag{12.2}$$

and thus

$$x \Rightarrow y = (-I_1 \cap -D_2 \cup D_1 \cup I_2, -D_1 \cap D_2) \tag{12.3}$$

It remains to check that

$$x \wedge z \leq y \Leftrightarrow z \leq x \Rightarrow y \tag{12.4}$$

We assume that $z = (C, D)$ and thus $x \wedge z \leq y$ amounts to

$$(I_1 \cap C, D_1 \cup D) \leq (I_2, D_2) \tag{12.5}$$

which is equivalent to the following conditions

(v) $I_1 \cap C \subseteq I_2$

and

(vi) $D_2 \subseteq D_1 \cup D$

Similarly, $z \leq x \Rightarrow y$ amounts to

$$(C, D) \leq (-I_1 \cap -D_2 \cup D_1 \cup I_2, -D_1 \cap D_2) \qquad (12.6)$$

which is equivalent to the following conditions

(vii) $C \subseteq -I_1 \cap -D_2 \cup D_1 \cup I_2$

and

(viii) $-D_1 \cap D_2 \subseteq D$

We may observe that conditions (vi) and (viii) are equivalent. Indeed, from (vi): $D_2 \subseteq D_1 \cup D$ it follows that

$$-D_1 \cap D_2 \subseteq -D_1 \cap D_1 \cup -D_1 \cap D = -D_1 \cap D \subseteq D \qquad (12.7)$$

i.e. (viii) is implied by (vi).
Conversely, assuming (viii): $-D_1 \cap D_2 \subseteq D$, we have

$$D_2 = D_2 \cap (D_1 \cup -D_1) = D_2 \cap D_1 \cup D_2 \cap -D_1 \subseteq D_1 \cup D \qquad (12.8)$$

i.e. (vi) is implied by (viii).

Now, we observe that (vii) implies (v); indeed, assuming (vii): $C \subseteq -I_1 \cap -D_2 \cup D_1 \cup I_2$, we have – as $I_1 \cap D_1 = \emptyset$ that

$$I_1 \cap C \subseteq I_1 \cap I_2 \subseteq I_2 \qquad (12.9)$$

i.e. (vii) follows.

We have shown that conditions (vii), (viii) imply conditions (v), (vi). Now, for the converse.

We assume (v), (vi) hence we know that (viii) holds as well and only (vii) needs a proof. First, we have by (viii) and the condition $C \cap D = \emptyset$ that

(ix) $C \subseteq -(-D_1 \cap D_2) = D_1 \cup -D_2$

Now, we have

$$C = C \cap (I_1 \cup -I_1) = C \cap I_1 \cup C \cap -I_1 \qquad (12.10)$$

from which it follows by (v) that

$$C \subseteq I_2 \cup C \cap -I_1 \qquad (12.11)$$

and now it follows by (ix) that

$$C \subseteq I_2 \cup (D_1 \cup -D_2) \cap -I_1 \subseteq I_2 \cup D_1 \cup -I_1 \cap -D_2 \qquad (12.12)$$

i.e. (vii) follows.

Thus, conditions (v), (vi) are equivalent to conditions (vii), (viii) and we have proved (cf. [Pagliani98^b])

Proposition 12.3. *The operation* $\Rightarrow$ *defined as*

$$x \Rightarrow y = \neg x \wedge \neg \sim y \vee \sim \neg \sim x \vee y$$

is a relative pseudo–complementation in the bounded lattice $RS(U)$ of rough sets, making it into a pseudo–Boolean (Heyting) algebra.

We know from Chapter 8 that the specialization $x \Rightarrow 0$ of relative pseudo–complementation is the pseudo–complement to x. We denote the pseudo–complement to x with the symbol † viz.

$$\dagger x = x \Rightarrow 0.$$

As with $\Rightarrow$, we first find the internal structure of $\dagger x$.

We assume that $x = (I, D)$ and then we have

$$\dagger x = x \Rightarrow 0 = (I, D) \Rightarrow (\emptyset, U) = (D, -D) \qquad (12.13)$$

Let us observe that there exists a unique $X \subseteq U$ with the property that $r(X) = \dagger x$ viz. $X = D$. Thus, the function r restricted to the algebra $\mathcal{E}(U)$ of exact sets maps $\mathcal{E}(U)$ onto the set $\dagger(RS(U))$ and there exists the inverse

$$r_{\dagger}^{-1} : \dagger(RS(U)) \to \mathcal{E}(U)$$

to the restricted function

$$r_{\dagger} = r | \mathcal{E}(U) \to \dagger(RS(U)).$$

In order to express properties of † more fully, we invoke the already familiar from Chapter 9 notion of a Stone algebra.

12.2.3 Stone algebras of rough sets

We recall that a bounded lattice

$$(L, \vee, \wedge, 0, 1)$$

with a pseudo–complement $\star$ is a Stone algebra if and only if the condition

$$(ST) \ x^{\star} \vee x^{\star\star} = 1$$

holds. Analogously, a bounded lattice

$$(L, \vee, \wedge, 0, 1)$$

with a unary operation $+$ is a *dual pseudo–complemented lattice* if and only if the following condition

$$(DST) \ x \vee y = 1 \Leftrightarrow x^+ \leq y$$

holds. Then we say that a dual pseudo–complemented latttice

$$(L, \vee, \wedge, +, 0, 1)$$

is a *dual Stone algebra* if and only if the condition

$$(DSA) \ x^+ \wedge x^{++} = 0$$

holds.

These properties may be satisfied simultaneously in a lattice viz. we say that a bounded lattice $(L, \vee, \wedge, 0, 1)$ endowed with a pseudo–complementation $\star$ and a dual pseudo–complementation $+$ is a *double pseudo–complemented lattice*. Consequently, a lattice which is a Stone algebra as well as a dual Stone algebra is said to be a *double Stone algebra*.

Pseudo–complementation $\star$ and dual pseudo–complementation $+$ in general need not be related by any requirement; in case they satisfy the condition

$$(RSA) \ x \wedge x^+ \leq y \vee y^\star$$

for each pair $x, y \in L$, the double Stone algebra L is said to be *regular*.

We already have found that $\dagger$ is a pseudo–complementation in the bounded lattice $RS(U)$; we now check that $\neg$ is a dual pseudo–complementation in $RS(U)$. To this end, we have to verify that

$$x \vee y = 1 \Leftrightarrow \neg x \leq y \qquad (12.14)$$

For $x = (I_1, D_1), y = (I_2, D_2)$, we have $x \vee y = 1$ if and only if

(x) $I_1 \cup I_2 = U$
(xi) $D_1 \cap D_2 = \emptyset$.

On the other hand, the condition $\neg x \leq y$ amounts to

(xii) $-I_1 \subseteq I_2$
(xiii) $D_2 \subseteq I_1$.

Clearly, (x) is equivalent to (xii) and (x) implies (xiii) while (xi) is equivalent to (xiii). Thus conditions (x) and (xi) are equivalent to conditions (xii) and (xiii) and our claim that $\neg$ is a dual pseudo–complementation is proved.

As $RS(U)$ with $\dagger, \neg$ is a double pseudo–complemented algebra, the question may arise whether it is a double Stone algebra. We answer this question in the affirmative (cf. [Pomykala88], [Iwinski87].)

Proposition 12.4. $RS(U)$ *is a regular double Stone algebra when endowed with pseudo–complementations* $\dagger$ *and* $\neg$.

Proof. We have to verify that $\dagger$ and $\neg$ satisfy the respective dual requirements of a Stone algebra. So we have with $x = (I, D)$ that $\dagger x = (D, -D)$ and $\dagger\dagger x = (-D, D)$ and thus

$$\dagger x \vee \dagger\dagger x = (D \cup -D, -D \cap D) = (U, \emptyset) = 1 \qquad (12.15)$$

Similarly

$$\neg x \wedge \neg\neg x = (-I, I) \wedge (I, -I) = (-I \cap I, I \cup -I) = (\emptyset, U) = 0 \qquad (12.16)$$

Finally, we verify the regularity condition with $x, y = (I_1, D_1)$; we have

$$x \wedge \neg x = (I, D) \wedge (-I, I) = (\emptyset, I \cup D) \qquad (12.17)$$

and

$$y \vee \dagger y = (I_1, D_1) \vee (D_1, -D_1) = (I_1 \cup D_1, \emptyset) \qquad (12.18)$$

hence, as

$$\emptyset \subseteq I_1 \cup D_1, I \cup D \qquad (12.19)$$

the regularity condition holds. $\qquad\qquad\qquad\qquad\qquad\qquad\qquad\qquad\qquad$ □

A further examination of properties of $\dagger, \neg$ may be carried out in this setting. First, we may notice that as

$$\sim \dagger \sim = \neg \qquad (12.20)$$

results about $\dagger$ may be translated via this duality into results about $\neg$.

So we look at $\dagger$. Now, it is manifest that $\dagger$ as a function on $RS(U)$ is not any homomorphism as it inverts components of the r–representation; it follows from this remark, however, that the composition of $\dagger$ with itself may be a good candidate for a homomorphism. This idea proves a correct one as witnessed by the following

Proposition 12.5. *The function* $\dagger \circ \dagger : RS(U) \to RS(U)$ *is a homomorphism of the bounded lattice* $RS(U)$ *into itself. By duality, the function* $\neg \circ \neg$ *is also a homomorphism of the bounded lattice* $RS(U)$ *into itself.*

Proof. For $x = (A, B), y = (C, D)$, we have

$$\dagger \circ \dagger(x \vee y) = \dagger\dagger(A \cup C, B \cap D) = (-(B \cap D), B \cap D)$$

and

$$\dagger \circ \dagger(x) \vee \dagger \circ \dagger(y) = (-B, B) \vee (-D, D) = (-B \cup -D, B \cap D)$$

so the two sets are identical to each other; replacing $\vee$ with $\wedge$, we get the same result about identity of

$$\dagger \circ \dagger(x \wedge y)$$

and

$$\dagger \circ \dagger(x) \wedge \dagger \circ \dagger(y).$$

Finally, we have

$$\dagger \circ \dagger(1) = \dagger\dagger(U, \emptyset) = (-\emptyset, \emptyset) = (U, \emptyset) = 1$$

and

$$\dagger \circ \dagger(0) = \dagger\dagger(\emptyset, U) = (-U, U) = (\emptyset, U) = 0.$$

$\square$

We may observe that $\dagger\dagger(I, D) = (-D, D) = (C, -C)$ where $C = ClX$ is the closure of a set $X \subseteq U$ such that $r(X) = (I, D)$ so $\dagger \circ \dagger$ is related to the closure operator; similarly, $\neg \circ \neg(I, D) = (I, -I)$ where $I = IntX$ is the interior of a set $X \subseteq U$ such that $r(X) = (I, D)$. These observations may be rendered formally in the following

Proposition 12.6. *The function* $\dagger \circ \dagger$ *is a closure operation on the algebra* $RS(U)$ *while the function* $\neg \circ \neg$ *is an interior operator on this algebra.*

Proof. We have

$$x \leq \dagger\dagger x :$$

for $x = (I, D)$, we have $\dagger\dagger x = (-D, D)$ and $I \subseteq -D$. Also

$$\dagger\dagger x = \dagger\dagger\dagger\dagger x = (-D, D)$$

and finally letting $y = (I_1, D_1)$, we verify that $x \leq y$ implies

$$\dagger\dagger x \leq \dagger\dagger y :$$

to this end, we observe that

$$(I, D) \leq (I_1, D_1)$$

amounts to

(xiv) $I \subseteq I_1$

(xv) $D_1 \subseteq D$

On the other hand,
$$\dagger\dagger(I, D) \leq \dagger\dagger(I_1, D_1)$$
amounts to

(xvi) $-D \subseteq -D_1$

(xvii) $D_1 \subseteq D$

Obviously from (xiv), (xv), conditions (xvi) and (xvii) follow.

The dual statement about $\neg \circ \neg$ follows directly or by means of duality mentioned above. □

Finally, we will characterize the exact set algebra $\mathcal{E}(U)$ within the algebra $RS(U)$. To this end, we recall the notion of a *center* of a lattice L cf. Sect. 8.7.

An element $x \in L$ is said to be *complemented* if there exists an element $-x$ with the property that $x \wedge -x = 0, x \vee -x = 1$ and the center of L is the set of complemented elements of L, denoted $Ctr(L)$.

We reveal internal structure of elements of $Ctr(RS(U))$. So, we begin with an $x = (I, D)$ and we will search for $y = (A, B)$ such that

(xviii) $x \wedge y = 0$

and

(xix) $x \vee y = 1$

From (xviii), we obtain conditions $I \cap A = \emptyset, D \cup B = U$ and from (xix) it follows that
$$I \cup A = U, D \cap B = \emptyset.$$
Thus, $I = -A, A = -I, B = -D, D = -B$. As $I \cap D = \emptyset$, we have $-A \cap -B = \emptyset$ i.e. $A \cup B = U$.

It follows that $A = -B, B = -A$ and in consequence $I = -D, D = -I$. We obtain finally that

$$(I, D) = (I, -I) \in r(\mathcal{E}(U)), (A, B) = (-I, I) \in r(\mathcal{E}(U)).$$

We have thus (cf. [Pagliani98^a])

Proposition 12.7. *The center $Ctr(RS(U))$ of the algebra $RS(U)$ coincides with the image $r(\mathcal{E}(U))$ of the Boolean algebra of exact subsets of U. Moreover, on the set $Ctr(RS(U))$, the complement $-$ coincides with $\sim$ viz. $-(I,-I)$ $= (-I,I) =\sim (I,-I)$.*

12.3 Łukasiewicz algebras of rough sets

We have already encountered many – valued logics of Łukasiewicz (cf. Chapter 4) ; we recall that an n–valued logic of Łukasiewicz is a logical calculus in which logical values of propositions may take any one of n–values, usually collected in the set $L_n = \{0, \frac{1}{n-1}, \frac{2}{n-1}, ..., \frac{n-2}{n-1}, 1\}$. In particular, we will be concerned with the 3–valued logic with admissible logical values $\{0, \frac{1}{2}, 1\}$.

Łukasiewicz algebras are Lindenbaum-Tarski algebras of respective many–valued logics of Łukasiewicz. We will consider the set L_n of logical values along with the Łukasiewicz interpretation of logical connectives viz.

1. $Cxy = min\{1, 1 - x + y\}$ where we as usual denote with the symbol C the functor of implication

2. $Nx = 1 - x$ where N denotes the functor of negation

3. $x \vee y = max\{x, y\}$

4. $x \wedge y = min\{x, y\}$.

In order to render algebraically properties of logical calculi over L_n, we introduce operators $\phi_1, ..., \phi_{n-1}, \phi'_1, ..., \phi'_{n-1}$ on L_n defined via the formulae

$$\phi_k\left(\frac{i}{n-1}\right) = 0$$

in case $i + k < n$,

$$\phi_k\left(\frac{i}{n-1}\right) = 1$$

otherwise, and

$$\phi'_i = N\phi_i.$$

These operators reflect the ordered structure of the set L_n of logical values.

We may check that the structure

$$\mathbf{L}_n = (L_n, \vee, \wedge, N, 0, 1, \{\phi_i : i = 1, 2, ..., n-1\}, \{\phi'_i : i = 1, 2, ..., n-1\})$$

does satisfy the following

(LUK1) $(L_n, \vee, \wedge, N, 0, 1)$ is a De Morgan algebra i.e. $(L_n, \vee, \wedge, 0, 1)$ is a bounded distributive lattice and N is a dual involutive homomorphism of L_n into itself (meaning: $N(x \vee y) = Nx \wedge Ny$, $N(x \wedge y) = Nx \vee Ny$, and $NNx = x$)

(LUK2) ϕ_i is a $0, 1$–homomorphism of L_n into itself for each i

(LUK3) $\phi_i \wedge \phi'_i = 0$; $\phi_i \vee \phi'_i = 1$ for each i

(LUK4) $\phi_i \circ \phi_j = \phi_j$ for each pair i, j

(LUK5) $i \leq j$ implies $\phi_i \leq \phi_j$

(LUK6) $\phi'_i = N\phi_i$ for each i

(LUK7) $\phi_i(Nx) = N\phi_{n-i}(x)$ for each i and each $x \in L_n$

One may additionally require that the condition

(LUK8) $\phi_i(x) = \phi_i(y)$ for $i = 1, 2, ..., n-1$ imply $x = y$

is satisfied (this condition is called the *determination principle*). We have

Proposition 12.8. *The structure* **L** $_n$ *with operations* $\vee, \wedge, N, \{\phi_i\}$, $\{\phi'_i\}$ *introduced above does satisfy conditions (LUK1)–(LUK8).*

Proof. Clearly, (LUK1) and (LUK6) hold obviously, and (LUK3) follows directly from (LUK6).

The proof of (LUK2) is by direct checking; clearly, $0, 1$ are preserved by each of ϕ's. We verify that

$$(i)\ \phi_i(x \wedge y) = \phi_i(x) \wedge \phi_i(y).$$

We have with $x = \frac{j}{n-1}, y = \frac{k}{n-1}$ where we assume that $j \leq k$ that

$$\phi_i\left(\frac{j}{n-1} \wedge \frac{k}{n-1}\right) = \phi_i\left(\frac{j}{n-1}\right)$$

takes on the value 0 in case $i + j < n$ and the value 1 in case $i + j \geq n$. On the other hand,

$$\phi_i\left(\frac{j}{n-1}\right) \wedge \phi_i\left(\frac{k}{n-1}\right)$$

takes on the value 0 when either $i + j < n$ or $i + k < n$ hence when $i + j < n$; the value 1 is taken when both $i + j, i + k \geq n$ i.e. when $i + j \geq n$. Thus (i) holds.

In similar vein, we would check $\vee$–preservation verifying (LUK2). Condition (LUK4) follows from (LUK2) as each ϕ_i takes on only values $0, 1$ which are preserved by each ϕ_j.

Condition (LUK5) holds by the very definition of ϕ_i, ϕ_j viz. $i + k \geq n$ implies $j + k \geq n$ for each $x = \frac{k}{n-1} \in L_n$ and $i \leq j$ hence $\phi_i(x) = 1$ implies $\phi_j(x) = 1$ and thus $\phi_i \leq \phi_j$.

Now, only (LUK7) is in need of verification; let $x = \frac{k}{n-1}$. We have

$$\phi_i(Nx) = \phi_i(1 - \frac{k}{n-1}) = \phi_i(\frac{n-1-k}{n-1})$$

so $\phi_i(Nx) = 0$ in case

$$(ii)\ i - k < 1$$

and $\phi_i(Nx) = 1$ in case $i - k \geq 1$.

On the other hand, $N\phi_{n-i}(\frac{k}{n-1})$ takes on the value 0 if and only if $\phi_{n-i}(\frac{k}{n-1})$ takes on the value 1 which takes place if and only if $n - i + k \geq n$ i.e. if and only if

$$(iii)\ k - i \geq 0.$$

As (ii) and (iii) are equivalent, (LUK7) follows.

Finally, the determination principle (LUK8) is clearly observed as $x = \frac{i}{n-1} < y = \frac{k}{n-1}$ implies $\phi_{n-1-i}(\frac{i}{n-1}) = 0$ while $\phi_{n-1-i}(\frac{k}{n-1}) = 1$. The proof is thus concluded.   □

A bounded distributive lattice $(L, \vee, \wedge, 0, 1)$ endowed with operations $N, \{\phi_i\}, \{\phi_i'\}$ in such manner that (LUK1)-(LUK8) are satisfied is called a *Łukasiewicz algebra* (cf. [Boicescu91] for a detailed reference to this subject).

We have already mentioned that our special interest would lay with the Łukasiewicz algebra $\mathbf{L}_3$ in which case the set of logical values L_3 consists of $0, \frac{1}{2}, 1$ and we have two homomorphisms ϕ_1, ϕ_2 related to each other by means of the specialization of (LUK7) viz.

$$\phi_1(x) = N\phi_2(Nx)$$

with $\phi_1 \leq \phi_2$.

We may observe here an analogy with the previously discussed case of a Nelson algebra representing rough sets over a universe U: in that case we have dealt with two operations on a set $X \subseteq U$ viz. the interior $IntX = I$ and the closure $ClX = U \setminus D$ related via $I \subseteq U \setminus D$. This analogy points to the possibility of interpreting rough sets as a Łukasiewicz algebra $\mathbf{L}_3$. That this indeed is a fruitful idea is a subject of verification on which we now embark ourselves.

First of all, we collect again axioms of $\mathbf{L}_3$ taking into account the above mentioned relation between ϕ_1 and ϕ_2. We thus call a 3–valued Łukasiewicz algebra a structure $(L, \vee, \wedge, N, 0, 1, \phi_2)$ in which the following requirements are fulfilled

(LUK1)$_3$ $(L, \vee, \wedge, N, 0, 1)$ is a De Morgan algebra

(LUK2)$_3$ $\phi_2(x \wedge y) = \phi_2(x) \wedge \phi_2(y); \phi_2(0) = 0$

(LUK3)$_3$ $\phi_2(x \vee y) = \phi_2(x) \vee \phi_2(y); \phi_2(1) = 1$

(LUK4)$_3$ $\phi_2(x) \wedge N\phi_2(x) = 0; \phi_2(x) \vee N\phi_2(x) = 1$

(LUK5)$_3$ $\phi_2 \circ \phi_2 = \phi_2$

(LUK6)$_3$ $\phi_2 \circ N\phi_2 = N\phi_2$

(LUK7)$_3$ $N\phi_2 N \leq \phi_2$

(LUK8)$_3$ $\phi_2(x) = \phi_2(y)$ and $\phi_2(Nx) = \phi_2(Ny)$ imply $x = y$

It is manifest that (LUK1)$_3$–(LUK8)$_3$ are paraphrases of (LUK1)–(LUK8) which come by replacing ϕ_1 with $N\phi_2 N$.

12.3.1 Wajsberg algebras

We now return to the 3–valued Łukasiewicz logic. We recall from Chapter 4 its axiomatization due to Wajsberg [Wajsberg31]. The list of axioms proposed by Wajsberg is recalled below

(W1) $CpCqp$
(W2) $CCpqCCqrCpr$
(W3) $CCCpNppp$
(W4) $CCNpNqCqp$

We remember that the derivation rule is detachment (modus ponens). We now invoke the Tarski–Lindenbaum algebra of this logic. We recall that this algebra is constructed by starting with the set P of meaningful expressions of the logic on which the equivalence relation $\cong$ is defined: $p \cong q$ if and only if Cpq and Cqp are theses. On the quotient set $P/\cong$, the following operations are introduced:

(1) $[p]_\cong \hookrightarrow [q]_\cong = [Cpq]_\cong$

(2) $N[p]_\cong = [Np]_\cong$

and the constant $1 = [p]_{\cong}$ where p is a thesis is also defined.

We will check that the algebra $(P/\cong, \hookrightarrow, N, 1)$ does satisfy the following requirements

(WA1) $x \hookrightarrow (y \hookrightarrow x) = 1$

(WA2) $(x \hookrightarrow y) \hookrightarrow ((y \hookrightarrow z) \hookrightarrow (x \hookrightarrow z)) = 1$

(WA3) $((x \hookrightarrow Nx) \hookrightarrow x) \hookrightarrow x = 1$

(WA4) $(Nx \hookrightarrow Ny) \hookrightarrow (y \hookrightarrow x) = 1$

(WA5) $1 \hookrightarrow x = 1$ implies $x = 1$

(WA6) $x \hookrightarrow y = 1$ and $y \hookrightarrow x = 1$ imply $x = y$

Indeed, (WA1)–(WA4) are direct consequences of axioms (W1)–(W4), (WA5) follows by detachment, and (WA6) is a rendering of the relation $\cong$.

An algebra $(L, \hookrightarrow, N, 1)$ which satisfies (WA1)–(WA6) is called a *Wajsberg algebra*.

We now should relate 3–valued Łukasiewicz algebras to Wajsberg algebras. We obviously would anticipate that these notions are equivalent; this is actually true, and we indicate the arguments in favor of this statement (cf. [Becchio72, 78], [Boicescu91]).

Proposition 12.9. *(Becchio) Letting*
(i) $x \vee y = (x \hookrightarrow y) \hookrightarrow y$
(ii) $x \wedge y = N(Nx \vee Ny)$
(iii) $\phi_2(x) = Nx \hookrightarrow x$
(iv) $0 = N1$

establishes in the Wajsberg algebra W the structure of the 3– Łukasiewicz algebra $\mathbf{L}_3$.
   Conversely, letting

(v) $x \hookrightarrow y = (\phi_2(Nx) \vee y) \wedge (\phi_2(y) \vee Nx)$

makes the 3– Łukasiewicz algebra $\mathbf{L}_3$ into the Wajsberg algebra W.

Proof. As usual the proof is a series of derivations of one structure within the other and for completeness sake we will give a fairly exact account of these derivations in which we follow [Becchio72, 78] and [Boicescu91]. We consider the Wajsberg algebra W first so we adopt (i)–(iv) as definitions of operations in the Lukasiewicz algebra and we will show that $(LUK1)_3$–$(LUK8)_3$ are satisfied with the interpretation (i)–(iv).

We begin by introducing a partial order into W. To this end, we let

(vi) $x \leq y$ if and only if $x \hookrightarrow y = 1$

It is fairly evident that $\leq$ is a partial order: $x \leq y, y \leq z$ mean $x \hookrightarrow y = 1 = y \hookrightarrow z$ hence (WA2) along with (WA5) implies $x \hookrightarrow z = 1$ i.e. $x \leq z$; substituting $y/Nx \hookrightarrow x$ in (WA1) yields

$$(vii) \quad x \hookrightarrow ((x \hookrightarrow Nx) \hookrightarrow x) = 1$$

i.e.

$$x \leq (x \hookrightarrow Nx) \hookrightarrow x$$

and (WA3) implies

$$(viii) \quad (x \hookrightarrow Nx) \hookrightarrow x \leq x$$

i.e. $x \leq x$ by already proved transitivity. From (WA6) it follows directly that $x \leq y, y \leq x$ imply $x = y$.

Now we check the following statements (a)–(w) which together give the proof that $(W, \leq)$ is a bounded (distributive) lattice:

(a) 1 is the greatest element in $(L, \leq)$

By (WA5), $1 \leq x$ implies $1 = x$ so 1 is maximal; substituting in (WA1) $x/1, y/x$ we get

$$1 \hookrightarrow (x \hookrightarrow 1) = 1$$

i.e. $1 \leq x \hookrightarrow 1$ so $x \hookrightarrow 1 = 1$ i.e. $x \leq 1$.

(b) $x \leq y$ implies $y \hookrightarrow z \leq x \hookrightarrow z$

Indeed, by (WA2) from $x \hookrightarrow y = 1$ it follows via (WA5) that

$$(y \hookrightarrow x) \hookrightarrow (x \hookrightarrow z) = 1$$

i.e.

$$y \hookrightarrow z \leq x \hookrightarrow z.$$

(c) $x = (x \hookrightarrow Nx) \hookrightarrow x$

We have

$$x \hookrightarrow ((x \hookrightarrow Nx) \hookrightarrow x) = 1$$

shown in (vii) and

$$(x \hookrightarrow Nx) \hookrightarrow x \leq x$$

shown in (viii), whence (c) follows.

(d) $Nx \leq x \hookrightarrow y$

Indeed, substitution $x/Nx, y/Ny$ in (WA1) yields $Nx \leq Ny \hookrightarrow Nx$ so by (WA4): $Ny \hookrightarrow Nx \leq x \hookrightarrow y$ we get (d).

(e) $NNx = x$

Substituting in (d) $x/Nx, y/N(x \hookrightarrow Nx)$ we get

$$NNx \leq Nx \hookrightarrow N(x \hookrightarrow Nx) \leq (x \hookrightarrow Nx) \hookrightarrow x \leq x$$

where the last two inequalities follow by (WA4) and (WA3), in that order. Thus $NNx \leq x$ which implies $NNNx \leq Nx$ i.e. $NNNx \hookrightarrow Nx = 1$ and now (WA4) yields $x \hookrightarrow NNx = 1$ i.e. $x \leq NNx$ so finally $x = NNx$.

(f) $Ny \hookrightarrow Nx = x \hookrightarrow y$

We have

$$Ny \hookrightarrow Nx \leq x \hookrightarrow y$$

by (WA4) and

$$x \hookrightarrow y = NNx \hookrightarrow NNy \leq Ny \hookrightarrow Nx$$

by (e) and (WA4).

(g) $x \leq y$ if and only if $Ny \leq Nx$

Directly from (f) and (vi).

(h) $Nx \hookrightarrow y = x \hookrightarrow Ny$

Indeed, by (e),

$$Nx \hookrightarrow y = Nx \hookrightarrow NNy = x \hookrightarrow Ny$$

where the last equality follows by (f).

(i) $1 \hookrightarrow x = x$

By (WA1), $x \hookrightarrow (1 \hookrightarrow x) = 1$ i.e. $x \leq 1 \hookrightarrow x$. On the other hand, $x \hookrightarrow Nx \leq 1$ by (a) so by (b) we have

$$1 \hookrightarrow x \leq (x \hookrightarrow Nx) \hookrightarrow x = x$$

where the last equality is by (c).

(j) $x \le x \vee y$

By (i) and (WA2), with substitutions $x/1, y/x, z/y$, we have

$$x = 1 \hookrightarrow x \le (x \hookrightarrow y) \hookrightarrow (1 \hookrightarrow y) = (x \hookrightarrow y) \hookrightarrow y = x \vee y.$$

(k) $y \le x \vee y$

Substituting $x/y, y/x \hookrightarrow y$ in (WA1), we get

$$y \hookrightarrow ((x \hookrightarrow y) \hookrightarrow y) = 1$$

i.e. $y \le x \vee y$.

(l) $x \le y$ if and only if $x \vee y = y$

If $x \le y$ then

$$x \vee y = (x \hookrightarrow y) \hookrightarrow y = 1 \hookrightarrow y = y$$

by (i). If $x \vee y = y$ then $x \le y$ by (j).

(m) $x \hookrightarrow (y \hookrightarrow z) = y \hookrightarrow (x \hookrightarrow z)$

Indeed, by (k), in the form $y \le y \vee z$, we have

$$y \hookrightarrow (y \hookrightarrow z) \hookrightarrow z = 1$$

hence (WA2) implies after substitution $y/y \hookrightarrow z$ that

$$x \hookrightarrow (y \hookrightarrow z) \le (((y \hookrightarrow z) \hookrightarrow z) \hookrightarrow (x \hookrightarrow z)) \le y \hookrightarrow (x \hookrightarrow z)$$

and

$$y \hookrightarrow (x \hookrightarrow z) \le x \hookrightarrow (y \hookrightarrow z)$$

follows by symmetry.

(n) $x \le y \hookrightarrow z$ if and only if $y \le (x \hookrightarrow z)$

It follows directly from (m).

(o) $x \le y$ implies $z \hookrightarrow x \le z \hookrightarrow y$

As $x \hookrightarrow y = 1$ we have by (i) that

$$(x \hookrightarrow y) \hookrightarrow (z \hookrightarrow y) = z \hookrightarrow y$$

and substituting in (WA2) $x/z, y/x, z/y$ yields

$$(z \hookrightarrow x) \hookrightarrow (z \hookrightarrow y) = 1.$$

(p) $x \leq z, x \hookrightarrow y \leq z, Ny \leq z$ imply $z = 1$

As $Ny \leq z$, we have $Nz \leq NNy = y$ by (g) and (e) hence (b) applied to $x \leq z$ and (o) imply

$$z \hookrightarrow Nz \leq x \hookrightarrow Nz \leq x \hookrightarrow y \leq z$$

and thus by (b) and (c)

$$1 = z \hookrightarrow z \leq (z \hookrightarrow Nz) \hookrightarrow z = z$$

and finally by (a) we have $1 = z$.

(q) $x \leq x \vee y \hookrightarrow y \vee x$

Substituting in (WA2) first $x/x \hookrightarrow y$ and next $y/(x \hookrightarrow y) \hookrightarrow x$ we get

$$x \vee y = (x \hookrightarrow y) \hookrightarrow y \leq$$
$$(y \hookrightarrow x) \hookrightarrow ((x \hookrightarrow y) \hookrightarrow x) \leq$$
$$(((x \hookrightarrow y) \hookrightarrow x) \hookrightarrow x) \hookrightarrow ((y \hookrightarrow x) \hookrightarrow x) \leq$$
$$x \hookrightarrow ((y \hookrightarrow x) \hookrightarrow x) = x \hookrightarrow y \vee x.$$

Thus

$$x \vee y \leq x \hookrightarrow y \vee x$$

hence by (n) we have

$$x \leq x \vee y \hookrightarrow y \vee x.$$

(r) $x \hookrightarrow y \leq x \vee y \hookrightarrow y \vee x$

By (o),
$$x \vee y = (x \hookrightarrow y) \hookrightarrow y \leq (x \hookrightarrow y) \hookrightarrow x \vee y$$
so we have by (k) as $y \leq x \vee y$ that
$$x \hookrightarrow y \leq x \vee y \hookrightarrow y \vee x$$
by (n).

(s) $Ny \leq x \vee y \hookrightarrow y \vee x$

As
$$x \vee y = (x \hookrightarrow y) \hookrightarrow y \leq Nx \hookrightarrow y$$

by (b) and (d), it follows by (f) and (e) that

$$x \vee y \leq Ny \hookrightarrow x$$

thus

$$Ny \leq x \vee y \hookrightarrow x \leq x \vee y \hookrightarrow y \vee x$$

by (m), (j) and (o).

(t) $x \vee y = y \vee x$

Substituting $z/x \vee y \hookrightarrow y \vee x$ in (p), we get

$$x \vee y \hookrightarrow y \vee x = 1$$

i.e. $x \vee y \leq y \vee x$. By symmetry, $y \vee x \leq x \vee y$.

(u) $x \leq z, y \leq z$ imply $x \vee y \leq z$

We have

$$z \hookrightarrow y \leq x \hookrightarrow y$$

by (b) thus

$$x \vee y = (x \hookrightarrow y) \hookrightarrow y \leq (z \hookrightarrow y) \hookrightarrow y$$
$$= z \vee y = y \vee z = (y \hookrightarrow z) \hookrightarrow z = 1 \hookrightarrow z = z$$

where we exploited (b), (t), and (i).

It follows that $\vee$ is a lattice join operation. Moreover,

(w) $0 \leq x$

Indeed, $0 = N1 \leq NNx = x$ by (vi).

Now, $x \wedge y = N(Nx \vee Ny)$ along with (e) and (g) implies dual statements about $\wedge$ showing that $(W, \vee, \wedge, N, 0, 1)$ is a bounded lattice.

Distributivity may be shown along similar lines (cf. [Becchio72, 78], [Boic-escu91]). We pass now to the specific structural properties of the Łukasiewicz algebra L. In order to reveal them in W, we need some auxiliary statements.

(I) $\phi_2(x) \wedge Nx = x \wedge Nx$

Passing to negations, we have by (e), (f), (iii) and (c) that

$$N(\phi_2(x) \wedge Nx) = x \vee N\phi_2(x)$$

$$= (x \hookrightarrow N\phi_2(x)) \hookrightarrow N\phi_2(x) = \phi_2(x) \hookrightarrow N(\phi_2(x) \hookrightarrow Nx)$$

$$= (Nx \hookrightarrow x) \hookrightarrow N((Nx \hookrightarrow x) \hookrightarrow Nx)$$

$$= (Nx \hookrightarrow x) \hookrightarrow NNx = (Nx \hookrightarrow x) \hookrightarrow x$$

$$= Nx \vee x = N(x \wedge Nx)$$

thus (I) follows by (e).

(II) $\phi_2(x) \vee Nx = 1$

Indeed, by (i), (iii), (c), and (e) we have

$$\phi_2(x) \vee Nx = (\phi_2(x) \hookrightarrow Nx) \hookrightarrow Nx$$

$$= ((Nx \hookrightarrow x) \hookrightarrow Nx) \hookrightarrow Nx$$

$$= Nx \hookrightarrow Nx = 1.$$

(III) $x \leq \phi_2(x)$

By (II),

$$x = x \wedge (\phi_2(x) \vee Nx) = x \wedge \phi_2(x) \vee (x \wedge Nx) \leq \phi_2(x).$$

(IV) $x \wedge Nx \leq y \vee Ny$

By (d), $Nx \leq x \hookrightarrow y$ thus by (k) and (o)

$$Nx \leq x \hookrightarrow Ny \vee y$$

and hence (n) implies that

$$x \leq Nx \hookrightarrow y \vee Ny$$

and thus by (b) we have

$$x \leq x \wedge Nx \hookrightarrow y \vee Ny.$$

Moreover,

$$x \hookrightarrow y = NNx \hookrightarrow y \leq x \wedge Nx \hookrightarrow y \vee Ny$$

by (o) and (b).

Finally, by (WA1) with substitutions $x/Ny, y/x \wedge Nx$ and (o) we have

$$Ny \leq x \wedge Nx \hookrightarrow Ny \leq x \wedge Nx \hookrightarrow y \vee Ny$$

and thus it follows by (p) with $z/x \wedge Nx \hookrightarrow y \vee Ny$ that

$$x \wedge Nx \hookrightarrow y \vee Ny = 1$$

i.e. $x \wedge Nx \leq y \vee Ny$.

We now may deduct properties of operators ϕ encoded by $(\text{LUK2})_3 - (\text{LUK7})_3$.

(V) $\phi_2(0) = 0$

Indeed, $\phi_2(0) = \phi_2(0) \wedge N0 = 0 \wedge N0 = 0$ by (I).

(VI) $N\phi_2(x) \wedge x = 0$

Indeed, $N\phi_2(x) \wedge x = N(\phi_2(x) \vee Nx) = N1 = 0$ by (II).

(VII) $\phi_2(N\phi_2(x)) \wedge \phi_2(x) = 0$

It follows by (V) and (VI), as

$$\phi_2(N\phi_2(x)) \wedge \phi_2(x) = \phi_2(N\phi_2(x) \wedge x)$$
$$= \phi_2(0) = 0.$$

(VIII) $\phi_2(N\phi_2(x)) \vee \phi_2(x) = 1$

Indeed, by (II), (I) and (e),

$$\phi_2(N\phi_2(x)) \vee \phi_2(x) = \phi_2(N\phi_2(x)) \vee NN\phi_2(x) = \phi_2(N(\phi_2(x) \wedge Nx))$$
$$= \phi_2(N(Nx \wedge x)) = \phi_2(1) = 1.$$

(IX) $\phi_2(N\phi_2(x)) = \phi_2'(x)$

(IX) follows by (VII) and (VIII).

(X) $N\phi_2(x) = \phi_2'(x)$

By (III) with $x/N\phi_2(x)$ we have

$$N\phi_2(x) \leq \phi_2(N\phi_2(x))$$

thus

$$\phi_2(x) \wedge N\phi_2(x) \leq \phi_2(x) \wedge \phi_2(N\phi_2(x))$$
$$= \phi_2(x \wedge N\phi_2(x)) = \phi_2(0) = 0$$

by (VI). Also,

$$\phi_2(x) \vee N\phi_2(x) = N(N\phi_2(x) \wedge \phi_2(x)) = N0 = 1.$$

It follows that $N\phi_2(x) = \phi_2'(x)$.

(LUK6)$_3$ $\phi_2 \circ N\phi_2 = N\phi_2$

This follows from (IX) and (X).

(XI) $N\phi_2(Nx) \vee N\phi_2(x) \vee (\phi_2(x) \wedge \phi_2(Nx)) = 1$

Indeed,
$$N\phi_2(Nx) \vee N\phi_2(x) \vee (\phi_2(x) \wedge \phi_2(Nx))$$
$$= (N\phi_2(Nx) \vee N\phi_2(x) \vee \phi_2(x)) \wedge (N\phi_2(Nx) \vee N\phi_2(x) \vee \phi_2(Nx)) = 1 \wedge 1 = 1$$
by (II).

(LUK5)$_3$ $\phi_2 = \phi_2 \circ \phi_2$

By (V), (IX), and (X),
$$\phi_2'(x) \wedge \phi_2(\phi_2(x)) = \phi_2(N\phi_2(x) \wedge \phi_2(\phi_2(x)))$$
$$= \phi_2(N\phi_2(x) \wedge \phi_2(x)) = \phi_2(0) = 0$$
hence by distributivity
$$\phi_2(x) = \phi_2(x) \vee (\phi_2'(x) \wedge \phi_2(\phi_2(x)))$$
$$= \phi_2(x) \vee \phi_2(\phi_2(x)) = \phi_2(\phi_2(x))$$
by (III).

(LUK4)$_3$ $\phi_2(x) \wedge N\phi_2(x) = 0$

By (LUK6)$_3$, (LUK5)$_3$ and (XI) along with (e) we have
$$N\phi_2(x) \vee \phi_2(x)$$
$$= N\phi_2(x) \vee \phi_2(x) \vee (\phi_2(x) \wedge N\phi_2(x))$$
$$= N\phi_2(N\phi_2(x)) \vee (\phi_2(\phi_2(x)) \wedge \phi_2(N\phi_2(x))) \vee N\phi_2(\phi_2(x)) = 1$$
so (LUK4)$_3$ follows by duality.

(LUK7)$_3$ $N\phi_2(Nx) \leq \phi_2(x)$

From (III) it follows that $Nx \leq \phi_2(Nx)$ and thus by (g) and (e) we have
$$N\phi_2(Nx) \leq NNx = x \leq \phi_2(x)$$

by (III).

(XII) $Nx \wedge y \leq x$ implies $y \leq \phi_2(x)$

Indeed, by (II),

$$y \leq y \vee \phi_2(x) = (y \vee \phi_2(x)) \wedge (\phi_2(x) \vee Nx)$$

$$= \phi_2(x) \vee (Nx \wedge y) \leq x \vee \phi_2(x) = \phi_2(x)$$

by (III).

$(LUK2)_3$ $\phi_2(x \wedge y) = \phi_2(x) \wedge \phi_2(y)$

By (I)

$$Nx \wedge \phi_2(x \wedge y) \leq N(x \wedge y) \wedge \phi_2(x \wedge y) \leq x \wedge y \leq x$$

and thus (XII) implies that $\phi_2(x \wedge y) \leq \phi_2(x)$.
    Analogously, $\phi_2(x \wedge y) \leq \phi_2(y)$ so

$$(*) \ \phi_2(x \wedge y) \leq \phi_2(x) \wedge \phi_2(y).$$

By (I), (III), (IV),

$$Nx \wedge \phi_2(x) \wedge \phi_2(y) = Nx \wedge x \wedge \phi_2(y)$$

$$= x \wedge (Nx \wedge x) \wedge \phi_2(y) \leq x \wedge (y \vee Ny) \wedge \phi_2(y)$$

$$= x \wedge [y \vee (Ny \wedge \phi_2(y)] = x \wedge y;$$

similarly,

$$Ny \wedge \phi_2(x) \wedge \phi_2(y) \leq x \wedge y.$$

It follows that

$$N(x \wedge y) \wedge \phi_2(x) \wedge \phi_2(y) \leq x \wedge y$$

so (XII) yields

$$(**) \ \phi_2(x) \wedge \phi_2(y) \leq \phi_2(x \wedge y).$$

From (*) and (**), $(LUK2)_3$ follows.

$(LUK3)_3$ $\phi_2(x) \vee \phi_2(y) = \phi_2(x \vee y)$

Clearly,

$$\phi_2(x) \vee \phi_2(y) \leq \phi_2(x \vee y)$$

follows by the fact that $x \leq y$ implies $\phi_2(x) \leq \phi_2(y)$, a direct consequence of $(LUK2)_3$. The converse comes by dualization via (e) and (g), $(LUK6)_3$, and $(LUK2)_3$:

$$\phi_2(x \vee y) \leq \phi_2(\phi_2(x \vee y))$$

$$= \phi_2(N(N\phi_2(x) \wedge N\phi_2(y))$$
$$= \phi_2 N(\phi_2((N\phi_2(x)) \wedge \phi_2(N\phi_2(y))))$$
$$= \phi_2(N\phi_2(N\phi_2(x) \wedge N\phi_2(y)))$$
$$= N(\phi_2(N\phi_2(x)) \wedge \phi_2(N\phi_2(y)))$$
$$= N(N\phi_2(x) \wedge N\phi_2(y)) = \phi_2(x) \vee \phi_2(y).$$

(XIII) $(N\phi_2(Nx) \vee Nx) \wedge \phi_2(x) = x$

Indeed, by (e), and (I),

$$(N\phi_2(Nx) \vee Nx) \wedge \phi_2(x)$$

$$= N(Nx \wedge x) \wedge \phi_2(x) = (x \vee Nx) \wedge \phi_2(x)$$
$$= (x \wedge \phi_2(x)) \vee (Nx \wedge \phi_2(x) = x \vee (x \wedge Nx) = x.$$

$(LUK8)_3$ (the determination principle)

Assume that $\phi_2(x) = \phi_2(y)$ and $\phi_2(Nx) = \phi_2(Ny)$. Then by (XIII) and $(LUK2)_3$,

$$x \vee y = (N\phi_2(N(x \vee y)) \vee N(x \vee y)) \wedge \phi_2(x \vee y)$$

$$= (N(\phi_2(Nx) \wedge \phi_2(Ny)) \vee (Nx \wedge Ny)) \wedge (\phi_2(x) \vee \phi_2(y))$$

$$= (N\phi_2(Nx) \vee Nx) \wedge (N\phi_2(Ny) \vee Ny) \wedge \phi_2(x) \wedge \phi_2(y) = x \wedge y$$

where again the final conclusion follows by (XIII).

In this way we have shown that any Wajsberg algebra carries a structure of a 3– Lukasiewicz algebra so 3– Lukasiewicz algebras are in this sense Tarski–Lindenbaum algebras of 3–valued Lukasiewicz calculi.

The proof of the converse goes along similar lines, we assume the definition (v) of $x \hookrightarrow y$ as $(\phi_2(Nx) \vee y) \wedge (\phi_2(y) \vee Nx)$.

Then for instance

(iii) follows: $Nx \hookrightarrow x = (\phi_2(NNx) \vee x) \wedge (\phi_2(x) \vee NNx) = \phi_2(x) \vee x = \phi_2(x)$ by (III).

We sketch proofs of (i), (ii), (iv) and (WA1)–(WA6) in Exercises, below. $\square$

Now, we return to rough set representation as a Nelson algebra; recalling our discussion of operators of interior and closure as counterparts of operators ϕ_1, ϕ_2, we may also turn our attention to operators in the Nelson algebra of rough sets which return us the interior, respectively, closure component of a rough set representation viz. operators $\neg\neg$ and $\dagger\dagger$.

We apply the relation $\phi_1 = N\phi_2 N$ and we adopt: $\neg\neg$ as ϕ_1, $\dagger\dagger$ as ϕ_2 and $\sim$ as N. Then we may state our conjecture

Proposition 12.10. *Under the above representation of ϕ_1, ϕ_2, N, the disjoint representation of rough sets with operations $\vee, \wedge$ and constants $0, 1$ becomes a 3 Lukasiewicz algebra.*

Proof. We first check that the relation $\phi_1 = N\phi_2 N$ holds so we may avail ourselves of the axiomatization in terms of ϕ_2 of a 3– Lukasiewicz algebra.

Indeed, we have $N\phi_2(N(I, D)) = N\phi_2(D, I) = N(-I, I) = (I, -I) = \phi_1(I, D)$.

Now, we may check that all axioms are fulfilled.
1. $\phi_2((I, D)) \wedge N\phi_2((I, D)) = (-D, D) \wedge (D, -D) = (\emptyset, U) = 0$.

2. $\phi_2 \circ \phi_2 = \phi_2$ as $\phi_2 = \dagger\dagger$ is a closure operator.

3. $\phi_2(N\phi_2((I, D)) = \phi_2(N(-D, D)) = \phi_2((D, -D)) = (D, -D) = N\phi_2((I, D))$.

4. $N\phi_2(N(I, D)) = (I, -I) \leq (-D, D) = \phi_2(N(I, D))$.

We know (cf. Proposition 12.8) that ϕ_2 is a $0, 1$–homomorphism and the determination principle follows immediately as $\phi_2, \phi_2 \circ N$ return the D, respectively the I component of (I, D). We may thus interpret rough sets in 3–valued logic.   □

12.3.2   Post algebra representation of rough sets

A 3– Lukasiewicz algebra L is *centered* if and only if there exist $d_1, d_2 \in L$ with the properties that

(1) $\phi_i(d_j) = 1$ in case $j \leq i$ and $\phi_i(d_j) = 0$ in case $j > i$ for $i, j = 1, 2$

(2) For each choice $a_1 \leq a_2$ of elements a_1, a_2 in the centre $Ctr(L)$ of L, we have $x = a_1 \wedge d_1 \vee a_2 \wedge d_2 \in L$ and the representation of x in this form is unique for every $x \in L$.

Let us check that 3–valued Lukasiewicz algebra of rough sets is centered. We are searching for $d_1 = (I_1, D_1), d_2 = (I_2, D_2)$ satisfying (1), (2). From (1) it follows that

(3) $\neg\neg(I_1, D_1) = (I_1, -I_1) = (U, \emptyset)$

hence $I_1 = U$ a fortiori $D_1 = \emptyset$. Thus $1 = (U, \emptyset)$ is a candidate for d_1.

(4) $\dagger\dagger 1 = 1$ so $d_1 = 1$ fulfilling the second requirement for d_1

(5) $\neg\neg(I_2, D_2) = (I_2, -I_2) = 0 = (\emptyset, U)$ hence $I_2 = \emptyset$

(6) $\dagger\dagger(\emptyset, D_2) = (-D_2, D_2) = 1 = (U, \emptyset)$ hence $D_2 = \emptyset$ and $d_2 = (\emptyset, \emptyset)$

We have found possible d_1, d_2 and it remains to verify the requirement (2).
We recall that the centre $Ctr(L)$ of L consists of exact set representations.
    Assume $a_1 = (I_1, -I_1) \leq a_2 = (I_2, -I_2)$ are elements of the centre, and
consider $x = a_1 \wedge d_1 \vee a_2 \wedge d_2 = (I_1, -I_2)$; as $I_1 \subseteq I_2$, we have $I_1 \cap -I_2 = \emptyset$
so x is a disjoint representation of a rough set i.e. $x \in L$. Clearly, this
representation of x is unique.
    We have found

Proposition 12.11. *3–valued Lukasiewicz algebra of rough sets is centered.*

    A 3–valued *Post algebra* $(L, \vee, \wedge, 0, 1, c_1, c_2)$ is a structure satisfying the
following requirements

(P1) $(L, \vee, \wedge, 0, 1)$ is a distributed bounded lattice

(P2) $c_1 \leq c_2$

(P3) each $x \in L$ has a unique representation in the form $x = b_1 \wedge c_1 \vee b_2 \wedge c_2$
with some $b_1 \geq b_2$ where $b_1, b_2 \in Ctr(L)$.

Post algebras were introduced (cf. [Rosenbloom42]) as Lindenbaum–Tarski
algebras of $n-$ valued Post logics introduced by Emil Post [Post21].
    Let us observe that letting $c_1 = d_2, c_2 = d_1$ and $b_1 = a_2, b_2 = a_1$ does
establish a one–to –one correspondence between Post algebras and centered
Lukasiewicz algebras. Thus we have

Proposition 12.12. *The algebra* $(L, \vee, \wedge, 0, 1, c_1, c_2)$ *obtained from 3– Luka-*
siewicz centered algebra of rough sets $(L, \vee, \wedge, 0, 1, d_1, d_2)$ *by letting* $c_1 =$
$d_2, c_2 = d_1$ *is a 3–valued Post algebra.*

    We have obtained in this way a characterization of rough sets as a Post
algebra.

At this point we conclude our survey of algebraic structures representing
rough sets and we pass to some selected logics expressing rough set theoretic
notions and developed within information systems.

We begin with a logical rendering of functional dependencies (cf. Chapter 10) due to C. Rauszer [Rauszer84, 85^{a,b}]; as shown by Rauszer, the logic of indiscernibility relations in an information system (hence a fortiori the logic of functional dependence (cf. Chapter 10, Exercise 20)) is equivalent to the fragment of intuitionistic logic expressed by means of functors of conjunction and implication.

12.4 A logic of indiscernibility

We recall Exercises 19–26 of Chapter 10, in which an algebra was pointed to built with the set of indiscernibility relations on the universe of an information systems. Given an information system $\mathcal{A} = (U, A)$, we consider the set $\mathcal{I}_A = \{IND_C : C \subseteq A\}$ (observe that we identify here sets of attributes having identical indiscernibility relations) with the set theoretical intersection $\cap$ ordered by set theoretical inclusion and having the unit element $1_A = IND_\emptyset$. The relative pseudocomplement $IND_C \Rightarrow IND_D$ is defined as the greatest among elements IND_E with the property that $IND_C \cap IND_E \leq IND_D$.

A subset $\mathcal{M} \subseteq \mathcal{I}_A$ closed with respect to $\Rightarrow$, $\cap$, and containing the unit 1_A is called an FD-algebra. FD-algebras will play a decisive part in development of semantics for the announced logic. With this end in mind, we are going to construct a canonical FD-algebra $\mathcal{M}^c$.

In order to construct $\mathcal{M}^c$, we begin with a development of the logic in question, the FD-logic which will be presented here according to Rauszer.

12.4.1 The syntax of FD–logic

The language of the FD-logic of the information system $\mathcal{A}$ contains *terms* and *formulas* as two basic components. Following [Rauszer84], we will denote subsets of the set A of attributes with small letters $a, b, c, \ldots$.

The set T of terms is the smallest set containing each a for $a \subseteq A$ and closed with respect to set-theoretic operations of $\cup$, $\cap$, and $\setminus$. Thus expressions like a, b, $a \cup b$, $a \cap b$, $a \setminus b \cap c$ etc. etc. are terms. Terms of the form a where $a \subseteq A$ are called *elementary*. According to [Rauszer84], we denote the empty term (the empty subset of A) with the symbol $\perp$.

Formulas of the FD-logic are built from terms by means of a unary function symbol ι: for a term $t \in T$, the expression $\iota(t)$ is an *atomic formula*.

The set F of formulas is the smallest set containing all atomic formulas and closed with respect to logical functors of conjunction $\wedge$ and implication $\rightarrow$ (please note that we here depart for technical reasons from our notation of C for implication). The intended meaning of the formula $\iota(a)$ will be $IND_{\{a\}}$. Formulas will be denoted with Greek letters $\alpha, \beta, \ldots$.

We use the sequent formalization of classical as well as intuitionistic sentential calculus. Thus, a *sequent of terms* is an expression of the form $\Gamma \vdash \Delta$ where Γ, Δ are finite sequences of term symbols. Similarly, a *sequent of formulas* is an expression of the form $\Phi \supset \alpha$ where Φ is a finite sequence of formulas and α a formula.

The axioms for FD–logic are as follows

(A1) $a \vdash a$

(A2) $\perp \vdash a$

and we admit substitution so (A1), (A2) are axiom schemes.

Inference rules for terms are the Gentzen style rules (cf. Chapter 2) for the sentential calculus (cf. also [Gentzen34]). Observe that in the list of rules given below, we have included the cut rule which may be eliminated from usage (cf. the Hauptsatz, Chapter 2, Exercise 43) as well as some auxiliary rules; their introduction simplifies some derivations. Rules for implication are not given here as they are dispensable. Thus we have the following inference rules for terms (we apply the notational conventions of [Rauszer84])

(str) $\dfrac{\Gamma \vdash \Delta}{a, \Gamma \vdash \Delta b}$; $\dfrac{a, b, \Gamma \vdash \Delta, c, d}{b, a, \Gamma \vdash \Delta, d, c}$; $\dfrac{a, a, \Gamma \vdash \Delta, b, b}{a, \Gamma \vdash \Delta, b}$

(cut) $\dfrac{\Gamma \vdash \Delta, a; a, \Theta \vdash \Sigma}{\Gamma, \Theta \vdash \Delta, \Sigma}$

($\cup \vdash$) $\dfrac{a, \Gamma \vdash \Delta; b, \Gamma \vdash \Delta}{a \cup b, \Gamma \vdash \Delta}$

($\vdash \cup$) $\dfrac{\Gamma \vdash \Delta, a, b}{\Gamma \vdash \Delta, a \cup b}$

($\cap \vdash$) $\dfrac{a, b, \Gamma \vdash \Delta}{a \cap b, \Gamma \vdash \Delta}$

($\vdash \cap$) $\dfrac{\Gamma \vdash \Delta, a; \Gamma \vdash \Delta, b}{\Gamma \vdash \Delta, a \cap b}$

($- \vdash$) $\dfrac{\Gamma \vdash \Delta, a}{-a, \Gamma \vdash \Delta}$

($\vdash -$) $\dfrac{a, \Gamma \vdash \Delta}{\Gamma \vdash \Delta, -a}$

The inference rules for formulas are the Gentzen rules for the fragment of intuitionistic sentential calculus expressible in terms of functors of conjunction $\wedge$ and implication $\rightarrow$ (cf. Chapter 9, [Gentzen34], [Rauszer84])

(str) $\dfrac{\beta, \beta, \gamma, \delta, \Phi \supset \alpha}{\eta, \beta, \delta, \gamma, \Phi \supset \alpha}$

(cut) $\dfrac{\Phi \supset \alpha; \alpha, \Psi \supset \beta}{\Phi, \Psi \supset \beta}$

($\wedge \supset$) $\dfrac{\alpha, \beta, \Phi \supset \gamma}{\alpha \wedge \beta, \Phi \supset \gamma}$

($\supset \wedge$) $\dfrac{\Phi \supset \alpha; \Phi \supset \beta}{\Phi \supset \alpha \wedge \beta}$

$(\to\supset)$ $\dfrac{\Phi\supset\alpha;\beta,\Phi\supset\gamma}{\alpha\to\beta,\Phi\supset\gamma}$

$(\supset\to)$ $\dfrac{\alpha,\Phi\supset\beta}{\Phi\supset\alpha\to\beta}$

The link between term inferences and formula inferences is provided by the specific inference rule

(FD) $\dfrac{a\vdash b,\Delta}{\Delta^{*},\iota(b)\supset\iota(a)}$

where Δ^{*} is the sequence $\iota(c_1),...,\iota(c_k)$ for Δ being the sequence $c_1,..,c_k$.

The notions of a proof and a thesis are as usual. A proof of a sequent S: $\Phi\supset\alpha$ is a finite sequence $S_0,S_1,..,S_m$ of sequents where S_0 is an axiom instance, S_m is S and such that each S_i is obtained from one or two preceding sequents by means of an inference rule. A sequent having a proof is said to be a *thesis*. We say that a formula α is a thesis in case the sequent $\supset\alpha$ is a thesis.

12.4.2 Semantics of FD–logic

Consider an FD–algebra $\mathcal{M}$ for the information system $\mathcal{A}$. We define a valuation on formulas of FD–logic with values in $\mathcal{M}$. This valuation will be defined in two steps. First, we define a valuation v on terms as follows. Let $A' = \{a \subseteq A : \iota(a) \in \mathcal{M}\}$ and let $v : T \to A'$ be a homomorphism of T into A'.

Having v defined, we define a valuation w_v on formulas as follows: for an atomic formula $\iota(a)$ with $a \subseteq A$, we let $w_v(\iota(a)) = \iota(v(a))$ and we extend w_v in the usual way i.e. $w_v(\alpha \wedge \beta) = w_v(\alpha) \cap w_v(\beta)$, $w_v(\alpha \to \beta) = w_v(\alpha) \Rightarrow w_v(\beta)$.

Valuations v, w_v extend over sequents of, respectively, terms and formulas as follows. For a sequence Γ of terms, we denote by the symbol δ_Γ the union of all terms in Γ, and the symbol γ_Γ will denote the intersection of all terms in Γ; similarly, for a sequence Φ of formulas, the symbol σ_Φ will denote the conjunction of all formulas, members of Φ. Then we let

$$v(\Gamma \vdash \Delta) = -v(\gamma(\Gamma) \cup v(\delta(\Delta)) \tag{12.21}$$

and

$$w_v(\Phi \supset \alpha) = w_v(\sigma_\Phi) \Rightarrow w_v(\alpha) \tag{12.22}$$

A sequent $\Gamma \vdash \Delta$ is true in case $v(\Gamma \vdash \Delta) = A$ for every valuation v. A sequent $\Phi \supset \alpha$ is true in case $w_v(\Phi \supset \alpha) = 1_\mathcal{M}$ for every valuation w_v in every FD–algebra $\mathcal{M}$ for the information system $\mathcal{A}$.

Let us observe that inference rules for FD–logic preserve truth i.e. the conclusion of the rule is true whenever its premises are true, for each rule. We know from Chapter 2 that this is the case with inference rules for sentential calculi so the only rule which needs checking is (DF). To see that (DF) has this property, assume that we have (DF) in the form

$$\frac{a \vdash b, \Delta}{\Delta^*, \iota(b) \supset \iota(a)} \tag{12.23}$$

Assume that the premise $a \vdash b, \Delta$ is true hence for every valuation v we have

$$-v(a) \cup v(b) \cup v(c_1) \cup v(c_2) \cup ... \cup v(c_k) = A \tag{12.24}$$

where Δ is the sequence $c_1, c_2, .., c_k$; we fix a valuation v.
For the associated valuation w_v on the conclusion $\Delta^*, \iota(b) \supset \iota(a)$ we have

$$\begin{aligned} w_v(\Delta^*, \iota(b) \supset \iota(a)) = \\ IND_{v(c_1)} \cap ... \cap IND_{v(c_k)} \cap IND_{v(b)} \Rightarrow IND_{v(a)} \end{aligned} \tag{12.25}$$

As the condition

$$-v(a) \cup v(b) \cup v(c_1) \cup v(c_2) \cup ... \cup v(c_k) = A \tag{12.26}$$

implies that

$$IND_{v(c_1)} \cap ... \cap IND_{v(c_k)} \cap IND_{v(b)} \subseteq IND_{v(a)} \tag{12.27}$$

we have

$$\begin{aligned} IND_{v(c_1)} \cap ... \cap IND_{v(c_k)} \cap IND_{v(b)} \Rightarrow \\ IND_{v(a)} = IND_\perp = 1_M \end{aligned} \tag{12.28}$$

i.e. the conclusion $\Delta^*, \iota(b) \supset \iota(a)$ is true under w_v. It follows that the conclusion is true.
We arrive at the

Proposition 12.13. *For every formula α, if α is a thesis then α is true.*

The converse i.e. the statement about completeness of the FD–logic turns out to be also true. Before we enter a discussion of this subject, we define a particular FD–algebra, the *canonical FD–algebra*. First we return to the already familiar idea of the Lindenbaum–Tarski algebra and we offer a closer look at the respective Lindenbaum–Tarski algebras of the term and formula calculi of the FD–logic.

The Lindenbaum–Tarski algebra $\mathcal{L}_T$ of the term calculus is the standard Lindenbaum – Tarski algebra of the sentential calculus i.e. it is a Boolean algebra (cf. Chapter 9) of equivalence classes $|t|$ for $t \in T$ where the equivalence relation $\sim$ on T is defined as usual i.e. $t_1 \sim t_2$ if and only if $t_1 \vdash t_2$ and $t_2 \vdash t_1$ are provable (i.e. are theses). We recall that in this algebra the zero element is the class $|\perp|$ and the unit element 1_T is the class $|t|$ where t is a thesis i.e. the formula $\vdash t$ is provable. We know that the sentential calculus is complete (cf. Chapter 2). Let us observe that completeness of the sentential calculus means essentially the following: any formula satisfied in the Boolean algebra $\{0, 1\}$ is satisfied in any Boolean algebra. In particular, we have that a term t is a thesis if and only if $v(t) = A$ for any valuation v. In particular, we have

Proposition 12.14. $a \vdash b$ is a thesis if and only if $a \subseteq b$ if and only if $IND_{\{b\}} \subseteq IND_{\{a\}}$

Indeed, as v is a homomorphism it suffices to consider the canonical valuation $v(a) = a$ in which case we have $a \vdash b$ is a thesis if and only if $v(a \vdash b) = A$ i.e. $-a \cup b = A$ i.e. $a \subseteq b$ i.e. $IND_{\{b\}} \subseteq IND_{\{a\}}$.

The Lindenbaum–Tarski algebra $\mathcal{L}_F$ is defined in a similar way: for formulas α, β, we let $\alpha \sim \beta$ if and only if $\supset \alpha \to \beta$, $\supset \beta \to \alpha$ are provable and we let $\mathcal{L}_F = F/\sim$. We denote by the symbol $|\alpha|$ the equivalence class of α. In a standard way we prove that the ordering $|\alpha| \leq |\beta|$ if and only if $\supset \alpha \to \beta$ is a thesis induces the meet $\cap$: $|\alpha| \cap |\beta| = |\alpha \cap \beta|$ and the pseudo-complementation $|\alpha| \Rightarrow |\beta| = |\alpha \to \beta|$.

The link between the two algebras is provided by the syntactic property of the FD–logic which follows from the above listed inference rules viz. $a \vdash b$ is a thesis if and only if $\supset \iota(b) \to \iota(a)$ is a thesis. Thus,

Proposition 12.15. $\supset \iota(b) \to \iota(a)$ is a thesis if and only if $IND_{\{b\}} \subseteq IND_{\{a\}}$.

Now, we define a canonical FD–algebra as the algebra

$$\mathcal{A}_0 = \{IND_{\{a\}} : a \subseteq A\}$$

with $\cap, \Rightarrow$ defined as $IND_{\{a\}} \cap IND_{\{b\}} = IND_{a \cup b}$, $IND_{\{a\}} \Rightarrow IND_{\{b\}} = IND_{--a \cap b}$ and the unit element $IND_{\perp}$.

Let us observe that we may define a function $h : \mathcal{L}_F \to \mathcal{A}_0$ from the Lindenbaum–Tarski algebra of formulas into the canonical FD–algebra by letting $h(|\iota(a)|) = IND_{\{a\}}$. It turns out moreover that

Proposition 12.16. The function h is an isomorphism of $\mathcal{L}_F$ onto $\mathcal{A}_0$ i.e. the following properties are observed
1. $|\iota(a)| \neq |\iota(b)|$ implies $h(|\iota(a)|) \neq h(|iota()|)$
2. $h(|\iota(a)| \cap |\iota(b)|) = h(|\iota(a)|) \cap h(|\iota(b)|)$

3. $h(|\iota(a)| \Rightarrow |\iota(b)|) = h(|\iota(a)|) \Rightarrow h(|\iota(b)|)$

Proof. For (1), assume that $h(|\iota(a)|) = h(|\iota(b)|)$ i.e. $IND_{\{a\}} = IND_{\{b\}}$. Thus, both $\iota(a) \to \iota(b)$ and $\iota(b) \to \iota(a)$ are theses (cf. Proposition 12.15) hence $|\iota(a)| = |\iota(b)|$.

In case (2), one checks that formulas $\iota(a) \cap \iota(b) \to \iota(a \cup b)$ and $\iota(a \cup b) \to \iota(a) \cap \iota(b)$ are theses. We reproduce a proof of the thesis $\iota(a) \cap \iota(b) \to \iota(a \cup b)$, the other proof goes on similar lines: from axiom instances $a \vdash a$, $b \vdash b$ one obtains $a \vdash a, b$, $b \vdash a, b$ by applying the rule (str) whence via ($\cup \vdash$) it follows that $a \cup b \vdash a, b$ and the rule (DF) implies that $\iota(a), \iota(b) \sup -set\iota(a \cup b)$. Applying the rule ($\cap \supset$) yields $\iota(a) \cap \iota(b) \supset \iota(a \cup b)$ and finally $\supset \iota(a) \cap \iota(b) \to \iota(a \cup b)$ follows by the rule ($\supset \to$). Thus $h(|\iota(a)| \cap |\iota(b)|) = h(\iota(a \cup b)) = IND(a \cup b) = IND(a) \cap IND(b) = h(|\iota(a)|) \cap h(|\iota(b)|)$.

The claim (3) follows immediately by (2) and Propositions 12.14, 12.15.

□

We are prepared now for the completeness theorem for FD–logic. Let us observe that if α is not a thesis of the FD–logic, then $|\alpha| \neq 1_{L_F}$ hence by Proposition 12.16 it follows that $h(|\alpha|) \neq IND(\perp)$. Letting a valuation w_v to be the composition $\alpha \to |\alpha| \to h(|\alpha|)$ we arrive at the conclusion that $w_v(\alpha) \neq 1_{A_0}$ i.e. α is not true in the canonical FD–algebra hence it is not true. A fortiori it follows that each true formula of the FD–logic is a thesis of this logic.

Proposition 12.17. *(Rauszer) The FD–logic is complete.*

The correspondence between functional dependence and provability in the FD–logic may be now stated clearly on the basis of the facts established above:

Proposition 12.18. *(Rauszer) The following are equivalent for each pair $a, b \subseteq A$ of attribute sets*
1. b depends functionally on a
2. the formula $\supset \iota(a) \to \iota(b)$ is a thesis of the FD–logic
3. for each valuation v on terms of the FD–logic it holds that $v(b) \subseteq v(a)$
4. for each valuation w_v on formulas of the FD–logic in an FD–algebra $\mathcal{M}$ it holds that $w_v(\iota(a) \to \iota(b)) = 1_{\mathcal{M}}$, the unit element in $\mathcal{M}$.

12.5 Information logics

Information logics have been constructed with the aim of giving a formal rendering of reasoning about objects in terms of attributes; they a fortiori refer to knowledge representation in the form of information systems. We will give here an account of an information logic IL due to [Vakarelov89] which does encompass and generalize logics $DIL, NIL, INDL$ constructed in [Orlowska–Pawlak84, 89].

The first step in constructing an information logic is to consider a *many–valued information system* $\mathcal{A} = (U, A)$ where U is a set of *objects* and A is a set of *attributes*; contrary to the cases discussed earlier, we assume that each attribute $a \in A$ is a *many–valued* function $a : U \to 2^V$ which maps objects to subsets of a *value set* V. In case $|a(x)| = 1$ for each $a \in A$ and each $x \in U$, we have the case of an information system.

For a many–valued information system $\mathcal{A}$, one may introduce some natural relations on the set U of objects. First of all, we have the already familiar to us relation of *indiscernibility* which we denote here with the symbol I:

$$xIy \ iff \ \forall a \in A(a(x) = a(y)) \tag{12.29}$$

As values of attributes are now sets, we have the *informational inclusion* relation C defined as follows:

$$xCy \ iff \ \forall a \in A(a(x) \subseteq a(y)) \tag{12.30}$$

Finally, we have the *informational tolerance* relation T defined (cf. Proposition 31, Chapter 6) via

$$xTy \ iff \ \forall a \in A(a(x) \cap a(y) \neq \emptyset \tag{12.31}$$

It will be useful to introduce a constant subset of the set U viz.

$$(Const) \ D = \{x \in U : \forall a \in A(|a(x)| = 1)\}.$$

The set D is the set of *deterministic objects*. The following proposition brings forth the basic properties of relations I, C, T and the set D.

Proposition 12.19. *Given a many–valued information system*
$\mathcal{A}=(U, A)$, *the following are true*
(S1) xCx
(S2) xCy, yCz *imply* xCz
(S3) xTy *implies* yTx
(S4) xTy *implies* xTx
(S5) xTy, xCz *imply* zTy
(S6) $y \in D, xCy$ *imply* $x \in D$
(S7) $x \in D, xTy$ *imply* xCy
(S8) xIx
(S9) xIy *implies* yIx
(S10) xIy, yIz *imply* xIz
(S11) xIy *implies* xCy
(S12) $x \in D, y \in D, xTy$ *imply* xIy

The reader has undoubtedly recognized in (S8), (S9), (S10) reflexivity, symmetry and transitivity properties of an equivalence relation I, similarly (S1), (S2) are reflexivity and transitivity of a partial ordering C. Other properties are equally obvious.

By a *relational structure* we will understand a tuple $(U, R_1, ..., R_k)$ where U is a set and $R_1, ..., R_k$ are relations on U. By *type* of a relational structure $(U, R_1, ..., R_k)$ we will understand a tuple $(m_1, .., m_k)$ where m_i is the arity of the relation R_i for $i = 1, 2, ..., k$.

We define now a *generalized information system* as a relational system (U, I, C, T, D) of type $(2, 2, 2, 1)$ such that (S1)–(S12) hold.

12.5.1   The logic IL

We define the information logic IL. First, we discuss its syntax. The language of IL is built from
1. propositional variables which constitute a countable set Var
2. propositional functor symbols $\neg, \vee, \wedge$
3. modal functors $[I], [C], [T]$
4. propositional constant D
5. auxiliary symbols: dots, commas, parentheses etc.etc.

The set *FORM* of *formulae* is defined as the least set containing all propositional variables and the propositional constant D and closed on propositional functors and modal functors.

Before embarking on formal development of the logic IL i.e. its axiomatization, we should catch a glimpse of its semantics.

Assume that a generalized information system $M = (U, I, C, T, D)$ is given. An M–*valuation* is then a function $v : Var \to 2^U$; thus, an M–valuation assigns to each propositional variable a subset of the set U. Given a valuation v, we define the satisfiability relation $x \vdash_v \alpha$ of α by x relative to v, where $x \in U$ and $\alpha \in FORM$, by a standard induction on formula complexity viz.

$$
\begin{array}{c}
x \vdash_v p \ iff \ x \in v(p) \ for \ each \ variable \ p \\
x \vdash_v D \ iff \ x \in D \\
x \vdash_v \alpha \vee \beta \ iff \ either \ x \vdash_v \alpha \ or \ x \vdash_v \beta \\
x \vdash_v \alpha \wedge \beta \ iff \ x \vdash_v \alpha \ and \ x \vdash_v \beta \\
x \vdash_v \neg\alpha \ iff \ not \ x \vdash_v \alpha \\
x \vdash_v [R]\alpha \ iff \ \forall y \in U(xRy \Rightarrow y \vdash_v \alpha)
\end{array}
\tag{12.32}
$$

Then the notion of truth is introduced as usual, i.e. a formula α is *true* in M, v if and only if $x \vdash_v \alpha$ for every $x \in U$; a fortiori, α is true in M if and only if α is true in M, v for every valuation v, and α is true if and only if α is true in every generalized information system M.

Let us also observe that the last line in (12.32) conforms to the rules in Kripke semantics for modal logic, discussed in Chapter 5. We thus recognize in $[R]$ the modal operator of necessity, relativized here to a relation $R \in \{I, C, T\}$.

We now are ready for the principal result bridging syntax and semantics of IL viz. we render properties (S1)–(S12) stated above in the form of formulae $\alpha_1 - \alpha_{12}$ with the property that the formula α_i is true if and only if the corresponding (Si) holds in a given relational system. We made this vague statement precise.

Proposition 12.20. *(Vakarelov) Assume that $M = (U, I, C, T, D)$ is a relational structure of type $(2, 2, 2, 1)$ and $\alpha_1 - \alpha_{12}$ are the following formulae (as on many occasions earlier, we denote implication functor with the symbol $\Rightarrow$. We define the dual modal functor $< R >$ as $\neg[R]\neg$ and we use the shortcut $\top$ for $\alpha \vee \neg\alpha$.)*

$\alpha_1 : [C]\alpha \Rightarrow \alpha$

$\alpha_2 : [C]\alpha \Rightarrow [C][C]\alpha$

$\alpha_3 : \alpha \vee [T]\neg[T]\alpha$

$\alpha_4 : < T > \top \wedge [T]\alpha \Rightarrow \alpha$

$\alpha_5 : < C > [T]\alpha \Rightarrow [T]\alpha$

$\alpha_6 : < C > D \Rightarrow D$

$\alpha_7 : D \wedge [C]\alpha \Rightarrow [T]\alpha$

$\alpha_8 : [I]\alpha \Rightarrow \alpha$

$\alpha_9 : \alpha \vee [I]\neg[I]\alpha$

$\alpha_{10} : [I]\alpha \Rightarrow [I][I]\alpha$

$\alpha_{11} : [C]\alpha \Rightarrow [I]\alpha$

$\alpha_{12} : D \wedge [I]\alpha \Rightarrow [T](D \Rightarrow \alpha)$

Then, the formula α_i is true in M if and only if the corresponding property (Si) is satisfied by M for $i = 1, 2, ..., 12$.

Proof. We consider the pair (S1) and the corresponding α_1. Assume that (S1): xCx holds in U and we check that α_1: $[C]\alpha \Rightarrow \alpha$ is satisfied in M. Thus, let us consider a valuation v and $x \in U$ and assume that $x \vdash_v [C]\alpha$; this means that for each $y \in U$ such that xCy we have $y \vdash_v \alpha$. In particular, as (S1) is satisfied, we have xCx hence $x \vdash_v \alpha$. It follows that $x \vdash_v ([C]\alpha \Rightarrow \alpha)$ for each $x \in U$ and each valuation v i.e. α_1 is satisfied in M.

To prove the converse, we assume that xCx does not hold in U i.e. for some $x_0 \in U$ we have that x_0Cx_0 does not hold. We select a propositional variable p, and we define a valuation v as follows: for $B = \{y \in U : yCx_0\}$, we let v(p)=B, and for any propositional variable $q \neq p$ we let $v(q) = \emptyset$. For this valuation v, we have $x_0 \vdash_v p$ as each $y \in U$ such that yCx_0 does satisfy $y \in v(p)$ while it is not true that $x_0 \vdash_v p$ as $x_0 \notin v(p)$. Thus the pair x_0, v does falsify α_1. $\qquad\square$

The proof given above in case $i = 1$ sets a pattern for proofs of other equivalences for $i = 2, 3, ..., 12$. We relegate these proofs to Exercise section (cf. Exercise 19, below).

Proposition 12.20 gives means of rendering properties of relational systems like M in a logical form. The logic IL accounts for all properties (S1)–(S12) by including in its axiom schemes all of α_1–α_{12}. We may now present an axiomatization of the logic IL.

Axiomatization of IL

We adopt as axiom schemes of IL the following

1. axiom schemes for the sentential calculus (e.g. (T1)–(T3) of Chapter 2)

2. schemes of the form $[R](\alpha \Rightarrow \beta) \Rightarrow ([R]\alpha \Rightarrow [R]\beta)$ where $R \in \{I, C, T\}$

3. schemes α_1–α_{12} listed in Proposition 12.20

Let us observe that schemes listed in 2, are variants of the modal formula (K) (cf. Chapter 5) while α_1, α_2 are axiom schemes for the modal logic S4 (the logic of information containment); similarly, $\alpha_8, \alpha_9, \alpha_{10}$ are axiom schemes for the modal logic S5 (the logic of indiscernibility).

The derivation (inference) rules of IL are:

Detachment: *if $\alpha, \alpha \Rightarrow \beta$ are theses then β is a thesis*

Necessitation: *if α is a thesis then $[R]\alpha$ is a thesis for $R \in \{I, C, T\}$*

Substitution: in its usual form already familiar to us from discussions in Chapters 2, 4, 5, 9.

We denote by the symbol $Th(IL)$ the set of theses of IL. Our goal now will be completeness proof of IL. To this end, we exploit the Lemmon–Scott technique of canonical models already applied in Chapter 5 towards completeness proofs of modal logics. We depart slightly from that discussion in that we make usage here of the language of filters. It may be proved directly that a proper filter in the sense 1–4 below is a consistent set in the sense defined in Chapter 5; also, $Th(IL) \subseteq X$ for each proper filter X.

12.5.2 A canonical model

We consider the set $FORM$ of formulae of IL. We recall that (cf. our discussion of the Lindenbaum–Tarski algebra in Chapter 9) a pre–order $\leq$ may be introduced in $FORM$ via $\alpha \leq \beta$ in case $\alpha \Rightarrow \beta \in Th(IL)$. We consider proper filters with respect to this pre–order i.e. sets $X \subseteq FORM$ such that

1. $\top \in X$ (each filter contains the unit 1)

2. $\alpha \in X, \alpha \Rightarrow \beta \in Th(IL)$ imply $\beta \in X$ (each filter is closed on $\leq$)

3. $\alpha, \beta \in X$ imply $\alpha \wedge \beta \in X$ (each filter is closed on $\wedge$)

4. $\perp \notin X$ where $\perp$ is the shortcut for $\alpha \wedge \neg\alpha$, i.e. $\perp$ is the null element in $FORM$ ordered by $\leq$

Let us observe that it follows by theses $C(p \wedge q)p, C(p \wedge q)q$ of the sentential calculus that the converse to requirement 3 holds as well: $\alpha \wedge \beta \in X$ implies $\alpha \in X, \beta \in X$.

For $R \in \{I, C, T\}$, and a proper filter $X \subseteq FORM$, we let $[R]X = \{\alpha \in FORM : [R]\alpha \in X\}$. Then

Proposition 12.21. $[R]X$ *is a (proper) filter.*

Indeed, by Necessitation, we have $\top \Rightarrow [R]\top \in Th(IL)$ hence by 2, $[R]\top \in X$ hence $\top \in [R]X$ fulfilling requirement 1.

An obvious paraphrase of proof of Proposition 1 in Chapter 5 (that $Lp \wedge Lq \Rightarrow L(p \wedge q)$ follows from (K)) shows that (∗) $[R]\alpha \wedge [R]\beta \Rightarrow [R](\alpha \wedge \beta) \in Th(IL)$. Thus, in case $\alpha, \beta \in [R]X$, we have $[R]\alpha, [R]\beta \in X$ hence by 3 we have $[R]\alpha \wedge [R]\beta \in X$ and it follows by 2 and (∗) that $[R](\alpha \wedge \beta) \in X$ i.e. $\alpha \wedge \beta \in [R]X$ which fulfills requirement 3.

To show that requirement 2 is fulfilled, assume that $\alpha \in [R]X, \alpha \Rightarrow \beta \in Th(IL)$. Thus, by Necessitation, $[R](\alpha \Rightarrow \beta) \in Th(IL)$ and it follows by the instance $[R](\alpha \Rightarrow \beta) \Rightarrow ([R]\alpha \Rightarrow [R]\beta)$ of the appropriate axiom scheme for IL and detachment that $[R]\alpha \Rightarrow [R]\beta \in Th(IL)$. As $\alpha \in [R]X$, we have

$[R]\alpha \in X$ and thus by requirement 2, satisfied by X, it follows that $[R]\beta \in X$ i.e. $\beta \in [R]X$ witnessing that $[R]X$ does fulfill requirement 2.

Requirement 4 is clearly fulfilled by $[R]X$.

Thus $[R]X$ is a filter once X is a filter.

We now recall the notion of a *prime filter* (cf. e.g. Proposition 12 in Chapter 8); a filter X is *prime* if and only if $\alpha \vee \beta \in X$ implies either $\alpha \in X$ or $\beta \in X$. The reasoning in Chapter 8 leading to Proposition 12 by which we have demonstrated that distinct x, y may be separated by a prime filter, may be repeated here to show that

Proposition 12.22. *If X a filter and $\alpha \notin X$ then there exists a prime filter Y with the properties that $X \subseteq Y$ and $\alpha \notin Y$.*

Let us mention here an easy to be verified fact that

Proposition 12.23. *For a prime filter X the following are equivalent for each pair β, γ of formulae*

1. $\beta, \gamma \in X$

2. $\beta \vee \gamma \in X$

3. $\beta \wedge \gamma \in X$

4. $\beta \in X$ if and only if $\neg\beta \notin X$

We denote by the symbol Pr the family of all prime filters on $(FORM, \leq)$ and we finally let for $R \in \{I, C, T\}$ and filters X, Y:

$$XRY \Leftrightarrow [R]X \subseteq Y \qquad (12.33)$$

Then we have

Proposition 12.24. *For $X \in Pr$ and $\alpha \in$ FORM: $[R]\alpha \in X$ if and only if $XRY \Rightarrow \alpha \in Y$ for every $Y \in Pr$.*

Proof. Assume first $[R]\alpha \in X$; then $\alpha \in [R]X$ and for $Y \in Pr, XRY$ it follows obviously that $\alpha \in Y$. To prove the converse, assume that the requirement $XRY \Rightarrow \alpha \in Y$ for every $Y \in Pr$ holds; was $[R]\alpha \in X$ not the case, we would have by Proposition 12.22 a prime filter Y with $[R]X \subseteq Y$ and $[R]\alpha \notin Y$ hence $\alpha \notin Y$, contradicting our assumption. Thus the proposition follows. □

We take care also of the propositional constant D and we let

$$\mathbf{D} = \{X \in Pr : D \in X\} \qquad (12.34)$$

We have arrived at the relational system $\mathcal{M} = (Pr, \mathbf{I}, \mathbf{C}, \mathbf{T}, \mathbf{D})$ and it will be now our aim to show that $\mathcal{M}$ is a generalized information system.

Proposition 12.25. *(Vakarelov) The relational system $\mathcal{M}$ is a generalized information system i.e. it does satisfy properties (S1)–(S12).*

Proof. As hinted at in the statement of the proposition, the proof is in checking that (S1)–(S12) hold in $\mathcal{M}$. Proofs for each of (Si)' follow a similar pattern, so we include here, following [Vakarelov89], a proof in case of (S7), relegating to Exercises proofs of other cases.

Assume to the contrary that (S7) does not hold in $\mathcal{M}$. Then, we have prime filters X, Y with (i) $X \in \mathbf{D}$ (ii) $X\mathbf{T}Y$ (iii) $X\mathbf{C}Y$ not true. By (i), $D \in X$, and (iii) implies that for some formula α we have $\alpha \in [C]X \setminus Y$ hence $[C]\alpha \in X$ which implies that $D \wedge [C]\alpha \in X$. Now we may invoke α_7 whose instance $D \wedge [C]\alpha \Rightarrow [T]\alpha$ implies that $[T]\alpha \in X$ i.e. $\alpha \in [T]X$ so by (ii) $\alpha \in Y$, a contradiction. Thus (S7) is satisfied in $\mathcal{M}$. $\qquad\square$

We are ready now for a completeness theorem for the logic IL.

IL is complete

As $\mathcal{M}$ is a generalized information system, it remains to define a canonical valuation and to this end we let v^c, the canonical valuation, to be defined as follows

$$(CV) \ v^c(\alpha) = \{X \in Pr : \alpha \in X\}.$$

The fundamental property of canonical models, witnessed by us in Chapter 5, is observed here.

Proposition 12.26. *(Vakarelov) For each $X \in Pr$ and each $\alpha \in$ FORM the following equivalence takes place:*

$$X \vdash_{v^c} \alpha \Leftrightarrow \alpha \in X.$$

Proof. As usual, the proof is by induction on complexity of the formula α. In case α is a propositional variable, the equivalence in question is satisfied by definition of $\vdash_v$. If α is one of $\beta \vee \gamma$, $\beta \wedge \gamma$, $\neg\gamma$, the equivalence follows easily by Proposition 12.23 and the definition of $\vdash_v$. Finally, in case α is $[R]\beta$, the equivalence follows immediately by Proposition 12.24. $\qquad\square$

We state now a completeness theorem for IL.

Proposition 12.27. *(Vakarelov) The following are equivalent for a formula α*

1. α is a thesis of IL i.e. $\alpha \in Th(IL)$
2. α is true in every generalized information system

Proof. As usual implication from 1 to 2 is true as an easy check shows that all axiom instances are true and derivation rules preserve truth. It remains to show that 2 implies 1, and the best way to do it is to argue from the contrary. So assume that α is not any thesis i.e. $\alpha \notin Th(IL)$. As $Th(IL)$ is obviously a filter, we may find by Proposition 12.22 a prime filter X with $Th(IL) \subseteq X$ and $\alpha \notin X$. It follows from Proposition 12.26 that it is not true that $X \vdash_{v^c} \alpha$ so α is not true in the canonical generalized information system $\mathcal{M}$. This concludes the argument for completeness of the information logic IL. $\qquad\square$

Some other results on algebraic as well as logical approaches to rough sets and information systems may be found in works on the subject mentioned in Historic remarks, below.

Historic remarks

In addition to works already mentioned in the text, we would like to provide here clues to other works, complementary as well as demonstrating other ideas about algebraization and logicization of rough sets. The subject of algebraic and relational characterizations of rough sets was discussed among others in [Obtulowicz85], [Pagliani98^b], [Duentsch94, 98, 00], [Cattaneo 97, 98], [Comer93]. Logical analysis of rough sets was also undertaken among others in [Rasiowa–Skowron84], [Orlowska 84, 85, 89], [Nakamura98], [Banerjee–Chakraborty98], [Archangelsky–Taitslin97].

Exercises

In Exercises 1–7 , we propose to look at Wajsberg as well as Łukasiewicz algebras. In particular, we outline the proof that Łukasiewicz algebras are Wajsberg algebras under the interpretation (v).

1. Prove that under (v), the fact (a) $x \leq y \Leftrightarrow x \hookrightarrow y = 1$ holds [Hint: observe that $x \hookrightarrow y = 1$ if and only if $\phi_1(x \hookrightarrow y) = 1$ if and only if $(\phi_2(Nx) \vee \phi_1(y)) \wedge (\phi_2(y) \vee \phi_1(Nx)) = 1$ if and only if $\phi_1(x) \leq \phi_1(y)$ and $\phi_2(x) \leq \phi_2(y)$ if and only if $x \leq y$]

2. Deduce from results of Exercise 1, that axioms (WA5), (WA6) hold under interpretation (v).

3. Prove that the axiom (WA3) holds under (v). [Hint: observe that $(x \hookrightarrow Nx) \hookrightarrow x = \phi_2(Nx) \hookrightarrow x = N\phi_1(x) \hookrightarrow x = \phi_2(\phi_1(x) \wedge x) \vee \phi_1(x) \vee x = x$]

4. Prove that the axiom (WA4) holds under (v). [Hint: observe that $Nx \hookrightarrow Ny = (\phi_2(x) \vee Ny) \wedge (\phi_2(Ny) \vee x) = y \hookrightarrow x$]

5. Verify that under (v) the following holds: $\phi_1(x \hookrightarrow y) = \phi_2(Nx \wedge y) \vee \phi_1(Nx \vee y)$.

6. Verify that under (v), the following holds: $\phi_2(x \hookrightarrow y) = \phi_2(Nx \wedge y) \vee \phi_2(Nx \vee y) = \phi_2(Nx \vee y)$.

7. Apply results of Exercises 5 and 6 to verify that under (v) the relationships (1), (2) hold. [Hint: prove, respectively, using Exercises 5,6, that $\phi_1((x \hookrightarrow y) \hookrightarrow y) = \phi_1(x \vee y)$, $\phi_2((x \hookrightarrow y) \hookrightarrow y) = \phi_2(x \vee y)$ as well as that $\phi_1((y \hookrightarrow z) \hookrightarrow (x \hookrightarrow z)) \geq \phi_1(x \hookrightarrow y)$ and $\phi_2((y \hookrightarrow z) \hookrightarrow (x \hookrightarrow z)) \geq \phi_2(x \hookrightarrow y)$ so the results follow by the determination principle]

In Exercises 8–14, we propose to prove some facts about Wajsberg algebras corresponding to theses (XIV)–(XX) of the Wajsberg system for the 3–valued Łukasiewicz logic (cf. Chapter 4).

8. Prove that in the Wajsberg algebra the following $N(x \hookrightarrow y) \hookrightarrow x = 1$ holds. Deduce from this the thesis (XVI) (loc. cit.) [Hint: apply (g) to (d) to get $N(x \hookrightarrow y) \leq NNx$ and finally invoke (e)]

9. Verify that in the Wajsberg algebra the following $N(x \hookrightarrow y) \hookrightarrow N(y) = 1$ holds. Deduce from this the thesis (XVII) (loc. cit.) [Hint: in (f), substitute $x/y, y/x \hookrightarrow y$ and apply (WA1) in which substitutions $x/y, y/x$ have been made]

10. Prove the following property of the Wajsberg algebra: $x \hookrightarrow (Ny \hookrightarrow N(x \hookrightarrow y)) = 1$ and deduce from it the thesis (XVIII) (loc.cit.) [Hint: by (f), one has $x \hookrightarrow (Ny \hookrightarrow N(x \hookrightarrow y)) = x \hookrightarrow ((x \hookrightarrow y) \hookrightarrow y)$ and then by (m) one gets $x \hookrightarrow ((x \hookrightarrow y) \hookrightarrow y) = (x \hookrightarrow y) \hookrightarrow (x \hookrightarrow y) = 1$]

11. Verify that in the Wajsberg algebra $((x \hookrightarrow Nx) \hookrightarrow N(x \hookrightarrow Nx)) \hookrightarrow x = 1$ holds and deduce therefrom the thesis (XV) (loc.cit.) [Hint: observe that $(x \hookrightarrow Nx) \hookrightarrow N(x \hookrightarrow Nx) = \phi_2(N\phi_2(Nx)) = \phi_2(\phi_1(x)) = \phi_1(x) \leq x$]

12. Prove the following property of the Wajsberg algebra: $(x \hookrightarrow (x \hookrightarrow Nx)) \hookrightarrow (x \hookrightarrow Nx) = 1$ and verify by this the thesis (XIV) (loc.cit.) [Hint: verify that $x \hookrightarrow (x \hookrightarrow Nx) = \phi_2(Nx \wedge (x \hookrightarrow Nx)) \vee Nx \vee (x \hookrightarrow Nx) = x \hookrightarrow Nx$]

13. Prove that $(x \hookrightarrow Nx) \hookrightarrow ((Ny \hookrightarrow NNy) \hookrightarrow (x \hookrightarrow y)) = 1$ in the Wajsberg algebra and deduce from this the thesis (XIX) (loc.cit.) [Hint: apply (v) to verify that $(Ny \hookrightarrow NNy) \hookrightarrow (x \hookrightarrow y) = \phi_2(y) \hookrightarrow (x \hookrightarrow y) \geq N\phi_2(y) \vee (x \hookrightarrow y) \geq N\phi_2(y) \vee \phi_2(Nx \wedge y) \geq \phi_2(Nx) = x \hookrightarrow Nx$]

14. Prove that $(x \hookrightarrow (x \hookrightarrow (y \hookrightarrow z))) \hookrightarrow ((x \hookrightarrow (x \hookrightarrow y)) \hookrightarrow (x \hookrightarrow (x \hookrightarrow z))) = 1$ and verify by this the thesis (XX) (loc.cit.) [Hint: apply results of Exercises 5,6 along with (d) to check that $\phi_1((x \hookrightarrow (x \hookrightarrow y)) \hookrightarrow (x \hookrightarrow (x \hookrightarrow z)) = \phi_1(x \hookrightarrow (x \hookrightarrow (y \hookrightarrow z))$ and $\phi_2((x \hookrightarrow (x \hookrightarrow y)) \hookrightarrow (x \hookrightarrow (x \hookrightarrow z)) \geq \phi_2(x \hookrightarrow (x \hookrightarrow (y \hookrightarrow z))$ so the result follows by the determination principle]

Exercises 15– 18 are devoted to relations between Łukasiewicz and pseudo–Boolean (Heyting) algebras as well as Post algebras and Stone algebras. General results on correspondence between Łukasiewicz and Heyting algebras may be found in [Moisil 42, 63, 64], [Cignoli69], [Iturrioz77], in Exercise 15 we refer to [Moisil63] (cf. [Boicescu91]). That every Post algebra is a Heyting algebra was proved in [Rousseau70]. Correspondence between regular double Stone algebras and 3– Łukasiewicz algebras was described in [Moisil60] and [Varlet68].

15. [Moisil63] Prove that in a 3– Łukasiewicz algebra L, the operator $\Rightarrow$ defined via $x \Rightarrow y = y \vee (N\phi_1(x) \vee \phi_1(y)) \wedge (N\phi_2(x) \vee \phi_2(y))$ is a pseudocomplement operator turning L into a Heyting algebra. [Hint: apply the determination principle; to this end, calculate $\phi_1(x \Rightarrow y) = \phi_1(y) \vee (N\phi_1(x) \vee \phi_1(y)) \wedge (\phi_1(Nx) \vee \phi_2(y))$ and verify that $\phi_1(x \wedge x \Rightarrow y) \leq \phi_1(y)$; similarly, calculate $\phi_2(x \Rightarrow y) = \phi_2(y) \vee N\phi_2(x)$ and verify that $\phi_2(x \wedge x \Rightarrow y) \leq \phi_2(y)$. Deduce from the determination principle that $x \wedge x \Rightarrow y \leq y$. Consider z with $x \wedge z \leq y$ and deduce that $\phi_i(x \wedge z) = \phi_1(x) \wedge \phi_i(z) \leq \phi_i(y)$ for $i = 1, 2$ hence $\phi_i(z) \leq N\phi_i(x) \vee \phi_i(y)$ for $i = 1, 2$. Infer from previous calculations that $\phi_i(z) \leq \phi_i(x \Rightarrow y)$ for $i = 1, 2$ hence $z \leq x \Rightarrow y$]

16. Deduce from the result of Exercise 15 that every 3–Post algebra is a Heyting algebra and describe explicitly the pseudocomplement operator.

17. Recall that a Stone algebra S is a pseudocomplemented lattice in which the (multiplicative) pseudocomplement x^* satisfies the condition: (i) $x^* \vee x^{**} = 1$ for every x. Dually, the dual pseudocomplement x^+ satisfies the requirement $x \vee y = 1 \Leftrightarrow x^+ \leq y$; S is a double Stone algebra in case in addition to (i) the condition (ii) $x \wedge x^{++} = 0$ holds. The double Stone algebra is regular in case from $x^+ = y^+$ and $x^* = y^*$ it follows that $x = y$.

Prove that an algebra L is a 3– Łukasiewicz algebra if and only if it is a regular double Stone algebra. [Hint: to introduce in a 3– Łukasiewicz algebra the structure of a regular double Stone algebra, let $x^* = N\phi_1(x)$, $x^+ = N\phi_0(x)$. Conversely, let $\phi_2(x) = x^{**}$, $Nx = (x \wedge x^+) \vee x^*$]

18. [Pomykala88], [Iwinski88] Apply results of Exercise 17, to show that letting $[I, D]^* = [D, D]$ and $[I, D]^+ = [-I, -I]$ establishes the structure of a regular double Stone algebra in the representation of rough sets of Proposi-

tions 12.2, 12.4. Describe the structure $(C(S), \uplus, \cap)$ for this representation of rough sets (cf. Sect. 8.9).

In Exercises 19, 20, we refer to the information logic IL of Vakarelov.

19. In the context of Proposition 12.20, provide proofs that the property (Si) is satisfied in U if and only if the formula α_i is true in M for every relational system $M = (U, I, C, T, D)$ of type $(2, 2, 2, 1)$ and $i = 2, 3, ..., 12$.

20. In the context of Proposition 12.25, prove that the canonical system $\mathcal{M}$ satisfies properties (S1)–(S6) and (S8)–(S12).

In Exercises 21,22, we address the relationship between modal logics and rough sets on the level of Lindenbaum–Tarski algebras. We refer to Chapter 9 for the general introduction into the subject of Lindenbaum – Tarski algebras. We recall from Section 10 in Chapter 8 the notion of a topological Boolean algebra. Finally, we refer to Chapter 5 for an introduction to modal logics.

21. Prove that the Lindenbaum–Tarski algebra of modal logic (S5) is a clopen topological Boolean algebra hence it has the property that the intersection of each family of open sets is open. [Hint: letting $L[\alpha] = [L\alpha]$, interpret L as the interior operator I, i.e. $I[\alpha] = [L\alpha]$. Then observe that by (K) one has $Ia \le Ib$ whenever $a \le b$, by (T) we have $Ia \le a$, from (K) via Proposition 1 in Chapter 4 it follows that $I(a \cap b) = Ia \cap Ib$, (S4) implies that $Ia \le IIa$ hence $Ia = IIa$, and finally (S5) implies that $Ca = ICa$ where C is the closure operator defined dually to I via $C[\alpha] = [M\alpha]$. From the equality $Ca = ICa$ it follows that every closed set is open a fortiori every open set is closed hence the intersection of every family of open sets is open]

22. Prove that the topological Lindenbaum–Tarski algebra $\mathcal{B}$ of modal logic $S5$ is a topological space of an approximation space $(\mathcal{B}, R)$ for an appropriate equivalence relation R. [Hint: apply Proposition 38 in Chapter 7]

23. We return here to Lukasiewicz algebras. We recall that in the set $\mathbf{L}_n$ of truth values $\{0, \frac{1}{n-1}, ..., \frac{n-2}{n-1}, 1\}$ the following Lukasiewicz formulae define negation, alternation and disjunction (cf. Chapter 4): $Nx = 1 - x$, $x \wedge y = min\{x, y\}$, $x \vee y = max\{x, y\}$. We also recall functions ϕ_i defined in this chapter via $\phi_i(\frac{j}{n-1}) = 0$ in case $i + j < n$ and $\phi_i(\frac{j}{n-1}) = 1$ in case $i + j \ge n$.

We would like to establish a relationship between functions ϕ_i and functions J_i of Rosser and Turquette (cf. Chapter 4) defined on the set of values $\{0, 1, ..., n - 1\}$. (i) and (ii) below establish on one hand a structure of a Lukasiewicz algebra in the formalization of n–logic due to Rosser and Turquette, and on the other hand, they suggest an implicit proof of completeness of the Rosser and Turquette axiomatization of n–logic.

24. Prove that (i) given functions ϕ_i, we have

$$J_i(j) = \phi_{n-i}\left(\frac{j}{n-1}\right) \wedge N\phi_{n-i-1}\left(\frac{j}{n-1}\right)$$

(ii) given functions J_i, we have $\phi_i\left(\frac{j}{n-1}\right) = \bigvee_{k=1}^{i} J_{n-k}(j)$.

Works quoted

[Archangelsky–Taitslin97] D. A. Archangelsky and M. A. Taitslin, *A logic for information systems*, Studia Logica, 58 (1997), pp. 3–16.

[Banerjee–Chakraborty98] M. Banerjee and M. K. Chakraborty, *Rough logics: a survey with further directions*, in: E. Orlowska (ed.), *Incomplete Information: Rough Set Analysis*, Studies in Fuzziness and Soft Computing, vol. 13, Physica Verlag, Heidelberg, 1998, pp. 579–600.

[Becchio78] D. Becchio, *Logique trivalente de Lukasiewicz*, Ann. Sci. Univ. Clermont–Ferrand, 16 (1978), pp. 38–89.

[Becchio72] D. Becchio, *Nouvelle démonstration de la complétude du systéme de Wajsberg axiomatisant la logique trivalente de Lukasiewicz*, C. R. Paris, 275 (1972), pp. 679–681.

[Boicescu91] V. Boicescu, A. Filipoiu, S. Georgescu and S. Rudeanu, *Lukasiewicz–Moisil Algebras*, North Holland, Amsterdam, 1991.

[Cattaneo98] G. Cattaneo, *Abstract approximation spaces for rough theories*, in: [Orlowska98], pp. 59–98.

[Cattaneo97] G. Cattaneo, *Generalized rough sets. Preclusivity fuzzy–intuitionistic (BZ) lattices*, Studia Logica, 58 (1997), pp. 47–77.

[Cignoli69] R. Cignoli, *Algebras de Moisil de orden n*, Doctoral Thesis, Univ. Nacional del Sur, Bahia Blanca, Brasil, 1969.

[Comer93] S. Comer, *On connections between information systems, rough sets and algebraic logic*, in: *Algebraic Methods in Logic and Computer Science*, Banach Center Publ., 28, Warszawa, 1993.

[Duentsch00] I. Duentsch, *Logical and algebraic techniques for rough set data analysis*, in: [Polkowski–Tsumoto–Lin00], pp. 521–544.

[Duentsch98] I. Duentsch, *Rough sets and algebras of relations*, in: [Orlowska98], pp. 95–108.

[Duentsch94] I. Duentsch, *Rough relation algebras*, Fundamenta Informaticae, 21(1994), pp. 321–331.

[Epstein60] G. Epstein, *The lattice theory of Post algebras*, Trans. Amer. Math. Soc., 95 (1960), 300–317.

[Gentzen34] G. Gentzen, *Untersuchungen über das logische Schliessen. I, II.*, Mathematische Zeitschrift 39 (1934–5), pp. 176–210, 405–431.

[Iturrioz77] L. Iturrioz, *Lukasiewicz and symmetrical Heyting algebras*, Zeit. Math. Logik u. Grundl. Math., 23(1977), pp. 131–136.

[Iwinski88] T. B. Iwiński, *Rough orders and rough set concepts*, Bull. Polish Acad. Ser. Sci. Math., 37 (1988), pp. 187–192.

[Iwinski87] T. B. Iwiński, *Algebraic approach to rough sets*, Bull. Polish Acad. Ser. Sci. Math., 35 (1987), pp. 673–683.

[Moisil64] Gr. C. Moisil, *Sur les logiques de Lukasiewicz á un nombre fini de valeurs*, Rev. Roumaine Math. Pures Appl., 9 (1964), pp. 905–920, 583–595.

[Moisil63] Gr. C. Moisil, *Les logiques non–chrysippiennes et leurs applications*, Acta Phil. Fennica, 16 (1963), pp. 137–152.

[Moisil60] Gr. C. Moisil, *Sur les idéaux des algébres lukasiewicziennes trivalentes*, An. Univ. C. I. Parhon, Acta logica, 3 (1960), pp. 83–95, 244–258.

[Moisil42] Gr. C. Moisil, *Logique modale*, Disquisitiones Math. Phys., 2 (1942), pp. 3–98, 217–328, 341–441.

[Nakamura98] A. Nakamura, *Graded modalities in rough logic*, in: L. Polkowski and A. Skowron (eds.), *Rough Sets in Knowledge Discovery. Methodology and Applications*, Studies in Fuzziness and Soft Computing, vol. 18, Physica Verlag, Heidelberg, 1998, pp. 192–208.

[Nelson49] D. Nelson, *Constructible falsity*, The Journal of Symbolic Logic, 14 (1949), pp. 16–26.

[Obtulowicz85] A. Obtułowicz, *Rough sets and Heyting algebra valued sets*, Bull. Polish Acad. Sci. Math., 33(1985), pp. 454–476.

[Orlowska98] E. Orłowska, ed., *Incomplete Information: Rough Set Analysis*, Studies in Fuzzines and Soft Computing vol. 13, Physica Verlag, Heidelberg, 1998.

[Orlowska89] E. Orłowska, *Logic for reasoning about knowledge*, Z. Math. Logik u. Grund. d. Math., 35(1989), pp. 559–572.

[Orlowska85] E. Orłowska, *Logic approach to information systems*, Fundamenta Informaticae, 8(1985), pp. 359–378.

[Orlowska84] E. Orłowska, *Modal logics in the theory of information systems*, Z. Math. Logik u. Grund.d. Math., 30(1984), pp. 213–222.

[Orlowska–Pawlak84^a] E. Orłowska and Z. Pawlak, *Logical foundations of knowledge representation*, Reports of the Comp. Centre of the Polish Academy of Sciences, 537, 1984.

[Orlowska–Pawlak84^b] E. Orłowska and Z. Pawlak, *Representation of non-deterministic information*, Theor. Computer Science, 29 (1984), pp. 27–39.

[Pagliani98^a] P. Pagliani, *Rough set theory and logic–algebraic structures*, in: [Orlowska98], pp. 109–192.

[Pagliani98^b] P. Pagliani, *A practical introduction to the modal–relational approach to approximation spaces*, in: [Polkowski–Skowron98^a], pp. 209–232.

[Pagliani96] P. Pagliani, *Rough sets and Nelson algebras*, Fundamenta informaticae, 27(1996), pp. 205–219.

[Pawlak87^b] Z. Pawlak, *Rough logic*, Bull. Polish Acad. Sci. Tech., 35 (1987), pp. 253–258.

[Pawlak82^b] Z. Pawlak, *Rough sets, algebraic and topological approach*, Int. J. Inform. Comp. Sciences, 11(1982), pp. 341–366.

[Pawlak81^b] Z. Pawlak, *Information systems–theoretical foundations*, Information Systems, 6(1981), pp. 205–218.

[Pomykala88] J. Pomykała and J. A. Pomykała, *The Stone algebra of rough sets*, Bull. Polish Acad. Ser. Sci. Math., 36 (1988), 495–508.

[Post21] E. Post, *Introduction to a general theory of elementary propositions*, Amer. J. Math., 43 (1921), pp. 163–185.

[Rasiowa74] H. Rasiowa, *An Algebraic Approach to Non-Classical Logics*, PWN-Polish Scientific Publishers – North-Holland, Warszawa–Amsterdam, 1974.

[Rasiowa–Skowron86^a] H. Rasiowa and A. Skowron, *Rough concept logic*, LNCS vol. 208, Springer Verlag, Berlin, 1986, pp. 288–297.

[Rasiowa–Skowron86^b] H. Rasiowa and A. Skowron, *The first step towards an approximation logic*, J. Symbolic Logic, 51 (1986), p. 509.

[Rasiowa–Skowron86^c] H. Rasiowa and A. Skowron, *Approximation logic*, Proc. Conf. on Mathematical Methods of Specification and Synthesis of Software Systems, Akademie Verlag, Berlin, 1986, pp. 123–139.

[Rasiowa–Skowron84] H. Rasiowa and A. Skowron, *A rough concept logic*, in: A. Skowron (ed.), *Proc. the 5th Symposium on Comp. Theory*, Lecture Notes in Computer Science, vol. 208 (1984), pp. 197–227.

[Rauszer85^a] C. M. Rauszer, *Dependency of attributes in information systems*, Bull. Polish Acad. Sci. Math., 33(1985), pp. 551–559.

[Rauszer85^b] C. M. Rauszer, *An equivalence between theory of functional dependencies and a fragment of intuitionistic logic*, Bull. Polish Acad. Sci. Math., 33(1985), pp. 571–579.

[Rauszer84] C. M. Rauszer, *An equivalence between indiscernibility relations in information systems and a fragment of intuitionistic logic*, in: A. Skowron (ed.), *Proc. the 5th Symp. Comp. Theory*, Lecture Notes in Computer Science, vol. 208, Springer Verlag, Berlin, 1984, pp. 298–317.

[Rosenbloom42] P. Rosenbloom, *Post algebras.I. Postulates and general theory*, Amer. J. Math., 64 (1942), pp. 167–183.

[Rousseau70] G. Rousseu, *Post algebras and pseudo–Post algebras*, Fund. Math., 67 (1970), pp. 133–145.

[Traczyk63] T. Traczyk, *Axioms and some properties of Post algebras*, Colloq. Math., 10 (1963), pp. 193–209.

[Vakarelov89] D. Vakarelov, *Modal logics for knowledge representation systems*, Lecture Notes in Computer Science, vol. 363 (1989), pp. 257–277.

[Varlet68] J. Varlet, *Algébres de Lukasiewicz trivalentes*, Bull. Soc. Roy.

Sci. Liége, 36 (1968), pp. 394–408.

[Wajsberg31] M. Wajsberg, *Axiomatization of the three–valued sentential calculus* (in Polish, Summary in German), C. R. Soc. Sci. Lettr. Varsovie, 24(1931), 126–148.

PART 4: ROUGH vs. FUZZY

Chapter 13

Infinite–valued Logical Calculi

Shortly shall all my labors end, and thou shall have the air at freedom. For a little do follow me, and do me service

Shakespeare, *The Tempest*, IV

13.1 Introduction

In Chapter 4, many–valued logics were discussed. In that case, sets of truth values were finite and in Chapter 12 we have demonstrated that rough sets may be interpreted as Łukasiewicz algebras of 3–valued Łukasiewicz calculi. However, the Łukasiewicz semantics, introduced in [Łukasiewicz18] and discussed also in [Łukasiewicz20, 30], and presented in Chapter 4, does allow for infinite sets of truth values. In this case, one may admit as a set of truth values any infinite set $T \subseteq [0, 1]$ which is closed under functions

$$c : min(1, 1 - x + y), n : 1 - x$$

in the sense that $c(x, y), n(x) \in T$ whenever $x, y \in T$; we recall that the function c is the meaning of implication C and the function n is the meaning of negation N in the Łukasiewicz semantics. One shows (cf. [McNaughton51] quoted in [Rose–Rosser58]) that any such T needs to be dense in $[0, 1]$; it is manifest that $0, 1 \in T$ for each admissible T. Thus, sets $Q_0 = Q \cap [0, 1]$ of rational numbers in $[0,1]$ as well as the whole unit interval $[0, 1]$ may be taken as sets of truth values leading to respective infinite valued logical calculi. The additional merit of these two sets of truth values is that both $Q, \mathbf{R}^1$ are from algebraic point of view, *algebraic fields* so they may be regarded as vector spaces over themselves, and we will make use of this fact later on.

It is well–known that [0,1]–valued logics are fundamentally important to fuzzy calculi. With this in mind, we discuss in this Chapter infinite valued in particular [0,1]–valued logics with the Łukasiewicz semantics and we outline their completeness following a syntactic argument in [Rose–Rosser58]. Then, we introduce the Łukasiewicz residuated lattice and we sketch the Pavelka proposal for a complete fuzzy sentential calculus.

We intend this Chapter as a first step toward the discussion of the rough versus fuzzy approaches to uncertainty.

In Chapter 4, we introduced the idea of designated truth values in the sense that if the truth value of a statement is designated then the statement is accepted and otherwise it is rejected. We adopt here the convention of Łukasiewicz that the value 1 be designated and all others values in T be non–designated. Thus, a statement be accepted if and only if its truth value be 1.

We will formalize infinite valued logical calculus of Łukasiewicz taking implication C and negation N as basic logical functors and admitting the following definitions of other functors

Def.1 Apq is $CCpqq$

Def.2 Kpq is $NANpNq$

Def.3 Bpq is $CNpq$

Def.4 Lpq is $NCpNq$

Def.5 Epq is $LCpqCqp$

In [Łukasiewicz30[b]], a conjecture by Łukasiewicz is stated that the following set of axioms gives a complete axiomatization of infinite valued logical calculus with the Łukasiewicz semantics in case when the only designated value is 1

(Ł$_\infty$1) $CpCqp$

(Ł$_\infty$2) $CCpqCCqrCpr$

(Ł$_\infty$3) $CApqCAqp$

(Ł$_\infty$4) $CCNpNqCqp$

(Ł$_\infty$5) $ACpqCqp$

Completeness of this axiomatization was announced in [Wajsberg35] with-

out any proof; the completeness proof was given in [Rose–Rosser58] and here
we follow their development. Let us remark also that the axiom $(Ł_\infty 5)$ was
shown to be redundant (i.e. following from $(Ł_\infty 1)$–$(Ł_\infty 4)$) in [Meredith58]
and independently in [Chang58^a] (cf. Exercise 1, below).

According to the Łukasiewicz semantics, truth functions c of C and n of N
are given by the formulae quoted above i.e. $c(x,y) = min\{1, 1-x+y\}, n(x) = 1 - x$. Then truth functions a, k, b, l, e of derived functors, respectively,
A, K, B, L, E are given by the following formulae (cf. Exercise 2)

$$a(x,y) = max\{x,y\}$$

$$k(x,y) = min\{x,y\}$$

$$b(x,y) = min\{1, x+y\}$$

$$l(x,y) = max\{0, x+y-1\}$$

$$e(x,y) = min\{1 - x + y, 1 - y + x\}$$

We may check easily that

Proposition 13.1. *The following hold*
1. $a(x,y) = 1$ *if and only if either* $x = 1$ *or* $y = 1$
2. $k(x,y) = 1$ *if and only if* $x = 1 = y$
3. $b(x,y) = 1$ *if and only if* $x + y \geq 1$
4. $l(x,y) = 1$ *if and only if* $x = 1 = y$
5. $e(x,y) = 1$ *if and only if* $x = y$

It follows that A, B are many–valued counterparts to classical alternation,
K, L are many–valued counterparts to classical conjunction, and E is a many–
valued counterpart to classical equivalence.

As with any logical calculus, also in case of infinite valued logical calculus,
its syntax, its semantics, and their interplay are subject of study. We will
present results of this study here. We will simply assume that the set of truth
values T is either $Q \cap [0,1]$ or $[0,1]$ and 1 is the only designated value. The
resulting calculus is denoted $Ł_\infty$.

13.2 Syntax of $Ł_\infty$

The axioms are $(Ł_\infty 1)$–$(Ł_\infty 4)$ and inference rules are detachment and
substitution. We will use as usual the acceptance ("yields") symbol $\vdash$ and for
a formula α as in earlier Chapters, the symbol $\vdash \alpha$ will mean that α is a *thesis*
of the calculus i.e. there exists a proof of α from axioms via detachment and
substitution in the usual sense of Chapters 2, 3 and 4. Similarly, the symbol
$\Gamma \vdash \alpha$ will mean that α is provable from a set Γ of formulae.

We follow closely [Rose–Rosser58] and we give now a number of theses and syntactic features of L_∞; we provide after [Rose–Rosser58] some of proofs; other proofs will be sketched in Exercises that follow.

(T1) $\Gamma \vdash Cpq, \Gamma \vdash Cqr$ imply $\Gamma \vdash Cpr$

Indeed, this is a direct consequence of ($L_\infty 2$) and detachment. We let $p \Leftrightarrow q$ to denote that Cpq and Cqp hold.

(T2) $\Gamma \vdash p \Leftrightarrow q$ and $\Gamma \vdash q \Leftrightarrow r$ imply $\Gamma \vdash p \Leftrightarrow r$

By (T1) and definition of $\Leftrightarrow$.

(T3) $Cpq, Cqr \vdash Cpr$

By (T1).

(T4) $\vdash Apq \Leftrightarrow Aqp$

By ($L_\infty 3$).

(T5) From $\Gamma \vdash p \Leftrightarrow q$ it follows that $\Gamma \vdash Cpr \Leftrightarrow Cqr$

Indeed, by ($L_\infty 2$) and detachment we have $Cpq \vdash CCqrCpr$ and $Cqp \vdash CCprCqr$.

(T6) $\vdash CpAqp$

By substituting q/Cqp in ($L_\infty 1$).

(T7) $\vdash CpApq$

From (T6) and ($L_\infty 3$).

(T8) $\vdash CCpCqrCqCpr$

Indeed, $\vdash CqAqr$ by (T7) and ($L_\infty 2$) implies

$$(i) \quad \vdash CCAqrCprCqCpr$$

and substituting q/Cqr and applying ($L_\infty 2$) yields

$$(ii) \quad \vdash CCpCqrCAqrCpr.$$

From (i), (ii) one gets (T8) via (T1).

(T9) $CCqrCCpqCpr$

From (T8) and ($L_\infty 2$).

(T10) $\Gamma \vdash p \Leftrightarrow q$ implies $\Gamma \vdash Crp \Leftrightarrow Crq$

It follows by substituting $p/q, q/p$ in (T8) which yields

$$(i) \quad \vdash CpCqr \Leftrightarrow CqCpr$$

and using ($L_\infty 2$) in which substitutions $p/q, q/r, r/p$ have been made.

(T11) $\vdash Cpp$

Indeed, substituting q/p in (T8) and applying ($L_\infty 1$) one gets (i) $\vdash CqCpp$. Letting q to be any already proved thesis and applying detachment yields (T11).

(T12) $\vdash p \Leftrightarrow p$

Immediate from (T11).

(T13) $p \vdash q \Leftrightarrow Cpq$

This follows from ($L_\infty 1$) where substitution $p/q, q/p$ has been made and from (T7) in which A has been replaced by the right–hand side of Def. 1.

(T14) $\vdash q \Leftrightarrow Aqq$

Indeed, substitute p/Cqq in (T14) and apply (T11).

(T15) $Cpr, Cqr \vdash CApqr$

Cf. Exercise 3.

(T16) From $p_1, .., p_k, p \vdash r$ and $q_1, ..., q_n, q \vdash r$ it follows that

$$p_1, ..., p_k, q_1, ..., q_n, Apq \vdash r$$

Cf. Exercise 4.

(T17) $CNNpp$

Cf. Exercise 5.

(T18) $\vdash CpNNp$

Cf. Exercise 6.

(T19) $\vdash p \Leftrightarrow NNp$

(T19) follows by (T17) and (T18).

(T20) $\vdash Cpq \Leftrightarrow CNqNp$

Substituting p/q in (T19) gives via (T10) (i) $Cpq \Leftrightarrow CpNNq$. Substituting q/Nq in (iv) in the proof of (T18) yields (ii) $CpNNq \Leftrightarrow CNqNp$. From (i) and (ii) (T20) follows.

(T21) $\Gamma \vdash p \Leftrightarrow q$ implies $\Gamma \vdash Np \Leftrightarrow Nq$

Directly from (T20).

(T19) and (T20) imply directly

(T22) $\vdash Lpq \Leftrightarrow Lqp$

(T23) $\vdash Bpq \Leftrightarrow Bqp$

(T24) $\vdash CpBpq$

Indeed, substituting q/Nq in ($Ł_\infty 1$) gets us (i) $\vdash CpBqp$ whence we get (T24) by (T23).

(T25) $\vdash CLqpp$

Similarly, from (T22).

(T26) $\vdash CLpqp$

(T27) $\vdash CpCqr \Leftrightarrow CLpqr$

From (T20) it follows that (i) $\vdash CpCqr \Leftrightarrow CpCNrNq$ and (T8) implies (ii) $\vdash CpCqr \Leftrightarrow CNrCpNq$ so (T27) follows whence by (T23).

(T28) $\vdash CpCqLpq$

It is enough to substitute in (T27) r/Lpq and invoke the thesis $\vdash CLpqLpq$.

(T29) $p \vdash q \Leftrightarrow Lpq$

By (T25) and (T29).

(T30) $\Gamma \vdash p \Leftrightarrow q$ is equivalent to $\Gamma \vdash Epq$

From (T28) one infers (i) $Cpq, Cqp \vdash Epq$ and (T25), (T26) imply (ii) $Epq \vdash p \Leftrightarrow q$.

(T31) $\vdash Epp$

It suffices to substitute q/p in (T30).

(T32) $Epq, Eqr \vdash Epr$

Directly from (T30).

(T31) $Epq \vdash ENpNq$

Directly from (T30) and (T21).

(T32) $Epq, Ert \vdash ECprCqt$

It follows as above by (T5) and (T10).

(T33) $p, q \vdash p \Leftrightarrow q$

By (T11) and ($Ł_\infty 4$) we get (i) $\vdash CpCrr$ so by (T28) it follows that we have (ii) $\vdash CCCrrpEpCrr$ and it follows from ($Ł_\infty 1$) that (iii) $CpCCrrp$ (substitute q/Crr) so by (T30), one gets (iv) $p \vdash p \Leftrightarrow Crr$ and (replacing p/q in (iv)) (v) $q \vdash q \Leftrightarrow Crr$ whence by (T32) the thesis (T33) follows.

(T34) $Np, Nq \vdash p \Leftrightarrow q$

By the thesis $p \vdash p \Leftrightarrow Crr$ ((iv) in (T33)), (T19), and (T21), it follows that (i) $Np \vdash p \Leftrightarrow NCrr$ and so (T34) follows.

(T35) $CANpqCpq$

By (T24) and (L_∞2), (i) $CCBpqqCpq$ follows which may be also written in the form (T35).

(T36) $\vdash CNpCpq$

Indeed, (L_∞1) implies (i) $\vdash CNpCNqNp$ whence by (T20) the thesis (T36) follows.

We now close our discussion of syntax of L_∞ with a proof of the Deduction Theorem for this calculus. As we will see, in this case the Deduction Theorem acquires a more general form than the classical Deduction Theorem. However, the proof in case of L_∞ follows the lines of the classical proof with some modifications.

We will need some preliminary facts. To state them, we introduce a symbol $\Gamma_{i=1}^{m} r_i s$ (cf. [Rosser–Turquette52], [Rose–Rosser58]). The definition is inductive on m and goes as follows.

$$\Gamma_{i=1}^{m} r_i s = \begin{array}{ll} s & \text{in case } m < 1 \\ Cr_m \Gamma_{i=1}^{m-1} r_i s & \text{otherwise} \end{array} \qquad (13.1)$$

In case all r_i are a formula r, we write $(Cr)^m s$ instead of $\Gamma_{i=1}^{m} r_i s$.

Proposition 13.2. *[Rosser–Turquette] If $r_1, ..., r_m$ are formulae and $q_1, .., q_k$ with $k \leq m$ are among $r_1, ..., r_m$ then*

$$\vdash C\Gamma_{i=1}^{k} q_i s \Gamma_{i=1}^{m} r_i s.$$

Proof. The proof is by induction on m. In case $m = 0$ also $k = 0$ and the thesis reduces to $\vdash Css$ i.e. to (T11).

Assuming the thesis for m, we consider $m + 1$.

First, let $k = 0$; then by inductive assumption, (i) $\vdash Cs\Gamma_{i=1}^{m} r_i s$. ($L_\infty$1) gives us

$$(ii) \quad \vdash C\Gamma_{i=1}^{m} r_i s \Gamma_{i=1}^{m+1} r_i s.$$

The thesis follows by (L_∞2).

Now, let $k > 0$; then q_k is r_u. By the inductive assumption,

$$(iii) \quad \vdash C\Gamma_{i=1}^{k-1} q_i s \Gamma_{i=u+1}^{m+1} r_i \Gamma_{i=1}^{u-1} r_i s$$

so (T9) implies

$$(iv) \quad \vdash C\Gamma_{i=1}^{k} q_i s Cq_u \Gamma_{i=u+1}^{m+1} r_i \Gamma_{i=1}^{u-1} r_i s.$$

There are two cases.

Case 1. $u = k + 1$ in which the thesis follows from (iv).

Case 2. $u < k + 1$ so by (T8)

$$(v) \quad \vdash C(Cq_u \Gamma_{i=u+1}^{m+1} r_i \Gamma_{i=1}^{u-1} r_i s C r_{m+1} C r_u \Gamma_{i=u+1}^{m+1} r_i) \Gamma_{i=1}^{u-1} r_i s.$$

The inductive assumption implies

$$(vi) \quad \vdash C(C r_u \Gamma_{i=u+1}^{m} r_i \Gamma_{i=1}^{u-1} r_i s) \Gamma_{i=1}^{m} r_i s$$

whence (T9) yields

$$(vii) \quad \vdash C(C r_{m+1} C r_u \Gamma_{i=u+1}^{m} r_i s) \Gamma_{i=1}^{u-1} r_i s.$$

Now, $(\text{L}_\infty 2)$ applied to (iv), (v), (vii) gives the thesis. This ends the proof. $\qquad\square$

In a similar way the following proposition may be proved.

Proposition 13.3. *[Rosser–Turquette52]* $\vdash CCqrC\Gamma_{i=1}^{m} p_i q \Gamma_{i=1}^{m} p_i r.$

Proof. Again, we induct on m. In case $m = 0$ the thesis becomes $\vdash Crr$ i.e. (T11).

The inductive step from m to $m + 1$ involves (T9) which yields

$$(i) \quad \vdash C(C\Gamma_{i=1}^{m} p_i q \Gamma_{i=1}^{m} p_i r)(C(\Gamma_{i=1}^{m+1} p_i q) \Gamma_{i=1}^{m+1} p_i r).$$

It suffices now to apply the inductive assumption and $(\text{L}_\infty 2)$. $\qquad\square$

Proposition 13.4. *[Rose–Rosser58]*
$\Gamma_{i=1}^{m} p_i q, \Gamma_{i=1}^{n} s_i Cqr \vdash \Gamma_{i=1}^{m} p_i \Gamma_{i=1}^{n} s_i r.$

Proof. Assume (i) $\Gamma_{i=1}^{n} s_i Cqr$ and apply to it (T8) to get (ii) $Cq\Gamma_{i=1}^{n} s_i r.$ Proposition 13.3 yields

$$(iii) \quad C\Gamma_{i=1}^{m} p_i q \Gamma_{i=1}^{m} p_i \Gamma_{i=1}^{n} s_i r.$$

$\qquad\square$

Letting p_i, q_i to be p we get a corollary (cf. [Novák 90], [Hajek97])

Proposition 13.5. $\vdash C(C((Cp)^m q)((Cp)^n Cqr))((Cp)^{n+m} r).$

We now are in position to prove the Deduction theorem for L_∞.

Proposition 13.6. *(the Deduction Theorem) [Rose–Rosser58], [No–vák87, 90], [Hajek97]* If $\Gamma, p \vdash q$ then $\Gamma \vdash (Cp)^n q$ for a natural number $n \geq 0$.

Proof. We refer to the proof of the Deduction Theorem in the case of classical propositional calculi (cf. Chapter 2). We carry out that proof: observe that the proof goes through in our new case here except in the case when α_n is gotten by detachment from $C\alpha_r\alpha_n$ with some $r < n$, in which case we apply Proposition 13.5. $\square$

Here the survey ends of basic theses and syntactic properties of the infinite valued calculus L_∞. In the part concerned with semantics, some further syntactic theses emerge when necessary. Now, we pass to semantics of L_∞ with the intention of giving an outline of the completeness proof by [Rose–Rosser58].

13.3   Semantics of L_∞

We begin with some preliminary results; the idea of the completeness proof rests on the separation properties in finite–dimensional vector spaces. We simplify the arguments in [Rose–Rosser58] by replacing matrix dichotomies with Propositions 7–9 below. Our first lemma is as follows.

Proposition 13.7. *Consider an algebraic infinite field $F \subseteq \mathbf{R}^1$. We are concerned with either $F = Q$, the field of rational numbers or $F = \mathbf{R}^1$. Consider linear functionals $f_1, ..., f_k, g$ on F^n with values in F. If (*) $f_1(x) = 0, ..., f_k(x) = 0$ imply $g(x) = 0$ for every $x \in F^n$ then there exist $\lambda_1, .., \lambda_k \in F$ with the property that $g = \sum_{i=1}^k \lambda_i f_i$.*

Proof. Consider $W = \{(f_1(x), ..., f_k(x)) : x \in F^n\}$. Then $W \subseteq F^k$ is a vector subspace of F^k. We define a functional h on W by letting $h((f_1(x), ..., f_k(x)))$ to be $g(x)$. Clearly, h is well–defined by our assumption (*) and then h extends to a functional $H : F^k \to F$. There exists the matrix of H say $[\lambda_1, ..., \lambda_k]$ i.e.

$$H(x) = [\lambda_1, ..., \lambda_k]x^T = \sum_{i=1}^k \lambda_i x_i.$$

By restricting H to W, we obtain a particular result that $g = \sum_{i=1}^k \lambda_i f_i$. $\square$

A simple corollary follows

Proposition 13.8. *In the notation of Proposition 13.7, if (**) $f_i(x) \geq 0$ for $i = 1, 2, ..., k$ imply $g(x) \geq 0$ for each $x \in F^n$, then $g = \sum_{i=1}^k \lambda_i f_i$ for some $\lambda_1 \geq 0, .., \lambda_k \geq 0 \in F$.*

Indeed, the assumption (*) follows easily from (**). That λ's be non-negative is also forced by (**).

We now consider a more general case when our functionals are *affine functions* i.e.

$$f_i(x) = c_i + \sum_{i=1}^{n} a_{ij}x_j$$

and

$$g(x) = d + \sum_{i=1}^{n} b_i x_i$$

where $c_i, a_{ij}, d, b_j \in F$. As clearly, $f_i - c_i, g - d$ are linear functionals, it follows from already proved cases that

Proposition 13.9. *If affine functions $f_1, .., f_k, g$ satisfy the condition that $f_1(x) \geq 0, .., f_k(x) \geq 0$ imply $g(x) \geq 0$ for every $x \in F^n$, then $g = \mu + \sum_{i=1}^{k} \lambda_i f_i$ with $\mu \geq 0, \lambda_1 \geq 0, ..., \lambda_k \geq 0 \in F$.*

13.3.1   Polynomials, polynomial formulae

The crux of the proof is in approximating general formulae with formulae of a special sort, called *polynomial formulae*.

Here, by a *polynomial*, a function of the form $c + \sum_{i=1}^{k} b_i x_i$ is meant with integral coefficients a, b_i. For each such polynomial f, a set $PF(f)$ of formulae will be defined. To this end, we define the *norm* $|f|$ of f by letting $|f| = \sum_{i=1}^{k} |b_i|$. The definition of $PF(f)$ splits into few cases.

(PF1) In case $|f| = 0$, we let: when $c \geq 1$ then $PF(f) = \{Cx_i x_j\}$ where x_i, x_j are variables, and when $c \leq 0$ then $PF(f) = \{NCx_i x_j\}$ where x_i, x_j are variables

In case $|f| > 0$, we assume that $PF(g)$ has been already defined for every g with $|g| < |f|$. Again, there are some subcases:
(PF2) When there is $b_i > 0$, we let $PF(f) = \{LBqx_j r\}$ where $q \in PF(f - x_j)$, $r \in PF(f + 1 - x_j)$
(PF3) When there is $b_i < 0$, we let $PF(f) = \{LBqNx_j r\}$ where $q \in PF(f + x_j - 1), r \in PF(f + x_j)$

For a polynomial formula p, we denote by $v(p)$ the truth value of p stemming from truth values of variables x_i. Actually then, $v(p)$ is a truth function. Let also sgn be the *sign function* defined via $sgn(x) = 1$ in case $x > 1$, $sgn(x) = 0$ in case $x < 0$ and $sgn(x) = x$, otherwise. Then we have

Proposition 13.10. *For each polynomial formula $p \in PF(f)$ and every set $x_i \in [0, 1]$ of variable values, we have $v(p) = sgn(f)$.*

Proof. It goes by inducting on $|f|$. In case $|f| = 0$ both sides are 1. Assuming truth of Proposition in case of all g with $|g| < |f|$, we consider $p \in PF(f)$.

Again, there are some cases to consider, according to definitions of polynomial formulae given above.

1. Some $b_i > 0$ so p is $LBqx_ir$ with some $q \in PF(f - x_i), r \in PF(f + 1 - x_i)$.

   Then we have four sub–cases viz. (a) $f > 1 + x_i$ in which sub–case $sgn(f - x_i), sgn(f + 1 - x_i) = 1$ hence $v(q) = v(r) = 1$ by the inductive assumption and thus by semantics of B, L it follows that $v(p) = 1$. Clearly, $sgn(f) = 1$.

   Other sub–cases (b) $x_i \le f < 1 + x_i$ (c) $-1 + x_i \le f < x_i$ (d) $f < -1 + x_i$ are discussed in analogous way

2. Some $b_i < 0$ so p is $LBqNx_ir$ with $q \in PF(f + x_i - 1), r \in PF(f + x_i)$. We have four sub–cases viz. (a) $f + x_i > 2$ (b) $1 \le f + x_i \le 2$ (c) $0 \le f + x_i < 1$ (d) $f + x_i < 0$ in which we argue as in Case 1.

   $\square$

This ends the basic introduction to polynomial formulae. Now, we list following [Rose–Rosser58] some deeper properties of polynomial formulae necessary in the proof of completeness. We would omit the tedious technical lemmas intervening in these proofs which otherwise would obscure the main ideas of the proof and we refer the interested reader to [Rose–Rosser58; Thms. 10.1–10.7, 13.1 – 13.18] for the proofs of these properties. For the reader convenience however, we list in the Exercise section the necessary syntactic facts and we propose to prove the results concerning polynomial formulae (1) – (10), listed in the proposition just below, in Exercises 18–34.

Proposition 13.11. *[Rose–Rosser58] The following are among properties of polynomial formulae*

1. $\vdash p \Leftrightarrow q$ whenever $p, q \in PF(f)$

2. $\vdash ANpq$ whenever $p \in PF(f), q \in PF(f + 1)$

3. $\vdash p \Leftrightarrow Nq$ whenever $p \in PF(f), q \in PF(1 - f)$

4. $\vdash Apq$ whenever $p \in PF(f), q \in PF(2 - f)$

5. $t \vdash CpCCqrs$ whenever $q \in PF(f), r \in PF(g), s \in PF(1 - f + g), t \in PF(2 - f), p \in PF(g + 1)$

6. $\vdash CpCqr$ whenever $q \in PF(f), r \in PF(g), p \in PF(1 - f + g)$

7. $p_1, .., p_k \vdash q$ whenever $p_i \in PF(1 + f_i), q \in PF(1 + \sum_{i=1}^{k} f_i)$ for some $k > 0$

8. $\vdash p$ whenever $p \in PF(1 + x_i)$

9. $\vdash p$ whenever $p \in PF(2 - x_i)$

   10. $p_1, ..., p_m \vdash q$ *whenever* $p_i \in PF(1 + f_i), q \in PF(1 + g), f_1, ..., f_m,$
      g have integral coefficients, and $f_1(x) \geq 0, ..., f_m(x) \geq 0$ *imply* $g(x) \geq 0$
      for every $x \in [0, 1]^n$ *whose each coordinate is rational*

Assuming truth of statements in Proposition 13.11, we now may proceed with concluding steps in completeness proof.

For a formula p and a polynomial f, a new formula $\alpha(p, f)$ is defined as follows. For arbitrarily chosen $q \in PF(f), r \in PF(2 - f), t \in PF(1 + f)$, we let $\alpha(p, f)$ to be $LLEpqrs$. Let us observe that any other choice of q, r, s subject to conditions yields an equivalent formula $\alpha(p, f)$ due to Proposition 13.11(1).

With help of this new class of formulas, one may prove the following principal statement.

Proposition 13.12. *[Rose–Rosser58] For each formula p, there exist $m \geq 0$, polynomial formulae $q_1, ..., q_m, r_1, ..., r_m$ and polynomials $f_1, ..., f_{2m}$ with the properties*

   1. $\vdash Aq_i r_i$ *for* $i = 1, 2, ..., m$

   2. for each subset $\{j_1, .., j_k\} \subseteq \{1, 2, ..., m\}$ *with the complementing subset* $\{i_1, .., i_{m-k}\}$ *there exists* $1 \leq k \leq 2^m$ *with the property that*

$$q_{j_1}, ..., q_{j_k}, r_{i_1}, ..., r_{i_{m-k}} \vdash \alpha(p, f).$$

Proof. We induct on complexity of p.

In case p consists of a single symbol, it is of the form x_i; consider $m = 0$ so only f_1 needs to be defined and choose $f_1(x) = x_i$. Then (i) $\vdash r$ (ii) $\vdash t$ by Proposition 13.11 (8, 9). By Proposition 13.11 (1), we have

$$\vdash q \Leftrightarrow LBNCssx_iCss$$

which is

$$\vdash q \Leftrightarrow LBNCsspCss.$$

By (T11) and (T33) we get that

$$\vdash q \Leftrightarrow BNCssp$$

whence $\vdash q \Leftrightarrow p$ follows by (T11), (T19), and (T13). Then $\vdash Epq$ follows by (T30) and (i), (ii), (T28), and Proposition 13.11 (6), imply $\vdash \alpha(p, f)$.

To get through the inductive step, assume that the proposition is true for all t with complexity less than l, and consider p with l symbols. There are two cases in need of consideration: 1. p is Nt 2. p is Ctr.

Case 1. There exist an integer n, polynomial formulae $q_1, ..., q_n, r_1, ..., r_n$ and polynomials $g_1, ..., g_{2^n}$ satisfying 1,2, in Proposition 13.12 with t. Keeping $m = n$, we let $f_i = 1 - g_i$ for $1 \le i \le 2^n$.

Assuming q^*, r^*, s^* were chosen from, respectively, $PF(g_i)$, $PF(2 - g_i)$, $PF(1 + g_i)$ and choosing q, r, s from respectively $PF(f_i), PF(2 - f_i), PF(1 + f_i)$ we have by Proposition 13.11 (1, 3), that

(i) $\vdash r \Leftrightarrow s^*$

(ii) $\vdash s \Leftrightarrow r^*$

(iii) $\vdash q \Leftrightarrow Nq^*$.

Now, we have $Etq^* \vdash ENtNq^*$ by (T31) i.e. (iv) $Etq^* \vdash Epq$.

We now consider a subset $\{j_1, ..., j_k\}$; we have by inductive assumption that

$$q_{j_1}, ..., q_{j_k}, r_{i_1}, ..., r_{i_{m-k}} \vdash \alpha(t, g_i)$$

for some i. It follows from (i), (ii), (iii), (iv), (T25), (T26), and (T28) that

$$\alpha(t, g) \vdash \alpha(p, f)$$

which concludes the proof in this case.

Case 2. As p is Ctr, the inductive assumption yields us $m, q_1, ..., q_m$, $r_1, ..., r_m$, $g_1, ..., g_{2^m}$ satisfying conditions 1,2 of the statement of Proposition 13.12 with t and

$n, q_1^*, ..., q_n^*, r_1^*, ..., r_n^*, h_1, ..., h_{2^n}$ satisfying those conditions with r.

It remains to glue these constructs together to produce the respective set for p.

We let $k = m + n + 2^{m+n}$. We make use of q_i's, q_i^*'s, r_i's, and r_i^*'s whose number is $m + n$ so we need 2^{m+n} new polynomial formulae. To construct them, for $1 \le u \le 2^m$ and $1 \le v \le 2^n$ we select $s_{uv} \in PF(1 - g_u + h_v)$, $s_{uv}^*(1 + g_u - h_v)$. As $\vdash As_{uv}s_{uv}^*$ by Proposition 13.11 (4), sets

$$\{q_1, ..., q_m, r_1, ..., r_n, s_{11}, ..., s_{2^m 2^n}\}$$

and

$$\{q_1^*, ..., q_m^*, r_1^*, ..., r_n^*, s_{11}^*, ..., s_{2^m 2^n}^*\}$$

satisfy condition 1 of Proposition 13.12. We denote elements of the first set with symbols a_i and elements of the second set with symbols a_i^*.

Now we need polynomials satisfying condition 2 with new sets of formulae. So we select a subset $\{j_1, ..., j_w\}$ of the set $\{1, ..., k\}$ and we consider the set $M = \{a_{j_1}, ..., a_{j_w}, a_{i_1}^*, ..., a_{i_{k-w}}^*\}$.

Let Γ be the subset of M consisting of formulae of the form of q_i, q_i^* and Δ be the subset of r_i, r_i^* of M. By inductive assumption, we have $u \le m$, $v \le n$ such that

(i) $\Gamma \vdash \alpha(t, g_u)$

(ii) $\Delta \vdash \alpha(r, h_v)$

Applying Proposition 13.11 (6), (T25), and (T26), we infer from (i), (ii) that

(iii) $\Gamma, \Delta \vdash Ett_1, t_2, t_3, Err_{1,r}, r_{2,r}, r_{3,r}$

where $\alpha(t, g_u)$ is $LLEtt_1 t_2 t_3$ and $\alpha(r, h_v)$ is $LLErr_{1,r} r_{2,r} r_{3,r}$.

By (iii) and (T32) it follows that

(iv) $\Gamma, \Delta \vdash EpCt_1 r_{1,r}$

By Proposition 13.11 (6), we obtain

(v) $\vdash Cs_{uv} Ct_1 r_{1,r}$

and Proposition 13.11 (5), implies via (v) that

(vi) $t_2 \vdash Cr_{3,r} CCt_1 r_{1,r} s_{uv}$

Now it follows from (iii), (iv) that

(vii) $\Gamma, \Delta \vdash Eps_{uv}$

It is time now to show the existence of the appropriate polynomial f_M for this case. There are two cases to discuss: either $s_{uv} \in M$ (Case 1) or $s_{uv} \notin M$ (Case 2).

Case 1. Let f_M be identically 1. By (T11) and (T33), we have

(viii) $s_{uv} \vdash Es_{uv} Czz$ for an arbitrarily chosen formula z

whence by (vii) and (T32)

(ix) $\Gamma, \Delta, s_{uv} \vdash EpCzz$

By Proposition 13.11 (1), and the definition of f_M, we have for $p_1 \in PF(f_M), p_2 \in PF(2 - f_M), p_3 \in PF(1 + f_M)$ that $p_1 \vdash Czz$, $p_2 \vdash Czz$, and $p_3 \vdash Czz$ and thus by (T11) and (T28) we arrive at $\Gamma, \Delta, s_{uv} \vdash \alpha(p, f_M)$

concluding the proof in this case.

Case 2. We have $s^*_{uv} \in M$. Let $f_M = 1 - g_u + h_v$ and select p_1, p_2, p_3 appropriately in respective PF's. By Proposition 13.11 (1), $\vdash p_1 \Leftrightarrow s_{uv}$ and $\vdash p_2 \Leftrightarrow s^*_{uv}$ which imply via (vii) and (T28) that

(x) $\Gamma, \Delta, s^*_{uv} \vdash LEpp_1 p_2$

By Proposition 13.11 (7), it follows that

(xi) $t_2, r_{3,r} \vdash p_3$

Finally, by (iii) and (x), one infers that $\Gamma, \Delta, s^*_{uv} \vdash \alpha(p, f_M)$

concluding the proof in Case 2 and hence the proof of the proposition. $\square$

We only need one more auxiliary statement.

Proposition 13.13. *For each formula p and each polynomial f the following hold*

1. $\Gamma \vdash \alpha(p, f)$ implies $\Gamma, p \vdash p_1$

2. $\Gamma \vdash \alpha(p, f)$ implies $\Gamma \vdash Cp_1 p$

Proof. Concerning (1), we begin with $\Gamma \vdash LLEpp_1 p_2 p_3$ whence by (T22) first $\Gamma \vdash Lp_3 LEpp_1 p_2$ and next $\Gamma \vdash Lp_3 Lp_2 Epp_1$ follow. Assuming p, we have $\Gamma, p \vdash Lp_3 Lp_2 p_1$ and applying twice (T25) we get $\Gamma, p \vdash p_1$.

In case (2), we begin as in (1) until $\Gamma \vdash Lp_3Lp_2Epp_1$ and at this point we apply (T25) twice getting $\Gamma \vdash Epp_1$ from which $\Gamma \vdash Cp_1p$ follows.   □

We now may state the completeness theorem. Clearly, $\mathbf{L}_\infty$ is sound: whenever a formula p is a thesis, it is a theorem as well i.e. its value $v(p)$ is identically 1. The converse is also true.

Proposition 13.14. *(The Completeness Theorem [Wajsberg35], [Rose– Rosser58]) For each formula p, if p is a theorem i.e. the value $v(p)$ is identically 1 then p is a thesis.*

Proof. The idea of the proof is a distant echo of the proof idea due to Kalmár in the case of classical propositional calculus (cf. Chapter 2).

Assume that $v(p)$ is identically 1 for a formula p. By Proposition 13.12 there exist $m \geq 0$, polynomial formulae $q_1, ..., q_m, r_1, ..., r_m$ and polynomials $g_1, ..., g_m, h_1, ..., h_m$ as well as $f_1, ..., f_{2^m}$ with the properties following from Proposition 13.13

(i) $\vdash Aq_ir_i$ for $i = 1, 2, ..., m$

(ii) for each subset $\{j_1, .., j_k\} \subseteq \{1, 2, ..., m\}$ with the complementing subset $\{i_1, ..., i_{m-k}\}$ we have a polynomial f_k such that

(a) $q_{j_1}, ..., q_{j_k}, r_{i_1}, ..., r_{i_{m-k}}, p \vdash p_1$

(b) $q_{j_1}, ..., q_{j_k}, r_{i_1}, ..., r_{i_{m-k}} \vdash Cp_1p$

with some $p_1 \in PF(f_k)$ and $q_i \in PF(g_i), r_i \in PF(h_i)$ for $i = 1, 2, ..., m$.

From (a), by Proposition 13.10 it follows that whenever $g_i(x), h_i(x) \geq 1$, then $f_k(x) \geq 1$ for each truth value vector x with coordinates in $[0, 1]$. Then by Proposition 13.11 (10), we have that

(iii) $q_{j_1}, ..., q_{j_k}, r_{i_1}, ..., r_{i_{m-k}} \vdash p_1$

whence by (b) and detachment

(iv) $q_{j_1}, ..., q_{j_k}, r_{i_1}, ..., r_{i_{m-k}} \vdash p$

Now, we have eliminated p_1 hence f_k a fortiori no more do we need formulae $\alpha(p, f_i)$ and we are left only with the fact that (iv) holds for each choice of $\{j_1, ..., j_k\} \subseteq \{1, 2, ..., m\}$.

Using (T16) along with (i) an appropriate number of times yields

(v) $\vdash p$

which concludes the proof.   □

Thus, the calculus $\mathbf{L}_\infty$ is complete. Please observe that the proof above covers cases when the set of truth values is an infinite subset of $[0, 1]$ (with additional above mentioned requirements) e.g. the set of rational values, or simply the segment $[0, 1]$.

13.4   Fuzzy logics of sentences

One may ask – and justly – what actually is a fuzzy logic? With fuzziness, one associates usually an approach to vagueness, due to [Zadeh65] in which statements known from classical set theory (cf. Chapter 6) of the form $x \in A$ may neither be true nor false but they may possess a degree of truth to a certain degree. Assuming that the set of truth values L is a bounded lattice (cf. Chapter 9), we may express fuzziness of the set $A \subseteq X$ by admitting that the *characteristic function* $\chi_A(x)$ on X which in the classical case takes only values 0 (in case $x \in A$) and 1 (in case $x \notin A$) may now take any value from L i.e. $\chi_A : X \to L$. In case $\chi_A(x) = a \in L$, one may say that x is an element of the set A to degree a. The function χ_A is called the *fuzzy characteristic (membership) function* of the *fuzzy set* A.

It is one of subjects of Fuzzy Set Theory to elaborate on algebra of fuzzy membership functions and on ways to ascribe degrees of membership to more complex constructs then sets e.g. to relations, functions etc. etc. Such activities are often styled *fuzzy logic*, but in the light of our experiences with logic it would be more advisable to keep the label of logic for a formal mechanism of inference in which fuzziness would play an essential syntactic and semantic role.

Thus, we should explore what ideas related to fuzziness and in what ways should enter the mechanism of many–valued logics presented already in this and previous Chapters. We have here few possibilities.

Referring to a logic, we usually set a language $\mathcal{L}$ consisting of certain symbols along with a syntactic mechanism, usually in the form of a set A of *axiom schemes* and *inference rules*. Now, the set of formulae, F, is formed and formulae are subject to a semantic evaluation.

Introducing fuzziness may be effected by making A into a fuzzy set over a bounded lattice L. In this way, each axiom instance acquires a degree of truth; accordingly, each inference rule R should consist of two components: $R = < R_1, R_2 >$, of which R_1 should be a syntactic inference rule in the sense we are acquainted with very well by now and R_2 a semantic counterpart to R_1 measuring the effect of R_1 on truth degrees (cf. [Goguen67, 69]).

It follows that a fuzzy logic in the genuine sense is a more complicated construct then ordinary many–valued logic as its semantic mechanism is more intricate and it is more closely coupled to the syntactic part. Also the role of the value set L is more important and essential, as it turns out (cf. [Pavelka 79^{a,b,c}]) that properties of a fuzzy logic depend essentially on properties of L.

We will see below essential excerpts from a study of fuzzy logic of propositions due to [Pavelka79^{a,b,c}].

13.4.1 Basic ingredients of a fuzzy logic

Following the scheme proposed in [Pavelka79[a]], we assume, as usual with logics of sentences, an infinite set V of *sentential variables* and sentential functors $\lor, \land, \dagger, \Rightarrow$.

From variables by means of functors, meaningful expressions (formulae) are constructed. The set F of meaningful expressions is the smallest of all sets X which satisfy the following properties

1. $v \in X$ for each $v \in V$

2. $\alpha \lor \beta \in X$, $\alpha \land \beta \in X$, $\alpha \dagger \beta \in X$, $\alpha \Rightarrow \beta \in X$ whenever $\alpha, \beta \in X$.

Once the set F of formulae is defined, we may choose a set $A \subseteq F$ of *axioms* (or, *axiom schemes* depending on our usage of the substitution rule) and fuzzify it over a chosen bounded lattice L as a fuzzy set $A : F \to L$ (called the *L–set of (logical) axioms*).

We need now a syntactic mechanism to generate theses from axioms by means of *inference rules*. We have already observed that inference rules are pairs of syntactic and semantic components. Accordingly, an *inference rule* R is a pair (R_1, R_2) such that for some positive integer k, we have
(1) $R_1 : dom R_1 \to F$ is a function on the domain $dom R_1 \subseteq F^k$ with values in F
(2) $R_2 : L^k \to L$ is a function which is join–semi–continuous i.e. the condition

$$(C) \quad R_2(a_1, ..., a_{i-1}, \bigvee_u a_{iu}, a_{i+1}, ..., a_k) =$$

$$\bigvee_u R_2(a_1, ..., a_{i-1}, a_{iu}, a_{i+1}, ..., a_k)$$

holds for each $i = 1, 2, ..., k$ and each choice of $\{a_{iu} : u \in U\}$ with $U \neq \emptyset$.

We may observe that (2) implies that R_2 is an isotone function (cf. Chapter 6).

Thus, an inference rule R at the same time converts meaningful expressions via R_1 and evaluates truth of results via R_2.

The behaviour of $X \in L^F$ with respect to R may be very intricate and hence some regularity conditions are desired. Following [Pavelka79[a]], we adopt a natural one viz.

$$(R) \quad \chi_X(R_1(x_1, ..., x_k)) \geq_L R_2(\chi_X(x_1), ..., \chi_X(x_k))$$

for every $(x_1, ..., x_k) \in dom R_1$.

We had had the need of an interlude on semantics prior to a further discussion of syntax and now we return to syntactic consequence.

We may fix an L–set of axioms A for a *complete lattice* L and a set $\mathcal{R}$ of inference rules. Given the pair $\mathcal{S} = (A, \mathcal{R})$, we may define (cf. [Pavelka79[a]]) the notion of syntactic consequence. Let us observe that the set L^F of all functions from F into L (i.e. the set of all L–sets of axioms) is a partially ordered set with the ordering $\leq$ induced coordinate–wise from the ordering $\leq_L$ of L i.e. for $A, B \in L^F$ we have

$$A \leq B \Leftrightarrow \forall x (\chi_A(x) \leq_L \chi_B(x)) \tag{13.2}$$

Then L^F is a completely ordered set (viz. $\chi_{\bigwedge Z}(x) = \bigwedge \{\chi_X(x) : X \in Z\}$) and one may define for each $X : F \to L$ a set $C_S(X) : F \to L$ as the set satisfying the following demands

1. $A \vee X \leq C_S(X)$

2. $C_S(X)$ does satisfy (R) for every $R \in \mathcal{R}$

3. $C_S(X)$ is the smallest set from among all sets Y satisfying 1, 2 above

The existence of $C_S(X)$ follows from the fact (cf. [Pavelka79[a]]) that the family of all sets satisfying (R) is closed with respect to all meets; indeed, given a family $\mathcal{M}$ of sets satisfying (R), and $x = (x_1, x_2, .., x_n) \in dom R_1$, we have

$$\begin{aligned}
\chi_{\bigwedge \mathcal{M}}(R_1(x)) = \bigwedge \{\chi_Y(R_1(x)) : Y \in \mathcal{M}\} \geq \\
\bigwedge \{R_2(\chi_Y(x_1), ..., \chi_Y(x_n)) : Y \in \mathcal{M}\} \geq \\
R_2(\bigwedge \{\chi_Y(x_1) : Y \in \mathcal{M}\}, ..., \bigwedge \{\chi_Y(x_n) : Y \in \mathcal{M}\}) = \\
R_2(\chi_{\bigwedge \mathcal{M}}(x_1), \\
..., \chi_{\bigwedge \mathcal{M}}(x_n))
\end{aligned} \tag{13.3}$$

where the last inequality comes by the fact that R_2 is isotone.

Thus, to produce $C_S(X)$ it suffices to take the meet of all sets $Y \in L^F$ that satisfy 1 and 2.

We now define the notion of the *syntactic consequence to a degree* viz. we let the symbol

$$X \vdash_{S,a} x$$

to denote the fact that

$$\chi_{C_S(X)}(x) \geq_L a$$

for $X \in L^F$, $x \in F$ and $a \in L$.

In case $X \vdash_{S,a} x$ holds, we say that x is a *syntactic consequence of X to degree at least a.*

Proposition 13.15. *(Pavelka) If* $X \vdash_{S,a_i} x_i$ *for* $i = 1, 2, ..., n$ *and* $(x_1, ..., x_n) \in dom R_1$, *then* $X \vdash_{S,R_2(a_1,...,a_n)} R_1(x_1, ..., x_n)$

Indeed, the thesis of Proposition 13.15 follows by definition of C_S.

Now, for the semantic factor. By a *semantics* on the set F, we will mean (cf.

[Pavelka79^a]) a family $\mathcal{E} \subseteq L^F$ of L–fuzzy sets; as with syntactic factor, we associate with $\mathcal{E}$ the *semantic consequence* function $C_{\mathcal{E}} : L^F \to L^F$ by letting

$$C_{\mathcal{E}}(X) =$$

$$\bigwedge \{Y \in \mathcal{E} : X \leq Y\}$$

We define the notion of a *semantic (logical) consequence to a degree* viz. we use the symbol

$$X \models_{\mathcal{E},a} x$$

to denote the fact that

$$\chi_{C_{\mathcal{E}}(X)}(x) \geq_L a$$

for $X \in L^F$, $x \in F$, and $a \in L$.

Proposition 13.16. *(Pavelka)* $X \models_{\mathcal{E},a} x$ *if and only if* $Y \in \mathcal{E}, Y \geq X$ *imply* $\chi_Y(x) \geq_L a$.

Indeed, the proof follows easily by definition of $C_{\mathcal{E}}$.

With notions of syntactic as well as semantic consequences, we may give a new meaning to the classical notions of soundness and completeness; we recall that a logical system is sound in case each thesis of the system is a theorem of the system. This means that the degree of semantic truth of each meaningful expression should be not less than the degree of its syntactic truth. So we will say that a syntax $S = (A, \mathcal{R})$ is *sound* with respect to a semantics $\mathcal{E} \subseteq L^F$ if and only if $C_S(X) \leq C_{\mathcal{E}}(X)$ for every $X \in L^F$.

Completeness in turn means in classical cases that – conversely – each theorem of a system is a thesis of the system. Thus, assuming soundness of a fuzzy logic as a prerequisite for its completeness, we will say that a fuzzy logic with a syntax $S = (A, \mathcal{R})$ is *complete* with respect to a semantics $\mathcal{E} \subseteq L^F$ if and only if $C_S(X) = C_{\mathcal{E}}(X)$ for every $X \in L^F$ i.e. if and only if $C_S = C_{\mathcal{E}}$. This means that syntactic and semantic consequences coincide in this case.

13.4.2   The Łukasiewicz residuated lattice

It is now time to discuss the relevant lattices L; as already pointed out, properties of L bear strongly on properties of the induced logic (cf. Exercise section for relevant results).

We consider a bounded lattice $(L, 0, 1, \vee, \wedge)$; a *pair of adjoints* is a pair $(\otimes, \to)$ of binary operations on L with the properties

(Ad1) $\otimes$ is an isotone function on $L \times L$

(Ad2) $\to$ is anti–isotone in the first coordinate and isotone in the second coordinate

(Ad3) $x \otimes y \leq z$ if and only if $x \leq y \to z$

The operation $\otimes$ is called *multiplication* and $\to$ is *residuation*.

An example of a pair of adjoints is $(\wedge, \to)$ in a pseudo–Boolean (Heyting) lattice (cf. Chapter 9) where $\to$ is the relative pseudo–complementation. We recall from Chapter 9 that in case L is a Boolean algebra, we have $x \to y = x' \vee y$ where x' is the complement to x in L.

A *residuated lattice* L is a bounded lattice L with a pair of adjoints and the property that $\otimes$ makes L into a commutative monoid i.e. $\otimes$ is commutative and associative with the lattice unit 1 as a monoid unit i.e. $1 \otimes x = x$ for every $x \in L$.

We will be concerned here with a particular L viz. we consider the unit interval $L = [0,1]$ with $x \vee y = max\{x, y\}, x \wedge y = min\{x, y\}$ and $0, 1$ as the null, respectively, unit elements.

In $L = [0,1]$, we consider multiplication $x \otimes y = max\{0, x + y - 1\}$ and residuation $x \to y = min\{1, 1 - x + y\}$. We will check that

Proposition 13.17. *The pair* $(\otimes, \to)$ *is a pair of adjoints on* $[0,1]$.

Indeed, in case $x + y - 1 < 0$, from $x \otimes y \leq z$ it follows that $0 \leq z$ and $x < 1 - y$ hence $x \leq min\{1, 1 - y + z\}$ i.e. $x \leq y \to z$; similarly, from $x \leq y \to z$ i.e. $x \leq min\{1, 1 - y + z\}$ it follows that $x + y - 1 \leq z$ i.e. $max\{0, x + y - 1\} \leq z$ i.e. $x \otimes y \leq z$. In case $x + y - 1 \geq 0$ the proof is similar.

We will call the tuple $L(0,1) = ([0,1], 0, 1, \vee, \wedge, \otimes, \to)$ the *Lukasiewicz residuated lattice*. We may easily observe that $\otimes$ is a commutative monoid operation on $L(0,1)$.

We introduce a new operation $\leftrightarrow$ in $L(0,1)$ via

$$x \leftrightarrow y = (x \to y) \wedge (y \to x);$$

clearly

$$x \leftrightarrow y = \\ min\{min\{1, 1 - x + y\}, min\{1, 1 - y + x\}\} = min\{1, 1 - |x - y|\} \quad (13.4)$$

We define the n–th power x^n as the product $x \otimes ... \otimes x$ of n–copies of x.

We observe here an important fact to be exploited in the sequel, viz., nilpotency of $L(0,1)$.

Proposition 13.18. *For every* $x \in [0,1)$*, there exists* $n \geq 1$ *with the property that* $x^n = 0$ *i.e. x is a nilpotent element.*

Proof. As $x^n = max\{0, 1 - (n-1)x\}$ it follows that $x^n = 0$ for $n \geq \frac{1}{1-x}$. $\square$

Then, we consider an operation $\phi : [0,1]^n \to [0,1]$; we will say after [Pavelka79[b]] that ϕ is *acceptable* (to $L(0,1)$) in case there exists non-negative integers $m_1, ..., m_n$ with the property that

$$(Ac) \ (x_1 \leftrightarrow y_1)^{m_1} \otimes (x_2 \leftrightarrow y_2)^{m_2} \otimes ..... \otimes (x_n \leftrightarrow y_n)^{m_n} \leq_L$$

$$\phi(\overline{x}) \leftrightarrow \phi(\overline{y})$$

for each $\overline{x} = (x_1, .., x_n), \overline{y} = (y_1, ..., y_n) \in [0,1]^n$.

To give an example, we recall that a function $f : I^n \to I$ is *lipschitzian* if for a constant $C \in (0, \infty)$ we have

$$(L) \ \rho(f(\overline{x}), f(\overline{y})) \leq C \cdot \rho(\overline{x}, \overline{y})$$

where ρ is a metric on I^n, say, $\rho(\overline{x}, \overline{y}) = \sum_{i=1}^n |x_i - y_i|$. Then

Proposition 13.19. *[Pavelka79[b]] Each lipschitzian operation $f : I^n \to I$ is acceptable to $L(0,1)$.*

Proof. Clearly, by definition of $\otimes$, we have (i) $(x \leftrightarrow y)^m = max\{0, 1 - m \cdot |x - y|\}$ for $x, y \in I$, $m \geq 0$. From the Lipschitz property (L) of the function f it follows that for $m \geq C$ we have (ii) $\sum_{i=1}^n (1 - m \cdot |x_i - y_i|) - (m - 1) \geq 1 - |\phi(\overline{x}) - \phi(\overline{y})|$.

Thus, from (i), (ii), we obtain

$$(x_1 \leftrightarrow y_1)^m \otimes (x_2 \leftrightarrow y_2)^m \otimes ..... \otimes (x_n \leftrightarrow y_n)^m =$$

$$max\{0, \sum_{i=1}^n max\{0, 1 - m \cdot |x_i - y_i|\}\} - (m - 1) \leq$$

$$max\{0, 1 - |\phi(x_1, ..., x_n) - \phi(y_1, ..., y_n)|\} =$$

$$\phi(x_1, ..., x_n) \leftrightarrow \phi(y_1, ..., y_n)$$

(13.5)

concluding the proof. □

It follows from Proposition 13.19 that we have a plethora of acceptable operations on $L(0,1)$, in particular every differentiable f is such.

In particular, operations $\vee, \wedge, \otimes, \to$ are acceptable (cf. Exercise section). Acceptable operations are also closed under composition.

Proposition 13.20. *Assume that $f : L^n \to L$ and $g_i : L^{n_i} \to L$, where $i = 1, 2, ..., n$, are acceptable. Then $f(g_1, ..., g_n)$ is acceptable.*

Proof. We assume that f satisfies the acceptability condition with integers $k_1, ..., k_n$ and each g_i does the same with coefficients $k_{i1}, ..., k_{in_i}$. Let $s_i =_1$

$+...+n_i$, each i .Then we have for given $\overline{x} = (x_1,...,x_{s_n}), \overline{y} = (y_1,...,y_{s_n})$
that

$$((x_1 \leftrightarrow y_1)^{k_{11}} \otimes ... \otimes (x_{n_1} \leftrightarrow y_{n_1})^{k_{1n_1}})^{k_1} \otimes ..$$

$$... \otimes ((x_{s_{n-1}+1} \leftrightarrow y_{s_{n-1}+1})^{k_{n1}} \otimes .... \otimes (x_{s_n} \leftrightarrow y_{s_n})^{k_{nnn}})^{k_n} \leq$$

$$(g_1(\overline{x}) \leftrightarrow g_1(\overline{y}))^{k_1} \otimes ... \otimes (g_n(\overline{x}) \leftrightarrow g_n(\overline{y})^{k_n} \leq$$

$$f(g_1,...,g_n)(\overline{x}) \leftrightarrow f(g_1,...,g_n)(\overline{y})$$

(13.6)

$\square$

The result of the last Proposition is that we may extend our lattice

$$L(I) = (L(0,1), 0, 1, \vee, \wedge, \otimes, \rightarrow)$$

to the algebra

$$L(I) = (L(0,1), 0, 1, \vee, \wedge, \otimes, \rightarrow, \{\delta_u : u \in U\})$$

where $\{\delta_u : u \in U\}$ is the set of all operations derived from the primitives $\vee, \wedge, \otimes, \rightarrow$ by means of composition. All these operations are acceptable in the sense of our definition.

13.4.3   Filters on residuated lattices

The proof of completeness we are going to sketch makes use of algebraic techniques of [Rasiowa–Sikorski63] (cf. Chapter 9) hence we need a discussion of filters on our algebra $L(0,1)$. According to a general idea of a filter (cf. Chapters 6, 8, 9), we call a subset $\mathcal{F} \subseteq L(0,1)$ a *filter* in case the following requirements are satisfied

(F1) $1 \in \mathcal{F}$
(F2) $x \otimes y \in \mathcal{F}$ whenever $x, y \in \mathcal{F}$
(F3) $y \in \mathcal{F}$ whenever $x \in \mathcal{F}$ and $x \leq y$

Let us observe (cf. Exercise 16) that $x \otimes y \leq_{L(0,1)} x \wedge y$ hence

(F4) $x \wedge y \in \mathcal{F}$ whenever $x, y \in \mathcal{F}$

and thus a filter on $L(0,1)$ is a filter on the lattice $(L(0,1), 0, 1, \vee, \wedge)$ in the sense of Chapter 8.

We may remember from our earlier chapters the important role played by Lindenbaum–Tarski algebras which were defined as quotient algebras of respective algebras of formulae by congruences induced by respective bidirectional implications. This scheme will be followed here as well so we now introduce relevant constructs into our discussion.

Given a filter $\mathcal{F}$, we define a congruence $\sim_{\mathcal{F}}$ on $L(I)$ via

$$(CF) \quad x \sim_{\mathcal{F}} y \ iff \ x \leftrightarrow y \in \mathcal{F} \tag{13.7}$$

That $\sim_{\mathcal{F}}$ is an equivalence relation follows from the following properties (cf. Exercise 12) of bi–residuation $\leftrightarrow$

$$\begin{aligned} &(i) \ x = y \ iff \ x \leftrightarrow y = 1 \ hence \ x \leftrightarrow x = 1 \\ &(ii) \ x \leftrightarrow y = y \leftrightarrow x \\ &(iii) \ x \leftrightarrow y \otimes y \leftrightarrow z \leq_{L(0,1)} x \leftrightarrow z \end{aligned} \tag{13.8}$$

That it is a congruence i.e. that from $x \sim_{\mathcal{F}} y, u \sim_{\mathcal{F}} v$ relations $x \bowtie u \sim_{\mathcal{F}} y \bowtie v$ where $\bowtie$ is respectively $\vee, \wedge, \otimes, \rightarrow$ follow is provided by the following further properties (cf. Exercise 12)

$$\begin{aligned} &(iv) \ (x \leftrightarrow y) \wedge (u \leftrightarrow v) \leq_{L(0,1)} (x \vee u) \leftrightarrow (y \vee v) \\ &(v) \ (x \leftrightarrow y) \wedge (u \leftrightarrow v) \leq_{L(0,1)} (x \wedge u) \leftrightarrow (y \wedge v) \\ &(vi) \ (x \leftrightarrow y) \otimes (u \leftrightarrow v) \leq_{L(0,1)} (x \otimes u) \leftrightarrow (y \otimes v) \\ &(vii) \ (x \leftrightarrow y) \otimes (u \leftrightarrow v) \leq_{L(0,1)} (x \leftrightarrow u) \leftrightarrow (y \leftrightarrow v) \end{aligned} \tag{13.9}$$

In this way each filter induces a congruence; conversely, given a congruence $\sim$ on $L(I)$, we may define a filter $\mathcal{F}_{\sim}$ via

$$x \in \mathcal{F}_{\sim} \ iff \ x \sim 1 \tag{13.10}$$

That $\mathcal{F}_{\sim}$ is indeed a filter follows from

$$\begin{aligned} &(viii) \ 1 \sim 1 \ by \ reflexivity \ of \ \sim \ hence \ 1 \in \mathcal{F}_{\sim} \ ((F1)) \\ &(ix) \ x \sim 1, \ x \leq_{L(0,1)} y \ imply \ y = x \vee y \sim 1 \vee y = 1 \\ &\qquad i.e. \ y \in \mathcal{F}_{\sim} \ ((F2)) \\ &(x) \ x \sim 1, y \sim 1 \ imply \ x \otimes y \sim 1 \otimes 1 = 1 \\ &\qquad i.e. \ x \otimes y \in \mathcal{F}_{\sim} \ ((F3)) \end{aligned} \tag{13.11}$$

The correspondence between filters and congruences revealed above is canonical i.e. $\mathcal{F}_{\sim_{\mathcal{F}}} = \mathcal{F}$ (cf. Exercise 17).

A filter $\mathcal{F}$ on $L(0,1)$ does induce a quotient algebra $L(0,1)/\mathcal{F}$ along with the canonical homomorphism $q_F : L(0,1) \rightarrow L(0,1)/\mathcal{F}$ sending x to $[x]_{\sim_{\mathcal{F}}}$. Then

Proposition 13.21. $q_F(x) \leq_{L(0,1)/\mathcal{F}} q_F(y)$ *if and only if* $x \rightarrow y \in \mathcal{F}$.

Proof. Indeed, $q_F(x) \leq_{L(0,1)/\mathcal{F}} q_F(y)$ means that $q_F(x) = q_F(x \wedge y)$ which is equivalent to $x \leftrightarrow x \wedge y \in \mathcal{F}$. Observe that $x \leftrightarrow x \wedge y = x \rightarrow y$. $\square$

By Proposition 13.20 and the above correspondence between filters and congruences it follows that congruences are preserved by all derived operations δ_u.

13.4.4 Syntax and semantics of a fuzzy sentential calculus

It is our purpose now to introduce formally syntactic and semantic ingredients into a general scheme presented above. We begin with a syntax of the fuzzy sentential calculus due to [Pavelka79^{a,b,c}].

So we begin with an infinite set V of *sentential variables* and with the Łukasiewicz algebra $\mathbb{L}(I) = (\mathbb{L}(0,1), 0, 1, \vee, \wedge, \otimes, \rightarrow, \{\delta_u : u \in U\})$.

We admit functors $\vee, \wedge, \dagger, \Rightarrow$ as sentence–forming, for each $a \in [0,1]$ we introduce the constant $\bar{a}$ corresponding to a, and for each δ_u we introduce a functor d_u of the same arity as δ_u.

Then we define the set $F(V, \mathbb{L}(I))$ (which we shall abbreviate to F from now on) called the *algebra of formulae over V, $\mathbb{L}(I)$* defined as the smallest of all sets X such that

(Form1) $v \in X$ whenever $v \in V$
(Form2) $\bar{a} \in X$ whenever $a \in [0,1]$
(Form3) $\alpha \vee \beta, \alpha \wedge \beta, \alpha \dagger \beta, \alpha \Rightarrow \beta \in X$ whenever $\alpha, \beta \in X$
(Form4) $d_u(\alpha_1, ..., \alpha_n) \in X$ whenever $\alpha_1, ..., \alpha_n \in X$ and arity of d_u is n

Let us add in passing that $\vee, \wedge, \Rightarrow$ are counterparts to classical functors of *alternation, conjunction, implication* while $\dagger$ is called *context*.

The bi–implication $\Leftrightarrow$ is defined as $\Rightarrow \wedge \Leftarrow$ and α^n will mean the shortcut for $\alpha \dagger ... \dagger \alpha$ where α is taken n times.

In ordinary order of things we should now single out axiom schemes and derivation rules; however, as we recall, fuzzy derivation rules do encompass both syntactic and semantic operations, hence we have now to make a short incursion into semantics prior to completing the description of syntax.

We extend once more the algebra $\mathbb{L}(I)$ by adding to it constants of the form a where $a \in (0,1)$ (note that we do not add $0,1$ as they are already singled out in $\mathbb{L}(I)$; let the resulting algebra be denoted with the symbol $\mathbb{L}(I)_a$.

According to the general idea of fuzzy semantics introduced above, we call a semantics on the algebra F the set

$$\mathcal{E} \subseteq \mathbb{L}(0,1)^F$$

such that

(FSEM) $T \in \mathcal{E}$ iff $T : F \rightarrow \mathbb{L}(I)_a$ is a homomorphism

Let us disentangle this definition; the demand that each T in $\mathcal{E}$ be a homomorphism means that T preserves functors on F, a fortiori, once T is defined on V it extends over F in a unique way viz.

(Sem1) $T(\bar{a}) = a$ for every $a \in [0,1]$

(Sem2) $T(\alpha \bowtie \beta) = T(\alpha) \bowtie T(\beta)$ where $\bowtie$ is $\vee$ or $\wedge$

(Sem3) $T(\alpha \dagger \beta) = T(\alpha) \otimes T(\beta)$

(Sem4) $T(\alpha \Rightarrow \beta) = T(\alpha) \rightarrow T(\beta)$

(Sem5) $T(d_u(\alpha_1, ..., \alpha_n)) = \delta_u(T(\alpha_1), ..., T(\alpha_n))$ whenever arity of δ_u is n

Clearly, we have also

(Sem6) $T(\alpha \Leftrightarrow \beta) = T(\alpha) \leftrightarrow T(\beta), T(\alpha^n) = (T(\alpha))^n$

Continuing with semantic aspects, we single out some formulae schemes (the reader will be pleased to notice that a majority of them are theses of the [0,1]–valued sentential calculus)

(Th1) $\bar{a} \wedge \bar{b} \Rightarrow \overline{a \wedge b}; \ \bar{a} \vee \bar{b} \Rightarrow \overline{a \vee b}$

(Th2) $\bar{a} \Rightarrow \bar{b} \Rightarrow \overline{a \rightarrow b}$

(Th3) $d_u(\alpha_1, ..., \alpha_n) \Leftrightarrow \overline{\delta_u(T(\alpha_1), ..., T(\alpha_n))}$ whenever arity of δ_u is n

(Th4) $\alpha \Rightarrow \bar{I}$

(Th5) $\alpha \Rightarrow \alpha$

(Th6) $(\beta \Rightarrow \gamma) \Rightarrow ((\alpha \Rightarrow \beta) \Rightarrow (\alpha \Rightarrow \gamma))$

(Th7) $(\alpha \Rightarrow (\beta \Rightarrow \gamma)) \Rightarrow (\beta \Rightarrow (\alpha \Rightarrow \gamma))$

(Th8) $\alpha \wedge \beta \Rightarrow \alpha$

(Th9) $\alpha \wedge \beta \Rightarrow \beta$

(Th10) $(\alpha \Rightarrow \beta) \Rightarrow ((\alpha \Rightarrow \gamma) \Rightarrow (\alpha \Rightarrow \beta \wedge \gamma))$

(Th11) $(\alpha \Rightarrow \beta) \Rightarrow ((\gamma \Rightarrow \beta) \Rightarrow (\alpha \vee \gamma \Rightarrow \beta))$

(Th12) $\alpha \Rightarrow (\alpha \vee \beta$

(Th13) $\beta \Rightarrow \alpha \vee \beta$

(Th14) $\alpha \Rightarrow (\beta \Rightarrow \alpha \dagger \beta)$

(Th15) $(\alpha \Rightarrow (\beta \Rightarrow \gamma)) \Rightarrow ((\alpha \Rightarrow \beta) \Rightarrow \gamma)$

(Th16) $\alpha \dagger \bar{I} \Leftrightarrow \alpha$

(Th17) $(\alpha_1 \Leftrightarrow \beta_1)^{k_1} \dagger \ldots \dagger (\alpha_n \Leftrightarrow \beta_n)^{k_n} \Rightarrow (d_u(\alpha_1, .., \alpha_n) \Leftrightarrow d_u(\beta_1, .., \beta_n))$
whenever arity of δ_u is n and parameters in the acceptability condition of δ_u
are $k_1, ..., k_n$

(Th18) $(\alpha \Rightarrow \beta) \vee (\beta \Rightarrow \alpha)$

(Th19) $(\alpha \vee \beta)^n \Rightarrow (\alpha^n \vee \beta^n)$ whenever $n \geq 1$

(Th20) $((\bar{a} \Rightarrow \alpha)^n \Rightarrow \bar{b}) \Rightarrow ((\bar{c} \Rightarrow \alpha)^n \Rightarrow \bar{d})$ whenever $0 \leq c < a$ and
$b < d < 1$, and $n \cdot a + b \geq n \cdot c + d$ for $n \geq 1$.

(Th21)
$$((\alpha \Rightarrow \bar{a})^n \Rightarrow \bar{b}) \Rightarrow ((\alpha \Rightarrow \bar{c})^n \Rightarrow \bar{d})$$
whenever $a < c \leq 1$ and $b < d < 1$, and $n \cdot a - b \leq n \cdot c - d$ for $n \geq 1$.

Now, we may return to syntax and give its basic components: an $L(0,1)$
fuzzy set A of axiom schemes and a set R of inference rules.

Having schemes (Th1-Th21) proposed above (the reader has undoubtedly
foreseen that these schemes would have turned out to be among theorems of
the system), we are going to define the set A of axiom schemes.

We thus define the set A as follows:

$$\chi_A(\alpha) = \begin{cases} a & \text{in case } \alpha \text{ is } \bar{a} \text{ for } \in [0,1] \\ a \otimes b & \text{in case } \alpha \text{ is } \bar{a} \dagger \bar{b} \text{ for } a, b \in [0,1] \\ a \to b & \text{in case } \alpha \text{ is } \bar{a} \Rightarrow \bar{b} \text{ for } a, b \in [0,1] \text{ } c \\ 1 & \text{in case } \alpha \text{ is one of } (Th1 - -Th21) \\ 0 & \text{otherwise} \end{cases} \qquad (13.12)$$

This defines the axiom scheme set and now we need a set of inference
rules to complete our description of syntax.

Following [Pavelka79^c], we introduce inference rules:

(FD) (*fuzzy detachment*) $D = (D_1, D_2)$ where $D_1(\alpha, \alpha \Rightarrow \beta) = \beta$ and
$D_2(x, y) = x \otimes y$. After our usage in previous chapters, we may write down
this rule in the form of a double fraction

$$(FD) \ (\frac{\alpha, \alpha \Rightarrow \beta}{\beta}; \frac{x, y}{x \otimes y})$$

(L_a) (*lifting by a*) $L_a = (L_{a1}, L_{a2})$ where $L_{a1}(\alpha) = \bar{a} \Rightarrow \alpha$ and $L_{a2}(x) = a \to$
x which we may write down in the fractional form as

$$(L_a) \ (\frac{\alpha}{\bar{a} \Rightarrow \alpha}; \frac{x}{a \to x})$$

We recall that an inference rule $R = (R_1, R_2)$ is sound with respect to
the semantics $\mathcal{E}$ whenever for each homomorphism T in $\mathcal{E}$, the requirement

(SND) $\chi_T(R_1(\alpha_1, ..., \alpha_n)) \geq_{L(0,1)} R_2(\chi_T(\alpha_1), ..., \chi_T(\alpha_n))$

is satisfied.

We may expect that inference rules FD and L_a are sound with respect to the semantics $\mathcal{E}$ for each a. It is so indeed.

Proposition 13.22. *[Pavelka79^c] Inference rules DF, L_a where $a \in [0,1]$ are sound with respect to the semantics $\mathcal{E}$.*

Proof. In case of FD, we have for a given pair α, β:

$$\chi_T(D_1(\alpha, \alpha \Rightarrow \beta)) = \chi_T(\beta) \geq \chi_T(\alpha) \otimes (\chi_T(\alpha) \to \chi_T(\beta)) = \\ \chi_T(\alpha) \otimes \chi_T(\alpha \Rightarrow \beta) = D_2(\chi_T(\alpha), \chi_T(\alpha \Rightarrow \beta)) \qquad (13.13)$$

This proves soundness of FD.

In case of L_a with $a \in [0,1]$, its soundness follows from

$$\chi_T(L_{a1}(\alpha)) = \chi_T(\bar{a} \Rightarrow \alpha) = a \to \chi_T(\alpha) = L_{a2}(\chi_T(\alpha)) \qquad (13.14)$$

As both D_2, L_{a2} are continuous, they satisfy the condition of upper $-$ semi $-$ continuity i.e. join $-$ preservation so this remark concludes the proof. $\square$

The axiom scheme set A along with the inference rule set $\mathcal{R} = \{FD\} \cup \{L_a : a \in [0,1]\}$ gives us the syntax $\mathcal{S} = (A, \mathcal{R})$.

Now, we need a deeper look into the interplay of syntax and semantics.

13.4.5 Syntax and semantics at work

We may recall ourselves that given the syntax $\mathcal{S}$, we define for each X : $F \to L(0,1)$ a set $C_{\mathcal{S}}(X) : F \to L(0,1)$ as the set satisfying the following demands

1. $A \vee X \leq C_{\mathcal{S}}(X)$

2. $C_{\mathcal{S}}(X)$ does satisfy the condition (R) for every $R \in \mathcal{R}$

3. $C_{\mathcal{S}}(X)$ is the smallest set from among all sets Y satisfying 1, 2 above

Then, we define the syntactic consequence to a degree by letting $X \vdash_{\mathcal{S},a}$ (α) if and only if $\chi_{C_{\mathcal{S}}(X)}(\alpha) \geq_{L(0,1)} a$. We will abbreviate this to the symbol $X \vdash_a \alpha$ as $\mathcal{S}$ is known to us.

The semantics $\mathcal{E}$, as we recall from our preliminary discussion, does induce the semantic consequence $C_{\mathcal{E}} : L(0,1)^F \to L(0,1)^F$ by letting

$$C_{\mathcal{E}}(X) = \bigwedge \{Y \in \mathcal{E} : X \leq Y\}.$$

We define the notion of the *semantic (logical) consequence to a degree* viz. we use the symbol

$$X \models_{\mathcal{E},a} \alpha$$

to denote the fact that

$$\chi_{C_{\mathcal{E}}(X)}(\alpha) \geq_{L(0,1)} a$$

for $X \in L(0,1)^F$, $\alpha \in F$, and $a \in L$. Again, we abbreviate this symbol to $X \models_a \alpha$ as $\mathcal{E}$ is already fixed.

Let us observe that

Proposition 13.23. *The syntax S is sound with respect to the semantics $\mathcal{E}$ i.e. $C_S \leq C_{\mathcal{E}}$.*

Indeed, it follows from definition of C_S that $C_S(X) \leq T$ whenever $X \leq T \in \mathcal{E}$ and thus $C_S(X) \leq \bigwedge\{T \in \mathcal{E} : X \leq T\}$ implying $C_S(X) \leq C_{\mathcal{E}}(X)$ as $C_{\mathcal{E}}(X) = \bigwedge\{T \in \mathcal{E} : X \leq T\}$ by definition.

Let us put for the record the following fact.

Proposition 13.24. $\chi_{C_{\mathcal{E}}(X)}(\alpha) = \bigvee\{a \in [0,1] : X \models_a \alpha\}.$

This follows by definition of $C_{\mathcal{E}}$.

Our semantic consequence was defined relative to a fuzzy set $X \in L(0,1)^F$ therefore its values were calculated on the basis of semantic functors $T \geq X$ only; in order to define theorems of the system, we should take into consideration the whole of $\mathcal{E}$ henceforth we should let $X = \overline{0}$ where $\chi_{\overline{0}}(x) = 0$ for every $x \in F$. Accordingly, $C_S(\overline{0}) = \bigwedge \mathcal{E}$.

We may define the notion of a theorem of our system, and we thus say that a formula $\alpha \in F$ is a *theorem* if and only if $\chi_{C_S(\overline{0})}(\alpha) = 1$.

Example 13.1. *As an example, let us mention that $\alpha \Rightarrow \beta$ is a theorem if and only if $\chi_T(\alpha) \leq_{L(0,1)} \chi_T(\beta)$ for every $T \in \mathcal{E}$.*

Indeed, $\overline{0} \models_1 \alpha \Rightarrow \beta$ if and only if $1 = T(\alpha \Rightarrow \beta) = T(\alpha) \to T(\beta)$ hence $T(\alpha) = 1 \otimes T(\alpha) \leq_{L(0,1)} T(\beta)$ for every $T \in \mathcal{E}$ by definition of residuation.

Let us also observe that

Proposition 13.25. *[Pavelka79ᶜ] Every formula of the form (Th1–Th21) is a theorem of the system. Moreover, the following properties are satisfied*
(Th22) $\overline{0} \models_a \overline{a}$
(Th23) $\overline{0} \models_{a \otimes b} \overline{a} \dagger \overline{b}$
(Th24) $\overline{0} \models_{a \to b} \overline{a} \Rightarrow \overline{b}$

We give some proofs here to indicate the general idea of proof. Clearly, formulae of the form (Th1) or (Th2) are theorems by (Sem1), (Sem4) and Example above.

Each formula of the form (Th5) is a theorem as $T(\alpha \Rightarrow \alpha) = T(\alpha) \to T(\alpha) = 1$.

For a formula $(\alpha \Rightarrow \beta) \vee (\beta \Rightarrow \alpha)$ of type (Th18), we have $T(\alpha \Rightarrow \beta) \vee (\beta \Rightarrow \alpha) = T(\alpha) \to T(\beta) \vee T(\beta) \to T(\alpha)$ and as $[0,1]$ is linearly ordered either

$T(\alpha) \leq T(\beta)$ or $T(\beta) \leq T(\alpha)$ so either $T(\alpha) \to T(\beta) = 1$ or $T(\beta) \to T(\alpha) = 1$ hence $T(\alpha \Rightarrow \beta) \vee (\beta \Rightarrow \alpha) = 1$, each $T \in \mathcal{E}$.

Similarly, for a formula α of type (Th19), by the isotonicity of $\otimes$ in each coordinate and the fact that $x \vee y \in \{x, y\}$ one has $(x \vee y)^n = x^n \vee y^n$ hence $T(\alpha) = 1$, every $T \in \mathcal{E}$.

Concerning formulae of types (Th20) and (Th21), one refers to the property $(x \to y)^n \to z = min\{1, z + max\{0, n(x-y)\}\}$ which may be directly verified by definition of $\to$ and which implies both $(a \to x)^n \to b \leq (c \to x)^n \to d$ in case of (Th20) and $(x \to a)^n \to b \leq (x \to c)^n \to d$ in case of (Th21).

Properties (Th22–Th24) follow immediately from the definition of semantics, notably from (Sem1), (Sem3), (Sem4).

We are ready for a sketch of completeness of the fuzzy sentential calculus based on L(I).

13.4.6 Completeness of the fuzzy sentential calculus

Let us first observe that by the definition of the axiom scheme set A it follows immediately that

Proposition 13.26. *(Pavelka) The following are true given the axiom scheme set A defined above*
1. $\emptyset \vdash_a \bar{a}$
2. $\emptyset \vdash_{a \otimes b} \bar{a} \dagger \bar{b}$
3. $\emptyset \vdash_{a \to b} \bar{a} \Rightarrow \bar{b}$
4. $\emptyset \vdash_a \alpha$ *implies* $X \vdash_a \alpha$ *for every* $X \in L(0,1)^F$
5. $\emptyset \vdash_1 \alpha$ *whenever α is a formula of a type belonging to (Th1)–(Th21).*

Clearly, (1), (2), (3), (5) follow by the very choice of A and (4) follows by definition of syntactic consequence.

We now enter the path leading to a completeness proof based on filter behavior in the Lindenbaum–Tarski algebra of our calculus. We begin at the algebra F of formulae and for a set $X \in L(0,1)^F$, we define a relation $\preceq$ on F via

$$\alpha \preceq \beta \; iff \; X \vdash_1 \alpha \Rightarrow \beta \tag{13.15}$$

Then we have

Proposition 13.27. *(Pavelka) The relation $\preceq$ is a pre–order on F i.e. it is reflexive and transitive. Moreover the property*
6. $X \vdash_a \alpha$ *if and only if $\bar{a} \preceq \alpha$*
   holds.

Proof. By Proposition 13.26, (4), (5), we have $X \vdash_1 \alpha \Rightarrow \alpha$ as an instance of (Th5) hence $\alpha \preceq \alpha$ i.e. reflexivity of $\preceq$ is verified.

We assume now that $\alpha \preceq \beta \preceq \gamma$ hold; thus $X \vdash_1 \beta \Rightarrow \gamma$. By Proposition 13.26, (4), (5), and (Th6) we have $X \vdash_1 (\beta \Rightarrow \gamma) \Rightarrow ((\alpha \Rightarrow \beta) \Rightarrow (\alpha \Rightarrow \gamma))$.

Applying fuzzy detachment FD, to $(\beta \Rightarrow \gamma, (\beta \Rightarrow \gamma) \Rightarrow ((\alpha \Rightarrow \beta) \Rightarrow (\alpha \Rightarrow \gamma))$ we get $X \vdash_1 ((\alpha \Rightarrow \beta) \Rightarrow (\alpha \Rightarrow \gamma))$ and repeating this for the pair $(\alpha \Rightarrow \beta, (\alpha \Rightarrow \beta) \Rightarrow (\alpha \Rightarrow \gamma))$ yields $X \vdash_1 \alpha \Rightarrow \gamma$ i.e. $\alpha \preceq \gamma$ proving transitivity of $\preceq$.

To settle (6), we assume first that $\overline{a} \preceq \alpha$ i.e. $X \vdash_1 \overline{a} \Rightarrow \alpha$. As $X \vdash_a \overline{a}$ by Proposition 13.26, (1), (4), applying FD yields $X \vdash_{D_2(a,1)} D_1(\overline{a}, \overline{a} \Rightarrow \alpha)$ i.e. $X \vdash_a \alpha$.

Conversely, assuming $X \vdash_a \alpha$, we may apply L_a to get $X \vdash_{L_{a2}(a)} L_{a1}(\alpha)$ i.e. $X \vdash_a \overline{a} \Rightarrow \alpha$. $\square$

By the Schröder theorem (cf. Chapter 6, Exercise 28), the relation $\approx$ defined via $\alpha \approx \beta$ if and only if $\alpha \preceq \beta \wedge \beta \preceq \alpha$ is an equivalence relation on F. Moreover

Proposition 13.28. *(Pavelka) The relation $\approx$ is a congruence on F.*

The proof follows on standard lines; to give an exemplary argument, assume that $\alpha_1 \approx \beta_1, \alpha_2 \approx \beta_2$ hence $X \vdash_1 \alpha_1 \Rightarrow \beta_1, \alpha_2 \Rightarrow \beta_2$. Then by (Th8), (Th9), and detachment FD, we get $X \vdash_1 \alpha_1 \wedge \alpha_2 \Rightarrow \beta_1, \beta_2$ hence by (Th10) and FD it follows that $X \vdash_1 \alpha_1 \wedge \alpha_2 \Rightarrow \beta_1 \wedge \beta_2$. By symmetry, it follows that $X \vdash_1 \beta_1 \wedge \beta_2 \Rightarrow \alpha_1 \wedge \alpha_2$ hence $\alpha_1 \wedge \alpha_2 \approx \beta_1 \wedge \beta_2$. In similar manner we may prove that $\approx$ is preserved by other functors.

Having the congruence $\approx$ on the algebra of formulae F, we may form the quotient set $F_X = F/\approx$ whose elements are classes $[\alpha]_\approx$ for $\alpha \in F$. Clearly, F_X is the underlying set for the Lindenbaum – Tarski algebra of F. It remains to verify that F_X becomes in a natural way the algebra of the same type as F i.e. with the corresponding operations of same arity and accessibility parameters.

Proposition 13.29. *(Pavelka) The Lindenbaum – Tarski algebra*

$$L(0,1)_X = (F_X, [0]_\approx, [1]_\approx, \vee, \wedge, \otimes, \rightarrow, \{d_u : u \in U\})$$

of F is a residuated lattice of same type as $L(0,1)$, where $\vee, \wedge, \otimes, \rightarrow$ are logical functors $\vee, \wedge, \dagger, \Rightarrow$ factored through $\approx$.

Proof. The proof consists in verifying step by step that all requirements for a residuated lattice are satisfied and the reduced operations on F_X preserve parameters of their counterparts on F. We define a pre–ordering $\prec$ on F_X via $[\alpha]_\approx \prec [\beta]_\approx$ if and only if $\alpha \preceq \beta$.

As $X \vdash_0 \alpha$, for every $\alpha \in F$, it follows by Proposition 13.27, (6), that $X \vdash_1 \overline{0} \Rightarrow \alpha$ i.e. $[0]_\approx \prec [\alpha]_\approx$. Similarly, by (Th4), $[\alpha]_\approx \prec [1]_\approx$ for each $\alpha \in F$. Thus $[0]_\approx, [1]_\approx$ are respectively the null and the unit elements in F_X.

It follows from (Th8), (Th9), that $[\alpha]_{\approx}, [\beta]_{\approx} \prec [\alpha \wedge \beta]_{\approx}$ and in case $[\gamma]_{\approx} \prec [\alpha]_{\approx}, [\beta]_{\approx}$ we have $X \vdash_1 \gamma \Rightarrow \alpha, \gamma \Rightarrow \beta$ so by (Th10) via detachment we arrive at $X \vdash_1 \gamma \Rightarrow \alpha \wedge \beta$ i.e. at $[\gamma]_{\approx} \prec [\alpha \wedge \beta]_{\approx}$.

Thus letting $[\alpha]_{\approx} \wedge [\beta]_{\approx} = [\alpha \wedge \beta]_{\approx}$, we obtain the result that $[\alpha \wedge \beta]_{\approx}$ is the meet of $[\alpha]_{\approx}, [\beta]_{\approx}$.

Similarly one may prove that $[\alpha \vee \beta]_{\approx}$ is the join of $[\alpha]_{\approx}, [\beta]_{\approx}$. As $\preceq$ is a congruence, it follows that $(F_X, [0]_{\approx}, [1]_{\approx}, \vee, \wedge)$ is a bounded lattice.

We now check that $(\otimes, \to)$ where $[\alpha]_{\approx} \otimes [\beta]_{\approx} = [\alpha \dagger \beta]_{\approx}$ and $[\alpha]_{\approx} \to [\beta]_{\approx} = [\alpha \Rightarrow \beta]_{\approx}$ is a pair of adjoints on $L(0,1)_X$.

We leave a verification that $\Rightarrow$ is antitone in the first coordinate and isotone in the second which follows by (Th6) in the former and by (Th6), (Th7) in the latter case. Thus the requirement (Ad2) is satisfied. Similarly we may prove that $\dagger$ is isotone in both coordinates fulfilling (Ad1). It remains to verify the adjointness condition (Ad3).

Consider $\alpha, \beta, \gamma \in F$; assuming $\alpha \preceq \beta \Rightarrow \gamma$ i.e. $X \vdash_1 \alpha \Rightarrow (\beta \Rightarrow \gamma)$ we may invoke (Th14), (Th15) with substitution $\gamma / \alpha \dagger \beta$ and detachment to get $\alpha \dagger \beta \preceq \gamma$. This is one half of (Ad3).

To get the other half, assume that $\alpha \dagger \beta \preceq \gamma$; by (Th14) we have $\alpha \preceq \beta \Rightarrow \alpha \dagger \beta$ and by the already verified fact that $\Rightarrow$ is isotone in the second coordinate our assumption yields $\alpha \preceq \beta \Rightarrow \gamma$ completing the proof that $(\dagger, \Rightarrow)$ is a pair of adjoints. As $\dagger, \Rightarrow$ preserve $\preceq$, the pair $(\otimes, \to)$ is a pair of adjoints on $L(0,1)_X$.

To conclude that we get a quotient residuated lattice, it suffices to check that $\otimes$ makes $L(0,1)_X$ into a monoid with $[1]_{\approx}$ as the unit element. We prove for example that $\otimes$ is commutative; it follows by (Th7) that $[\alpha]_{\approx} \prec [\beta]_{\approx} \to [\gamma]_{\approx}$ if and only if $[\beta]_{\approx} \prec [\alpha]_{\approx} \to [\gamma]_{\approx}$ implying that $[\alpha]_{\approx} \otimes [\beta]_{\approx} \prec [\gamma]_{\approx}$ if and only if $[\beta]_{\approx} \otimes [\alpha]_{\approx} \prec [\gamma]_{\approx}$ for each $\gamma \in F$. This implies that $[\alpha]_{\approx} \otimes [\beta]_{\approx} = [\beta]_{\approx} \otimes [\alpha]_{\approx}$.

It follows that $L(0,1)_X$ is a residuated lattice.

It remains to check that $L(0,1)_X$ with operations indicated in its full description is a residuated algebra of same type as $L(0,1)$.

We continue by considering the factored through $\approx$ operations d_u. First, let us observe that

$$(D_n) \; (D_{n1}(\alpha_1, ..., \alpha_n) = \alpha_1 \dagger ... \dagger \alpha_n, D_{n2}(x_1, ..., x_n) = x_1 \otimes ... \otimes x_n)$$

is an inference rule with respect to the syntax $\mathcal{S}$.

Indeed, for $n = 1$, (D_1) holds trivially; assume (D_n) holds and consider $X \vdash_{a_i} \alpha_i$ for $i = 1, 2, ..., n+1$. By the assumption we have that $X \vdash_{a_1 \otimes ... \otimes a_n} \alpha_1 \dagger ... \dagger \alpha_n$ and $X \vdash_{a_{n+1}} \alpha_{n+1}$; by (Th14) applied to $\alpha_1 \dagger ... \dagger \alpha_n$ and α_{n+1} we obtain applying detachment twice that $X \vdash_{a_1 \otimes ... \otimes a_n \otimes a_{n+1}} \alpha_1 \dagger ... \dagger \alpha_n \dagger \alpha_{n+1}$.

We consider now an operation d_u of arity n with parameters $k_1, .., k_n$; we assume that $\alpha_i \approx \beta_i$ for $i = 1, 2, ..., n$. Then

$$X \vdash_1 (\alpha_1 \Leftrightarrow \beta_1)^{k_1} \dagger ... \dagger (\alpha_n \Leftrightarrow \beta_n)^{k_n} \tag{13.16}$$

by an application of $(D_{\sum k_i})$. Now we invoke (Th17) and from there via detachment we get

$$X \vdash_1 d_u(\alpha_1, ..., \alpha_n) \Leftrightarrow d_u(\beta_1, ..., \beta_n) \tag{13.17}$$

i.e.

$$d_u(\alpha_1, ..., \alpha_n) \approx d_u(\beta_1, ..., \beta_n) \tag{13.18}$$

This means that d_u preserve $\approx$ and thus letting

$$[d_u]([\alpha_1]_\approx, ..., [\alpha_n]_\approx) = [d_u(\alpha_1, ..., \alpha_n)]_\approx \tag{13.19}$$

we have that

$$([\alpha_1]_\approx \leftrightarrow [\beta_1]_\approx)^{k_1} \otimes ... \otimes ([\alpha_n]_\approx \leftrightarrow [\beta_n]_\approx)^{k_n} \prec$$
$$\tag{13.20}$$
$$([d_u]([\alpha_1]_\approx, ..., [\alpha_n]_\approx) \leftrightarrow [d_u]([\beta_1]_\approx, ..., [\beta_n]_\approx)$$

It follows that $L(0,1)_X$ is a residuated algebra of the same type as $L(0,1)$. Our proposition is proved.   □

We now need some means to relate the lattice $L(0,1)$ to the algebra $L(0,1)_X$; the natural choice is a function $i_X : L \to L(0,1)_X$ defined via the condition

$$i_X(a) = [\bar{a}]_\approx.$$

Then we have the following

Proposition 13.30. *(Pavelka) The following properties hold for i_X*
(h1) i_X is a homomorphism which preserves joins
(h2) i_X is either an injection or $L(0,1)_X$ is trivially null
(h3) If $L(0,1)_X$ is not trivially null then
   (i) every filter $\mathcal{F}$ on $L(0,1)_X$ such that $(*)$ $\mathcal{F} \cap i_X([0,1]) = \{i_X(1)\}$ extends to a filter $\mathcal{G}$ which is maximal with respect to $(*)$
   (ii) every filter $\mathcal{G}$ maximal with respect to $(*)$ satisfies: $x \notin \mathcal{G}$ if and only if $(**)$ $x^n \otimes u \prec i_X(a)$ for some $u \in \mathcal{G}$, some $a \in L$ and some $n \geq 1$
   (iii) every filter $\mathcal{G}$ maximal with respect to $(*)$ is prime (cf. Chapter 8) i.e. $x \vee y \in \mathcal{G}$ if and only if either $x \in \mathcal{G}$ or $y \in \mathcal{G}$

We include a proof of this auxiliary proposition.

Proof. First, we show that i_X is isotone; so we assume that $x \leq_{L(0,1)} y$. As $X \vdash_y \bar{y}$ by the definition of the set A, we have also $X \vdash_x \bar{y}$ hence by Proposition 13.27, (6), we have $\bar{x} \preceq \bar{y}$.

From this fact we infer that $i_X(x \wedge y) \prec i_X(x) \wedge i_X(y)$; on the other hand, from the definition of the axiom set A and specifically from the condition that $\chi_A(\bar{x} \wedge \bar{y} \Rightarrow \overline{x \wedge y}) = 1$ it follows that $X \vdash_1 \bar{x} \wedge \bar{y} \Rightarrow \overline{x \wedge y}$ and thus $[\bar{x} \wedge \bar{y}]_{\approx} \prec [\overline{x \wedge y}]_{\approx}$ i.e. $i_X(x) \wedge i_X(y) \prec i_X(x \wedge y)$. Thus i_X preserves $\wedge$ and in similar way one shows that it does preserve $\vee, \otimes, \rightarrow$. By (Th3) it follows that i_X preserves operations d_u i.e. i_X is a homomorphism.

By Proposition 13.24, and Proposition 13.27, we have that if $\bar{x} \preceq \alpha$ for every $x \in C \subseteq [0,1]$, then $X \vdash_x \alpha$ for every $x \in c$ and thus $X \vdash_{supC} \alpha$ i.e. $\overline{supC} \preceq \alpha$. This implies by arbitrariness of α that i_X does preserve joins. Thus (h1) is verified.

To verify (h2), we recall that each element $x < 1$ in $[0,1]$ is nilpotent cf. Proposition 13.18 i.e. for some n we have $x^n = 0$ where $x^n = x \otimes ... \otimes$ over n copies of x. Assuming that i_X is not any injection so there are $a \neq b \in [0,1]$ with $i_X(a) = i_X(b)$. Thus $a \leftrightarrow b < 1$ and $i_X(a \leftrightarrow b) = i_X(a) \leftrightarrow i_X(b) = 1$. As $a \leftrightarrow b < 1$, we have $n \geq 1$ for which $(a \leftrightarrow b)^n = 0$ and thus

$$0 = i_X(0) = i_X((a \leftrightarrow b)^n) = (i_X(a) \leftrightarrow i_X(b))^n = 1^n = 1 \qquad (13.21)$$

so this means $L(0,1)_X$ is trivially null.

Assume now that X is such that $L(0,1)_X$ is not trivially null (e.g. in the most interesting case to us viz. $X = \bar{0}$). Consider a filter $\mathcal{F}$ on $L(0,1)_X$ with the property that $(*)$ $\mathcal{F} \cap i_X([0,1]) = \{i_X(1)\}$. The existence of $\mathcal{G}$ maximal with respect to $(*)$ follows directly by the Maximum principle as given a chain Ψ of filters each of which satisfies $(*)$, it follows immediately that

$$\bigvee \Psi \cap i_X([0,1]) = \bigvee \mathcal{K} \cap i_X([0,1]) = \bigvee\{\{i_X(1)\}\} = i_X(1) \qquad (13.22)$$

Thus letting $\mathcal{G} = \bigvee \Psi$ concludes the proof of (h3)(i).

Considering (h3)(ii), we may observe that if the condition $(**)$ is satisfied with x then necessarily $x \notin \mathcal{G}$ as in case $x \in \mathcal{G}$, $(**)$ implies $i_X(a) \in \mathcal{G}$, contrary to $(*)$. Conversely, assuming that for a given x no u, a, n as requested satisfy (3ii) with x, we may define a filter

$$\mathcal{K} = \{z : \exists u \in \mathcal{G}. \exists n. z \geq x^n \otimes u\} \qquad (13.23)$$

Clearly, $\mathcal{K}$ contains $\mathcal{G} \cup \{x\}$ and it satisfies $(*)$ so by maximality of $\mathcal{G}$ we have $\mathcal{K} = \mathcal{G}$ hence $x \in \mathcal{G}$.

Finally, to prove (h3)(iii), we begin with recalling that by (Th18), (Th19) and the definition of the axiom scheme set A, we have

$$X \vdash_1 (\alpha \Rightarrow \beta) \vee (\beta \Rightarrow \alpha)$$
$$X \vdash_1 (\alpha \vee \beta)^n \Rightarrow (\alpha^n \vee \beta^n \qquad (13.24)$$

which translate into properties

$$(a)\ (x \to y) \vee (y \to x) = i_X(1)$$
$$(b)\ x^n \vee y^n = (x \vee y)^n \tag{13.25}$$

of $L(0,1)_X$.

It suffices to prove that $x \vee y \in \mathcal{G}$ implies $x \in \mathcal{G}$ or $y \in \mathcal{G}$. So assume to the contrary that $x, y \notin \mathcal{G}$. By (h3)(ii) there exist $u, v \in \mathcal{G}, n, m \geq 1, a < 1, b < 1$ with

$$(c)\ x^n \otimes u \leq i_X(a)$$
$$(d)\ y^m \otimes v \leq i_X(b) \tag{13.26}$$

Combining (c) and (d), we get with e.g. $max\{n, m\} = n$ that

$$(x \vee y)^n \otimes (u \otimes v) = (x^n \vee y^n) \otimes u \otimes v = x^n \otimes u \otimes v \vee y^n \otimes u \otimes v \leq_{L(0,1)_X}$$

$$(x^n \otimes u) \vee (y^m \otimes v) \leq_{L(0,1)_X}$$

$$i_X(a) \vee i_X(b) = i_X(a \vee b)$$

and as $a \vee b < 1$, it follows by (h2) that $x \vee y \notin \mathcal{G}$ proving (h3). $\square$

We may attempt at the completeness theorem for the fuzzy sentential calculus.

Proposition 13.31. *[Pavelka79^c] (The completeness theorem) Given the syntax $S = (A, \{FD\} \cup \{L_a : a \in [0,1]\})$ and the semantics $\mathcal{E}$, the fuzzy sentential calculus is complete i.e. $C_S = C_{\mathcal{S}}$ i.e. $\chi_{C_{S(X)}} = \chi_{C_{\mathcal{E}(X)}}$ for every fuzzy set $X \in L(0,1)^F$ i.e. (c) $\chi_{C_{S(X)}}(\alpha) = \chi_{C_{\mathcal{E}(X)}}(\alpha)$ for every $\alpha \in F$.*

We outline the proof of this theorem.

Proof. We have to verify (c) for every pair (X, α). By soundness, Proposition 13.22, it suffices to consider the case when $\chi_{C_{S(X)}}(\alpha) = a < 1$; given $b > a$, we should point to a homomorphism T in $\mathcal{E}$ such that $T(\alpha) < b$. If we succeed, then $\chi_{C_{\mathcal{E}(X)}}(\alpha) = \bigwedge\{T(\alpha) : T \in \mathcal{E}\} = a$ proving the equality (c).

So we fix an element $b > a$; as $L(0,1)_X$ is not trivially null (having $a < 1$), we infer from Proposition 13.30, (h2) that i_X is an injection. We also have by Proposition 13.24 that $a = \bigvee\{b : i_X(b) \prec [\alpha]_\approx\}$.

We look at the filter

$$\mathcal{F} = \{z : \exists n \geq 1. z \geq_{L(0,1)_X} ([\alpha]_\approx \to i_X(b))^n\} \tag{13.27}$$

We will check that $\mathcal{F}$ satisfies

$$(*)\ \mathcal{F} \cap i_X([0,1]) = \{i_X(1)\}.$$

Assume to the contrary that for some $c < 1$ we have

$$i_X(c) \geq_{L(0,1)_X} ([\alpha]_\approx \to i_X(b))^n \qquad (13.28)$$

with some $n \geq 1$. By already exploited in the above proof of (h3) properties
   (a) $(x \to y) \vee (y \to x) = i_X(1)$
   (b) $x^n \vee y^n = (x \vee y)^n$
of $L(0,1)_X$, we have

$$
\begin{aligned}
1 = i_X(1) = (i_X(1))^n = \\
([\alpha]_\approx \to i_X(b))^n \vee (i_X(b) \to [\alpha]_\approx)^n \leq_{L(0,1)_X} \\
i_X(c) \vee (i_X(b) \to [\alpha]_\approx)
\end{aligned} \qquad (13.29)
$$

It follows that

$$i_X(c) \vee (i_X(b) \to [\alpha]_\approx) = 1 \qquad (13.30)$$

We make use again of nilpotency in $L(0,1)$: there is $m \geq 1$ such that $c^m = 0$ and for this m we have

$$
\begin{aligned}
1 = i_x(1) = (i_x(1))^m = \\
(i_X(c))^m \vee (i_X(b) \to [\alpha]_\approx)^m = \\
(i_X(b) \to [\alpha]_\approx)^m \leq_{L(0,1)_X} \\
i_X(b) \to [\alpha]_\approx
\end{aligned} \qquad (13.31)
$$

which clearly implies that $i_X(b) \prec [\alpha]_\approx$ henceforth contradicting the fact that

$$b > a = \bigvee \{z : z \prec [\alpha]_\approx\} \qquad (13.32)$$

So $\mathcal{F}$ satisfies (*) and by Proposition 13.30, (h3)(i) it extends to a filter $\mathcal{G}$ maximal with respect to (*). By same proposition, (h3)(iii), $\mathcal{G}$ is prime.

We know from our discussion of filters that $\mathcal{G}$ does induce a homomorphism

$$q_G : L(0,1)_X \to L(0,1)_X / \mathcal{G} \qquad (13.33)$$

which by the property (*) of $\mathcal{G}$ makes the composition

$$q_G \circ i_X : L(0,1) \to L(0,1)_X / \mathcal{G} \qquad (13.34)$$

injective.

It is our task now to prove more viz. that q_G is a surjection hence $q_G \circ i_X$ is an isomorphism.

To this end we need to show that given $\alpha \in F$, we can find $e \in [0,1]$ with the property

$$(s) \; i_X(e) = [\alpha]_\approx.$$

The argument for this is curious enough as it takes us to topology of the interval, viz., we will exploit the fact that $[0,1]$ is connected i.e. it cannot be represented as a disjoint union of two non–empty open sets (or, for that matter, of two non–empty closed sets). The argument for connectedness is simple, so we may insert it here for completeness: given two disjoint non–empty open sets M, N with $M \cup N = [0,1]$ assume $0 \in M$ and consider $q = supM$; if $q \in M$ then there is an interval $M \supseteq (u,v) \ni q$ so $q < supM$. Thus $q \in N$ and an analogous argument will show that then $q > supM$. So it must be $M \cap N \neq \emptyset$.

We define two subsets of $[0,1]$: $M_\alpha = \{z : i_X(z) \to [\alpha]_\approx \in \mathcal{G}\}$ and $N_\alpha = \{z : [\alpha]_\approx \to i_X(z) \in \mathcal{G}\}$. Then $0 \in M_\alpha, 1 \in N_\alpha$ and by properties (a), (b) above $M_\alpha \cup N_\alpha = [0,1]$. It remains to check that these two sets are closed. As $\to$ is continuous hence it does preserve suprema and infima it is sufficient to observe that M_α is downward closed i.e. $z \in M_\alpha$ and $z_1 < z$ imply $z_1 \in M_\alpha$ hence as M_α is closed on suprema it is closed; similarly, N_α is upward closed and closed on infima hence closed. It follows that there must be some $e \in M_\alpha \cap N_\alpha$ and clearly e satisfies (s).

It follows that $i_X \circ q_G$ is an isomorphism. We invoke the homomorphism $q_\approx : \mathrm{L}(0,1) \to \mathrm{L}(0,1)_X$ and we consider the homomorphism

$$T = (q_G \circ i_X)^{-1} \circ q_G \circ q_\approx : F \to \mathrm{L}(0,1) \qquad (13.35)$$

It is the idea of the proof now to verify that T is an element of semantics $\mathcal{E}$ and that $T(\alpha) \leq b$ witnessing the completeness condition with α, a.

As composition of homomorphisms, T is a homomorphism itself so it preserves all functors. We need to check that $T(\overline{x}) = x$ for each $x \in [0,1]$. But as $x = ([\overline{x}]_\approx)_G$ which identification is made here in virtue of the isomorphism $q_G \circ i_X$, we have $\chi_T(\overline{x}) = (q_G \circ i_X)^{-1} \circ (q_G \circ i_X)(x) = x$.

For each $\alpha \in F$, it follows by definition of semantics that $X \vdash_{\chi_X(\alpha)} \alpha$ hence $i_X(\chi_X(\alpha)) \prec [\alpha]_\approx$ implying that

$$
\begin{aligned}
T(\alpha) = (q_G \circ i_X)^{-1} \circ q_G \circ q_\approx(\alpha) = \\
(q_G \circ i_X)^{-1} \circ q_G)([\alpha]_\approx) \\
\geq_{\mathrm{L}(0,1)_X} \\
(q_G \circ i_X)^{-1} \circ q_G \circ i_X(\chi_X(\alpha)) = \\
\chi_X(\alpha)
\end{aligned}
\qquad (13.36)
$$

i.e. $T \geq X$. Thus, $T \in \mathcal{E}$.

By definition of the filter $\mathcal{F}$, we have $[\alpha]_\approx \to i_X(b) \in \mathcal{F}$ (with $n = 1$) and thus $[\alpha]_\approx \to i_X(b) \in \mathcal{G}$ hence by Proposition 13.21

$$\{[\alpha]_\approx\}_G \leq_{\mathrm{L}_{(0,1)_X/G}} \{[i_X(b)]_\approx\}_G.$$

This implies that

$$T(\alpha) =$$
$$(q_G \circ i_X)^{-1} \circ q_G \circ q_{\approx}(\alpha) =$$
$$(q_G \circ i_X)^{-1} \circ q_G)([\alpha]_{\approx}) \leq \qquad (13.37)$$
$$((q_G \circ i_X)^{-1} \circ q_G \circ i_X)(b) = b$$

We have shown that T fulfills its duty in showing that the semantic degree of truth of α is less or equal to b. As $b > a$ was chosen arbitrarily it follows that the semantic degree of truth of α is a i.e. it is equal to the syntactic degree of consequence. □

The fuzzy sentential calculus due to Pavelka over the Lukasiewicz residuated lattice L$(0, 1)$ is complete.

13.4.7 Discrete Lukasiewicz residuated lattices

The Lukasiewicz lattice L$(0, 1)$ has discrete counterparts. For a natural number m, we consider a chain $L(m) = \{0 = a_0 < a_1 < .... < a_m = 1\}$ of $m + 1$ elements. We will consider $L(m)$ with lattice operations $\vee(x, y) = max\{x, y\}, \wedge(x, y) = min\{x, y\}$. The adjoint pair $(\otimes, \rightarrow)$ will be defined as the discrete counterpart of continuous adjoints i.e. we will adopt $a_x \otimes a_y = a_{max\{0, x+y-m\}}$ and $a_x \rightarrow a_y = a_{min\{m, m-x+y\}}$.

We may verify in the same way as with the continuous case that requirements for a pair of adjoints are satisfied. We denote the resulting residuated lattice with the symbol L(m).

L(m) is nilpotent in the sense that $x^m = 0$ for every $x < 1$ which may be verified in a similar way as with the continuous case.

The analogy with the continuous case may be carried further viz. inspecting the proof of completeness given above in the continuous case we may verify that it may be carried out in the discrete case with syntax, semantics, and inference rules defined as in the continuous case.

There are some modifications however, on which we comment briefly. We may observe that in the discrete case the following formula is a theorem

(Th25) $(\alpha \Rightarrow \overline{a_k}) \vee (\overline{a_{k+1}} \Rightarrow \alpha)$

Indeed, given $x \in L(m)$ there is k with either $x \leq a_k$ hence $x \rightarrow a_k = 1$ or $a_{k+1} \leq x$ so that $a_{k+1} \rightarrow x = 1$ and finally $x \rightarrow a_k \vee a_{k+1} \rightarrow x = 1$.

We make use of (Th25) in the final part of the completeness proof, viz., given sets M_α, N_α defined as in the continuous case and the claim that $M_\alpha \cap N_\alpha \neq \emptyset$ we prove this claim as follows.

Assuming that $M_\alpha \cap N_\alpha = \emptyset$, we argue as in the continuous case to prove that M_α is a lower segment of the form $[a_0, a_k]$ and N_α is the upper segment of the form $[a_{k+1}, a_m]$ for some k. Thus, a_k is the greatest element in M_α and a_{k+1} is the least element in N_α. By (Th25), we have

$$([\alpha]_\approx \to [\overline{a_k}]_\approx) \vee ([\overline{a_{k+1}}]_\approx \to [\alpha]_\approx) = 1 \qquad (13.38)$$

so either

$$[\alpha]_\approx \to [\overline{a_k}]_\approx = 1 \qquad (13.39)$$

in which case $a_k \in N_\alpha$ or

$$[\overline{a_{k+1}}]_\approx \to [\alpha]_\approx = 1 \qquad (13.40)$$

in which case $a_{k+1} \in M_\alpha$. In either case, we have a contradiction which proves that $M_\alpha \cap N_\alpha \neq \emptyset$ and each $e \in M_\alpha \cap N_\alpha$ witnesses that $T(\alpha) = b$ in the notation in the continuous case.

In this way one obtains complete fuzzy sentential calculi over discrete Łukasiewicz residuated lattices of the form Ł(m).

In practical applications, given a family C of subsets of a universe U, we may take as variables in V statements of the form $\lceil x \in A \rceil$ for $A \in C$ and $x \in U$ and develop a complete fuzzy sentential calculus to be used e.g. in a fuzzy controller. However, in practical applications, this calculus is often insufficient and better results are obtained with functors defined in a different way cf. [Driankov93].

13.4.8   3 – Łukasiewicz algebras vs. Lindenbaum–Tarski algebras of the fuzzy sentential calculus

We now return to Wajsberg algebras (cf. Chapter 4, Chapter 12) and Łukasiewicz algebras (loc.cit., loc.cit.). We recall the Wajsberg axioms for the 3–valued Łukasiewicz sentential calculus (cf. Chapter 4):

(W1) $q \Rightarrow (p \Rightarrow q)$

(W2) $(p \Rightarrow q) \Rightarrow ((q \Rightarrow r) \Rightarrow (p \Rightarrow r))$

(W3) $((p \Rightarrow Np) \Rightarrow p) \Rightarrow p$

(W4) $(Nq \Rightarrow Np) \Rightarrow (p \Rightarrow q)$

We define negation here via $N\alpha = \alpha \Rightarrow \overline{0}$ (cf. Exercise 15), and we depart from our convention of denoting implication with the symbol C in Chapter 4 in order to reach agreement with the notation of this Chapter.

We verify that all formulae of form (W1)–(W4) are theorems of the fuzzy sentential calculus based on Ł$(2) = \{0 < \frac{1}{2} < 1\}$ (we refrain here from the general case).

Proposition 13.32. *Every formula of type (W1)–(W4) is a theorem in the fuzzy sentential calculus over L(2).*

Proof. Let us choose an arbitrary homomorphism $T \in \mathcal{E}$ and we calculate the semantic degree of truth of formulae in question with respect to T.

Concerning (W1), we need to verify, by Example 1, that $T(q) \leq T(p \Rightarrow q) = T(p) \to T(q)$ which follows easily as $T(q) \otimes T(p) \leq T(q)$.

In case of (W2), it is of the form of (Th6) so we know it is a theorem (cf. Proposition 13.25).

Now, for (W3); using our definition of negation N, we may write down a formula of type of (W3) as $((p \Rightarrow (p \Rightarrow \bar{0})) \Rightarrow p) \Rightarrow p$ and as with (W1) we need to verify that $((T(p) \to (T(p) \to 0)) \to T(p) \leq T(p)$; it may be confirmed by a direct calculation that actually we have the equality of both sides.

Finally, we address (W4); we write down this formula as $((q \Rightarrow \bar{0}) \Rightarrow (p \Rightarrow \bar{0})) \Rightarrow (p \Rightarrow q)$ and again we verify that $(T(q) \to 0) \to (T(p) \to 0) \leq T(p) \to T(q)$.

It follows that all formulae in question here are theorems. $\square$

By completeness of the fuzzy calculus already proved, we know that all those formulae are theses of this calculus as well.

We may observe that the fuzzy detachment rule FD applied to (W1)–(W4) produces pairs of the form $(\alpha, 1)$ where α is a thesis of the 3 – Lukasiewicz logic. Also, if $\alpha \Rightarrow \beta$ is a thesis in the 3 Lukasiewicz logic, then also $X \vdash_1 \alpha \Rightarrow \beta$. Denoting by $\approx_W$ the congruence in the 3– Lukasiewicz logic, we infer that $\alpha \approx_W \beta$ implies $\alpha \approx \beta$ hence $[\alpha]_{\approx_W} \subseteq [\alpha]_{\approx}$ and in consequence the 3– Lukasiewicz algebra embeds into the Lindenbaum – Tarski algebra $L(2)_X$.

Historic remarks

The semantics for many valued implication and negation was proposed in [Lukasiewicz opera cit.]. The completeness proof of the [0,1]–valued sentential calculus, based on axioms proposed by Jan Lukasiewicz and mentioned in [Lukasiewicz30[b]], which is presented here, comes from [Rose–Rosser58]. The completeness proof for this calculus was announced in [Wajsberg35] nevertheless it has not been ever published in the literature. An algebraic proof of completeness theorem for this calculus was presented in [Chang58[b], 59] with the help of MV–algebras. Other proofs were proposed by [Cignoli93] and [Panti95]; in [Cignoli–Mundici97] the Chang proof has been simplified. It was shown independently by [Meredith58] and [Chang58[a]] that the axiom (A5) is dependent on axioms (A1)–(A4) (cf. Exercise 1).

The idea of a fuzzy set was introduced in [Zadeh65]. A scheme for developing fuzzy sentential logic was proposed in [Goguen67, 69]; our presentation and the completeness proof is based on [Pavelka 79[a,b,c]]. Following the calculus due to Pavelka, a complete predicate fuzzy calculus was constructed in [Hajek97] (cf. also [Novák 87, 90]). The approach to infinite valued Łukasiewicz logics via Wajsberg algebras was proposed in [Font84].

Exercises

The first part of exercises (Exercises 1–6) concerns syntactic properties of L_∞ in particular the dependence of the axiom $(L_\infty 5)$ on remaining ones.

1. [Meredith58] Prove that the axiom $(L_\infty 5)$ is a consequence to $(L_\infty 1)$ – $(L_\infty 4)$. [Hint: in $(L_\infty 2)$, substitute $p/Cpq, q/Cqp, r/p$ to get
   (i) $\vdash CCCpqCqpCAqpCCpqp$ i.e.

$$(ii) \ \vdash CCCpqCqpCApqCCpqp$$

by (T4). Substitute
   $p/CCpqCqp, q/CApqCCpqp, r/CCqCpqCqp$ in $(L_\infty 2)$ so

$$(iii) \ \vdash CCCApqCCpqpCCqCpqCqpCCCpqCqpCCqCpqCqp$$

follows. By commutativity of A, (iv)$\vdash CCpqCrq \Leftrightarrow CCqpCrp$ follows and substitution in (iv) of $p/Np, q/Nq, r/Nr$ along with (T20) yields

$$(v) \ \vdash CCqpCqr \Leftrightarrow CCpqCpr.$$

Substitute $p/q, q/Cpq, r/p$ in (v) to get (vi)

$$CCCpqCqpCCqCpqCqp$$

and (T8) along with $(L_\infty 1)$ gives $\vdash ACpqCqp$ which is $(L_\infty 5)$]

2. Prove the formulae for semantic counterparts $a(x,y), b(x,y), k(x,y), l(x,y),$ $e(x,y)$ to respectively A, B, K, L, E.

3. Prove the thesis (T15) $Cpr, Cqr \vdash CApqr$. [Hint: First, observe that $(L_\infty 2)$ implies

$$(i) \vdash CCCqrCprCAprAqr$$

and $(L_\infty 2)$ applied to (i) yields

$$(ii) \vdash CCpqCAprAqr$$

from which it follows by $(L_\infty 3)$ that

$$(iii) \ \vdash CCpqCArpArq$$

which implies (iv) $CCprCApqArq$ and (v) $CCqsCArqArs$ by the commutativity of A so (iv) and (v) imply by Proposition 13.4 that

$$(vi) \vdash CCprCApqCCqsArs$$

so (T8) implies

$$(vii) \vdash CCprCCqsCApqArs$$

so (T14) gives (T15)]

4. Prove (T16) From $p_1, .., p_k, p \vdash r$ and $q_1, ..., q_n, q \vdash r$ it follows that $p_1, ..., p_k, q_1, ..., q_n, Apq \vdash r$. [Hint: Apply the Deduction Theorem (Prop.13.6) to infer from assumptions that (i) $p_1, .., p_k \vdash (Cp)^m r$ and (ii) $q_1, ..., q_n \vdash (Cq)^n r$. Now, for $0 < j \le m + n$, one proves by inducting on k that

$$(iii) p_1, ..., p_k, q_1, ..., q_n, Apq \vdash \Gamma_{i=1}^{m+n-j} t_i r$$

where each t_i is either p or q.

Consider two cases (1) $j = 1$ in which consider two subcases (1a) that among t_i there is less than m occurrences of p hence at least n occurrences of q and by Proposition13.2 it follows that $\vdash C(Cq)^n r \Gamma_{i=1}^{m+n-1} t_i r$ so the thesis follows by (ii), and (1b) that contrary to (1a) there are at least m occurrences of p clearly dual to (1a) and settled in the same way.

(2) is the inductive step from j to $j + 1$. Assuming the case for j true, we get from (iii) with j that

$$(iv) p_1, ..., p_k, q_1, ..., q_n, Apq \vdash Cp\Gamma_{i=1}^{m+n-j-1} t_i r$$

and

$$(v) p_1, ..., p_k, q_1, ..., q_n, Apq \vdash Cq\Gamma_{i=1}^{m+n-j-1} t_i r$$

and (T15) implies

$$(vi) \ p_1, ..., p_k, q_1, ..., q_n, Apq \vdash \Gamma_{i=1}^{m+n-j-1} t_i r$$

completing the inductive step. Letting $j = m + n$ gives (T16)]

5. Prove (T17) $CNNpp$. [Hint: first, (L$_\infty$1) implies

$$(i) \ CNNpCNNqNNp$$

via substitutions $p/NNp, q/NNq$. Next, (L$_\infty$4) with $p/Np, q/Nq$ applied to (i) yields (ii) $\vdash CNNpCNpNq$ and from (ii) by a similar application of (L$_\infty$4) (iii) $CNNpCqp$ follows. Then (T9) applied to (iii) gives (iv) $CqCNNpp$ and letting q to be Cpp ((T11)), we get $CNNpp$ by detachment]

6. Prove (T18) $\vdash CpNNp$ [Hint: from (T17), applying (T9), we obtain (i) $CCpNqCNNpNq$ and we apply to it (L$_\infty$4) to get (ii) $CCpNqCqNp$. Substitution $p/q, q/p$ in (T17) gives (iii) $CCqNpCqNp$ and (iv) $CpNq \Leftrightarrow CqNp$ by (T17). Substituting q/Np in (iv) yields by (T11) $\vdash CpNNp$]

In Exercises 7–17, we are concerned with the fuzzy sentential calculus and its underlying structures – residuated lattices.

7. Prove that operations $\vee$, $\wedge$ $\otimes$, and $\to$ are acceptable on $L(0,1)$.

8. [Pavelka79^b] Prove that the operation $\odot(x,y) = x \cdot y$ on $[0,1]^2$ is acceptable on $L(0,1)$. [Hint: check that $\odot$ is lipschitzian with constant 1 (to this end, observe:

$$|x_1 y_1 - x_2 y_2| = |x_1 y_1 - x_1 y_2 + x_1 y_2 - x_2 y_2| \leq y_2 \cdot |x_1 - x_2| + x_1 \cdot |y_1 - y_2| \leq$$

$$|x_1 - x_2| + |y_2 - y_2|)$$

]

9. Verify that the residuation $y \to_1 z$ (the *Goguen implication*) induced by $\odot$ takes on values
1 in case $y \leq z$
$\frac{z}{y}$ in case $y > z$.

10. [Pavelka79^b] Prove that any operation $(x_1, ..., x_n)$ acceptable by $L(0,1)$ is continuous. [Hint: assume H acceptable with parameters $k_1, .., k_n$ and let $k = max\{k_1, ..., k_n\}$. Then apply the acceptability condition and use $(x \leftrightarrow y)^m = 1 - m|x - y|$ to deduce therefrom that for a given $1 > \varepsilon > 0$ and $\delta = \frac{\varepsilon}{kn}$ we get in case $max|x_i - y_i| < \delta$ that

$$H(x_1, ..., x_n) \leftrightarrow H(y_1, ..., y_n) = 1 - |H(x_1, ..., x_n) - H(y_1, ..., y_n)| \geq$$

$$(x_1 \leftrightarrow y_1)^k \otimes ... \otimes (x_n \leftrightarrow y_n)^k =$$

$$\sum_{i=1}^{n}(1 - k|x_i - y_i|) - (n - 1) = 1 - k\sum_{i+1}^{n}|x_i - y_i| >$$

$$1 - kn\delta = 1 - \varepsilon$$

hence $|H(x_1, ..., x_n) - H(y_1, ..., y_n)| < \varepsilon$ proving continuity of H]

11. Prove that the residuation $\to_1$ is not acceptable by $L(0,1)$ [Hint: observe that $\to_1$ is not continuous in $(0,0)$ as the limit $lim_{y,z\to 0} \to_1 (y,z)$ does not exist]

12. Prove that residuation $\to$ and bi–residuation $\leftrightarrow$ on $L(0,1)$ as well as on $L(m)$ satisfy the following properties
(1) $x = y$ if and only if $x \leftrightarrow y = 1$
(2) $(x \leftrightarrow y) \otimes (y \leftrightarrow z) \leq x \leftrightarrow z$
(3) $(x \leftrightarrow y) \wedge (z \leftrightarrow w) \leq (x \wedge z) \leftrightarrow (y \wedge w)$
(4) $(x \leftrightarrow y) \wedge (z \leftrightarrow w) \leq (x \vee z) \leftrightarrow (y \vee w)$
(5) $(x \leftrightarrow y) \otimes (z \leftrightarrow w) \leq (x \otimes z) \leftrightarrow (y \otimes w)$

(6) $(x \leftrightarrow y) \otimes (z \leftrightarrow w) \leq (x \to z) \leftrightarrow (y \to w)$.

13. [Pavelka79^b] Prove that any operation of finite arity is acceptable on $L(m)$. [Hint: observe that $x^m = 0$ for every $x < 1$ in $L(m)$ and conclude that given $H(x_1, ..., x_n)$ on $L(m)$, either $(x_1, ..., x_n) = (y_1, ..., y_n)$ hence $H(x_1, ..., x_n) = H(y_1, ..., y_n)$ or $(x_1 \leftrightarrow y_1)^m \otimes ... \otimes (x_n \leftrightarrow y_n)^m = 0$ and in both cases the acceptability condition is satisfied]

14. Construct matrices of $\otimes$ and $\to$ on the Łukasiewicz lattice $L(2) = \{0, \frac{1}{2}, 1\}$. Verify that the residuation in $L(2)$ is identical to the implication in the Łukasiewicz 3–valued logic.

15. Verify that $x \to 0$ coincides with the negation operation in the Łukasiewicz many–valued logic viz. $x \to 0 = 1 - x$ where $\to$ is the residuation in either $L(0,1)$ or $L(m)$. Observe that on the basis of this fact one may introduce negation into the fuzzy sentential calculus as $N\alpha = \alpha \Rightarrow \overline{0}$.

16. Verify that $x \otimes y \leq_{L(0,1)} x \wedge y$ in $L(0,1)$ and in $L(m)$.

17. Prove that $\mathcal{F}_{\sim_{\mathcal{F}}} = \mathcal{F}$ for every filter $\mathcal{F}$ on residuated lattices $L(0,1)$, $L(m)$.

In the following Exercises 18 – 34, we return to the infinite valued logical calculus with the intention of giving the reader an opportunity to see the interplay of purely syntactic and topological arguments leading to the completeness proof of those calculi. We address here specifically, Proposition 13.11, (1)-(10) and we propose to prove these statements.

We list here the syntactic results which are applied in proofs of (1)–(10) of Proposition 13.11. Proofs of these results may be found in [Rose–Rosser58, Thms. 10.1 - 10.9].

(I) $\vdash ANpq$,

   $\vdash r \Leftrightarrow LBpzq$,

   $\vdash s \Leftrightarrow pzx$,

   $\vdash t \Leftrightarrow LBpyq$,

   $\vdash u \Leftrightarrow LBqyx$ imply

   $\vdash LBrys \Leftrightarrow LBtzu$

(II) $\vdash ANrs$

   $\vdash ANst$

   $\vdash p \Leftrightarrow LBrxs$

   $\vdash q \Leftrightarrow LBsxt$ imply

   $\vdash ANpq$

(III) $\vdash ANsm$,

$\vdash ANuv,$
$\vdash ANvw$
$\vdash ANyz$
$\vdash q \Leftrightarrow LBuxv$
$\vdash t \Leftrightarrow LBvxw$
$\vdash r \Leftrightarrow LByxz$
$s \vdash CvCBpuy$
$s \vdash CwCBpvz$
$m \vdash CvCBsuz$ imply
$s \vdash CtCBpqr$

(IV) $\vdash ANuv$
$\vdash ANyz$
$\vdash Apw$
$\vdash q \Leftrightarrow LBuxv$
$\vdash r \Leftrightarrow LByxz$
$\vdash CyBpu$
$\vdash CzBpv$
$w \vdash CzCCpyv$ imply
$\vdash CrBpq$

(V) $ANpq, ANqr \vdash LBLBpxqNxLBqxr \Leftrightarrow q$

We now propose to verify statements (1)–(10) of Proposition 13.11 assuming (I)–(V). Throughout all exercises, $f = a + \sum_{j=1}^{n} b_j x_j$ and $g = c + \sum_{j=1}^{n} d_j x_j$. All statements come from [Rose–Rosser58].

18.   Prove Proposition 13.11 (1) i.e. that $p \in PF(f), q \in PF(f)$ imply $\vdash p \Leftrightarrow q$ along with (2) $p \in PF(f), q \in PF(f+1)$ imply $\vdash ANpq$. [Hint [Rose–Rosser]: prove (1) , (2) in parallel by inducting on $|f|$. Begin with (1) and in case $|f| = 0$, use (T11), (T33) when $f \geq 1$ and (T11), (T19), (T34) when $f \leq 0$.

In case (2) with $|f| = 0$, use (T11), (T6) when $f \geq 1$ and (T11), (T19), (T7) when $f \leq 0$.

Assume now that (1), (2) are proved in case $|f| < m$ and consider (1) in the case $|f| = m$. There are two subcases according to definition of p.

In subcase 1, there is $b_j > 0$; choose $r, t \in PF(f - x_j)$ and $s, u \in PF(f + 1 - x_j)$. Then $p = LBrx_j s$ and $q = LBtx_j u$. By inductive assumption $\vdash r \Leftrightarrow t, \vdash s \Leftrightarrow u$ and thus $\vdash p \Leftrightarrow q$. In subcase 2, there is $b_j < 0$, proceed as in subcase 1.

For (2) in case $|f| = m$, the subcase 1 is $b_j > 0$ for some j. Take

$$r \in PF(f - x_j), s \in PF(f + 1 - x_j), t \in PF(f + 2 - x_j)$$

and observe that $\vdash ANrs, \vdash ANst$ by the inductive assumption and $LBrx_j s \in PF(f)$ so $\vdash p \Leftrightarrow LBrx_j s$, similarly, $\vdash q \Leftrightarrow LBx_j st$. Use (II) to get $\vdash ANpq$.

In subcase 2, $b_j < 0$ for some j, proceed analogously]

19. Prove (3) in Proposition 13.11 viz. $p \in PF(f), q \in PF(1-f)$ imply $\vdash p \Leftrightarrow Nq$. [Hint [Rose–Rosser58]: induct on $|f|$. In case $|f| = 0$, assume $f \geq 1$ and observe that $\vdash p$ by (T11) and $\vdash Nq$ by (T11), (T19) so $\vdash p \Leftrightarrow Nq$ by (T33).

Assume $f \leq 0$ and observe that $\vdash Np$ by (T11), $\vdash NNq$ by (T11) and (T19) so $\vdash p \Leftrightarrow Nq$ by (T34).

Assume (3) true in case $|f| < m$ and let $|f| = m$. In the subcase 1, $b_j > 0$ for some j, choose $r \in PF(f - x_j), s \in PF(f + 1 - x_j), t \in PF(x_j - f + 1)$. Then (i) $\vdash r \Leftrightarrow Nu$ (ii) $\vdash s \Leftrightarrow Nt$ by the inductive assumption and (iii) $\vdash p \Leftrightarrow LBrx_j s$ (iv) $\vdash q \Leftrightarrow LBtNx_j u$ by (1); also (v) $\vdash ANrs$ by (2) and (vi) $s \vdash Brx_j \Leftrightarrow BrLsx_j$ by (T29). Infer (vii) $s \vdash p \Leftrightarrow Brx_j$ by (T29), (iii) and commutativity of L and therefrom (viii) $s \vdash p \Leftrightarrow BrLsx_j$.

By (T13) (ix)$Np \vdash q \Leftrightarrow Bpq$ follows; infer (x) $Nr \vdash Lsx_j \Leftrightarrow BrLsx_j$ from (ix) and (xi) $Nr \vdash p \Leftrightarrow Lx_j s$ from (ix) and (iii) and apply commutativity of L to get (xii) $Nr \vdash p \Leftrightarrow BrLsx_j$. Use (T16), (v), (vii), (viii) to get (xiii) $\vdash p \Leftrightarrow BrLsx_j$. Use commutativity of B to get (xiv) $\vdash p \Leftrightarrow BLsx_j r$. Use (T19), (T20), to get (xv) $\vdash Bpq \Leftrightarrow NLNpNq$ and hence (xvi) $\vdash p \Leftrightarrow NLNLsx_j Nr$. From (T19), (T20) get (xvii) $\vdash p \Leftrightarrow NLNLsx_j Nr$, (xix) $\vdash p \Leftrightarrow NLNNBNsNx_j Nr$. Use (i), (ii), (T19) to get (xx) $\vdash p \Leftrightarrow NLBtNx_j u$ and from (iii) get $\vdash p \Leftrightarrow Nq$.

In the subcase 2, $b_j < 0$ for some j, reduce it to the subcase 1, by substituting $p/q, q/p$ and replacing f with $1 - f$. Conclude by the subcase 1 that $\vdash q \Leftrightarrow Np$ and use (T19), (T21) to get $\vdash p \Leftrightarrow Nq$]

20. Prove (4) in Proposition 13.11 i.e. $p \in PF(f), q \in PF(2-f)$ imply $\vdash Apq$. [Hint [Rose–Rosser58]: choose $r \in PF(1-f)$ and observe that $\vdash p \Leftrightarrow Nr$ by (3) and $\vdash ANrq$ by (2) hence $\vdash Apq$ follows]

21. Prove the following auxiliary statement (A) : $m \geq 0, p \in PF(f), q \in PF(m+f)$ imply $\vdash Cpq$. [Hint [Rose–Rosser58]: induct on m. In case $m = 0$, apply (1). In the inductive step with $m + 1$, take $q \in PF(f+m+1)$ and $r \in PF(m+f)$. Observe that $\vdash Cpr$ by the inductive assumption and $\vdash ANrq$ by (2), and infer $\vdash Crq$ from (T35) so $\vdash Crq$ follows]

22. Prove the following auxiliary statement (B): $m \geq 0, p \in PF(f), q \in PF(1 - m - f)$ imply $\vdash CpCqr$ for every formula r. [Hint [Rose–Rosser58]: for $s \in PF(1 - f)$, observe that $\vdash Cqs$ by (A) and $\vdash p \Leftrightarrow Ns$ by (3). Infer $\vdash CCsrCqr$ from $(L_\infty)2$ and $\vdash CpCsr$ by (T36), and deduce $\vdash CpCqr$]

23. Prove the following auxiliary statement (C): assuming that $b_j = 0$, and $\alpha(x_j) \in PF(f)$, we have $\vdash \alpha(x_j) \Leftrightarrow \alpha(q)$ for every formula q where substitution x_j/q has been made. [Hint [Rose–Rosser58] : induct on $|f|$.

In case $|f| = 0$, as $\alpha(x_j)$ is either Cx_jx_j or NCx_jx_j so apply either (T33) or (T34). In the inductive step, in case e.g. $b_u > 0$ for some u, choose $\beta(x_j) \in PF(f - x_u), \gamma(x_j) \in PF(f + 1 - x_u)$. Observe that $\vdash \alpha(x_j) \Leftrightarrow LB\beta(x_j)x_u\gamma(x_j)$ by (1) so by substitution x_j/r (i) $\vdash \alpha(r) \Leftrightarrow LB\beta(r)x_u\gamma(r)$, and that (ii) $\vdash \beta(x_j) \Leftrightarrow \beta(r)$ and (iii) $\vdash \gamma(x_j) \Leftrightarrow \gamma(r)$ by the inductive assumption. Deduce from (i), (ii), (iii) that $\vdash \alpha(x_j) \Leftrightarrow \alpha(r)$. In case $b_j < 0$ proceed similarly]

24. Prove the following auxiliary statement (D): assuming $b_j = 0 = b_k$, and $m \geq 0$, if $\alpha(x_k) \in PF(f + mx_k), q \in PF(f + m - mx_j)$ then $\vdash \alpha(Nx_j) \Leftrightarrow q$. [Hint [Rose–Rosser58]: induct on m. In case $m = 0$, $\vdash \alpha(x_k) \Leftrightarrow \alpha(Nx_j)$ by (B), and $\vdash \alpha(x_k) \Leftrightarrow q$ by (1) so $\vdash \alpha(Nx_j) \Leftrightarrow q$ follows. In the inductive step with $m + 1$, we have $\alpha(x_k) \in PF(f + mx_k + x_k), q \in PF(f + m + 1 - (m + 1)x_j)$. Pick $\beta(x_k) \in PF(f + mx_k), \gamma(x_k) \in PF(f + 1 + mx_k)$. Observe that $\vdash \alpha(x_k) \Leftrightarrow LB\beta(x_k)x_k\gamma(x_k)$ by (1) and deduce (i) $\vdash \alpha(Nx_j) \Leftrightarrow LB\beta(Nx_j)Nx_j\gamma(Nx_j)$. Pick $r \in PF(f+m-mx_j), s \in PF(f+1+m-mx_j)$ and deduce that (ii) $\vdash q \Leftrightarrow LBrNx_js$. Observe that (iii) $\vdash \beta(Nx_j) \Leftrightarrow r$ (iv) $\gamma(Nx_j) \Leftrightarrow s$ by the inductive assumption and deduce from (i)–(iv) the thesis of (C)]

25. Prove the following auxiliary statement (E): assuming $b_j = 0 = b_k$, and $m, n \geq 0$, if $\alpha(x_j, x_k) \in PF(f + mx_j + (m+n)x_k), q \in PF(f + m + n - mx_j)$, then $\alpha(x_j, Nx_j) \Leftrightarrow q$. [Hint [Rose–Rosser58]: induct on n. In case $n = 0$, (D) is (C). In the inductive step with $n+1$, select $\beta(x_j, x_k) \in PF(f + nx_j + (m+n)x_k), \gamma(x_j, x_k) \in PF(f + 1 + nx_j + (m+n)x_k), \delta(x_j, x_k) \in PF(f + 2 + nx_j + (m+n)x_k), p \in PF(f+m+n-mx_j), r \in PF(f+m+n+2-mx_j)$. Observe that by induction hypothesis (i) $\vdash \beta(x_j, Nx_j) \Leftrightarrow p$, (ii) $\vdash \gamma(x_j, Nx_j) \Leftrightarrow q$ (iii) $\delta(x_j, Nx_j) \Leftrightarrow r$.

Observe that

$$LB\beta x_j\gamma \in PF(f + (n + 1)x_j + (m + n)x_k),$$

$$LB\gamma x_j\delta \in PF(f + 1 + (n + 1)x_j + (m + n)x_k0$$

and deduce that

$$LBLB\beta x_j\gamma x_kLB\gamma x_j\delta \in PF(f + (n + 1)x_j + (m + n + 1)x_k)$$

from which infer by (1) that

$$(iv) \quad \vdash \alpha(x_j, x_k) \Leftrightarrow$$

$$LBLB\beta(x_j, x_k)x_j\gamma(x_j, x_k)x_kLB\gamma(x_j, x_k)x_j\delta(x_j, x_k).$$

Deduce from (iv) via (i), (ii), (iii) that

$$(v) \quad \alpha(x_j, Nx_j) \Leftrightarrow LBLBpx_jqNx_jLBqx_jr.$$

Observe that (vi) $\vdash ANpq$ (vii) $\vdash ANqr$ by (2) and deduce from (v), (vi), (vii) and (V) that $\vdash \alpha(x_j, Nx_j) \Leftrightarrow q$.

26. Prove the following auxiliary statement (F): $p \in PF(f), q \in PF(g), r \in PF(f+g), s \in PF(f+1), t \in PF(g+1)$ imply $s \vdash CtCBpqr$. [Hint [Rose–Rosser58]: induce on $|g|$. In case $|g| = 0$, either $g \geq 1$ in which case $s \vdash r$ by (A) so $s \vdash CtCBpgr$ via $(L_\infty)1$ applied twice, or $g = 0$ in which case $\vdash Nq$ by (T11), (T19) and $\vdash Cpr$ by (1)so infer by (T13) (ix)$Np \vdash q \Leftrightarrow Bpq$ and deduce that $s \vdash CBpqr$ and then infer $s \vdash CtCBpqr$ by $(L_\infty)1$, or $g \leq -1$ in which case $\vdash Nt$ so deduce $s \vdash CtCBpqr$ by (T36).

In the inductive step, consider subcases 1° $d_j > 0, d_j + b_j > 0$ for some j 2° $d_j > 0$ for some j but for no j is $b_j + d_j > 0$ 3° $d_j \leq 0$ for every j.

In 1°, pick $u \in PF(g - x_j), v \in PF(g + 1 - x_j), w \in PF(g + 2 - x_j), y \in PF(f + g + 1 - x_j), z \in PF(f + 2)$. Observe that (i) $\vdash q \Leftrightarrow LBux_jv$ (ii) $\vdash t \Leftrightarrow LBvx_jw$ (iii) $\vdash r \Leftrightarrow LByx_jz$ by (1) and (iv) $\vdash ANsm$ (v) $\vdash ANuv$ (vi) $ANvw$ (vii) $ANyz$ follow by (2) , and (viii) $s \vdash CvCBpuy$ (ix) $s \vdash CwCBpvz$ (x) $m \vdash CvCBsuz$ are true by the inductive assumption. To conclude, invoke (III)

In case 2°, let $d_j > 0$, so $-b_j \geq d_j$ and with k such that $b_k = d_k = 0$ (i.e. k is sufficiently large, possibly greater than degree of f and degree of g) pick $\alpha(x_j, x_k) \in PF(f + b_j - b_jx_j - b_jx_k), \beta(x_j, x_k) \in PF(f + g + b_j - b_jx_j - b_jx_k), \gamma(x_j, x_k) \in PF(f + 1 + b_j - b_jx_j - b_jx_k)$. Observe that by (E), (i) $\vdash \alpha(x_j, Nx_j) \Leftrightarrow p$ (ii) $\vdash \beta(x_j, Nx_j) \Leftrightarrow r$ (iii) $\vdash \gamma(x_j, Nx_j) \Leftrightarrow s$ and that Case 1 implies (iv) $\gamma(x_j, x_k) \vdash CtCB\alpha(x_j, x_k)q\beta(x_j, x_k)$ which by substitution yields

$$(v) \; \gamma(x_j, Nx_j) \vdash CtCB\alpha(x_j, Nx_j)q\beta(x_j, Nx_j).$$

Deduce from (i)–(iv) the thesis.

In case 3°, replace x_j by $1 - x_k$ and use (D) instead of (E) when concluding.]

27. Prove (5), Proposition 13.11: $p \in PF(f), q \in PF(g), r \in PF(1 - f + g), s \in PF(2 - f), t \in PF(g+1)$ imply $s \vdash CtCCpqr$. [Hint [Rose–Rosser58]: for $y \in PF(1 - f)$, observe that $p \Leftrightarrow Ny$ by (3), Proposition 13.11 (Exercise 19, above), and $s \vdash CtCBuqr$ by (E) and deduce $s \vdash CtCCpqr$]

28. Prove the statement (G): $p \in PF(f), q \in PF(g), r \in PF(f + g)$ imply $\vdash CrBpq$. [Hint [Rose–Rosser]: induct on $|f + g|$; in case $|f + g| = 0$, consider the subcase $f + g = c \geq 1$. Pick $s \in PF(1 - f)$ and observe that (i) $\vdash p \Leftrightarrow Ns$ by (3), Proposition 13.11 (Exercise 19, above) and (ii) $\vdash BNss$ by (T17), and (iii) $\vdash Csq$ by (A). Deduce (iv) $\vdash Bps$ from (i), (ii) and (v) $\vdash Bpq$ from (iii), (iv) via $\vdash CCpqCBrpBrq$ which follows from (T9). From (v), deduce $\vdash CrBpq$ by $(L_\infty)1$. In case $f + g < 0$, observe that $\vdash Nr$ by (T11), (T18) so (T36) implies the thesis.

In the inductive step, with $|f+g| = m$, consider two subcases $1°$ $b_j+d_j > 0$ for some j $2°$ $b_j + d_j < 0$. In case $1°$, assume e.g $d_j > 0$, otherwise substitute $p/q, q/p$. Select $u \in PF(g - x_j), v \in PF(g + 1 - x_j), w \in PF(2 - f), y \in PF(f+g-x_j), z \in PF(f+g+1-x_j)$. Observe that (vi) $\vdash q \Leftrightarrow LBux_jv$ (vii) $\vdash r \Leftrightarrow LByx_jz$ by (1), Proposition 13.11 (Exercise 18, above), (viii) $\vdash ANuv$ (ix) $\vdash ANyz$ by (2), Proposition 13.11 (Exercise 18, above),(x) $\vdash Apw$ by (4), Proposition 13.11 (Exercise 20, above), (xi) $w \vdash CzCCpyv$ by (F) (Exercise 27, above), and (xii) $\vdash CyBpu$, (xiii) $\vdash CzBpv$ by the inductive assumption. Apply (IV) to (vi)–(xiii) to conclude.

In Case $2°$ proceed analogously]

29. Prove (6), Proposition 13.11: $p \in PF(f), q \in PF(g), r \in PF(1 - f + g)$ imply $\vdash CrCpq$. [Hint [Rose–Rosser58]: with $s \in PF(1 - f)$, observe that $\vdash p \Leftrightarrow Ns$ by (3), Proposition 13.11, and $\vdash CrBsq$ by (6), Proposition 13.11]

30. Prove the auxiliary statement (H): $m \geq 1$ and $p \in PF(1 + f), q \in PF(1 + mf)$ imply $\vdash Cqp$. [Hint [Rose–Rosser58]: induct on m. In case $m = 1$, apply (1), Proposition 13.11. In the inductive step, with $m + 1$ and $q \in PF(1+(m+1)f)$ pick $r \in PF(1+mf)$ and $s \in PF(1-mf)$, and observe that (i) $\vdash Ars$ by (4), Proposition 13.11, (ii) $\vdash CsCqp$ by (6), Proposition 13.11, (iii) $\vdash Crp$ by the inductive assumption, (iv) $\vdash CpCqp$ by $(L_\infty)1$ so (v) $\vdash CrCqp$ by (iii), (iv). From (i), (ii), (v) deduce the thesis by (T15).]

31. Prove (7), Proposition 13.11: $m \geq 1$ and $p_i \in PF(1 + f_i), q \in PF(1 + \sum_{i=1}^m f_i)$ imply $p_1, ..., p_m \vdash q$. [Hint [Rose–Rosser58]: induct on m. In case $m = 1$, apply (1), Proposition 13.11. In the inductive step, with $m + 1$, pick $r \in PF(1 + \sum_{i=1}^m f_i)$. Observe that $\vdash CrCp_{m+1}q$ by (6), Proposition 13.11, and $p_1, ..., p_m \vdash r$ by the inductive assumption. Deduce therefrom the thesis]

32. Prove (8), Proposition 13.11: $\vdash p$ whenever $p \in PF(1 + x_j)$ and observe that an analogous argument proves (9), Proposition 13.11. [Hint [Rose–Rosser58]: pick $q \in PF(1), r \in PF(2)$ with $\vdash p \Leftrightarrow LBqx_jr$ (observe that q, r exist by definition of p in case some $b_j > 0$) and observe that $\vdash q, \vdash r$ by (T11). Deduce $\vdash p$ by (T24), (T28)]

33. Prove that if polynomials $f_1, ..., f_m, g$ have integral coefficients and for every vector $x \in [0,1]^n$ with rational coefficients from $f_1(x) \geq 0, ..., f_m(x) \geq 0$ it follows that $g(x) \geq 0$ then there exists non-negative rational numbers $\mu, \lambda_1, ..., \lambda_m$ with the property that $g = \mu + \sum_{i=1}^m \lambda_i f_i$. [Hint: consider the set $M = \{x \in [0,1]^n : f_i(x) \geq 0 \ (i = 1, 2, ..., n)\}$. By continuity of g and density of the set of vectors with rational coefficients in $[0,1]^n$ deduce from the assumptions that $g(x) \geq 0$ for every $x \in M$. Apply Proposition 13.9 and observe that by assumption that f_i, g have integral coefficients, $\mu, \lambda_1, ..., \lambda_m$ are rational]

34. Prove (10), Proposition 13.11: $p_1, ..., p_m \vdash q$ whenever $p_i \in PF(1 + f_i), q \in PF(1 + g)$, $f_1, ..., f_m, g$ have integral coefficients, and $f_1(x) \geq 0, ...,$ $f_m(x) \geq 0$ imply $g(x) \geq 0$ for every $x \in [0,1]^n$ whose each coordinate is rational. [Hint [Rose–Rosser58]: consider two cases $1°$ there is a rational vector $x \in [0,1]^n$ with $f_1(x) \geq 0, ..., f_m(x) \geq 0$ $2°$, otherwise.

In case $1°$, apply Exercise 33 to conditions $f_1(x) \geq 0, ..., f_m(x) \geq 0, x_1 \geq 0, ..., x_n \geq 0, 1 - x_1 \geq 0, ..., 1 - x_n \geq 0$ which render a part of assumptions and conclude that

$$(i) \quad g = \mu + \sum_{i=1}^{m} \lambda_i f_i + \sum_{j=1}^{n} \kappa_j x_j + \sum_{j=1}^{n} \delta_j (1 - x_j).$$

Convert (i) to

$$(ii) \quad Kg = M + \sum_{i=1}^{m} L_i f_i + \sum_{j=1}^{n} K_j x_j + \sum_{j=1}^{n} D_j (1 - x_j)$$

by appropriately multiplying (i) so all K, M, L_i, K_j, D_j are non–negative integers.

Pick $r_j \in PF(1 + x_j), s_j \in PF(2 - x_j)$ and $t \in PF(1 + Kg - M), u \in PF(1 + Kg)$. Observe that (iii) $\vdash r_j$, (iv) $\vdash s_j$ by (8), (9), Proposition 13.11. Apply (7), Proposition 13.11 (Exercise 31, above) with each p_i taken L_i times, each r_j taken K_j times and each s_j taken D_j times to t to deduce that

$$(v) \quad p_1, .., p_m, r_1, ..., r_n, s_1, ..., s_n \vdash t.$$

Deduce from (v), (iii), (iv) that

$$(vi) \quad p_1, ..., p_m \vdash t$$

and observe (vii) $\vdash Ctu$ by (A) (Exercise 21), (viii) $\vdash Cuq$ by (H) (Exercise 30). Deduce $p_1, ..., p_m \vdash q$ from (vi), (vii), (viii).

In case $2°$, assume that k is the last integer with rational vectors $x \in [0,1]^n$ such that (a) $f_1(x) \geq 0, ..., f_k(x) \geq 0$; thus $f_{k+1}(x) < 0$ whenever (a) holds hence (a) implies $-f_{k+1} > 0$. By Exercise 33, $f_{k+1} = -\mu + \sum_{i=1}^{n}(-\lambda_i)f_i$ with $\mu, \lambda_i > 0$ hence $-f_{k+1}(x) \geq \mu$ for every $x \in [0,1]^n$.

Let $\mu = \frac{N}{K}$ with N, K integers > 0. Observe that $-N - Kf_{k+1} \geq 0$ whenever (a) holds so apply Case $1°$ with $g = -N - Kf_{k+1}$ to $r \in PF(1 - N - Kf_{k+1})$ and $s \in PF(1 + Kf_{k+1})$ to get

$$(ix) \quad p_1, ..., p_k \vdash r.$$

Apply (7), Proposition 13.11 (Exercise 31, above) with p_{k+1} used K times to get

$$(x) \quad p_{k+1} \vdash s$$

and observe that (xi) ⊢ $CrCsq$ by (B) (Exercise 22, above). Deduce that

$$p_1, p_2, ..., p_m ⊢ q$$

from (ix), (x), (xi)]

Works quoted

[Borkowski70] *Jan Łukasiewicz. Selected Works*, L. Borkowski (ed.), North Holland–Polish Scientific Publishers, Amsterdam–Warsaw, 1970.

[Chang59] C. C. Chang, *A new proof of the completeness of the Łukasiewicz axioms*, Trans. Amer. Math. Soc., 93 (1959), pp. 74–80.

[Chang58^a] C. C. Chang, *Proof of an axiom of Łukasiewicz*, Trans. Amer. Math. Soc., 87 (1958), pp. 55–56.

[Chang58^b] C. C. Chang, *Algebraic analysis of many–valued logics*, Trans. Amer. Math. Soc., 88 (1958), pp. 467–490.

[Cignoli93] R. Cignoli, *Free lattice–ordered abelian groups and varieties of MV – algebras*, Notás de Lógica Matemática, Univ. Nac. del Sur, 38 (1993), pp. 113–118.

[Cignoli–Mundici97] R. Cignoli and D. Mundici, *An elementary proof of Chang's completeness theorem for the infinite–value calculus of Łukasiewicz*, Studia Logica, 58 (1997), pp. 79–97.

[Driankov93] D. Driankov, H. Hellendoorn, and M. Reinfrank, *An Introduction to Fuzzy Control*, Springer Verlag, Berlin, 1993.

[Font84] J. M. Font, A. J. Rodriguez, and A. Torrens, *Wajsberg algebras*, Stochastica, 8 (1984), pp. 5–31.

[Goguen69] J. A. Goguen, *The logic of inexact concepts*, Synthese, 18/19 (1968/ 1969), pp. 325–373.

[Goguen67] J. A. Goguen, *L–fuzzy sets*, J. Math. Anal. Appl., 18 (1967), pp. 145–174.

[Hajek97] P. Hajek, *Fuzzy logic and arithmetical hierarchy II*, Studia Logica, 58 (1997), pp. 129–141.

[Łukasiewicz30^a] J. Łukasiewicz, *Philosophische Bemerkungen zu mehrwertige Systemen des Aussagenkalkuls*, C. R. Soc. Sci. Lettr. Varsovie, 23(1930), 51–77 [English translation in [Borkowski], pp. 153–178]

[Łukasiewicz30^b] J. Łukasiewicz and A. Tarski, *Untersuchungen ueber den Aussagenkalkul*, C. R. Soc. Sci. Lettr. Varsovie, 23(1930), 39–50 [English translation in [Borkowski], pp. 130–152].

[Łukasiewicz20] J. Łukasiewicz, *On three–valued logic* (in Polish), Ruch Filozoficzny 5(1920), 170–171 [English translation in [Borkowski], pp. 87–88].

[Łukasiewicz18] J. Łukasiewicz, *Farewell Lecture by Professor Jan Łukasiewicz* (delivered in the Warsaw University Lecture Hall on March 7, 1918) [English translation in [Borkowski], pp. 84–86].

[McNaughton51] R. McNaughton, *A theorem about infinite–valued sentential logic*, J. Symbolic Logic, 16 (1951), pp. 1–13.

[Meredith58] C. A. Meredith, *The dependence of an axiom of Lukasiewicz*, Trans. Amer. Math. Soc., 87 (1958), p. 54.

[Novák90] V. Novák, *On the syntactico–semantical completeness of first–order fuzzy logic*, Kybernetika, 2 (1990), Part I pp. 47–62, Part II pp. 134–152.

[Novák87] V. Novák, *First–order fuzzy logic*, Studia Logica, 46 (1987), pp. 87–109.

[Panti95] G. Panti, *A geometric proof of the completeness of the calculus of Łukasiewicz*, J. Symbolic Logic, 60 (1995), pp. 563–578.

[Pavelka79^a] J. Pavelka, *On fuzzy logic I*, Zeit. Math. Logik Grund. Math., 25 (1979), pp. 45–52.

[Pavelka79^b] J. Pavelka, *On fuzzy logic II*, Zeit. Math. Logik Grund. Math., 25 (1979), pp. 119–134.

[Pavelka79^c] J. Pavelka, *On fuzzy logic III*, Zeit. Math. Logik Grund. Math., 25 (1979), pp. 447–464.

[Rasiowa–Sikorski63] H. Rasiowa and R. Sikorski, *The Mathematics of Metamathematics*, PWN–Polish Sci. Publishers, Warsaw, 1963.

[Rose–Rosser58] A. Rose and J. B. Rosser, *Fragments of many–valued statement calculi*, Trans. Amer. Math. Soc., 87 (1958), pp. 1–53.

[Rosser–Turquette52] J. B. Rosser and A. R. Turquette, *Many–valued Logics*,

North–Holland Publ. Co., Amsterdam, 1952.

[Wajsberg35] M. Wajsberg, *Beiträge zum Metaaussagenkalkül I*, Monat. Math. Phys., 42 (1935), pp. 221–242.

[Zadeh65] L. A. Zadeh, *Fuzzy sets*, Information and Control, 8 (1965), pp. 338–353.

Chapter 14

From Rough to Fuzzy

Nature can only raid Reason to kill; but Reason can invade Nature to take prisoners and even to colonize

C. S. Lewis, *Miracles*, 4

14.1 Introduction

We have witnessed the development of logical calculi with truth values ranging continuously from 0 to 1 and in the development of the fuzzy sentential logic we have treated the set of logical axioms as a fuzzy set. The fuzzy sentential calculus in Chapter 13 has been developed in the framework of residuated lattices and essential usage has been made of the adjoint pair $(\otimes, \rightarrow)$. In the wider perspective of fuzzy calculi on sets it has turned useful to extend the notion of an adjoint pair to the notion of a pair $(T, \rightarrow_T)$ where T is a *triangular norm* (or, t – *norm*) and $\rightarrow_T$ is the induced *residuated implication*. t – norms and duals of them, t – *conorms*, may be applied in the development of algebra of fuzzy sets viz. t – norms determine intersections of fuzzy sets according to the formula $\chi_{A \cap B}(x) = T(\chi_A(x), \chi_B(x))$ where T is a t – norm and χ_A is the fuzzy characteristic (membership) function of the fuzzy set A while t – conorms may be used in determining unions of fuzzy sets via $\chi_{A \cup B}(x) = S(\chi_A(x), \chi_B(x))$ where S is a t – conorm.

In the following section, we introduce t – norms and we study their properties; in particular, we provide characterization (representation) theorems for t – norms along with dual results for t – conorms, stemming from [Ling65] et al. We also study properties of residuated implications and we characterize t – norms (respectively t – conorms) whose residuated implications are continuous. The importance of those results follows from Chapter 13: as we recall, only such implications are acceptable. As a result, we obtain a remarkable result [Pavelka79[a,b,c]] that the Lukasiewicz adjoint pair $(\otimes, \rightarrow)$ is

up to equivalence the only adjoint pair making the fuzzy sentential calculus
in the Pavelka sense complete.

Although rough sets and fuzzy sets address distinct aspects of reasoning
under uncertainty, yet both these ideas can be combined into a hybrid approach. Actually they may be traced to a common origin which is a partition
in the rough case and a special partition viz. $\{A, X \setminus A\}$ (represented by the
characteristic function of the set A) in the fuzzy case which in the latter case
is given a graded substitute in the form of a fuzzy characteristic (membership) function. Thus we may say that both paradigms address – in distinct
languages – the problem of boundary of a non – crisp notion.

Following the scheme laid out in [Dubois – Prade92] we distinguish between *rough fuzzy sets* which are fuzzy sets filtered through a classical equivalence relations to fuzzy sets in resulting quotient spaces and *fuzzy rough
sets* which emerge in the purely fuzzy set – theoretic context by imitating
rough set approximations with fuzzy similarity relations in place of classical equivalence relations. Both these notions are examined in the following
sections.

In this study, we work with the *Zadeh lattice $Fuz(X, [0, 1])$* of fuzzy sets on
a universe X and valued in the interval $[0, 1]$ endowed with lattice operations
min, max. As pointed to, among others, in [Catteneo97], the Zadeh lattice
may be endowed with natural complementation operations viz. the *Kleene,
Brouwer* – complementations, making it into a *Brouwer – Zadeh lattices*.
By means of these complementations, similarly as in Chapter 12, rough set
approximations may be introduced to fuzzy sets. This approach to rough
fuzzy sets is examined in the last section of this Chapter.

14.2   Triangular norms

In Chapter 13, we witnessed an important role played in the develop-
ment of the fuzzy sentential logic by the adjoint pair $(\otimes, \rightarrow)$ in the lattice
$([0, 1], min, max)$. We exploited in that development properties of $\otimes$ that
it made $[0, 1]$ into a commutative monoid with the unit 1 as well as it was
lipschitzian.

Let us observe for the record that

$$T(x, y) = x \otimes y = max\{0, x + y - 1\}$$

does satisfy the following as a function from $[0, 1]^2$

(TN1) T is *associative*: $T(T(x, y), z) = T(x, T(y, z))$
(TN2) T is *commutative*: $T(x, y) = T(y, x)$
(TN3) T is *non–decreasing*: $T(z, y) \geq T(x, y)$ whenever $z \geq x$ and $T(x, z) \geq$
$T(x, y)$ whenever $z \geq y$
(TN4) $T(1, x) = x = T(x, 1)$

(TN5) $T(0,0) = 0$

Departing now from $\otimes$ and considering a general function T on the square $[0,1]^2$, into $[0,1]$, we say that T is a *triangular norm* (or, a t – *norm* for short) in the case when it does satisfy (TN1) – (TN5).

We have therefore an example of a t – norm viz. $\otimes$. As we know, $\otimes$ does satisfy one more requirement on T

(TN6) T is continuous

but it is easy to verify that $\otimes$ does not satisfy

(TN7) T is *increasing*: $T(z,y) > T(x,y)$ whenever $z > x > 0$ and $T(z,y) > T(x,y)$ whenever $z > y > 0$

A t – norm T which additionally satisfies (TN6), (TN7) is said to be *strict*.

It follows that $\otimes$ is not any strict t – norm. In order to produce an example of a strict t– norm, we need only to go to the arithmetical multiplication: $T(x,y) = Prod(x,y) = x \cdot y$ is clearly a strict t – norm.

Let us observe that $\otimes$ does satisfy the following

(TN8) $T(x,x) < x$ for each $x \in (0,1)$

Indeed, if $x \otimes x = 0$ it is obvious and if $x \otimes x = 2x - 1$ then $2x - 1 > 0$ but $2x - 1 < x$ as $x < 1$.

A t – norm $T(x,y)$ which satisfies (TN6), (TN8) is said to be *archimedean*. Let us observe that any strict t – norm satisfies (TN8) by (TN7), (TN6) and (TN4) thus any strict t – norm is archimedean.

Much effort was put into the task of recognizing the structure of t – norms. To this end, a notion of the *pseudo–inverse* to a given function was introduced with which we begin this topic.

Given a function $f : [a,b] \to [0,\infty]$ of which we assume it is continuous and increasing, we define its *pseudo – inverse* $g : [0,\infty] \to [a,b]$ by letting

$$g(x) = \begin{array}{l} a \ in \ case \ x \in [0, f(a)] \\ f^{-1}(x) \ in \ case \ x \in [f(a), f(b)] \\ b \ in \ case \ x \in [f(b), \infty] \end{array} \tag{14.1}$$

In the dual case when f is decreasing, the definition is similar, with the obvious changes viz. b takes the place of a and vice versa.

The correctness of this definition follows from the fact that f is injective on $[a,b]$ and that by continuity of f, the image $f([a,b])$ is the interval $[f(a), f(b)]$ (or, $[f(b), f(a)]$).

We assume now that a function $T : [0,1]^2 \to [0,1]$, not necessarily a t – norm, satisfies (TN1), (TN3), (TN4) in the weakened form $T(1,x) = x$, (TN6), (TN8). We will say in this case that T is *archimedean*. We prove a structure theorem about T. To this end, we establish first some properties of T. We introduce the symbol $T^n(x)$ defined inductively as follows

$$
\begin{aligned}
&(i)\ T^0(x) = 1 \\
&(ii)\ T^{n+1}(x) = T(T^n(x), x)
\end{aligned}
\tag{14.2}
$$

Proposition 14.1. *[Ling65] (the auxiliary structure theorem) Every archimedean function T has the following properties*

(A1) the sequence $(T^n(x))_n$ is non–increasing for every $x \in [0,1]$

Indeed, given $x \in [0,1]$, $T^{n+1}(x) = T(T^n(x), x) \leq T(T^n(x), 1) = T^n(x)$.

(A2) given $0 < y < x < 1$, we have $T^n(x) < y$ for some n; a fortiori, $lim_n T^n(x) = 0$

Assume to the contrary that there are x, y with $y < x, T^n(x) \geq y$ for each n; consider $z = inf T^n(x) \geq y$. By (TN6), (TN1) $T(z,z) = lim_n T(T^n(x), T^n(x))$ $= lim_n T^{2n}(x) = z$ contradicting (TN8).

(A3) 0 is the two – sided null element i.e. $T(0,x) = 0 = T(x,0)$ for every $x \in [0,1]$
    Indeed,

$$
T(0,x) = T(lim_n T^n(x), x) = lim_n T(T^n(x), x) = lim_n T^{n+1}(x) = 0
$$

by (A2); in the case of $T(x,0)$ the argument is analogous.

(A4) $T(x,1) = x$ for every $x \in [0,1]$
    Indeed, consider the function $x \to T(x,1)$; by continuity, it takes all values from $0 = T(0,1)$ to $1 = T(1,1)$ so for a given x there is y with $x = T(y,1)$ and $T(x,1) = T(T(y,1),1) = T(y,T(1,1)) = T(y,1) = x$.

(A5) $y < x,\ 0 < T^n(x)$ imply $T^n(y) < T^n(x)$
    First, observe $T(x,y) \leq T(1,y) = y$; if it was $T(x,y) = y$, we would have by induction that $T(T^n(x), y) = y$. But for some n_0 we have $T^{n_0}(x) < y$ by (A2) and thus $y = T(T^{n_0}(x), y) \leq T(y,y) < y$, a contradiction. We have proved

(i) $T(x,y) < y, T(x,y) < x$ for $0 < x, y < 1$

We consider now the function $z \to T(x,z)$ for a given x and let $y < x$; as this function is continuous and maps $[0,1]$ onto the interval $[0,x]$ so for some z_1 we have $T(x, z_1) = y$. Similarly, we prove that $T(z_2, x) = y$ for some z_2. We have verified

(ii) $x < y$ implies the existence of z_1, z_2 such that $T(x, z_1) = y$, $T(z_2, x) = y$

We return to the proof of (A5); assume to the contrary that for some x, y, n we have $y < x, 0 < T^n(x), T^n(x) = T^n(y)$. Then by (i), (ii), (TN1) we have $0 < T^n(x) = T^n(y) = T(T^{n-1}(y), y) = T(T^{n-1}(y), T(x, z)) = T(T(T^{n-1}(y), x), z) \leq T(T(T^{n-1}(x), x), z) = T(T^n(x), y) < T^n(x)$ by (i), a contradiction.

(A6) $0 < x < 1$, $0 < T^n(x)$ imply $T^{n+1}(x) < T^n(x)$
  This follows immediately from (i).

(A7) the set $T_n(x) = \{y \in [0,1] : T^n(y) = x\}$ is non – empty
  Indeed, as T^n is continuous and $T^n(0) = 0, T^n(1) = 1$, we have $T^n(y) = x$ with some y.
  We let $r_n(x) = \inf T_n(x)$. Then

(A8) $r_n(x) \in T^n(x)$ i.e. $T^n(r_n(x)) = x$
  Indeed, as $T_n(x)$ is a closed set it contains its extrema.

(A9) $r_n(x)$ is a continuous, decreasing function of x
  As by (A8) r_n is the right inverse to T^n which is continuous and decreasing by (A5), so r_n is also continuous and decreasing.

(A10) $r_n(x) < r_{n+1}(x)$ for every $0 < x < 1$
  Assume to the contrary that $0 < r_{n+1}(x) \leq r_n(x) < 1$ for some x, n so

$$0 < T^n(r_{n+1}(x)) \leq T^n(r_n(x)) = x < 1 \tag{14.3}$$

so

$$0 < x = T^{n+1}(r_{n+1}(x)) = T(T^n(r_{n+1}(x)), r_{n+1}(x)) \leq \\ T(x, r_{n+1}(x)) < x \tag{14.4}$$

by (A5), a contradiction.

(A11) $\lim_n r_n(x) = 1$ for every $0 < x < 1$
  By (A10), $(r_n(x))_n$ is increasing hence it has a limit $s = \lim_n r_n(x)$; thus, $s > r_n(x)$ for each n. On the other hand, $x = T^n(r_n(x)) < T^n(s)$ for each n. Was $s < 1$, we would have $T^n(s) < 1$ for each n and n, we would find by (A2) some m with $T^{mn}(s) = T^m(T^n(s)) < x$, a contradiction.

(A12) the superposition $T^m \circ r_n$ depends on $\frac{m}{n}$ only i.e. $T^m \circ r_n = T^{km} \circ r_{kn}$ for every $k = 1, 2, ...,$

Assume to the contrary that $T^m \circ r_n(x) \neq T^{km} \circ r_{kn}(x)$ for some x, m, n, k; then for $y = r_n(x)$ and $z = r_{kn}(x)$ we have $T^k(z) \geq y$ by definition of $r_n(x)$ and finally $T^k(z) > y$ by our assumption. By continuity of T^k we have $y = T^k(w)$ with some $w < z$ so $T^{kn}(w) = T^n(y) = x$ i.e. $w \in T_{kn}(x)$ contrary to $w < z$ and $z = r_{kn}(x)$.

We now are able to state and verify the principal structure theorem for archimedean functions. Motivations for this theorem go back to the 13th Hilbert problem. As noticed in [Sierpiński34], if a function $\phi(x, y)$ has the property that $\phi(x_1, y_1) = \phi(x_2, y_2)$ implies $x_1 = x_2, y_1 = y_2$ then for every function $f(x, y)$ there exists a function $g(t)$ with the property that $f(x, y) = g(\phi(x, y))$; in [Lindenbaum33] it was pointed out that $\phi(x, y)$ may be of the form $\phi_1(x) + \phi_2(y)$ where

$$\phi_1(x) = \sum_{n=1}^{\infty} \frac{\lfloor 2^n x \rfloor - 2 \lfloor 2^{n-1} x \rfloor}{2^{2^n}}, \phi_2(y) = \frac{1}{2}\phi_1(y) \qquad (14.5)$$

This very general result may be essentially improved in case $f(x, y)$ has sufficiently good properties e.g. it is a t – norm.

Proposition 14.2. *[Ling65] (the principal structure theorem) If a function* $T : [0, 1]^2 \to [0, 1]$ *is archimedean i.e. it does satisfy (TN1), (TN3), (TN4), (TN6), (TN8) then there exists a continuous decreasing function f on $[0, 1]$ with the property that*

$$(*) \quad T(x, y) = g(f(x) + f(y))$$

for each pair x, y where g is the pseudo – inverse to f. Moreover, every function $T(x, y)$ for which a representation of the form (*) exists is archimedean.

Proof. Selecting and fixing $a \in (0, 1)$, we define by (A12) a function $h(\frac{m}{n}) = T^m(r_n(a))$ on the set Q_0 of rational positive numbers. As both T^m, r_n are continuous decreasing, h is continuous and decreasing as well. Moreover

h does satisfy the functional equation

$$(**) \quad T(h(p), h(q)) = h(p + q)$$

on the set Q_0.

To verify (**), we consider $p = \frac{m}{n}, q = \frac{k}{l}$; finding a common denominator d we write down $x = \frac{c}{d}, y = \frac{f}{d}$ and calculate $T(h(p), h(q)) = T(T^c(r_d(a)), T^f(r_d(a))) = T^{c+f}(r_d(a)) = h(\frac{c+f}{d}) = h(p + q)$.

Having (**) verified, we now extend h in a unique way by continuity and density of the set Q_0 in the interval $[0, 1]$ to the continuous and decreasing function g. Clearly

$$(* * *) \quad T(g(x), g(y)) = g(x + y)$$

is satisfied on $[0,1]$.

Letting A to be the first argument with $g(A) = 0$, we have that g is a continuous decreasing function from $[0, A]$ onto $[0, 1]$ so it has the inverse f. Returning to the identity (***) and letting $g(x) = u, g(y) = v$, we have (***) in the form of

$$T(u, v) = g(f(u) + f(v))$$

for every pair $u, v \in [0, 1]$.

The second part viz. that (*) implies T archimedean may be verified by a straightforward verification that (TN1), (TN3), (TN4), (TN6), (TN8) hold (cf. [Schweizer – Sklar83]). □

As a corollary we may note as a direct consequence of the last theorem that as shown by Proposition 14.2

Proposition 14.3. *Each archimedean function T is commutative.*

It follows that any continuous t – norm admits a representation of the form (*). The function f occurring in the representation (*) is called a *generator* of T. Let us return to the well – known to us t – norm $\otimes$ and find its generator.

Example 14.1. *We are at f with the property that $x \otimes y = g(f(x) + f(y))$. Assuming that $f(1) = a$, we have $g(x) = 1$ for $x \in [0, a]$ and thus $1 = 1 \otimes 1 = g(f(1) + f(1)) = g(2a)$ hence $2a \in [0, a]$ so $a = 0$. As $f(1) = 0$, the candidate is $f(x) = 1 - x$; in this case the pesudo–inverse is $g(y) = 1 - y$ in case $y \leq 1$ and $g(y) = 0$ in case $y > 1$. Combining f, g into (*), we have*

$$g(f(x) + f(y)) = \begin{array}{l} 1 - (1 - x) - (1 - y) = x + y - 1 \text{ in case } x + y \geq 1 \\ 0 \text{ in case } x + y < 1 \end{array}$$

(14.6)

i.e. $x \otimes y = g(f(x) + f(y))$ holds with the indicated f, g. Thus the generator for $\otimes$ is $f(x) = 1 - x$ i.e. the negation in the sense of Łukasiewicz.

It turns out that assumptions (TN1), (TN3), (TN4), (TN6), (TN8) are essential; the example for that is provided by the t – norm *min* which does not satisfy (TN8). Actually

Proposition 14.4. *[Ling] In any representation*

$$(\ddagger) \ min(x, y) = g(f(x) + f(y))$$

the function f cannot be continuous or decreasing.

Proof. Assuming ($\ddagger$), as $g \circ f$ is an identity, f is injective. For $f(0) = a, f(1) = b$, we may e.g. assume $a > b$. As $0 = min(0, x) = g(f(0) + f(x))$ for every x, in case f is continuous, the value of g on $[a + b, 2a]$ is constantly 0. For x close enough to 0 we have $a + b < f(x) + f(x) \leq 2a$ so $x = min(x, x) = g(f(x) + f(x)) = 0$, a contradiction when $x \neq 0$. So f may not be continuous.

For the second part, assuming f decreasing, we may apply a well-known theorem of Riemann stating that any monotone real function on an interval has at most countably many points of discontinuity (these points may happen only where f experiences jumps and as each jump determines an open interval, the family of jumps may be at most countable). Thus there is a continuity point $a \in (0, 1)$ for f. Let $(a_n)_n$ be an increasing sequence with the limit a. It follows that $(f(a_n)_n$ is a decreasing sequence with the limit $f(a)$. As $f(a) > 0$, we have a_n with $f(a) < f(a_n) < 2 \cdot f(a)$ and applying g which is decreasing as the pseudo $-$ inverse to f, we get $a > a_n > g(f(a) + f(a)) = min(a, a) = a$, a contradiction. So there is no decreasing generator for min.   □

It will be useful to mention here the dual notion of a $t - conorm$. Given a $t - norm$ T, the associated $t - conorm$ S_T is produced via

$$S_T(x, y) = 1 - T(1 - x, 1 - y) \tag{14.7}$$

Clearly, associativity, commutativity, and continuity are retained by this operation. A closer inspection tells us that null elements for T (i.e. elements a such that $T(a, x) = a$ for every x) are changed into null elements $1 - a$ for S_T and the unit elements b for T (i.e. elements such that $T(b, x) = x$ for every x) are turned into unit elements $1 - b$ for S_T. Also S_T is non-decreasing as T is, and the inequality $T(x, x) < x$ is changed into inequality $S_T(x, x) > x$. Modifying accordingly (TN1)–(TN8) we obtain conditions defining a $t - conorm$ in general.

As an example we may calculate the $t - conorm$ $S_\otimes$. We have $S_\otimes(x, y) = 1 - (1 - x) \otimes (1 - y) = 1 - max\{0, 1 - x - y\} = min(x + y, 1)$.

The corresponding duality takes place also for representations of $t - norms$ and $t - conorms$. Letting $T(x, y) = g_T(f_T(x) + f_T(y))$ as a representation for a $t - norm$ T, we may produce the respective representation for the $t - conorm$ S_T via

$$S_T(x, y) = 1 - T(1 - x, 1 - y) = 1 - g_T(f_T(1 - x) + f_T(1 - y))$$

and thus letting

$$f_S(x) = f_T(1 - x), g_S(x) = 1 - g_T(x)$$

we obtain the representation

$$S_T(x, y) = g_S(f_S(x) + f_S(y)).$$

For instance, in case of $S_\otimes$ we have $f_S(x) = 1 - (1 - x) = x$ and $g_S(x) = 1 - g_T(x) = min(x, 1)$. Clearly, in the general case, f_S is increasing, and g_S is the pseudo–inverse to f_S.

Given a t – norm T in place of the Łukasiewicz product $\otimes$, we may replace as well the residuation $\rightarrow$ with a more general construct $\rightarrow_T$ called also in this case the *residuated implication* and defined via the condition

$$(RI)\ T(x,y) \leq z \Leftrightarrow x \leq y \rightarrow_T z.$$

Let us collect in the following proposition the elementary properties of the residuated implication following from its definition.

Proposition 14.5. *For every t – norm T, the residuated implication $\rightarrow_T$ satisfies the following*

(RI1) $T(x,y) \leq z \rightarrow_T u$ if and only if $x \leq T(y,z) \rightarrow_T u$
(RI2) $y \rightarrow_T (z \rightarrow_T u) = T(y,z) \rightarrow_T u$
(RI3) $y \rightarrow_T z = 1$ if and only if $y \leq z$
(RI4) $x \leq y \rightarrow_T u$ if and only if $y \leq x \rightarrow_T u$

Indeed, by associativity of T i.e. $T(T(x,y),z) = T(x,T(y,z))$ we have

$$x \leq y \rightarrow_T (z \rightarrow_T u) \Leftrightarrow T(x,y) \leq z \rightarrow_T u \Leftrightarrow T(T(x,y),z) \leq u \Leftrightarrow \\ T(x,T(y,z)) \leq u \Leftrightarrow x \leq T(y,z) \rightarrow_T u \tag{14.8}$$

whence (RI1,2) follow. (RI3) follows from definition of $\rightarrow_T$ and the fact that $T(x,y) \leq min(x,y)$ following from $T(x,1) = x$ and commutativity and monotonicity of T. Finally, commutativity of T i.e. $T(x,y) = T(y,x)$ implies

$$x \leq y \rightarrow_T u \Leftrightarrow T(x,y) \leq u \Leftrightarrow T(y,x) \leq u \Leftrightarrow y \leq x \rightarrow_T u \tag{14.9}$$

Passing to deeper properties of T, $\rightarrow_T$, let us observe that

(RI5) *for a fixed a, the function $T(x,a)$ does preserve suprema*

To verify (RI5), consider a set C with $c_0 = supC$; then

$$T(c_0,a) = T(supC,a) \geq T(c,a) \tag{14.10}$$

for every $c \in C$ hence $T(supC,a) \geq sup_C T(c,a)$. On the other hand

$$T(supC,a) \leq u \Leftrightarrow supC \leq a \rightarrow_T u \Leftrightarrow \forall c \in C.c \leq a \rightarrow_T u \Leftrightarrow \\ \forall c \in C.T(c,a) \leq u \tag{14.11}$$

thus with $u = supT(c,a)$ we get $T(supC,a) \leq sup_C T(c,a)$ and finally the equality $T(supC,a) = sup_C T(c,a)$ follows.

Clearly, the function $T(a,x)$ preserves suprema by commutativity of T.
   Returning to $\rightarrow_T$, let us observe that

Proposition 14.6. *The residuated implication* $\to_T$ *satisfies*

(RI6) $\to_T$ *is non–increasing in the first coordinate and non–decreasing in the second coordinate*

(RI7) $a \to_T x$ *for a fixed a preserves infima; consequently, $x \to_T a$ changes suprema into infima.*

Proof. (RI6) follows by monotonicity of T and definition of $\to_T$ e.g. for $u \leq v$ from $x \leq y \to_T u$ i.e. $T(x,y) \leq u$ it follows that $T(x,y) \leq v$ i.e. $x \leq y \to_T v$ for every x hence $y \to_T u \leq y \to_T v$. The proof for the first coordinate follows on the same lines.

The argument for (RI7) parallels that for (RI5); given a set C we have for every x that

$$x \leq inf_C(a \to_T c) \Leftrightarrow \forall c \in C.x \leq a \to_T c \Leftrightarrow \forall c \in C.T(x,a) \leq c \Leftrightarrow$$
$$T(x,a) \leq infC \Leftrightarrow x \leq a \to_T infC$$

$$(14.12)$$

whence $a \to_T infC = inf_C(a \to_T c)$ follows. $\qquad\square$

It turns out that these properties may be equivalently stated in topological terms. We recall that a function f between metric spaces is continuous (cf. Chapter 7) if it satisfies the condition $f(lim_n x_n) = lim_n f(x_n)$ for every sequence $(x_n)_n$. This condition may be too strong for some nevertheless important functions and therefore it has been ramified into two *semi – continuity* conditions. We introduce here two operations on sequences viz. given a sequence $(x_n)_n$, we define its

lower limit as $liminf x_n = inf\{lim x_{n_k}\}$ where the infimum is taken over all convergent sub–sequences $(x_{n_k})_k$ of the sequence $(x_n)_n$

upper limit as $limsup x_n = sup\{lim x_{n_k}\}$ where the supremum is taken over all convergent sub–sequences $(x_{n_k})_k$ of the sequence $(x_n)_n$

With respect to these two notions, two kinds of semi–continuity emerge

the lower–semicontinuity viz. a function f is lower–semicontinuous at x_0 if and only if for every sequence $(x_n)_n$ with $x_0 = lim x_n$ the inequality $f(x_0) \leq liminf f(x_n)$ holds. The function f is lower–semicontinuous if and only if it is lower–semicontinuous at every point

the upper–semicontinuity viz. a function f is upper–semicontinuous at x_0 if and only if for every sequence $(x_n)_n$ with $x_0 = lim x_n$ the inequality $f(x_0) \geq limsup f(x_n)$ holds. The function f is upper–semicontinuous if and only if it is upper–semicontinuous at every point

An important and not difficult to get characterization is the following

Proposition 14.7. *A function f is upper–semicontinuous (respectively lower–semi – continuous) if and only if for every number r the set $\{x : f(x) \geq r\}$ is closed (respectively for every number r the set $\{x : f(x) \leq r\}$ is closed).*

We may now characterize in topological terms t – norms and residuated implications.

Proposition 14.8. *1. Any t – norm T is lower–semicontinuous; moreover, any associative and commutative function T with $T(0,x) = 0, T(1,x) = x$ which is lower–semicontinuos is a t – norm. In this case the residuated implication $\to_T$ is given by the condition $y \to_T z = max\{x : T(x,y) \leq z\}$.*

2. Any residuated implication $\to_T$ is upper–semicontinuos and any function non–increasing in the first coordinate, non–decreasing in the second coordinate which is upper–continuous is of the form $\to_T$ for some t – norm T.

Proof. We consider as a matter for an exemplary argument, a t – norm T about which we would like to prove upper–semicontinuity.

So let us fix a real number r and consider the set

$$M = \{(x,y) : T(x,y) \leq r\} \tag{14.13}$$

We want to show that M is closed. Assume then to the contrary that $(x_0, y_0) \in [0,1]^2$ and a sequence $(x_n)_n$ in M exist such that

$$(i) \ T(x_0, y_0) > r$$
$$(ii) \ T(x_n, y_n) \leq r \ for \ each \ n \tag{14.14}$$
$$(iii) \ (x_0, y_0) = lim_n (x_n, y_n)$$

Then by monotonicity of T we have for each n that either $x_n < x$ or $y_n < y$; we may assume that the set $\{n : x_n < x_0\}$ is infinite and in consequence we may assume that $x_n < x_0$ for each n. Letting $\bar{x}_n = max\{x_1, .., x_n\}$ and $\underline{y}_n = min\{y_1, ..., y_n\}$ we have $T(\bar{x}_n, \underline{y}_n) \leq r$ so we may assume that the sequence $(x_n)_n$ is increasing to x_0 and the sequence $(y_n)_n$ is decreasing to 0. As $sup_n T(x_n, y_n) = T(x_0, y_1) \leq r$ it follows by monotonicity of T that $T(x_0, y_0) \leq r$, a contradiction. Thus M is closed and a fortiori T is lower–semicontinuous. In all remaining cases we may argue similarly.

The converse may be proved along the same lines. As a corollary to the lower–semicontinuity of a t – norm T, we have that the set $M = \{(x,y) : T(x,y) \leq z\}$ is closed hence the set $\{x : x \leq y \to z\}$ is closed hence it contains its greatest element and thus $y \to_T z = max\{x : T(x,y) \leq z\}$.

The proof of the second part follows along similar lines and we leave it as an exercise. $\square$

The question what t – norms may have continuous residuations was settled in [Menu–Pavelka76] whose solution was based on results in [Mostert –

Shields57] and [Faucett55] viz. on the result that any continuous t − norm with $0, 1$ as the only solutions to the equation $T(x, x) = x$ and at least one nilpotent element x (i.e. $T^n(x) = 0, x \neq 0$) is equivalent to $\otimes$ [Mostert–Shields57] and on the related result in [Faucett55] that any continuous t − norm with $0, 1$ as the only solutions to the equation $T(x, x) = x$ and no nilpotent element is equivalent to *Prod*.

We outline a proof of these results as consequences of the principal structure theorem. Following [Ling65], we produce an argument for the Mostert–Shields theorem in its dual form for the respective t − conorm $S_\otimes(x, y) = min\{x + y, 1\}$. The proof for $\otimes$ may be directly recovered form the given one via dualization, or it may be carried directly on parallel lines.

Let us comment briefly on nilpotent elements; in Chapter 13 we witnessed an important role of nilpotency in developing technical aspects of completeness proof for the fuzzy sentential logic. We recall once more that an element $x \neq 0$ is nilpotent with respect to T if $T^n(x) = 0$ for some n (respectively $x \neq 1$ is nilpotent with respect to S_T in case $S_T^n(x) = 1$ for some n). For instance, with $S = S_\otimes$, and $f_S(x) = x, g_S(x) = min\{x, 1\}$, we have by an easy calculation involving the identity $f(g(y)) = y$ (as g is the pseudo − inverse to f) that $S^n(x) = min\{n(x), 1\}$ hence $S^n(x) = 1$ for every $x \geq \frac{1}{n}$.

The difference between the existence and the non − existence of nilpotent elements may be captured in terms of the generator f and its pseudo–inverse g viz. we may observe that regardless of the precise and specific formula for the t–conorm S (or t − norm T), we have the inductively proved identity

$$(N) \quad S^n(x, x) = g(f(2x) + (n - 2)f(x))$$

due to the identity $f(g(x)) = x$ which bounds the generator f and its pseudo − inverse g. Thus, two possibilities arise: (i) f is bounded, a fortiori $g(x)$ returns to 1 sufficiently large values of x i.e. from (N) it follows that S has nilpotent elements (ii) f is unbounded so g is the inverse to f and there are no nilpotents.

We return to the problem of classification of t − norms and t − conorms. We sketch its solution. First we need the notion of *equivalent t − norms*. It is manifest that given a t − norm T and an increasing function $\phi : [0, 1] \rightarrow [0, 1]$ the function

$$(E) \quad T'(x, y) = \phi^{-1}(T(\phi(x), \phi(y)))$$

is a t − norm. We will say that T, T' are *equivalent*. The relation (E) bears also on residuated implications viz.

$$(ER) \quad y \rightarrow_{T'} z = \phi^{-1}(\phi(y) \rightarrow_T \phi(z)).$$

Indeed,

$$x \leq y \rightarrow_{T'} z \Leftrightarrow T'(x,y) \leq z \Leftrightarrow$$
$$\phi^{-1}(T(\phi(x),\phi(y))) \leq z \Leftrightarrow T(\phi(x),\phi(y)) \leq \phi(z) \Leftrightarrow$$
$$\phi(x) \leq \phi(y)) \rightarrow_T \phi(z) \Leftrightarrow x \leq$$
$$\phi^{-1}(\phi(y)) \rightarrow_T \phi(z)) \tag{14.15}$$

Thus we state the result due to [Ling65].

Proposition 14.9. *[Ling65] The theorem of Mostert–Shields:*

if $S(x,y)$ is associative and continuous on $[a,b]^2$ with values in $[a,b]$, with $S(a,x) = x$, and $S(b,x) = b$ for every x, and with at least one nilpotent element in the open interval (a,b) then S is equivalent to the t – conorm $S_\otimes$

is a consequence of the principal structure theorem for t – conorms.
The dual result follows for t – norms.

Proof. We have to verify first that S satisfies the conditions of the dual principal structure theorem i.e. that

(i) S is non – decreasing
(ii) $S(x,x) > x$ for every $x \in (a,b)$

To this end, we may observe that if $S(x,z_1) = S(x,z_2)$ then by associativity we have $z_1 = S(a,z_1) = S(S(a,x),z_1) = S(a,S(x,z_1)) = S(a,S(x,z_2)) = S(S(a,x),z_2)S(a,z_2) = z_2$ thus $S(x,.)$ is an injective function for every x. As $S(x,.)$ is continuous with $S(x,a) = x, S(x,b) = b$ it is increasing. Similar proof shows that $S(.,x)$ is increasing also. Thus (i) holds.

By (i), we have $S(x,x) > S(a,x) = x$ proving (ii).

Thus S satisfies assumptions of the principal structure theorem and there exist functions f,g with $S(x,y) = g(f(x) + f(y))$. As S has nilpotent elements, f is bounded, say, $f : [a,b] \rightarrow [0,M]$; then the pseudo – inverse $g : [0,\infty] \rightarrow [a,b]$.

We define a new function $h : [a,b] \rightarrow [0,1]$ via $h(x) = \frac{f(x)}{M}$ and we observe that $x = h(f^{-1}(Mx))$.

We denote by the symbol $g_\otimes(x) = min\{x,1\}$ the pseudo – inverse to the generator of $S_\otimes(x,y) = min\{x+y,1\}$. Then we have to verify the

Claim $g_\otimes(x) = h(g(Mx))$ for every $x \in [0,\infty]$

But indeed

$$g_\otimes(x) = h(f^{-1}(Mg_\otimes(x))) = h(f^{-1}(Mmin(x,1))) \tag{14.16}$$

and thus

$$g_\otimes(x) = \begin{array}{l} h(f^{-1}(Mx) = h(g(Mx)) \text{ in case } x \le 1 \\ h(f^{-1}(M)) = h(b) = 1 \text{ in case } x \ge 1 \end{array} \qquad (14.17)$$

and thus $g_\otimes(x) = h(g(Mx))$ for every $x \in [0, \infty]$.
Claim is verified so now

$$h(S(x,y)) = h(g(f(x) + f(y)) = h(g(M \cdot (\tfrac{f(x)}{M} + \tfrac{f(y)}{M}))) =$$
$$g_\otimes(\tfrac{f(x)}{M} + \tfrac{f(y)}{M}) = g_\otimes(h(x) + h(y)) = \qquad (14.18)$$
$$min\{h(x) + h(y), 1\} = S_\otimes(h(x), h(y))$$

and this means that S is equivalent to $S_\otimes$.   □

Now we pass to the problem of characterizing t – norms which have continuous residuated implications. We will need some additional properties of residuation.

Proposition 14.10. *The following properties hold for every t – norm T*

(RI8) $y \le (y \to_T z) \to_T z$

(RI9) $T(x,y) \to_T 0 = x \to_T (y \to_T 0)$

Indeed, as $y \to_T z \le y \to_T z$ we have $T(y \to_T z, y) \le z$ hence by commutativity of T it follows that $T(y, y \to_T z) \le z$ which implies $y \le (y \to_T z) \to_T z$ i.e. (RI8).

Associativity of T implies that $z \le T(x,y) \to_T 0$ if and only if $T(z, T(x,y)) \le 0$ i.e. $T(T(z,x),y) \le 0$ so equivalently $T(z,x) \le y \to_T 0$ which is equivalent to $z \le x \to_T (y \to_T 0)$. From the equivalence of the first and the last statements (RI9) follows.

It turns out that continuity of $\to_T$ implies continuity of T. This fact has remarkable consequences: in the case when $\to_T$ is continuous, the t – norm T is equivalent to $\otimes$.

Proposition 14.11. *[Menu–Pavelka76] For every t–norm T, continuity of $\to_T$ implies continuity of T. In this case, T is equivalent to $\otimes$.*

Proof. We assume that $\to_T$ is continuous; we first verify the following

Claim Consider the function $x \to_T a$ with $0 \le a < 1$. Then $y = (y \to_T a) \to_T a$ for each $y \in [a, 1]$.

To verify *Claim*, we recall that (i) $y \le (y \to_T a) \to_T a$ by (RI8); it remains to show that in case $y \ge a$, we have $y \ge (y \to_T a) \to_T a$. As the function $q(x) = x \to_T a$ is continuous decreasing with $q(a) = 1, q(1) = a$ there is $z \ge a$ with $y = z \to_T a$. Then

$$y = z \to_T a \ge (((z \to_T a) \to_T a) \to_T a) = \qquad (14.19)$$
$$(y \to_T a) \to_T a$$

by the fact that the superposition $(x \to_T a) \to_T a$ is increasing so
(ii) $y \geq (y \to_T a) \to_T a$ and combining (i) and (ii) gives Claim.

From *Claim* it follows with $a = 0$ that

$$(iii) \quad T(x,y) = (T(x,y) \to_T 0) \to_T 0 = ((x \to_T (y \to_T 0) \to_T 0$$

where the last equality holds by (RI9) and (iii) shows that T is definable in
terms of superpositions of $\to_T$ and thus T is continuous.

We now solve the equation (iv) $T(x,x) = x$; assume that (iv) holds with
$x \neq 0,1$. Take $c \leq x \leq d$ so there is u with $c = T(x,u)$. Then $T(x,c) =$
$T(x,T(x,u)) = T(T(x,x),u) = T(x,u) = c$ and we have $c = T(x,c) \leq$
$T(d,c) \leq T(1,c) = c$ i.e. $T(d,c) = c$ hence $d \to_T c = c$ for each pair d,c with
$c \leq x \leq d$.

This means that for $c < x$, the function $\to_T c$ is not any injection contrary
to the result in Claim implying that it is injective. We infer that the equation
$T(x,x) = x$ has only $0,1$ as solutions.

To conclude the proof one has to refer to the mentioned above results
of [Mostert–Shields57] and [Faucett55]. By these results, T is equivalent either to $\otimes$ or to *Prod*; however, *Prod* has to be excluded as its residuated
implication (the Goguen implication cf. Chapter 13, Exercise 9) is discontinuous. $\qquad\square$

Going back to Chapter 13, and recalling Exercise 10, we may conclude
(cf. [Pavelka79^c]) that the adjoint pair $(\otimes, \to)$ due to Łukasiewicz is up to
equivalence the only adjoint pair which could make the fuzzy sentential logic
as proposed by Pavelka complete.

14.3 Rough fuzzy and fuzzy rough sets

We know already from our discussion of the fuzzy sentential logic in Chapter 13, that L–fuzzy sets on a given universe X and with a given lattice L,
make a partially ordered set viz. given fuzzy sets A, B on X, represented by
fuzzy membership functions $\chi_A : X \to L$, $\chi_B : X \to L$ we let (cf. [Zadeh65])

$$A \leq B \Leftrightarrow \forall x \in X.\chi_A(x) \leq_L \chi_B(x) \tag{14.20}$$

We denote this partially ordered set of L–fuzzy sets on X with the symbol
$Fuz(X,L)$. Clearly, in case L is a complete lattice, $Fuz(X,L)$ is completely
ordered viz. given a family $G \subseteq Fuz(X,L)$ we define its infimum $\bigwedge G$ via

$$\chi_{\wedge G}(x) = \bigwedge\{\chi_C(x) : C \in G\} \tag{14.21}$$

and similarly we define the supremum $\bigvee G$ via

$$\chi_{\vee G}(x) = \bigvee\{\chi_C(x) : C \in G\} \tag{14.22}$$

The greatest element of the set $Fuz(X, L)$ is the fuzzy set $\bar{1}$ defined via $\chi_{\bar{1}}(x) = 1$ for every $x \in X$ and the least element in $Fuz(X, L)$ is the fuzzy set $\bar{0}$ defined via $\chi_{\bar{0}}(x) = 0$ for every $x \in X$ where $0, 1$ are respectively the null and the unit elements in L.

The structure of L may be rendered in $Fuz(X, L)$ more closely: given two fuzzy sets A, B in $Fuz(X, L)$ we define the meet $A \wedge B$ and the join $A \vee B$ as follows:

$$\begin{aligned}\chi_{A \wedge B}(x) &= \chi_A(x) \wedge_L \chi_B(x) \\ \chi_{A \vee B}(x) &= \chi_A(x) \vee_L \chi_B(x)\end{aligned} \qquad (14.23)$$

where $\wedge_L, \vee_L$ are respectively the meet and the join in the lattice L. Then clearly, $(Fuz(X, L), \bar{0}, \bar{1}, \wedge, \vee)$ is a lattice.

These operations may be exploited as semantic counterparts of the syntactic set operations of union and intersection (as proposed in [Zadeh65]) viz. we define the *union* $A \cup B$ of two fuzzy sets A, B and the *intersection* $A \cap B$ of the sets A, B admitting that

$$\begin{aligned}\chi_{A \cup B}(x) &= \chi_{A \vee B}(x) = max\{\chi_A(x), \chi_B(x)\} \\ \chi_{A \cap B}(x) &= \chi_{A \wedge B}(x) = min\{\chi_A(x), \chi_B(x)\}\end{aligned} \qquad (14.24)$$

It is manifest that operations of union and intersection defined in this way obey the standard laws of associativity, commutativity and distributivity known from the theory of lattices (cf. Chapter 9) as well as from elementary set theory (cf. Chapter 6). In particular we may observe that the null element $\bar{0}$ satisfies the identity $A \cup \bar{0} = A = \bar{0} \cup A$ showing it to be the counterpart of the empty set in the standard set algebra while the unit element $\bar{1}$ plays the role of the universe as it satisfies the identity $A \cap \bar{1} = A = \bar{1} \cap A$. In this setting the lattice ordering $\leq$ plays the role of classical containment $\subseteq$.

We will from now on focus on the lattice $Fuz(X, [0, 1])$ of fuzzy sets valued in the interval $[0, 1]$. In this case we may endow the set algebra $(Fuz(X, [0, 1]), \cup,$
$\cap, \bar{0}, \bar{1})$ with the complement operation based on the Łukasiewicz negation $N(x) = 1 - x$ and we may define the *complement* A^c to a fuzzy set A via

$$\chi_{A^c}(x) = N(\chi_A(x)) = 1 - \chi_A(x) \qquad (14.25)$$

Then again the operation $(.)^c$ obeys some basic laws governing its classical counterpart – the complement $X \setminus A$ e.g. $(A^c)^c = A$, $A \leq B$ implies $B^c \leq A^c$, while some other basic laws are not obeyed as a rule, e.g. it needs not to be true that $A \cap A^c = \bar{0}, A \cup A^c = \bar{1}$.

These examples tell us already that it is not a straightforward task to translate properties of point – constructed objects into properties of set – constructed objects and in any process of such translation some properties will be lost. It is a matter of a judicious choice which of properties/laws to retain and which to lose.

Here, the familiar to us constructs, t – norms, t – conorms, and various modifications of the Łukasiewicz negation come fore: replacing max with a t – conorm S, min with a t – norm T, the negation N with one of its modifications n leads to many calculi on fuzzy sets which may better suit a particular purpose. The reader will consult e.g. [Driankov93] in this matter.

As our interest here is in mathematical foundations a fortiori in coherent and rationally justified approaches, we will stay for a while with the Zadeh lattice of fuzzy sets defined above.

In this section we will pursue the topic of rough fuzzy sets along with that of fuzzy rough sets following the ideas for both exposed in [Dubois – Prade92].

Rough fuzzy sets will arise when in the universe X of the Zadeh algebra $Z(X) = (Fuz(X, [0,1]), \cup, \cap, N, \overline{0}, \overline{1})$ an equivalence relation R is given. As we know well, the relation R does induce a rough set algebra over X stratifying subsets of X into exact and rough, and it will be our task to survey possible effects of the action of R on elements of $Z(X)$.

But the other way is possible: one may try to define in the language of fuzzy sets the basic ingredients of rough set theory viz. partitions and equivalence relations, and having this done, to construct approximations to fuzzy sets with respect to introduced fuzzy partitions. In this way *fuzzy rough sets* will emerge.

We will be interested in constructions bringing forth these objects and in their properties confronted with properties of rough sets and rough set approximations.

14.3.1 Rough fuzzy sets

We begin with the algebra $Z(X)$ and with an equivalence relation R on the universe X. We know that the set $X/R = \{[x]_R : x \in X\}$ of equivalence classes of the relation R (cf. Chapter 6) and the set X are related via the natural quotient function $q_R : X \to x/R$ which assigns to every $x \in X$ its equivalence class $[x]_R$. We denote by the symbol $\omega(y)$ the function sending $y \in X/R$ to $\omega(y) = \{x : q_R(x) = y\}$; thus, ω assigns to every equivalence class as an element of the quotient set X/R this class as a subset of X.

Now, suppose a fuzzy set $A \in Z(X)$ is given; there are two tasks before us viz (1) to induce a fuzzy set by means of A in X/R (2) to approximate A with respect to R in the spirit of rough set theory.

Approaching the task (1), we define two fuzzy sets on X/R (cf. [Dubois – Prade92]) viz. $\underline{A}_R, \overline{A}_R$ by letting

$$\chi_{\underline{A}_R}(y) = inf\{\chi_A(x) : x \in \omega(y)\}$$
$$\chi_{\overline{A}_R}(y) = sup\{\chi_A(x) : x \in \omega(y)\} \tag{14.26}$$

Then clearly

(RF1) $\underline{A}_R \le \overline{A}_R$

and the pair $(\underline{A}_R, \overline{A}_R)$ is an R – *rough fuzzy set* on X/R induced by A.

This rough fuzzy set is an R – *exact fuzzy set* on X/R if and only if $\underline{A}_R = \overline{A}_R$ which happens to be the case if and only if the fuzzy set A is R–*constant* i.e. the function $\chi_A(x)$ is constant on equivalence classes of R.

Let us observe that $\underline{A}_R, \overline{A}_R$ preserve properties of lower and upper approximations in rough set theory:

(RF2) $\underline{A \cap B}_R = \underline{A}_R \cap \underline{B}_R; \overline{A \cup B}_R = \overline{A}_R \cup \overline{B}_R$
(RF3) $\underline{A^c}_R = \overline{A}_{R^c}; \overline{A^c}_R = \underline{A}_R^c$
(RF4) $\underline{A \cup B}_R \supseteq \underline{A}_R \cup \underline{B}_R; \overline{A \cap B}_R \subseteq \overline{A}_R \cap \overline{B}_R$

Indeed, it is enough to justify the first of the two identities in each of (RF2)–(RF4) as the other follows by duality. Concerning (RF2), we have

$$\begin{aligned}
\chi_{\underline{A \cap B}_R}(y) &= inf_{x \in y}\{min\{\chi_A(x), \chi_B(x)\}\} \\
\chi_{\underline{A}_R \cap \underline{B}_R}(x) &= min\{inf_{x \in y}\chi_A(x), inf_{x \in y}\chi_B(x)\}
\end{aligned} \qquad (14.27)$$

We have

$$min\{\chi_A(x), \chi_B(x)\} \ge min\{inf_{x \in y}\chi_A(x), inf_{x \in y}\chi_B(x)\} \qquad (14.28)$$

hence

$$inf_{x \in y}\{min\{\chi_A(x), \chi_B(x)\}\} \ge min\{inf_{x \in y}\chi_A(x), inf_{x \in y}\chi_B(x)\} \quad (14.29)$$

The converse inequality follows equally easily, so first part of (RF2) is verified and the second follows along similar lines. Analogous considerations prove (RF4).

In case of (RF3), the identities in question follow from the following easy to verify identities $inf_x\{1 - a_x\} = 1 - sup_x\{a_x\}$ and its dual $sup_x\{1 - a_x\} = 1 - inf_x\{a_x\}$.

We may now lift R–rough fuzzy sets $\underline{A}_R, \overline{A}_R$ to the universe X via the function ω; we define fuzzy sets $\underline{A}_{R,\omega}, \overline{A}_{R,\omega}$ on X as follows:

$$\begin{aligned}
\chi_{\underline{A}_{R,\omega}}(x) &= \chi_{\underline{A}_R}(q_R(x)) \\
\chi_{\overline{A}_{R,\omega}}(x) &= \chi_{\overline{A}_R}(q_R(x))
\end{aligned} \qquad (14.30)$$

The pair $(\underline{A}_{R,\omega}, \overline{A}_{R,\omega})$ is then the R – *rough fuzzy set* on X induced by A. Again, we have that the R – rough fuzzy set $(\underline{A}_{R,\omega}, \overline{A}_{R,\omega})$ is exact if and only if the fuzzy set A is R – constant.

Fuzzy sets $\underline{A}_{R,\omega}, \overline{A}_{R,\omega}$ may be regarded as rough fuzzy approximations to the set A as they obey the basic laws governing rough set approximations viz.

(RF5) $\underline{A}_{R,\omega} \le A \le \overline{A}_{R,\omega}$
(RF6) $\underline{A \cap B}_{R,\omega} = \underline{A}_{R,\omega} \cap \underline{B}_{R,\omega}; \overline{A \cup B}_{R,\omega} = \overline{A}_{R,\omega} \cup \overline{B}_{R,\omega}$

(RF7) $\underline{A^c}_{R,\omega} = \overline{A}^c_{R,\omega}; \overline{A^c}_{R,\omega} = \underline{A}^c_{R,\omega}$

(RF8) $\underline{A \cup B}_{R,\omega} \supseteq \underline{A}_{R,\omega} \cup \underline{B}_{R,\omega}; \overline{A \cap B}_{R,\omega} \subseteq \overline{A}_{R,\omega} \cap \overline{B}_{R,\omega}$

(RF9) $\underline{A}_{R,\omega}, \overline{A}_{R,\omega}$ are exact R – rough fuzzy sets

Proofs of (RF6)–(RF8) are the same as with (RF2)–(RF4), (RF5) follows by the very definition, and (RF9) holds as well because both $\underline{A}_{R,\omega}, \overline{A}_{R,\omega}$ are R – constant.

We may now recall the notions of a rough bottom equality and of rough top equality (cf. Chapter 10) and render them in our new context of fuzzy sets viz. we will say that fuzzy sets A, B are R – *rough bottom equal*, $A \sim_R B$ symbolically, if and only if $\underline{A}_{R,\omega} = \underline{B}_{R,\omega}$; similarly, A, B are *rough top equal* if and only if $\overline{A}_{R,\omega} = \overline{B}_{R,\omega}$, symbolically $A \simeq_R B$. The sets A, B are *rough equal*, symbolically $A \equiv_R B$ if and only if $A \simeq_R B$ and $A \approx_R B$ hold.

From these definitions the conditions in terms of the membership functions may be easily read off viz.

Proposition 14.12. *For fuzzy sets A, B on a universe X (i) $A \simeq_R B$ if and only if $inf_{x \in y} \chi_A(x) = inf_{x \in y} \chi_B(x)$ for every $y \in X/R$ (ii) $A \approx_R B$ if and only if $sup_{x \in y} \chi_A(x) = sup_{x \in y} \chi_B(x)$ for every $y \in X/R$.*

In particular, it follows that for two R – constant fuzzy sets A, B we have $A \equiv_R B$ if and only if $A = B$ if and only if $A \simeq_R B$ if and only if $A \approx_R B$. We also point to the fact that every rough membership function in the sense of Pawlak and Skowron (cf. Chapter 1) defined with respect to the partition induced by R is R–constant hence exact.

At this point our discussion of rough fuzzy sets ends. The reader may consult [Dubois – Prade92] for a discussion of relations of the notion of a rough fuzzy set to other approaches to uncertainty like the C – calculus of Caianiello [Caianiello87] or the evidence theory of Dempster – Shafer [Dempster67], [Shafer76].

14.3.2 Fuzzy rough sets

The dual notion of a *fuzzy rough set* does emerge in purely fuzzy context. Having the set $Fuz(X, [0,1])$ of fuzzy sets on a universe X valued in the interval $[0, 1]$, we would like to discuss the fuzzy counterpart of the notion of an equivalence relation and the resulting notion of a fuzzy partition.

The closest approximation in the language of fuzzy sets to the notion of an equivalence relation is the T – *fuzzy similarity relation* (cf. [Zadeh71]) which in the most general formulation may be defined as a fuzzy set E_T on the square $X \times X$ satisfying the following requirements

(FS1) $\chi_{E_T}(x, x) = 1$

(FS2) $\chi_{E_T}(x, y) = \chi_{E_T}(y, x)$

(FS3) $\chi_{E_T}(x,z) \geq T(\chi_{E_T}(x,y), \chi_{E_T}(y,z))$

Thus, (FS1) models reflexivity of an equivalence relation, (FS2) models its symmetry, while (FS3) renders transitivity with respect to a t − norm T. Let us observe that for each idempotent a of the t −norm T (i.e. for a with $T(a,a) = a$) the relation $R_{a,E}$ defined via

$$(x,y) \in R_{a,E} \Leftrightarrow \chi_{E_T}(x,y) \geq a \tag{14.31}$$

is an equivalence in the classical sense.

For instance, for *similarity relations* (cf. [Zadeh71]) defined with $T(x,y) = min(x,y)$ the relation $R_{a,E}$ is defined for every $a \in [0,1]$ and relations $R_{a,E}$ form the descending chain of equivalence relations

$$R_{1,E} \subseteq ... \subseteq R_{a,E} \subseteq .. \subseteq R_{0,E} \tag{14.32}$$

On the other hand, for *probabilistic similarity relations* defined with $T(x,y) = Prod(x,y)$ (cf. [Menger42], the only relations $R_{a,E}$ are $R_{1,E} \subseteq R_{0,E} = X \times X$.

The same happens to *likeness relations* (cf. [Ruspini91]) defined with $T(x,y) = x \otimes y$.

In the classical case, an equivalence relation R does induce a *pseudo − metric* (cf. Chapter 7) on the underlying set X viz. we let $d_R(x,y) = 0$ in case $(x,y) \in R$ and $d_R(x,y) = 1$, otherwise. One could expect from a fuzzy similarity relation E_T a similar service; it is indeed so viz. letting for $E = E_T$:

$$\chi_{d_E}(x,y) = 1 - \chi_{E_T}(x,y) \tag{14.33}$$

defines a fuzzy set d_E which we may call the *fuzzy metric associated with* E_T. That d_E retains properties of a metric follows from the following (cf. [Mantaras–Valverde88])

Proposition 14.13. *The fuzzy set d_E induced by a fuzzy similarity relation $E = E_T$ has the following properties*

(FM1) $\chi_{d_E}(x,y) = 0$ *if and only if* $(x,y) \in R_{1,E}$

(FM2) $\chi_{d_E}(x,y) = \chi_{d_E}(y,x)$

(FM3) $\chi_{d_E}(x,z) \leq \chi_{d_E}(x,y) + \chi_{d_E}(y,z)$ *in case* $T = min, Prod, \otimes$

Indeed, (FM1), (FM2) follow from the definition of d_E, and concerning the triangle inequality (FM3), we may paraphrase it to the equivalent inequality

(FM4) $1 + \chi_{E_T}(x,z) \geq \chi_{E_T}(x,y) + \chi_{E_T}(y,z)$

which by (FS3) follows from the inequality

(*) $1 + T(\alpha, \beta) \geq \alpha + \beta$

satisfied by $T = min, Prod, \otimes$ which may be verified directly.

An interesting relation between fuzzy metrics and fuzzy similarities was discovered in [Mantaras – Valverde88] by dualization of the notion of fuzzy distance. Given a t – conorm $S(x, y)$, we may introduce the notion of the *residuated dual implication* $x \leftarrow_S y$ via

$$(DRI) \quad x \leftarrow_S y \leq z \Leftrightarrow S(z, x) \geq y.$$

Then we define a a fuzzy set d_S via

$$\chi_{d_S}(x, y) = f(max\{x, y\}) \leftarrow_S f(min\{x, y\}) \tag{14.34}$$

where f is an arbitrarily chosen decreasing continuous function from the interval $[0, 1]$ onto itself. It turns out that

Proposition 14.14. *[Mantaras – Valverde88] For each t – conorm S, the fuzzy set d_S satisfies the following*

(SM1) $\chi_{d_S}(x, x) = 0$

(SM2) $\chi_{d_S}(x, y) = \chi_{d_S}(y, x)$

(SM3) $\chi_{d_S}(x, z) \leq S(\chi_{d_S}(x, z), \chi_{d_S}(y, z))$

Let us remark for the proof that (SM1), (SM2) follow by definition of d_S and properties of S, and (SM3) follows by direct calculations involving the definition of $\leftarrow_S$.

Thus, d_S is a pseudo – metric induced by the t – conorm S. Involving once more a continuous decreasing $f : [0, 1] \rightarrow [0, 1]$, we may convert d_S into a fuzzy similarity relation $E = E_T$ with $T(x, y) = f^{-1}(S(f(x), f(y))$ by letting

$$\chi_{E_T}(x, y) = f(\chi_{d_S}(x, y)) \tag{14.35}$$

In this way we obtain the correspondence between fuzzy similarity relations and pseudo – metrics on X as reversing the above construction we pass from a fuzzy similarity relation to a pseudo – metric.

Now we may address the problem of similarity (equivalence) classes for a fuzzy similarity relation $E = E_T$. In [Zadeh71] it is proposed to define the similarity class $[x]_E$ for $x \in X$ by letting

$$\chi_{[x]_E}(y) = \chi_E(x, y) \tag{14.36}$$

Then the following properties are observed (cf. [Höhle88], [Dubois – Prade88])

Proposition 14.15. *The similarity class* $[x]_E$ *satisfies the following*

(SC1) $\chi_{[x]_E}(x) = 1$

(SC2) $T(\chi_{[x]_E}(y), \chi_E(y, z)) \leq \chi_{[x]_E}(z)$

(SC3) $T(\chi_{[x]_E}(y), \chi_{[x]_E}(z)) \leq \chi_E(y, z)$

Proof. (SC1) follows by (FS1); (SC2) is a direct paraphrase of (FS3) and (SC3) paraphrases (FS3) via (FS2).   □

Properties (SC1–(SC3)) reflect properties of classical equivalence relations viz. (SC1) corresponds to the property that $x \in [x]_R$, (SC2) paraphrases the property that xRy implies $y \in [x]_R$, and (SC3) paraphrases the property that $y, z \in [x]_R$ imply $(y, z) \in R$.

We may call a *fuzzy equivalence class* of a fuzzy similarity relation E any fuzzy set A satisfying (SC1) – (SC3) in place of $[x]_E$. However, it turns out that we have already found all fuzzy equivalence classes (cf. [Höhle88], [Dubois – Prade92]). We denote with the symbol $A \times_T B$ the T – Cartesian product of fuzzy sets A, B defined via

$$\chi_{A \times_T B}(x, y) = T(\chi_A(x), \chi_B(y)) \tag{14.37}$$

Proposition 14.16. *If A is a fuzzy equivalence class of a fuzzy similarity relation E_T then A is of the form $[x]_{E_T}$ with some $x \in X$. Moreover*
  1. $\bigcup_{x \in X}[x]_{E_T} \times_T [x]_{E_T} = E_T$
  2. *in the case X is finite,* $\sup_y \{\min\{\chi_{[x]_{E_T}}(y), \chi_{[z]_{E,T}}(y)\}\} < 1$ *whenever* $[x]_{E_T} \neq [z]_{E,T}$.

Proof. Assuming A a fuzzy equivalence class, we have (i) $\chi_A(x) = 1$ with some x by (SC1), and thus by (RI5)

$$\chi_A(y) = T(\chi_A(y), \sup_z T(\chi_{[x]_{E_T}}(z), \chi_{[x]_{E_T}}(z))) = \\ \sup_z T(\chi_A(y), T(\chi_{[x]_{E_T}}(z), \chi_{[x]_{E_T}}(z))) \tag{14.38}$$

As

$$\sup_z T(\chi_A(y), T(\chi_{[x]_{E_T}}(z), \chi_{[x]_{E_T}}(z))) \leq \\ \sup_z T(T(\chi_A(y), \chi_A(z)), \chi_{[x]_{E_T}}(z)) \tag{14.39}$$

by (i) and associativity and monotonicity of T, and our assumption about A, and moreover

$$\sup_z T(T(\chi_A(y), \chi_A(z)), \chi_{[x]_{E_T}}(z)) \leq \\ \sup_z T(\chi_{E_T}(y, z), \chi_{[x]_{E_T}}(z)) \tag{14.40}$$

by (SC2); finally

$$\sup_z T(\chi_{E_T}(y, z), \chi_{[x]_{E_T}}(z)) \leq \chi_{[x]_{E_T}}(y) \tag{14.41}$$

by (SC3). Thus, $A \leq [x]_{E_T}$ and replacing roles of A and $[x]_{E_T}$ yields by symmetry that $[x]_{E_T} \leq A$ so finally $A = [x]_{E_T}$.

To settle (i), we observe that

$$\chi_{\bigcup_{[x]}}[x]_{E_T} \times_T [x]_{E_T}(y,z) = sup_{[x]}T(\chi_{[x]_{E_T}}(y), \chi_{[x]_{E_T}}(z)) =$$
$$sup_x T(\chi_{E_T}(x,y), \chi_{E_T}(x,z)) = \chi_{E_T}(y,z) \tag{14.42}$$

by (FS3). Thus (i) is verified.

To verify (ii), we may observe that by the already proved part, if $[x]_{E_T} \neq [z]_{E,T}$ then $min_y\{\chi_{[x]_{E_T}}(y), \chi_{[z]_{E_T}}(y)\} < 1$ hence (ii) follows. $\qquad\square$

Properties (i) and (ii) are counterparts to properties of classical equivalence relations viz. covering the universe by equivalence classes ((i)) and disjointness of equivalence classes ((ii)). It is justified to call collections $\mathcal{F}$ of fuzzy sets satisfying (i),(ii) and additional property (iii) $\exists x.\chi_A(x) = 1$ for every $A \in \mathcal{F}$ *fuzzy partitions*.

Given a fuzzy partition $\mathcal{F} = \{A_i : i \in I\}$, we can ask about the existence of a fuzzy similarity relation E_T which could approximate fuzzy sets A_i with its equivalence classes and at the best give sets A_i as its equivalence classes.

In order to address this problem, we consider a t – norm T and we define a relation R_T on the universe X by the formula

$$\chi_{R_T}(x,y) =$$
$$inf_i\{max\{\chi_{A_i}(x), \chi_{A_i}(y)\} \to_T min\{\chi_{A_i}(x), \chi_{A_i}(y)\}\} \tag{14.43}$$

Then we have

Proposition 14.17. *[Valverde85] The relation R_T satisfies (FS1) – (FS3) and thus R_T is a fuzzy similarity relation.*

Indeed, (FS1) is satisfied as $x \to_T x = 1$ for every x hence $\chi_{R_T}(x,x) = 1$ for every x.

Symmetry (FS2) follows immediately from definition of R_T. Finally, transitivity property (FS3) comes down after a suitable paraphrase to the property

(RI10) $y \to_T x \leq (z \to_T y) \to_T (z \to_T x)$

of residuated implications. To prove (RI10), we begin with a string of equivalent inequalities following from the definition of residuated implications starting with (RI10):

(i) $T(y \to_T x, z \to_T y) \leq z \to_T x$

(ii) $T(T(y \to_T x, z \to_T y), z) \leq x$

(iii) $T(y \to_T x, T(z \to_T y, z)) \leq x$

As (iv) $T(y \to_T x, y) \leq x$, and similarly (v) $T(z \to_T y, z) \leq y$, we obtain from (iv) and (v) the inequality (vi) $T(y \to_T x, T(z \to_T y, z)) \leq T(y \to_T x, y)) \leq x$ i.e. (iii) is proved verifying (RI10) and a fortiori (FS3).

As $\mathcal{F}$ was supposed to be a fuzzy partition, we have

$$\chi_{A_i}(x) = 1$$

with some x for every A_i. So (SC1) is satisfied with $\mathcal{F}$. We now verify that each A_i satisfies (SC2). We need to verify that

$$(vii) \ T(\chi_A(x), \chi_{R_T}(x, y)) \leq \chi_A(y)$$

with $A = A_i$. Replacing $\chi_{R_T}(x, y)$ with its definition and using the generic symbol B for an element of $\mathcal{F}$, we have

$$T(\chi_A(x), inf_B\{max\{\chi_B(x), \chi_B(y)\} \to_T min\{\chi_B(x), \chi_B(y)\}\}) \leq$$

$$T(\chi_A(x), \{max\{\chi_A(x), \chi_A(y)\} \to_T min\{\chi_A(x), \chi_A(y)\}\}) \leq$$

$$T(\max\{\chi_A(x), \chi_A(y)\}, \{max\{\chi_A(x), \chi_A(y)\} \to_T \\ min\{\chi_A(x), \chi_A(y)\}\}) \leq \tag{14.44}$$

$$min\{\chi_A(x), \chi_A(y)\} \leq \chi_A(y)$$

proving that (SC2) holds.

Now, for (SC3), we should verify, again using generic symbols A, B that

$$T(\chi_A(x), \chi_A(y)) \geq \chi_{R_T}(x, y) \tag{14.45}$$

Let us look at

$$(x) \ \chi_{R_T}(x, y) = \\ inf_B\{max\{\chi_B(x), \chi_B(y)\} \to_T min\{\chi_B(x), \chi_B(y)\} \tag{14.46}$$

as far as we have no other information about $\mathcal{F}$, then that it is merely a fuzzy partition, we are not able to prove or disprove (SC3). However, in case our fuzzy partition $\mathcal{F}$ is of the form $\{[x]_{E_T}\}$ i.e. it is generated by a fuzzy similarity relation E_T, we may say more viz.

Proposition 14.18. *[Valverde85] For the fuzzy partition $\mathcal{F} = \{[x]_{E_T} : x \in X\}$ induced by a fuzzy similarity relation $R = E_T$ with a t–norm T, the fuzzy similarity relation E_R induced by $\mathcal{F}$ according to the recipe of Proposition 14.17 coincides with E_T; a fortiori, $\mathcal{F}$ is a fuzzy partition with respect to E_R.*

Proof. Indeed, let us observe that by (FS3):

$$\chi_{E_T}(x,z) \geq T(\chi_{E_T}(x,y), \chi_{E_T}(y,z))$$

we have

$$(i)\ \chi_{E_T}(x,y) \leq \chi_{E_T}(y,z) \to_T \chi_{E_T}(x,z).$$

Going now to E_R, we consider the term

$$(ii)\ max\{\chi_B(x), \chi_B(y)\} \to_T min\{\chi_B(x), \chi_B(y)\}$$

with $B = [b]_{E,T}$, some b. By (i), we have

$$(iii)\ max\{\chi_B(x), \chi_B(y)\} \to_T min\{\chi_B(x), \chi_B(y)\} \geq \chi_{E_T}(x,y)$$

and letting $b = x$ we have that

$$(iv)\ inf_B\{max\{\chi_B(x), \chi_B(y)\} \to_T min\{\chi_B(x), \chi_B(y)\}\} = \chi_{E_T}(x,y)$$

which concludes our proof.   □

It follows that the construction presented above is canonical: starting with a fuzzy similarity R it induces a fuzzy partition which in turn returns the initial fuzzy similarity R. We have here a counterpart to the classical correspondence between equivalence relations and partitions (cf. Chapter 6).

Now it is time to give a final touch to our analysis and discuss the problem of rough set – modeled approximations to a given fuzzy set, induced by a given fuzzy partition. We may hardly expect as a regular theory here as with rough set approximations yet we present a sufficiently regular theory here.
We settle on the context: we assume the Zadeh lattice

$$(Fuz(X, [0,1]), min, max, N, \overline{0}, \overline{1})$$

as our standard playground and we restrict ourselves to the case when $T = min$. In this case the residual implication $\to_m$ is the following function:

$$y \to_m z = \begin{array}{l} 1\ in\ case\ y \leq z \\ z\ \ \ otherwise \end{array} \qquad (14.47)$$

We define for a given fuzzy partition $\mathcal{F} = \{A : A \in \mathcal{F}\}$ and a given fuzzy set B

the lower $\mathcal{F}$–approximation $\underline{B}_{\mathcal{F}}$ via

$$\chi_{\underline{B}_{\mathcal{F}}}(x) = inf_A\{min\{\chi_A(x), \chi_B(x)\}\};$$

the upper $\mathcal{F}$–approximation $\overline{B}_{\mathcal{F}}$ via

$$\chi_{\overline{B}_{\mathcal{F}}}(x) = sup_A\{\chi_A(x) \to_m \chi_B(x)\}.$$

It remains to verify that the basic properties of rough approximations are observed with these proposal for rough fuzzy approximations. We introduce a localized variant of the notion of an exact set viz. we will say that a fuzzy set B is $\mathcal{F}$ – *exact* at $x \in X$ if and only if $\chi_{\underline{B}_{\mathcal{F}}}(x) = \chi_{\overline{B}_{\mathcal{F}}}(x)$. We will say that the fuzzy partition $\mathcal{F}$ is *centered* at $x \in x$ if and only if $inf_A\{\chi_A(x) : A \in \mathcal{F}\} = 1$.

Proposition 14.19. *1. The lower approximation $\underline{B}_{\mathcal{F}}$ is an interior operator (opening) on $Fuz(X, [0, 1])$*

2. The upper approximation $\overline{B}_{\mathcal{F}}$ is a closure operator (closing) on $Fuz(X, [0, 1])$

3. A fuzzy set B is $\mathcal{F}$ – exact at $x \in X$ if and only if $\chi_B(x) = 1$ for every x at which $\mathcal{F}$ is centered or $B(x) \leq inf_A\{\chi_A(x) : A \in \mathcal{F}\}$ In particular, the lower approximation $\overline{B}_{\mathcal{F}}$ is exact.

Proof. We verify the consecutive elements in our statement. Concerning (i), it follows by definition that (O1) $\underline{\overline{0}} = \overline{0}$; as

$$min\{\chi_A(x), \chi_B(x)\} \leq \chi_B(x),$$

we have

(O2) $\underline{B}_{\mathcal{F}} \leq B$

By monotonicity of min

(O3) $B \leq C$ implies $\underline{B}_{\mathcal{F}} \leq \underline{C}_{\mathcal{F}}$

Finally, as

$$inf_A\{min\{\chi_A(x), \chi_B(x)\} = inf_A\{\chi_A(x), inf_A\{min\{\chi_A(x), \chi_B(x)\}\}\},$$

it follows that

(O4) $\underline{B}_{\mathcal{F}} = \underline{B}_{\mathcal{F}\mathcal{F}}$

(O1) – (O4) witness that the lower approximation is an interior mapping hence it satisfies all the basic demands on the rough set – theoretic lower approximation.

Concerning (ii), we proceed along similar but dual lines. Clearly, (C1) $\overline{\overline{1}}_{\mathcal{F}} = \overline{1}$ holds. As $y \to_m z \geq z$, we have

(C2) $B \leq \overline{B}_{\mathcal{F}}$

By monotonicity in the second coordinate of $\to_m$

(C3) $B \leq C$ implies $\overline{B}_{\mathcal{F}} \leq \overline{C}_{\mathcal{F}}$

Finally, one verifies directly that

(C4) $\overline{B}_{\mathcal{F}} = \overline{\overline{B}_{\mathcal{F}}}_{\mathcal{F}}$

As witnessed by (C1) – (C4), the upper approximation is a closure operation on $Fuz(X, [0, 1])$ thus it obeys all basic laws governing rough set – theoretic upper approximations.

We may yet ask about exact fuzzy sets: what is their characterization? To answer this question we may consider the equation

$$(?) \quad \overline{B}_{\mathcal{F}} = \underline{B}_{\mathcal{F}} \tag{14.48}$$

We may consider the three cases
(I) $\chi_B(x) \leq inf_A\{\chi_A(x) : A \in \mathcal{F}\}$
(II) $\chi_B(x) \leq sup_A\{\chi_A(x) : A \in \mathcal{F}\}$
(III) neither (I) nor (II).

In case (I), (?) is satisfied as both sides are equal to $\chi_B(x)$. In case (II), we get easily that the condition for equality in (?) is $inf_A\{\chi_A(x) : A \in \mathcal{F}\} = 1$ in which case $\chi_B(x) = 1$. Case (III) satisfies (?) if and only if $inf_A\{\chi_A(x) : A \in \mathcal{F}\} = 1$ in which case again $\chi_B(x) = 1$. Thus (?) follows. For the lower approximation $\overline{B}_{\mathcal{F}}$, we have $\chi_{\overline{B}_{\mathcal{F}}}(x) \leq inf_A\{\chi_A(x) : A \in \mathcal{F}\}$ i.e. case (I) so the lower approximation is exact. $\qquad\square$

We may observe that the upper approximation defined above need not be exact.

14.4 Brouwer–Zadeh lattices

We know already from our discussion in preceding sections, that L–fuzzy sets on a given universe X and with a given lattice L, make a partially ordered set viz. given fuzzy sets A, B on X, represented by fuzzy membership functions $\chi_A : X \to L$, $\chi_B : X \to L$ we let

$$A \leq B \Leftrightarrow \forall x \in X . \chi_A(x) \leq_L \chi_B(x) \tag{14.49}$$

We denote this partially ordered set of L–fuzzy sets on X with the symbol $Fuz(X, L)$. Clearly, in case L is a complete lattice, $Fuz(X, L)$ is completely ordered viz. given a family $G \subseteq Fuz(X, L)$ we define its infimum $\bigwedge G$ via

$$\chi_{\bigwedge G}(x) = \bigwedge\{\chi_C(x) : C \in G\} \tag{14.50}$$

and similarly we define the supremum $\bigvee G$ via

$$\chi_{\vee G}(x) = \bigvee \{\chi_C(x) : C \in G\} \tag{14.51}$$

The greatest element of the set $Fuz(X,L)$ is the fuzzy set $\overline{1}$ defined via $\chi_{\overline{1}}(x) = 1$ for every $x \in X$ and the least element in $Fuz(X,L)$ is the fuzzy set $\overline{0}$ defined via $\chi_{\overline{0}}(x) = 0$ for every $x \in X$ where $0, 1$ are respectively the null and the unit elements in L.

The structure of L may be rendered in $Fuz(X,L)$ more closely: given two fuzzy sets A, B in $Fuz(X,L)$ we define the meet $A \wedge B$ and the join $A \vee B$ as follows:

$$\begin{aligned} \chi_{A \wedge B}(x) &= \chi_A(x) \wedge_L \chi_B(x) \\ \chi_{A \vee B}(x) &= \chi_A(x) \vee_L \chi_B(x) \end{aligned} \tag{14.52}$$

where $\wedge_L, \vee_L$ are respectively the meet and the join in the lattice L. Then clearly, $(Fuz(X,L), \overline{0}, \overline{1}, \wedge, \vee)$ is a lattice.

Let us also observe that $Fuz(X,L)$ contains for each $a \in L$ the fuzzy set $\overline{a}$ with

$$\chi_{\overline{a}}(x) = a \tag{14.53}$$

for every $x \in X$. The properties of L translate via coordinate–wise definitions of operations in $Fuz(X,L)$ into properties of the latter, e.g., when L is a distributive lattice, then $Fuz(X,L)$ has also this property.

We now consider, following [Cattaneo97], some complementation operations in $Fuz(X,L)$ induced by respective complementation operations in the lattice L.

The reader will be asked to have in mind the lattice $[0,1]$ with usual operations min, max as the archetypical model for all objects we introduce here.

We first consider a lattice L endowed with the *Kleene – Zadeh* complementation i.e. with the operation $\prime$ having the properties

(KZ1) $x = x''$ for every $x \in L$

(KZ2) $x \leq_L y$ implies $y' \leq_L x'$ for each pair x, y in L

(KZ3) $x \leq_L x', y' \leq_L y$ imply $x \leq_L y$ for each pair x, y in L

It is easy to see that (KZ3) implies

(KZ4) $x \wedge_L x' \leq_L y \vee_L y'$ for each pair x, y in L.

Similarly, from (KZ1), (KZ2) the following hold in a standard way so we leave the proof as an exercise

(KZ4) $(x \wedge y)' = x' \vee y'$

(KZ5) $(x \vee y)' = x' \wedge y'$

We may observe an obvious fact that letting in the lattice $[0,1]$ $x' = 1 - x$ induces in $[0,1]$ the Kleene–Zadeh complementation.

A more remote example may be gotten by considering subspaces of a finite–dimensional real vector space V with the ordering $M \leq N$ if and only if $M \subseteq N$ for vector subspaces M, N of V. In this case we let $M' = M^{\perp}$, where $M^{\perp}$ is the orthogonal complement to M.

Let us also observe that $0' = 1, 1' = 0$ whenever $0, 1$ exist in L; indeed, in e.g. case of $0'$, for an arbitrary $b \in L$, we have $0 \leq_L b'$ hence by (KZ2) and (KZ1), $b = b'' \leq_L 0'$.

The Kleene–Zadeh complementation may be rendered in the lattice $Fuz(X, L)$ by letting A' to be the fuzzy set in $Fuz(X, L)$ defined via

$$\chi_{A'}(x) = (\chi_A(x))' \qquad (14.54)$$

In particular letting

$$\chi_{A'}(x) = 1 - \chi_A(x) \qquad (14.55)$$

defines the Kleene – Zadeh complementation in the lattice $Fuz(X, [0,1])$.

We put this observation for the record

Proposition 14.20. *The operation A' defined on $Fuz(X, L)$ by means of*

$$\chi_{A'}(x) = (\chi_A(x))'$$

is a Kleene–Zadeh complementation in $Fuz(X, L)$ whenever x' is a Kleene–Zadeh complementation in L.

An other type of complementation is provided by the *Brouwer* complementation $(.)^{\sim}$ which is subject to the following requirements

(B1) $x \leq_L x^{\sim\sim}$ for each $x \in L$

(B2) $x \leq_L y$ implies $y^{\sim} \leq_L x^{\sim}$ for each pair $x, y \in L$

(B3) $x \wedge x^{\sim} = 0$ for each $x \in L$

The reader will notice that complementation defined via $x^{\sim} = x \Rightarrow 0$ where

$\Rightarrow$ is the relative pseudo – complementation in a pseudo – Boolean lattice does satisfy (B1)–(B3) (cf. Chapter 9). Therefore (cf. Chapter 9) natural examples of such complementation are provided by lattices of open sets in a topological space viz. letting for any open $P \subseteq X$ in a topological space (X, τ)

$$P^\sim = Int(X \setminus ClP) \qquad\qquad (14.56)$$

defines a Brouwerian complementation.

We observe that

(B4) $x^\sim = x^{\sim\sim\sim}$

as (B1), (B2) imply $x^{\sim\sim\sim} \leq_L x^\sim$ and $x^{\sim\sim\sim} \geq x^\sim$ by (B1). Obviously, $0^\sim = 1$.

Let us also observe that under (B1), (B2) the following property also holds

(B5) $x^\sim \wedge y^\sim = (x \vee y)^\sim$

Indeed, from $x, y \leq_L x \vee y$ it follows that $(x \vee y)^\sim \leq_L x^\sim, y^\sim$ hence $(x \vee y)^\sim \leq_L x^\sim \wedge y^\sim$. On the other hand, $x, y \geq_L x \wedge y$ imply $x^\sim, y^\sim \leq (x \wedge y)^\sim$ hence $x^\sim \vee y^\sim \leq_L (x \wedge y)^\sim$ and thus $(x \wedge y)^{\sim\sim} \leq_L (x^\sim \vee y^\sim)^\sim$. Substituting $x/u^{\sim\sim}, y/v^{\sim\sim}$ and using (1), we get $(u^{\sim\sim} \wedge v^{\sim\sim})^{\sim\sim} \leq_L (u^\sim \vee v^\sim)^\sim$ whence by substitution $u^\sim/z, v^\sim/w$ we get by (B1) $z^\sim \wedge w^\sim \leq_L (z^\sim \wedge w^\sim)^{\sim\sim} \leq_L (z \vee w)^\sim$.

A *Brouwer–Zadeh lattice* is a lattice L with a Kleene–Zadeh complementation $'$ and a Brouwer complementation $\sim$ which are related to each other by means of the following property

(KB1) $x^{\sim\prime} = x^{\sim\sim}$

Let us observe that (KB1) implies

(KB2) $x^\sim \leq_L x'$

Indeed, by (B1) we have $x \leq_L x^{\sim\sim}$ hence (i) $x^{\sim\sim\prime} \leq_L x'$ by (KZ2) and thus assuming (KB1) we get by (1) and (i) that $x^\sim = x^{\sim\sim\sim} = x^{\sim\sim\prime} \leq_L x'$.

We define a new operation in L by letting $x^\dagger = x'^\sim{}'$ for $x \in L$. In a standard way (cf. Chapter 12...), we prove that the following properties hold

(DB1) $x^{\dagger\dagger} \leq_L x$

(DB2) $x \leq_L y$ implies $y^\dagger \leq_L x^\dagger$

(DB3) $x \vee x^\dagger = 1$

Indeed, as $x^{\dagger\dagger} = x'^{\sim''\sim'} = x'^{\sim\sim'}$ (DB1) follows from $x' \leq_L x'^{\sim\sim}$ which implies $x'^{\sim\sim'} \leq_L x'' = x$. Similarly, (DB2) follows directly from (KZ2) and (B2). For (DB3), as $x^\sim \wedge x'^\sim = 0$ by (B3) and (KB2) we have $x^{\sim\sim\sim} \wedge x'^{\sim\sim\sim} = 0$ so $(x^{\sim\sim} \vee x'^{\sim\sim})^\sim = 0$ by (B4) implying $x^{\sim\sim} \vee x'^{\sim\sim} = 1$ and thus $x \vee x'^{\sim\sim} = 1$ by (B1). Finally we may observe that $x \vee x'^{\sim\sim} = x \vee x'^{\sim'}$ by (KB1).

We now explore further the interplay of the two complementations in a Brouwer – Zadeh lattice by recalling the following

Proposition 14.21. *[Cattaneo–Nistico97] For any element x in a Brouwer – Zadeh lattice L the following statements are equivalent*

1. $x^\sim = x'$
2. $x = x'^\sim$
3. $x^{\sim\sim} = x'^\sim$
4. $x = x^{\sim'}$
5. $x = x^{\dagger\dagger}$
6. $x = x^{\sim\sim}$

Moreover, any of (1) – (6) implies each of equivalent statements

7. $x \wedge x' = 0$
8. $x \vee x' = 1$

The direct proof of equivalence of (1) – (6) we would like to leave to the reader, we point only to derivation of (7) from (1) (clearly, (7), (8) are equivalent by (KZ5,6)). So assuming (1), we have $x \wedge x' = x \wedge x^\sim = 0$ by (B3).

We may now proceed as in Chapter 12, and we may define the notion of an interior as well as of a closure functions. We recall that a *closure* function on a lattice L is a function $c : L \to L$ such that

(C1) $x \leq_l c(x)$
(C2) $c(c(x)) = c(x)$
(C3) $x \leq_L y$ implies $c(x) \leq_L c(y)$

and, dually an *interior* function on a lattice L is a function $o : L \to L$ with the properties

(O1) $o(x) \leq_l x$
(O2) $o(o(x)) = o(x)$
(O3) $x \leq_L y$ implies $o(x) \leq_L o(y)$

An element $a \in L$ is *closed* (respectively *open*) if and only if $a = c(a)$ (respectively $a = o(a)$).

The reader who read chapter 12 will not be surprised at the choice of candidates for closure respectively interior in a Brouwer – Zadeh lattice L: we verify that $c(x) = x^{\sim\sim}$ and $o(x) = x^{\dagger\dagger}$ are natural closure respectively interior on L.

Proposition 14.22. *[Cattaneo97, 98] The function $x \to x^{\dagger\dagger}$ does satisfy properties (O1)–(O3) of an interior and the function $x \to x^{\sim\sim}$ satisfies properties (C1)–(C3) of a closure. Moreover, in any Brouwer – Zadeh lattice L the two functions coincide.*

Proof. That (O1), (O3) are satisfied follows from (DB1), (DB2). Also by (DB1), (DB2) the identity $x^{\dagger} = x^{\dagger\dagger\dagger}$ follows which implies $x^{\dagger\dagger} = x^{\dagger\dagger\dagger\dagger}$ i.e. (O2). The proof that (C1)–(C3) are satisfied follows same lines with (B1), (B2) in place of (DB1), (DB2). By Proposition 14.21, an element in L is of the form $o(x)$ if and only if it is of the form $c(x)$.   □

We have therefore that *open* elements i.e. elements of the form $x^{\dagger\dagger}$ with some x are of the form $x^{\sim\sim}$ i.e. they are *closed*.

With the rough set ideology which we extensively exploited in Chapter 12, we will call the pair $(o(x), c(x))$ the *approximation* to an element $x \in L$, where $o(x)$ will be the *lower approximation* to x and $c(x)$ will be the *upper approximation* to x.

It remains to introduce into the lattice $Fuz(X, L)$ the lacking element viz. a Brouwer complementation. We focus ourselves on the most important case of $Fuz(X, [0, 1])$. As we know in this case the Kleene – Zadeh complementation A' may be defined via $\chi_{A'}(x) = 1 - \chi_A(x)$. To define the other complementation, let us consider for a given fuzzy set A the following sets :

(the necessity kernel) $(A)_1 = \{x \in X : \chi_A(x) = 1\}$

(the non–possibility kernel) $(A)_0 = \{x \in X : \chi_A(x) = 0\}$

(the possibility kernel) $(A)_{>0} = \{x \in X : \chi_A(x) > 0\}$

For a given fuzzy set A, we define its Brouwer complementation $A^{\sim}$ by letting $\chi_{A^{\sim}} = \delta_{(A)_0}$. It remains to verify that $A^{\sim}$ is the complement in the sense of Brouwer to A, indeed.

Concerning (B1), we have by applying the definition of $A^{\sim}$ one more time that $\chi_{A^{\sim\sim}}(x) = 1$ whenever $\chi_A(x) > 0$ and $\chi_{A^{\sim\sim}}(x) = 0$ whenever $\chi_A(x) = 0$ thus $A \leq A^{\sim\sim}$ follows.

In case $A \leq B$ for fuzzy sets A, B, we have $(B)_0 \subseteq (A)_0$ hence $A^{\sim\sim} \leq B^{\sim\sim}$ verifying (B2).

By definition of $A^{\sim}$, $A \wedge A^{\sim} = 0$: $\chi_{A^{\sim}}(x) = 1$ whenever $\chi_A(x) = 0$ and $\chi_{A^{\sim}}(x) = 0$ whenever $\chi_A(x) > 0$. Thus (B3) is satisfied.

Finally, the property (KB1) remains to be checked. But by definitions of two complementations we obtain easily that both $A^{\sim\prime}$ and $A^{\sim\sim}$ are fuzzy sets with the fuzzy membership function $\chi(x) = 1$ whenever $\chi_A(x) > 0$ and $\chi(x) = 0$ whenever $\chi_A(x) = 0$.

We may now find rough approximations to elements of $Fuz(X, [0,1])$. Given a fuzzy set A, we define its *crisp* counterpart A^{ex} via $\chi_{A^{ex}}(x) = 1$ in case $x \in A$ and $\chi_{A^{ex}}(x) = 0$ in case $x \notin A$. We denote $\chi_{A^{ex}}$ with the symbol δ_A in the sequel.

Proposition 14.23. *[Cattaneo98] For each fuzzy set A in the Brouwer – Zadeh lattice*

$$Fuz(X, [0,1]),$$

the rough approximation $(\underline{A}, \overline{A})$ to A is given as follows

1. $\chi_{\underline{A}} = \delta_{(A)_1}$
2. $\chi_{\overline{A}} = \delta_{A_{>0}}$

Proof. The proof consists in direct verifying that the proposed identities hold. For instance, in case (i), applying recipes for $(.)^{\sim}, (.)'$ to calculate $\chi_{A^{\dagger\dagger}}$, we find that $\chi_{A^{\dagger\dagger}}(x) = 1$ if and only if $\chi_A(x) = 1$ so (i) holds. Similarly, we may verify (ii). $\square$

It follows from the last proposition that an element A of the lattice $Fuz(X, [0,1])$ is *exact* if and only if $\delta_{(A)_1} = \delta_{A_{>0}}$ which is equivalent to $\chi_A \in \{0,1\}^X$ i.e. to A being a crisp (classical) set.

The rough approximation $(\underline{A}, \overline{A})$ admits an interpretation in terms of possibility and necessity: the upper approximation is related to possibility via its characteristic function while the lower approximation may be interpreted as the necessity associated with the fuzzy set A (cf. [Zadeh78], [Dubois–Prade88] for a discussion of these notions.)

It follows from the fact that open elements are closed as well that the induced here rough approximations do observe the general properties of rough approximations discussed in Chapter 10.

Historic remarks

Fuzzy sets were introduced in [Zadeh65]. David Hilbert conjectured that a continuous 3–ary functions need not be any superposition of continuous 2–ary functions which conjecture was disproved in [Arnold63]. In [Kolmogorov63] it was shown that any continuous n–ary function may be represented as a superposition of unary continuous functions and binary addition. For continuous, increasing and associative 2–ary functions, the representation in the form $S(x, y) = f^{-1}(f(x) + f(y))$ with f continuous and increasing was proposed in [Aczél49].

Triangular norms were first studied in [Menger42] in connection with *probabilistic metrics* (cf. [Schweizer – Sklar83]). In addition to ideas presented above, relationships between fuzzy and rough sets were studied, among others, in [Fariñas del Cerro86], [Nakamura88], [Nakamura–Gao91], [Pawlak85], [Pedrycz99], [Pal – Skowron99].

Works quoted

[Aczél49] J. Aczél, *Sur les operations défines pour les nombres réels*, Bull. Soc. Math. France, 76 (1949), pp. 59 – 64.

[Arnold63] V. I. Arnold, *On functions of three variables*, Amer. Math. Soc. Transl., 28 (1963), pp. 51 – 54.

[Caianiello87] E. R. Caianiello, *C – calculus: an overview*, in: E.R. Caianiello and M. A. Aizerman (eds.), *Topics in the General Theory of Structures*,Reidel, Dordrecht, 1987.

[Cattaneo98] G. Cattaneo, *Abstract approximation spaces for rough theories*, in: L. Polkowski and A. Skowron (eds.), *Rough Sets in Knowledge Discovery. Methodology and Applications*, vol. 18 in Studies in Fuzziness and Soft Computing, Physica Verlag, Heidelberg, 1998, pp. 59 – 98.

[Cattaneo97] G. Cattaneo, *Generalized rough sets. Preclusivity fuzzy – intuitionistic (BZ) lattices*, Studia Logica, 58 (1997), pp. 47–77.

[Cattaneo – Nistico89] G. Cattaneo and G. Nisticó, *Brouwer – Zadeh posets and three valued Łukasiewicz posets*, Fuzzy Sets Syst., 33 (1989), pp. 165 – 190.

[Dempster67] A. P. Dempster, *Upper and lower probabilities induced by a multiple – valued mapping*, Annals Math. Stat., 38 (1967), pp. 325 – 339.

[Driankov93] D. Driankov, H. Hellendoorn, and M. Reinfrank, *An Introduction to Fuzzy Control*, Springer Verlag, Berlin, 1993.

[Dubois – Prade92] D. Dubois and H. Prade, *Putting rough sets and fuzzy sets together*, in: R. Słowiński (ed.), *Intelligent Decision Support. Handbook of Applications and Advances of the Rough Sets Theory*, Kluwer, Dordrecht, 1992, pp. 203 – 232.

[Dubois – Prade88] D. Dubois and H. Prade (with coll.), *Possibility Theory: An Approach to Computerized processing of Uncertainty*, Plenum Press, New York, 1988.

[Fariñas del Cerro86] L. Fariñas del Cerro and H. Prade, *Rough sets, twofold fuzzy sets and modal logic – fuzziness in indiscernibility and partial information*, in: A. Di Nola and A. G. S. Ventre (eds.), *The Mathematics of Fuzzy Systems*, Verlag TÜV Rheinland, Köln, 1986.

[Faucett55] W. M. Faucett, *Compact semigroups irreducibly connected between two idempotents*, Proc. Amer. Math. Soc., 6 (1955), pp. 741 – 747.

[Höhle88] U. Höhle, *Quotients with respect to similarity relations*, Fuzzy Sets Syst., 27 (1988), pp. 31 – 44.

[Kolmogorov63] A. N. Kolmogorov, *On the representation of continuous functions of many variables by superposition of continuous functions of one variable and addition*, Amer. Math. Soc. Transl., 28 (1963), pp. 55 – 59.

[Lindenbaum33] A. Lindenbaum, *Sur les ensembles dans lesquels toutes les équations d'une famille donnée ont un nombre de solution fixé d'avance*, Fund. Math., 20 (1933), p. 20.

[Ling65] C. – H. Ling, *Representation of associative functions*, Publ. Math. Debrecen, 12 (1965), pp. 189 – 212.

[Mantaras – Valverde88] L. de Mántaras and L. Valverde, *New results in fuzzy clustering based on the concept of indistinguishability relation*, IEEE Trans. on Pattern Analysis and Machine intelligence, 10 (1988), pp. 754 – 757.

[Menger42] K. Menger, *Statistical metrics*, Proc. Nat. Acad. Sci. USA, 28 (1942), pp. 535 – 537.

[Menu – Pavelka76] J. Menu and J. Pavelka, *A note on tensor products on the unit interval*, 17 (1976), pp. 71 – 83.

[Mostert – Shields57] P. S. Mostert and A. L. Shields, *On the structure of semigroups on a compact manifold with boundary*, Ann. Math., 65 (1957), pp. 117 – 143.

[Nakamura88] A. Nakamura, *Fuzzy rough sets*, Notes on Multiple – Valued Logic in Japan, 9 (1988), pp. 1 – 8.

[Nakamura – Gao91] A. Nakamura and J. M. Gao, *A logic for fuzzy data analysis*, Fuzzy Sets Syst., 39 (1991), pp. 127 – 132.

[Pal – Skowron99] S. K. Pal and A. Skowron, *Rough – Fuzzy Hybridization. A New Trend in Decision – Making*, Springer Verlag, Singapore, 1999.

[Pavelka79[a,b,c]] J. Pavelka, *On fuzzy logic I, II, III* , Zeit. Math. Logik Grund. Math., 25 (1979), pp. 45–52, 119–134, 447–464.

[Pawlak85[c]] Z. Pawlak, *Rough sets and fuzzy sets*, Fuzzy Sets Syst., 17 (1985), pp. 99 – 102.

[Pedrycz99] W. Pedrycz, *Shadowed sets: bringing fuzzy and rough sets*, in: [Pal – Skowron], pp. 179 – 199.

[Ruspini91] E. H. Ruspini, *On the semantics of fuzzy logic*, Int. J. Approx. Reasoning, 5 (1991), pp. 45 – 88.

[Schweizer – Sklar83] B. Schweizer and A. Sklar, *Probabilistic Metric Spaces*, North – Holland, Amsterdam, 1983.

[Shafer76] G. Shafer, *A Mathematical Theory of Evidence*, Princeton U. Press, Princeton N. J., 1976.

[Sierpiński34] W. Sierpiński, *Remarques sur les fonctions de plusieurs variables réelles*, Prace Matematyczno – Fizyczne, 41 (1934), pp. 171 – 175.

[Valverde85] L. Valverde, *On the structure of F – indistinguishability operators*, Fuzzy Sets Syst., 17 (1985), pp. 313 – 328.

[Zadeh78] L. A. Zadeh, *Fuzzy sets as a basis for the theory of possibility*, Fuzzy Sets Syst., 1 (1978), pp. 3 – 28.

[Zadeh71] L. A. Zadeh, *Similarity relations and fuzzy orderings*, Information Sciences, 3 (1971), pp. 177 – 200.

[Zadeh65] L. A. Zadeh, *Fuzzy sets*, Information and Control, 8 (1965), pp. 338 – 353.

Bibliography

Works quoted in the book

[Aczél49] J. Aczél, *Sur les operations défines pour les nombres réels*, Bull. Soc. Math. France, 76 (1949), pp. 59 – 64.

[Alexandrov25] P. S. Alexandrov, *Zur Begründung der n–dimensionalen mengentheoretischen Topologie*, Math. Ann., 94(1925), pp. 296–308.

[Alexandrov–Urysohn29] P. S. Alexandrov and P. S. Urysohn, *Mémoire sur les espaces topologiques compacts*, Verh. Konink. Akad. Amsterdam, 14(1929).

[Archangelsky–Taitslin97] D. A. Archangelsky and M. A. Taitslin, *A logic for information systems*, Studia Logica, 58 (1997), pp. 3–16.

[Arnold63] V. I. Arnold, *On functions of three variables*, Amer. Math. Soc. Transl., 28 (1963), pp. 51 – 54.

[Baire899] R. Baire, Ann. di Math., 3(1899).

[Balcar– Štěpánek86] B. Balcar and P. Štěpánek, *Teorie Množin*, Academia, Praha, 1986.

[Banach22] S. Banach, *Sur les opérations dans les ensembles abstraits et leur application aux équations intégrales*, Fund. Math., 3(1922), pp. 133–181.

[Banerjee–Chakraborty98] M. Banerjee and M. K. Chakraborty, *Rough logics: a survey with further directions*, [Orlowska98], pp. 579–600.

[Barnsley88] M. F. Barnsley, *Fractals Everywhere*, Academic Press, 1988.

[Bazan00] J. Bazan, H.S. Nguyen, S. H. Nguyen, P. Synak, and J. Wróblewski, *Rough set algorithms in classification problems*, in: [Polkowski–Tsumoto–Lin], pp. 49–88.

[Bazan98] J.G. Bazan, Nguyen Hung Son, Nguyen Tuan Trung, A. Skowron, and J. Stepaniuk , *Decision rules synthesis for object classification*, in: [Orłowska98], pp. 23–57.

[Becchio78] D. Becchio, *Logique trivalente de Lukasiewicz*, Ann. Sci. Univ. Clermont–Ferrand, 16(1978), pp. 38–89.

[Becchio72] D. Becchio, *Nouvelle démonstration de la complétude du systeme de Wajsberg axiomatisant la logique trivalente de Lukasiewicz*, C.R. Acad. Sci. Paris, 275(1972), pp. 679–681.

[Bernays26] P. Bernays, *Axiomatische Untersuchung den Aussagenkalküls der "Principia Mathematica"*, Mathematische Zeitschrift, 25 (1926).

[Birkhoff67] G. Birkhoff, *Lattice Theory*, AMS, Providence, 1940 (3rd ed. 1967).

[Bocheński61] I. M. Bocheński, *A History of Formal Logic*, Notre Dame Univ. Press, 1961.

[Boicescu91] V. Boicescu, A. Filipoiu, G. Georgescu, and S. Rudeanu, *Lukasiewicz–Moisil Algebras*, North Holland, Amsterdam, 1991.

[Boole847] G. Boole, *The Mathematical Analysis of Logic*, Cambridge, 1847.

[Borel898] E. Borel, *Leçons sur la Théorie des Fonctions*, Paris, 1898.

[Borkowski70] *Jan Lukasiewicz. Selected Works*, L. Borkowski (ed.), North Holland–Polish Scientific Publishers, Amsterdam–Warsaw, 1970.

[Brouwer08] L. E. J. Brouwer, *De onbetrouwbaarheid der logische principes*, Tijdschrift voor wijsbegeerte 2(1908), pp. 152–158.

[Buszkowski–Orłowska86] W. Buszkowski and E. Orłowska, *On the logic of database dependencies*, Bull. Polish Acad. Sci. Math., 34(1986), pp. 345–354.

[Caianiello87] E. R. Caianiello, *C – calculus: an overview*, in: E.R. Caianiello and M. A. Aizerman (eds.), *Topics in the General Theory of Structures*,Reidel, Dordrecht, 1987.

[Cantor62] G. Cantor, *Gesammelte Abhandlungen mathematischen und philosophischen Inhalts*, Hildesheim, 1962.

[Cantor883] G. Cantor, Math. Ann., 21(1883).

[Cantor880] G. Cantor, Math. Ann., 17(1880).

[Carathéodory14] C.Carathéodory, *Über das lineare Mass von Punktmenge eine Verallgemeinerung des Längenbegriffs*, Nach. Gesell. Wiss. Göttingen, 1914, pp. 406-426.

[Cattaneo98] G. Cattaneo, *Abstract approximation spaces for rough theories*, in: [Orlowska98], pp. 59–98.

[Cattaneo97] G. Cattaneo, *Generalized rough sets. Preclusivity fuzzy–intuitionistic (BZ) lattices*, Studia Logica, 58 (1997), pp. 47–77.

[Cattaneo – Nistico89] G. Cattaneo and G. Nisticó, *Brouwer – Zadeh posets and three valued Lukasiewicz posets*, Fuzzy Sets Syst., 33 (1989), pp. 165 – 190.

[Chang59] C. C. Chang, *A new proof of the completeness of the Lukasiewicz axioms*, Trans. Amer. Math. Soc., 93 (1959), pp. 74–80.

[Chang58^a] C. C. Chang, *Proof of an axiom of Lukasiewicz*, Trans. Amer. Math. Soc., 87 (1958), pp. 55–56.

[Chang58^b] C. C. Chang, *Algebraic analysis of many–valued logics*, Trans. Amer. Math. Soc., 88 (1958), pp. 467–490.

[Cignoli93] R. Cignoli, *Free lattice-ordered abelian groups and varieties of MV – algebras*, Notás de Lógica Matemática, Univ. Nac. del Sur, 38 (1993), pp. 113–118.

[Cignoli69] R. Cignoli, *Algebras de Moisil de orden n*, Doctoral Thesis, Univ. Nacional del Sur, Bahia Blanca, Brasil, 1969.

[Cignoli–Mundici97] R. Cignoli and D. Mundici, *An elementary proof of Chang's completeness theorem for the infinite-value calculus of Lukasiewicz*, Studia Logica, 58 (1997), pp. 79–97.

[Comer93] S. Comer, *On connections between information systems, rough sets and algebraic logic*, in: *Algebraic Methods in Logic and Computer Sci-*

ence, Banach Center Publ., 28, Warszawa, 1993.

[Čech66] E. Čech, *Topologicke' prostory*, in: E. Čech, *Topological Spaces*, Academia, Praha, 1966.

[Dedekind881] R. Dedekind, *Was sind und was sollen die Zahlen*, Braunschweig, 1881.

[Dempster67] A. P. Dempster, *Upper and lower probabilities induced by a multiple – valued mapping*, Annals Math. Stat., 38 (1967), pp. 325 – 339.

[Driankov93] D. Driankov, H. Hellendoorn, and M. Reinfrank, *An Introduction to Fuzzy Control*, Springer Verlag, Berlin, 1993.

[Dubois – Prade92] D. Dubois and H. Prade, *Putting rough sets and fuzzy sets together*, in: R. Słowiński (ed.), *Intelligent Decision Support. Handbook of Applications and Advances of the Rough Sets Theory*, Kluwer, Dordrecht, 1992, pp. 203 – 232.

[Dubois – Prade88] D. Dubois and H. Prade (with coll.), *Possibility Theory: An Approach to Computerized processing of Uncertainty*, Plenum Press, New York, 1988.

[Duentsch00] I. Duentsch, *Logical and algebraic techniques for rough set data analysis*, in: [Polkowski–Tsumoto–Lin00], pp. 521–544.

[Duentsch98] I. Duentsch, *Rough sets and algebras of relations*, in: [Orlowska98], pp. 95–108.

[Duentsch94] I. Duentsch, *Rough relation algebras*, Fundamenta Informaticae, 21(1994), pp. 321–331.

[Epstein60] G. Epstein, *The lattice theory of Post algebras*, Trans. Amer. Math. Soc., 95 (1960), 300–317.

[Falconer90^a] K. J. Falconer, *The Geometry of Fractal Sets*, Cambridge U. Press, 1990.

[Falconer90^b] K. J. Falconer, *Fractal Geometry. Mathematical Foundations and Applications*, Wiley and Sons, 1990.

[Fariñas del Cerro86] L. Fariñas del Cerro and H. Prade, *Rough sets, twofold fuzzy sets and modal logic – fuzziness in indiscernibility and partial information*, in: A. Di Nola and A. G. S. Ventre (eds.), *The Mathematics of Fuzzy*

Systems, Verlag TÜV Rheinland, Köln, 1986.

[Faucett55] W. M. Faucett, *Compact semigroups irreducibly connected between two idempotents*, Proc. Amer. Math. Soc., 6 (1955), pp. 741 – 747.

[Feys37] R. Feys, *Les logiques nouvelles des modalités*, Revue Néoscholastique de Philosophie, 40(1937), pp, 517–553; 41(1937), pp. 217–252.

[Font84] J. M. Font, A. J. Rodriguez, and A. Torrens, *Wajsberg algebras*, Stochastica, 8 (1984), pp. 5–31.

[Fréchet28] M. Fréchet, *Les espaces abstraits*, Paris, 1928.

[Fréchet06] M. Fréchet, *Sur quelques points du Calcul fonctionnel*, Rend. Circ. Matem. di Palermo, 22(1906).

[Frege03] G. Frege, *Grundgesetze der Arithmetik 2*, Jena, 1903.

[Frege874] G. Frege, *Begriffsschrift, eine der mathematischen nachgebildete Formelsprache des Reinen Denkens*, Halle, 1874.

[Gentzen34] G. Gentzen, *Untersuchungen über das logische Schliessen. I, II.*, Mathematische Zeitschrift 39 (1934–5), pp. 176–210, 405–431.

[Głazek79] K. Głazek, *Some old and new problems of independence in mathematics*, Coll. Math., 17(1979), pp. 127–189.

[Goldberg89] D.E. Goldberg, *GA in Search, Optimisation, and Machine Learning*, Addison–Wesley, 1989.

[Goldberg–Leblanc–Weaver74] H. Goldberg, H. Leblanc, and G. Weaver, *A strong completeness theorem for 3–valued logic*, Notre Dame J. Formal Logic, 15(1974), 325–332.

[Goguen69] J. A. Goguen, *The logic of inexact concepts*, Synthese, 18/19 (1968/ 1969), pp. 325–373.

[Goguen67] J. A. Goguen, *L–fuzzy sets*, J. Math. Anal. Appl., 18 (1967), pp. 145–174.

[Gödel30] K. Gödel, *Die Vollständigkeit der Axiome des Logischen Funktionenkalküls*, Monats. Math. Phys., 37(1930), pp. 349–360.

[Greco99] S. Greco, B. Matarazzo, R. Słowiński, *Fuzzy dominance as basis for rough approximations*, in: Proceedings: the 4th Meeting of the EURO WG on Fuzzy Sets and 2nd Internat. Conf. on Soft and Intelligent Computing, (EUROFUSE-SIC'99), Budapest, Hungary, May 1999, pp. 273-278.

[Greco98] S. Greco, B. Matarazzo, and R. Słowiński, *On joint use of indiscernibility, similarity and dominance in rough approximation of decision classes*, in: Proceedings: the 5th International Conference of the Decision Sciences Institute, Athens, Greece, July 1999, pp. 1380–1382; also in: Research Report RA–012/98, Inst. Comp. Sci., Poznań Univ. Technology, 1998.

[Grzymala–Busse86] J. Grzymala–Busse, *On the reduction of knowledge representation systems*, in: Proceedings of the 6th Intern. Workshop on Expert Systems and Appl., Avignon, France, 1986, vol. 1, pp. 463–478.

[Hahn32] H. Hahn, *Reelle Funktionen I*, Leipzig, 1932.

[Hajek97] P. Hajek, *Fuzzy logic and arithmetical hierarchy II*, Studia Logica, 58 (1997), pp. 129–141.

[Hasenjaeger53] G. Hasenjaeger, *Eine Bemerkung zu Henkin's Beweis füer die Vollständigkeit des Prädikatenkalküls des Ersten Stufe*, J. Symb. Logic, 18(1953), pp. 42–48.

[Hausdorff19] F.Hausdorff, *Dimension und äusseres Mass*, Math. Annalen, 79(1919), pp. 157-179.

[Hausdorff14] F. Hausdorff, *Grundzüge der Mengenlehre*, Leipzig, 1914.

[Henkin49] L. Henkin, *The completeness of the first–order functional calculus*, J. Symb. Logic, 14(1949), pp. 159–166.

[Herbrand30] J. Herbrand, *Recherches sur la theórie de la démonstration*, Travaux de la Soc.Sci. Lettr. de Varsovie, III, 33 (1930), pp. 33–160.

[Heyting56] A. Heyting, *Intuitionism, an Introduction*, North Holland, Amsterdam, 1956.

[Höhle88] U. Höhle, *Quotients with respect to similarity relations*, Fuzzy Sets Syst., 27 (1988), pp. 31 – 44.

[Hughes–Creswell84] G. E. Hughes and M. J. Creswell, *A Companion to Modal Logic*, Methuen, London, 1984.

[Hughes–Creswell72] G. E. Hughes and M. J. Creswell, *An Introduction to Modal Logic*, Methuen, London, 1972.

[Hurewicz–Wallman41] W.Hurewicz and H. Wallman, *Dimension Theory*, Princeton U. Press, 1941.

[Hutchinson81] J. E. Hutchinson, *Fractals and self-similarity*, Indiana Math. Journal, 30(1981), pp. 713–747.

[Iturrioz77] L. Iturrioz, *Lukasiewicz and symmetrical Heyting algebras*, Zeit. Math. Logik u. Grundl. Math., 23(1977), pp. 131–136.

[Iwinski88] T. B. Iwiński, *Rough orders and rough set concepts*, Bull. Polish Acad. Ser. Sci. Math., 37 (1988), pp. 187–192.

[Iwinski87] T. B. Iwiński, *Algebraic approach to rough sets*, Bull. Polish Acad. Ser. Sci. Math., 35 (1987), pp. 673–683.

[Kalmár34] L. Kalmár, *Über die Axiomatisierbarkeit des Aussagenkalküls*, Acta Scientiarum Mathematicarum, 7 (1934–5), pp.222–243.

[Kanger57] S. Kanger, *Provability in Logic*, Acta Universitatis Stockholmiensis, Stockholm Studies in Philosophy I, 1957.

[Knaster28] B. Knaster, *Un théoréme sur les fonctions d'ensembles*, Ann. Soc. Polon. Math., 6(1928), pp. 133–134.

[Kolmogorov63] A. N. Kolmogorov, *On the representation of continuous functions of many variables by superposition of continuous functions of one variable and addition*, Amer. Math. Soc. Transl., 28 (1963), pp. 55 – 59.

[König27] D. König, *Über eine Schlussweise aus dem Endlichen ins Unendliche*, Acta litt. ac sc. univ. Franc. Josephinae, Sec. Sc. Math., 3(1927), pp. 121–130.

[Kripke63] S. A. Kripke, *Semantical analysis of modal logic I, normal propositional calculi*, Zeit. Math. Logik, 9(1963), pp. 67–96.

[Kuratowski22] C. Kuratowski, *Sur l'operation $\overline{A}$ de l'Analysis Situs*, Fund. Math., 3(1922), pp.182–199.

[Kuratowski–Mostowski65] K. Kuratowski and A. Mostowski, *Set Theory*, Polish Scientific Publ., Warsaw, 1965.

[Lebesgue05] H. Lebesgue, J. de Math., 6(1905).

[Lemmon–Scott63] E. J. Lemmon and D. S. Scott, *The "Lemmon Notes":
An Introduction to Modal Logic*, K. Segerberg, ed., Blackwell, Oxford, 1963.

[Lewis–Langford59] C. I. Lewis and C. H. Langford, *Symbolic Logic*, Dover,
New York, 1959 (2nd ed.).

[Lindenbaum33] A. Lindenbaum, *Sur les ensembles dans lesquels toutes les
équations d'une famille donnée ont un nombre de solution fixé d'avance*,
Fund. Math., 20 (1933), p. 20.

[Ling65] C. – H. Ling, *Representation of associative functions*, Publ. Math.
Debrecen, 12 (1965), pp. 189 – 212.

[Luxenburger98] M. Luxenburger, *Dependencies between many–valued attributes*, in: [Orłowska 98], pp. 316–346.

[Łukasiewicz70] J. Łukasiewicz, *On the history of the logic of propositions*,
in: [Borkowski70], pp. 197–217].

[Łukasiewicz63] J. Łukasiewicz, *Elements of Mathematical Logic*, Pergamon
Press – Polish Scientific Publishers, Oxford – Warsaw, 1963.

[Łukasiewicz57] Jan Łukasiewicz, *Aristotle's Syllogistic from the Standpoint
of Modern Formal Logic*, 2nd ed., Oxford, 1957.

[Łukasiewicz53] J. Łukasiewicz, *A system of modal logic*, The Journal of Computing Systems, 1(1953), 111–149.

[Łukasiewicz39] Jan Łukasiewicz, *On Aristotle's Syllogistic* (in Polish), Compt.
Rend. Acad. Polon. Lettr., Cracovie, 44 (1939).

[Łukasiewicz30[a]] J. Łukasiewicz, *Philosophische Bemerkungen zu mehrwertige Systemen des Aussagenkalkuls*, C. R. Soc. Sci. Lettr. Varsovie, 23(1930),
51–77 [English translation in [Borkowski70], pp. 153–178].

[Łukasiewicz30[b]] J. Łukasiewicz and A. Tarski, *Untersuchungen üeber den
Aussagenkalküls*, C. R. Soc. Sci. Lettr. Varsovie, 23(1930), 39–50 [English
translation in [Borkowski70], pp. 130–152].

[Łukasiewicz20] J. Łukasiewicz, *On three–valued logic* (in Polish), Ruch Filozoficzny 5(1920), 170–171 [English translation in [Borkowski70], pp. 87–88].

[Łukasiewicz18] J. Łukasiewicz, *Farewell Lecture by Professor Jan Łukasiewicz* (delivered in the Warsaw University Lecture Hall on March 7, 1918) [English translation in [Borkowski70], pp. 84–86].

[Łukasiewicz13] J. Łukasiewicz, *Die Logischen Grundlagen der Wahrscheinlichkeitsrechnung*, Cracow, 1913 [English translation in: [Borkowski70], pp. 16–63].

[Mac Neille37] H. M. Mac Neille, *Partially ordered sets*, Trans. Amer. Math. Soc., 42(1937), pp. 416–460.

[Mandelbrot75] B. Mandelbrot, *Les Objects Fractals: Forme, Hasard et Dimension*, Flammarion, Paris, 1975.

[Mantaras – Valverde88] L. de Mántaras and L. Valverde, *New results in fuzzy clustering based on the concept of indistinguishability relation*, IEEE Trans. on Pattern Analysis and Machine intelligence, 10 (1988), pp. 754 – 757.

[Marcus94] S. Marcus, *Tolerance rough sets, Čech topologies, learning processes*, Bull. Polish Acad. Sci. Tech., 42 (1994), pp. 471–487.

[Marczewski58] E. Marczewski, *A general scheme of independence in mathematics*, Bull. Polish Acad. Math. Sci., 6(1958), pp. 731–736.

[Marek–Rasiowa86] W. Marek and H. Rasiowa, *Approximating sets with equivalence relations*, Theor. Computer Sci. 48(1986), pp. 145–152.

[McKinsey–Tarski44] J. C. C. McKinsey and A. Tarski, *The algebra of topology*, Annals of Mathematics, 45(1944), pp. 141–191.

[McNaughton51] R. McNaughton, *A theorem about infinite–valued sentential logic*, J. Symbolic Logic, 16 (1951), pp. 1–13.

[Menger42] K. Menger, *Statistical metrics*, Proc. Natl. Acad. Sci. USA, 28 (1942), pp. 535 – 537.

[Menu – Pavelka76] J. Menu and J. Pavelka, *A note on tensor products on the unit interval*, 17 (1976), pp. 71 – 83.

[Meredith58] C. A. Meredith, *The dependence of an axiom of Łukasiewicz*, Trans. Amer. Math. Soc., 87 (1958), p. 54.

[Mitchell98] T. Mitchell, *Machine Learning*, McGraw–Hill, Boston, 1998.

[Moisil64] Gr. C. Moisil, *Sur les logiques de Lukasiewicz á un nombre fini de valeurs*, Rev. Roumaine Math. Pures Appl., 9 (1964), pp. 905–920, 583–595.

[Moisil63] Gr. C. Moisil, *Les logiques non–chrysippiennes et leurs applications*, Acta Phil. Fennica, 16 (1963), pp. 137–152.

[Moisil60] Gr. C. Moisil, *Sur les idéaux des algébres lukasiewicziennes trivalentes*, An. Univ. C. I. Parhon, Acta logica, 3 (1960), pp. 83–95, 244–258.

[Moisil42] Gr. C. Moisil, *Logique modale*, Disquisitiones Math. Phys., 2 (1942), pp. 3–98, 217–328, 341–441.

[Mostert – Shields57] P. S. Mostert and A. L. Shields, *On the structure of semigroups on a compact manifold with boundary*, Ann. Math., 65 (1957), pp. 117 – 143.

[Nakamura98] A. Nakamura, *Graded modalities in rough logic*, in: L. Polkowski and A. Skowron (eds.), *Rough Sets in Knowledge Discovery. Methodology and Applications*, Studies in Fuzziness and Soft Computing, vol. 18, Physica Verlag, Heidelberg, 1998, pp. 192–208.

[Nakamura88] A. Nakamura, *Fuzzy rough sets*, Notes on Multiple – Valued Logic in Japan, 9 (1988), pp. 1 – 8.

[Nakamura – Gao91] A. Nakamura and J. M. Gao, *A logic for fuzzy data analysis*, Fuzzy Sets Syst., 39 (1991), pp. 127 – 132.

[Nelson49] D. Nelson, *Constructible falsity*, The Journal of Symbolic Logic, 14 (1949), pp. 16–26.

[Nguyen Hung Son98] Nguyen Hung Son, *From optimal hyperplanes to optimal decision trees*, Fundamenta Informaticae, 34(1-2) (1998), pp. 145–174.

[Nguyen–Nguyen98^a] Nguyen Sinh Hoa and Nguyen Hung Son, *Pattern extraction from data*, Fundamenta Informaticae, 34(1-2) (1998), pp. 129–144.

[Nguyen–Nguyen98^b] Nguyen Hung Son and Nguyen Sinh Hoa, *Discretization methods in Data Mining*, in: [Polkowski –Skowron], pp. 451–482.

[Nguyen Sinh Hoa00] Nguyen Sinh Hoa, *Regularity analysis and its applications in Data Mining*, in: [Polkowski–Tsumoto–Lin] pp. 289–378.

[Nguyen–Skowron99] Nguyen Hung Son and A. Skowron, *Boolean reasoning scheme with some applications in Data Mining*, in: Proceedings: Principles of Data Mining and Knowledge Discovery PKDD'99, Prague, Czech Republic, September 1999, LNAI vol. 1704, Springer Verlag, Berlin, 1999, pp. 107–115.

[Novák90] V. Novák, *On the syntactico–semantical completeness of first-order fuzzy logic*, Kybernetika, 2 (1990), Part I pp. 47–62, Part II pp. 134–152.

[Novák87] V. Novák, *First-order fuzzy logic*, Studia Logica, 46 (1987), pp. 87–109.

[Novotný98^a] M. Novotný, *Dependence spaces of information systems*, in [Orłowska98], pp. 193–246.

[Novotný98^b] M. Novotný, *Applications of dependence spaces*, in [Orłowska 98], pp. 247–289.

[Novotný83] M. Novotný, *Remarks on sequents defined by means of information systems*, Fund. Inform., 6(1983), pp. 71–79.

[Novotný–Pawlak92] M. Novotný and Z. Pawlak, *On a problem concerning dependence spaces*, Fund. Inform., 16(1992), pp. 275–287.

[Novotný–Pawlak91] M. Novotný and Z. Pawlak, *Algebraic theory of independence in information systems*, Fund. Inform., 14(1991), pp. 454–476.

[Novotný–Pawlak90] M. Novotný and Z. Pawlak, *On superreducts*, Bull. Polish Acad. Sci. Tech., 38(1990), pp. 101–112.

[Novotný–Pawlak89] M. Novotný and Z. Pawlak, *Algebraic theory of independence in information systems*, Report 51, Institute of Mathematics of the Czechoslovak Academy of Sciences, 1989.

[Novotný–Pawlak88^a] M. Novotný and Z. Pawlak, *Partial dependency of attributes*, Bull. Polish Acad. Sci. Math., 36 (1989), pp. 453–458.

[Novotný–Pawlak88^b] M. Novotný and Z. Pawlak, *Independence of attributes*, Bull. Polish Acad. Sci. Math., 36(1988), pp. 459–465.

[Novotný–Pawlak87] M. Novotný and Z. Pawlak, *Concept forming and black boxes*, Bull. Polish Acad. Sci. Math., 35(1987), pp. 133–141.

[Novotný–Pawlak85^a] M. Novotný and Z. Pawlak, *Characterization of rough top equalities and rough bottom equalities*, Bull. Polish Acad. Sci. Math., 33

(1985), pp. 91–97.

[Novotný–Pawlak85^b] M. Novotný and Z. Pawlak, *On rough equalities*, Bull. Polish Acad. Sci. Math., 33 (1985), pp. 99-104.

[Novotný–Pawlak85^c] M. Novotný and Z. Pawlak, *Black box analysis and rough top equality*, Bull. Polish Acad. Sci. Math., 33 (1985), pp. 105-113.

[Novotný–Pawlak85^d] M. Novotný and Z. Pawlak, *Independence of attributes*, Bull. Polish Acad. Sci. Tech., 33 (1985), pp. 459-465.

[Novotný–Pawlak83] M. Novotný and Z. Pawlak, *On a representation of rough sets by means of information systems*, Fund. Inform., 6(1983), pp. 289-296.

[Obtulowicz85] A. Obtułowicz, *Rough sets and Heyting algebra valued sets*, Bull. Polish Acad. Sci. Math., 33(1985), pp. 454-476.

[Orłowska98] E. Orłowska, ed., *Incomplete Information: Rough Set Analysis*, Studies in Fuzziness and Soft Computing, vol. 13, Physica Verlag, Heidelberg, 1998.

[Orlowska90] E. Orłowska, *Kripke semantics for knowledge representation*, Studia Logica, 49 (1990), pp. 255-272.

[Orlowska89] E. Orłowska, *Logic for reasoning about knowledge*, Z. Math. Logik u. Grund. d. Math., 35(1989), pp. 559-572.

[Orlowska85] E. Orłowska, *Logic approach to information systems*, Fundamenta Informaticae, 8 (1985), pp. 359-378.

[Orlowska84] E. Orłowska, *Modal logics in the theory of information systems*, Z. Math. Logik u. Grund.d. Math., 30(1984), pp. 213-222.

[Orłowska83] E. Orłowska, *Dependencies of attributes in Pawlak's information systems*, Fund. Inform., 6(1983), pp. 247-256.

[Orlowska–Pawlak84^a] E. Orłowska and Z. Pawlak, *Logical foundations of knowledge representation*, Reports of the Comp. Centre of the Polish Academy of Sciences, 537, 1984.

[Orlowska–Pawlak84^b] E. Orłowska and Z. Pawlak, *Representation of non-deterministic information*, Theor. Computer Science, 29 (1984), pp. 27-39.

[Pagliani98^a] P. Pagliani, *Rough set theory and logic–algebraic structures*, in: [Orlowska98], pp. 109–192.

[Pagliani98^b] P. Pagliani, *A practical introduction to the modal–relational approach to approximation spaces*, in: [Polkowski–Skowron98^a], pp. 209–232.

[Pagliani96] P. Pagliani, *Rough sets and Nelson algebras*, Fundamenta Informaticae, 27(1996), pp. 205–219.

[Pal – Skowron99] S. K. Pal and A. Skowron, *Rough – Fuzzy Hybridization. A New Trend in Decision – Making*, Springer Verlag, Singapore, 1999.

[Panti95] G. Panti, *A geometric proof of the completeness of the calculus of Łukasiewicz*, J. Symbolic Logic, 60 (1995), pp. 563–578.

[Pavelka79^a] J. Pavelka, *On fuzzy logic I*, Zeit. Math. Logik Grund. Math., 25 (1979), pp. 45–52.

[Pavelka79^b] J. Pavelka, *On fuzzy logic II*, Zeit. Math. Logik Grund. Math., 25 (1979), pp. 119–134.

[Pavelka79^c] J. Pavelka, *On fuzzy logic III*, Zeit. Math. Logik Grund. Math., 25 (1979), pp. 447–464.

[Pawlak01] Z. Pawlak, *Combining rough sets and Bayes' rule*, Computational Intelligence: An Intern. Journal, 17, 2001, pp. 401–408.

[Pawlak91] Z. Pawlak, *Rough Sets. Theoretical Aspects of Reasoning about Data*, Kluwer, Dordrecht, 1991.

[Pawlak87^a] Z. Pawlak, *Decision tables–a rough set approach*, Bull. EATCS, 33 (1987), pp. 85–96.

[Pawlak87^b] Z. Pawlak, *Rough logic*, Bull. Polish Acad. Sci. Tech., 35 (1987), pp. 253–258.

[Pawlak86] Z. Pawlak, *On decision tables*, Bull. Polish Acad. Sci. Tech., 34 (1986), pp. 553–572.

[Pawlak85^a] Z. Pawlak, *On rough dependency of attributes in information systems*, Bull. Polish Acad. Sci. Tech., 33 (1985), pp. 551–559.

[Pawlak85^b] Z. Pawlak, *Rough sets and decision tables*, LNCS vol. 208, Springer Verlag, Berlin, 1985, pp. 186–196.

[Pawlak85^c] Z. Pawlak, *Rough sets and fuzzy sets*, Fuzzy Sets Syst., 17 (1985), pp. 99 – 102.

[Pawlak83] Z. Pawlak, *Rough classification*, Reports of the Computing Centre of the Polish Academy of Sciences, 506, Warsaw, 1983.

[Pawlak82^a] Z. Pawlak, *Rough sets*, Intern. J. Comp. Inform. Sci., 11 (1982), pp. 341–356.

[Pawlak82^b] Z. Pawlak, *Rough sets, algebraic and topological approach*, Int. J. Inform. Comp. Sciences, 11(1982), pp. 341–366.

[Pawlak81^a] Z. Pawlak, *Information Systems–Theoretical Foundations* (in Polish), PWN–Polish Scientific Publishers, Warsaw, 1981.

[Pawlak81^b] Z. Pawlak, *Information systems–theoretical foundations*, Information Systems, 6 (1981), pp. 205–218.

[Pawlak81] Z. Pawlak, *Information systems-theoretical foundations*, Inform. Systems, 6(1981), pp. 205–218.

[Pawlak–Rauszer85] Z. Pawlak and C. Rauszer, *Dependency of attributes in information systems*, Bull. Polish Acad. Sci. Math., 33 (1985), pp. 551–559.

[Pawlak–Skowron94] Z. Pawlak and A. Skowron, *Rough membership functions*, in: R.R. Yaeger, M. Fedrizzi, and J. Kacprzyk, eds., *Advances in the Dempster–Schafer Theory of Evidence*, Wiley, New York, 1994, pp. 251–271.

[Pedrycz99] W. Pedrycz, *Shadowed sets: bringing fuzzy and rough sets*, in: [Pal – Skowron], pp. 179 – 199.

[Poincaré05] H. Poincaré, *Science et Hypothèse*, Paris, 1905.

[Polkowski01] L. Polkowski, *On fractals defined in information systems via rough set theory*, in: Proceedings *RSTGC-2001*, Bulletin Intern. Rough Set Society 5(1/2)(2001), pp. 163–166.

[Polkowski99] L. Polkowski, *Approximation mathematical morphology*, in: S. K. Pal, A. Skowron (eds.), *Rough Fuzzy Hybridization. A New Trend in Decision Making*, Springer Verlag Singapore, 1999, pp. 151–162.

[Polkowski98] L. Polkowski, *Hit –or–miss topology*, in: *Encyclopaedia of Mathematics, Supplement 1*, Kluwer, Dordrecht, 1998, p.293.

[Polkowski94] L. Polkowski, *Concerning mathematical morphology of almost rough sets*, Bull. Polish Acad. Sci. Tech., 42(1994), pp. 141–152.

[Polkowski93^a] L. Polkowski, *Metric spaces of topological rough sets from countable knowledge bases*, Foundations of Computing and Decision Sciences, 18(1993), pp. 293–306.

[Polkowski93^b] L. Polkowski, *Mathematical morphology of rough sets*, Bull. Polish Acad. Sci. Math., 41(1993), pp. 241–273.

[Polkowski92] L. Polkowski, *On convergence of rough sets*, in: R. Słowiński (ed.), *Intelligent Decision Support. Handbook of Applications and Advances of the Rough Sets Theory*, Kluwer, Dordrecht, 1992, pp. 305–311.

[Polkowski–Polkowski00] L. Polkowski and M. Semeniuk–Polkowska, *Towards usage of natural language in approximate computation: a granular semantics employing formal languages over mereological granules of knowledge*, Scheda Informaticae (Sci. Fasc. Jagiellonian University), 10 (2000), pp. 131–146.

[Polkowski–Skowron01] L. Polkowski and A. Skowron, *Rough mereological calculi of granules: a rough set approach to computation*, Computational Intelligence: An Intern. Journal, 17 (2001), pp. 472–492.

[Polkowski–Skowron98^a] L. Polkowski and A. Skowron, *Rough Sets in Knowledge Discovery 1. Methodology and Applications*, Physica Verlag, Heidelberg, 1998.

[Polkowski–Skowron98^b] L. Polkowski and A. Skowron (eds.), *Rough Sets in Knowledge Discovery 2. Applications, Case Studies and Software Systems*, Studies in Fuzziness and Soft Computing, vol. 19, Physica Verlag, Heidelberg, 1998.

[Polkowski–Skowron–Zytkow94] L. Polkowski, A. Skowron, and J. Zytkow, *Tolerance based rough sets*, in: T. Y. Lin and M. Wildberger (eds.), *Soft Computing: Rough Sets, Fuzzy Logic, Neural Networks, Uncertainty Management, Knowledge Discovery*, Simulation Councils, Inc., San Diego, 1995, pp. 55–58.

[Polkowski–Tsumoto–Lin00] L. Polkowski, S. Tsumoto, and T. Y. Lin, eds., *Rough Set Methods and Applications. New Developments in Knowledge Discovery in Information Systems*, Studies in Fuzzines and Soft Computing, vol. 56, Physica Verlag, Heidelberg, 2000.

[Pompéju05] D. Pompéju, Ann. de Toulouse, 7(1905).

[Pomykala88] J. Pomykała and J. A. Pomykała, *The Stone algebra of rough sets*, Bull. Polish Acad. Ser. Sci. Math., 36 (1988), 495–508.

[Post21] E. Post, *Introduction to a general theory of elementary propositions*, Amer. J. Math., 43(1921), 163–185.

[Ramsey30] F. P. Ramsey, *On a problem of formal logic*, Proc. London. Math. Soc., 30(1930), pp. 264–286.

[Rasiowa74] H. Rasiowa, *An Algebraic Approach to Non–Classical Logics*, North Holland, 1974.

[Rasiowa53] H. Rasiowa, *On satisfiability and deducibility in non–classical functional calculi*, Bull. Polish Acad.Sci. Math., (Cl.III), 1(1953), pp. 229–231.

[Rasiowa51] H. Rasiowa, *A proof of the Skolem–Löwenheim theorem*, Fund. Math., 38 (1951), pp. 230–232.

[Rasiowa–Sikorski63] H. Rasiowa and R. Sikorski, *The Mathematics of Metamathematics*, PWN-Polish Scientific Publishers, Warszawa, 1963.

[Rasiowa–Sikorski50] H. Rasiowa and R. Sikorski, *A proof of the completeness theorem of Gödel*, Fund. Math., 37(1950), pp. 193–200.

[Rasiowa–Skowron86^a] H. Rasiowa and A. Skowron, *Rough concept logic*, LNCS vol. 208, Springer Verlag, Berlin, 1986, pp. 288–297.

[Rasiowa–Skowron86^b] H. Rasiowa and A. Skowron, *The first step towards an approximation logic*, J. Symbolic Logic, 51 (1986), p. 509.

[Rasiowa–Skowron86^c] H. Rasiowa and A. Skowron, *Approximation logic*, Proc. Conf. on Mathematical Methods of Specification and Synthesis of Software Systems, Akademie Verlag, Berlin, 1986, pp. 123–139.

[Rasiowa–Skowron84] H. Rasiowa and A. Skowron, *A rough concept logic*, in: A. Skowron (ed.), *Proc. the 5th Symposium on Comp. Theory*, Lecture Notes in Computer Science, vol. 208 (1984), pp. 197–227.

[Rauszer91] C. M. Rauszer, *Reducts in information systems*, Fund. Inform., 15(1991), pp. 1–12.

[Rauszer88] C. M. Rauszer, *Algebraic properties of functional dependencies*, Bull. Polish Acad. Sci. Math., 36(1988), pp. 561–569.

[Rauszer87] C. M. Rauszer, *Algebraic and logical description of functional and multi–valued dependencies*, Proceedings ISMIS'87, Charlotte, NC, North Holland, Amsterdam, 1987, pp. 145–155.

[Rauszer85^a] C. M. Rauszer, *Dependency of attributes in information systems*, Bull. Polish Acad. Sci. Math., 33(1985), pp. 551–559.

[Rauszer85^b] C. M. Rauszer, *An equivalence between theory of functional dependencies and a fragment of intuitionistic logic*, Bull. Polish Acad. Sci. Math., 33(1985), pp. 571–579.

[Rauszer84] C. M. Rauszer, *An equivalence between indiscernibility relations in information systems and a fragment of intuitionistic logic*, in: *Lecture Notes in Computer Science*, vol. 208, Springer Verlag, Berlin, 1984, pp. 298–317.

[Riesz09] F. Riesz, *Stetigskeitbegriff und abstrakte Mengenlehre*, Atti IV Congr. Int. Mat., Rome, 1909.

[Rissanen83] J. Rissanen, *A universal prior for integers and estimation by minimum description length*, The Annals of Statistics, 11 (1983), pp. 416–431.

[Rosenbloom42] P. Rosenbloom, *Post algebras.I. Postulates and general theory*, Amer. J. Math., 64 (1942), pp. 167–183.

[Rose–Rosser58] A. Rose and J. B. Rosser, *Fragments of many–valued statement calculi*, Trans. Amer. Math. Soc., 87 (1958), pp. 1–53.

[Rosser – Turquette58] J. B. Rosser and A. R. Turquette, *Many–valued Logics*, North Holland, Amsterdam, 1958.

[Rousseau70] G. Rousseu, *Post algebras and pseudo–Post algebras*, Fund. Math., 67 (1970), pp. 133–145.

[Ruspini91] E. H. Ruspini, *On the semantics of fuzzy logic*, Int. J. Approx. Reasoning, 5 (1991), pp. 45 – 88.

[Schröder895] E. Schröder, *Algebra der Logik*, Leipzig, 1895.

[Schweizer – Sklar83] B. Schweizer and A. Sklar, *Probabilistic Metric Spaces*,

North – Holland, Amsterdam, 1983.

[Shafer76] G. Shafer, *A Mathematical Theory of Evidence*, Princeton U. Press, Princeton N. J., 1976.

[Sierpiński34] W. Sierpiński, *Remarques sur les fonctions de plusieurs variables réelles*, Prace Matematyczno – Fizyczne, 41 (1934), pp. 171 – 175.

[Skowron89] A. Skowron, *The implementation of algorithms based on discernibility matrix*, manuscript, 1989.

[Skowron88] A. Skowron, *On topology in information systems*, Bull. Polish Acad. Sci. Math., 36 (1988), pp. 477–480.

[Skowron–Rauszer92] A. Skowron and C. Rauszer, *The discernibility matrices and functions in information systems*, in: R. Słowiński, ed., *Intelligent Decision Support. Handbook of Applications and Advances of the Rough Set Theory*, Kluwer, Dordrecht, 1992, pp. 311–362.

[Skowron–Stepaniuk96] A. Skowron and J. Stepaniuk, *Tolerance approximation spaces*, Fundamenta Informaticae, 27 (1996), pp. 245–253.

[Slowinski0x] R. Słowiński and D. Vanderpooten, *A generalized definition of rough approximations based on similarity*, IEEE Transactions on Data and Knowledge Engineering, to appear.

[Słupecki49] J. Słupecki, *On Aristotle's Syllogistic*, Studia Philosophica (Poznań), 4(1949–50), pp. 275–300.

[Słupecki36] J. Słupecki, *Der volle dreiwertige Aussagenkalkul*, C. R. Soc. Sci. Lettr. Varsovie, 29(1936), 9–11.

[Stepaniuk00] J. Stepaniuk, *Knowledge discovery by application of rough set model*, in: [Polkowski–Tsumoto–Lin00] , pp. 137–234.

[Stone36] M. H. Stone, *The theory of representations for Boolean algebras*, Trans. Amer. Math. Soc., 40(1936), pp. 37–111.

[Ślęzak00] D. Ślęzak, *Various approaches to reasoning with frequency based decision reducts: a survey*, in: [Polkowski–Tsumoto–Lin00], pp. 235–288.

[Tarski38] A. Tarski, *Der Aussagenkalkül und die Topologie*, Fund. Math., 31(1938), pp. 103–134.

[Tarski24] A. Tarski, *Sur les ensembles finis*, Fund. Math., 6(1924), pp. 45–95.

[Tikhonov35] A. N. Tikhonov, *Über einen Funktionenraum*, Math. Ann., 111(1935).

[Traczyk63] T. Traczyk, *Axioms and some properties of Post algebras*, Colloq. Math., 10 (1963), pp. 193–209.

[Vakarelov89] D. Vakarelov, *Modal logics for knowledge representation systems*, Lecture Notes in Computer Science, vol. 363 (1989), Springer Verlag, Berlin, pp. 257–277.

[Valverde85] L. Valverde, *On the structure of F – indistinguishability operators*, Fuzzy Sets Syst., 17 (1985), pp. 313 – 328.

[Varlet68] J. Varlet, *Algébres de Lukasiewicz trivalentes*, Bull. Soc. Roy. Sci. Liége, 36 (1968), pp. 394–408.

[Vaught52] R. L. Vaught, *On the equivalence of the axiom of choice and the maximal principle*, Bull. Amer. Math. Soc., 58(1952), p. 66.

[Vietoris21] L. Vietoris, Monat. Math. Ph., 31(1921), pp. 173–204.

[Wajsberg35] M. Wajsberg, *Beiträge zum Metaaussagenkalkül I*, Monat. Math. Phys., 42 (1935), pp. 221–242.

[Wajsberg31] M. Wajsberg, *Axiomatization of the three–valued sentential calculus* (in Polish, Summary in German), C. R. Soc. Sci. Lettr. Varsovie, 24(1931), 126–148.

[Wiweger88] A. Wiweger, *On topological rough sets*, Bull. Polish Acad. Sci. Math., 37(1988), pp. 89–93. [Zeeman65] E. C. Zeeman, *The topology of the brain and the visual perception* , in: *Topology of 3-manifolds and Selected Topics*, K. M. Fort (ed.), Prentice Hall, Englewood Cliffs, NJ, 1965, pp. 240–256.

[Von Wright] G. H. Von Wright, *An Essay in Modal Logic*, North Holland, Amsterdam, 1951.

[Zadeh78] L. A. Zadeh, *Fuzzy sets as a basis for the theory of possibility*, Fuzzy Sets Syst., 1 (1978), pp. 3 – 28.

[Zadeh71] L. A. Zadeh, *Similarity relations and fuzzy orderings*, Information Sciences, 3 (1971), pp. 177 – 200.

[Zadeh65] L. A. Zadeh, *Fuzzy sets*, Information and Control, 8 (1965), pp. 338–353.

[Zeeman65] E. C. Zeeman, *The topology of the brain and the visual perception*, in: *Topology of 3-manifolds and Selected Topics*, K. M. Fort (ed.), Prentice Hall, Englewood Cliffs, NJ, 1965, pp. 240–256.

[Zermelo08] E. Zermelo, *Untersuchungen über die Grundlagen der Mengenlehre I*, Math. Annalen, 65(1908), pp. 261–281.

[Zermelo04] E. Zermelo, *Beweiss das jede Menge wohlgeordnet werden kann*, Math. Annalen, 59(1904), pp. 514–516.

[Zorn35] M. Zorn, *A remark on method in transfinite algebra*, Bull. Amer. Math. Soc., 41(1935), pp. 667–670.

Index

List of Symbols